Freeman Laboratory Separates in Chemistry

The exercises listed below are available as Freeman Laboratory Separates. Each Separate is derived from the experiments in the fourth edition of this manual. The Separates are self-bound, self-contained exercises. They are 8½ inches by 11 inches in size, and are punched for a three-ring notebook. They can be ordered in any assortment or quantity. Order through your bookstore, specifying the ISBN and title. (For a complete listing of Freeman Laboratory Separates in Chemistry, see the last page of this manual.)

ISBN	Title
0-7167-9409-8	Student Handbook: Introduction and Appendix
0-7167-9410-1	Scientific Measurements
0-7167-9411-X	Mass and Volume Relationships
0-7167-9412-8	Separations of Food Dyes by Paper Chromatography. Separation of Lycopene and Beta-Carotene by Thin Layer Chromatography
0-7167-9413-6	Observing the Reactions of Household Chemicals. Deductive Chemical Reasoning
0-7167-9414-4	A Cycle of Copper Reactions
0-7167-9415-2	Synthesis of a Chemical Compound: Making Alum from Aluminum
0-7167-9416-0	The Chemistry of Oxygen: Basic and Acidic Oxides and the Periodic Table
0-7167-9417-9	Determination of a Chemical Formula: The Reaction of Iodine with Zinc
0-7167-9418-7	Determination of a Chemical Formula by Titration. The Reaction of Calcium with Water
0-7167-9457-8	Study Assignment A. The Language of Chemistry: Chemical Nomenclature
0-7167-9419-5	The Estimation of Avogadro's Number
0-7167-9420-9	The Molar Volume of Dioxygen and Other Gases
0-7167-9421-7	The Molar Mass of a Gas
0-7167-9422-5	The Reactivity of Metals with Hydrochloric Acid
0-7167-9425-X	The Heat Capacity of Metals
0-7167-9426-8	Enthalpy Changes in Chemical Reactions: Hess's Law
0-7167-9427-6	The Enthalpy of Combustion of a Vegetable Oil
0-7167-9428-4	Emission Spectra and the Electronic Structure of Atoms. The Hydrogen Atom Spectrum
0-7167-9429-2	Ionic and Covalent Bonding. Conductivity of Solutions of Ionic and Covalent Compounds
0-7167-9458-6	Study Assignment B. Writing Lewis Structures
0-7167-9430-6	Models of Molecular Shapes. VSEPR Theory and Orbital Hybridization
0-7167-9431-4	Intermolecular Forces. Slime Gel: Making and Killing Slime
0-7167-9432-2	Liquids and Solids. The Vapor Pressure and Enthalpy of Vaporization of Water. The Enthalpy of Fusion of Water
0-7167-9433-0	Colligative Properties. The Molar Mass of a Soluble Substance by Freezing Point Depression
0-7167-9434-9	Some Examples of Le Châtelier's Principle
0-7167-9435-7	Determination of an Equilibrium Constant by Spectrophotometry The Iron(III) - Thiocyanate Complex Ion
0-7167-9436-5	Factors Affecting the Rates of Chemical Reactions
0-7167-9437-3	The Rate of a Chemical Reaction. Chemical Kinetics
0-7167-9438-1	The pH Scale and Acid-Based Titrations of NaOH Solutions. Standardization of NaOH Solutions. The Titration of Vinegar and Stomach Antacids
0-7167-9439-X	Determination of the Molar Mass and Ionization Constant of a Weak Acid
0-7167-9440-3	The Chemistry of Natural Waters. Water Hardness by EDTA Titration. Water Softening by Ion Exchange
0-7167-9441-1	The Solubility Product Constant of Calcium Iodate, $Ca(IO_3)_2$
0-7167-9442-X	Identification of Silver, Lead, and Mercury(I) Ions
0-7167-9443-8	Identification of Ferric, Aluminum, and Zinc Ions
0-7167-9444-6	Identification of Alkaline Earth and Alkali Metal Ions
0-7167-9445-4	The Qualitative Analysis of Some Common Anions
0-7167-9446-2	Qualitative Analysis of Unlabeled Solutions: The N-Solution Problem
0-7167-9447-0	Paper Chromatography in the Separation of Iron(III), Cobalt(II), Nickel(II), and Copper(II) Ions
0-7167-9448-9	Oxidation-Reduction. Electron Transfer Reactions
0-7167-9449-7	Redox Titrations. The Oxidizing Capacity of a Household Cleanser or Liquid Bleach
0-7167-9450-0	Electrochemical Cells
0-7167-9451-9	Electrochemical Puzzles. Metal Corrosion and Anodic Protection. The Golden Penny Experiment
0-7167-9452-7	The Chemistry of Some Nonmetals: Nitrogen, Sulfur, and Chlorine
0-7167-9453-5	Equilibria of Coordination Compounds
0-7167-9454-3	The Chemistry of Vitamin C
0-7167-9455-1	Atmospheric Chemistry of Nitric Oxide. Smog and Acid Rain

W. H. Freeman and Company
41 Madison Avenue, New York, NY 10010
Houndmills, Basingstoke RG21 6XS, England

Chemistry in the Laboratory

FIFTH EDITION

James M. Postma

CALIFORNIA STATE UNIVERSITY, CHICO

Julian L. Roberts, Jr.

UNIVERSITY OF REDLANDS

J. Leland Hollenberg

UNIVERSITY OF REDLANDS

W. H. Freeman and Company
New York

Publisher: Michelle Russel Julet
Acquisitions Editor: Jessica Fiorillo
Marketing Manager: John A. Britch
Project Editor: Proof Positive/Farrowlyne Associates, Inc.
Cover Designer: Diana Blume
Text Designer: Black Dot Graphics
Production Coordinator: Susan Wein
Supplements Coordinator: Vivien Weiss
Composition: Black Dot Graphics
Manufacturing: Maple-Vail Book Manufacturing Group

Cover Photo: Lester Lefkowitz/Stock Market

Printed in the United States of America

First printing 2000

W. H. Freeman and Company
41 Madison Avenue, New York, NY 10010
Houndmills, Basingstoke RG21 6XS, England

To our students, who keep us motivated, challenged, and encouraged.

Contents

Note to Users of Jones and Atkins, *Chemistry: Molecules, Matter, and Change*, Fourth Edition

The table below shows the Experiments and Study Assignments in this manual that are suggested as companions to the chapters of *Chemistry: Molecules, Matter, and Change*, Fourth Edition, by Jones and Atkins.

Chapter Number	Experiment Number	Topic
1. Matter	3, Study Assignment A, 36	The elements, compounds, separation techniques; nomenclature of compounds
2. Measurements and Moles	1, 2, 8, 9, 10	The metric system, the mole concept, determination of chemical formulas
3. Chemical Reactions	4, 5, 6, 7, 29, 38, 43	Chemical equations; precipitation, acid-base, redox reactions, classifying chemical reactions
4. Reaction Stoichiometry	5, 6, 8, 9	Mole-to-mole, mass-to-mass calculations; limiting reactants; solutions
5. The Properties of Gases	11, 12, 13, 44	Nature of gases; the gas laws; kinetic model of gases; real gases
6. Thermochemistry	14, 15, 16, 21	Energy, heat, enthalpy; enthalpy of physical and chemical change
7. Atomic Structure and the Periodic Table	17, 7	Atomic spectra and energy levels; orbitals and energy levels; electron configuration of atoms; periodicity of atomic properties
8. Chemical Bonds	18, Study Assignment B, 44	Ionic and covalent bonds; writing Lewis structures; electronegativity; polarizability; atmospheric chemistry
9. Molecular Structure	19	VSEPR model; charge distributions in molecules; bond energy; bond length; valence bond and molecular orbital theory
10. Liquids and Solids	20, 21	Intermolecular forces, liquid and solid structures; metals, phase changes
11. Carbon-Based Materials	20	Hydrocarbons, functional groups, isomers, polymers, biopolymers
12. The Properties of Solutions	22	Solutes and solvent, solubility, colligative properties
13. The Rates of Reactions	25, 26	Concentration and rate, effects of temperature and catalysis; rate law and mechanism
14. Chemical Equilibrium	23, 24	Kinetics and equilibrium; equilibrium constants; heterogeneous and gaseous equilibria; response of equilibria to changing conditions
15. Acids and Bases	27, 28	Bronsted-Lowry definition; the pH scale; weak acids and bases; pH of solutions of weak acids and bases
16. Aqueous Equilibria	27, 28, 29, 30	Ionic equilibria; titrations, buffers, solubility; water hardness
17. The Direction of Spontaneous Change	23	Entropy and disorder; free energy and equilibrium; effect of temperature on equilibrium
18. Electrochemistry	37, 38, 39, 40	Half-reactions; balancing redox equations; galvanic cells; the electrochemical series; the Nernst equation; batteries; corrosion; electrolysis
19. The Elements: The First Four Main Groups	15, 9, 33	Periodic trends; hydrogen; the alkali metals, the alkaline earth metals; the boron family; the carbon family
20. The Elements: The Last Four Main Groups	41, 7, 8, 34, 35	The nitrogen family; the oxygen family; the halogens; the noble gases
21. The *d* Block: Metals in Transition	24, 42, 36	*d*-block elements: trends in physical and chemical properties; crystal field theory; effect of ligands on color
22. Nuclear Chemistry	—	Nuclear stability; radioactivity; nuclear energy

The table below shows the Experiments and Study Assignments in this manual that are suggested as companions to the chapters of *Chemical Principles: The Quest for Insight*, by Atkins and Jones.

Chapter Number	Experiment Number	Topic
A, B, C: Matter and Energy; Elements and Atoms; Compounds	2	Physical properties, density, energy, force; atoms, isotopes and the periodic table; molecular compounds; ionic compounds
D: The Nomenclature of Compounds	Study Assignment A	Names of cations and anions; naming ionic compounds; naming inorganic molecular compounds; naming organic compounds
E, F: Moles and Molar Masses; Determination of Chemical Formulas	10, 6, 8, 9	The mole concept; molar masses; mass percent composition; empirical and molecular formulas
G: Mixtures and Solutions	3, 36	Classifying mixtures; separation techniques; molarity; molality; dilution calculations
H: Chemical Equations	4, 5, 8, 9	Symbolizing chemical reactions; balancing chemical equations
I, J, K: Precipitation; Acid-Base; Redox Reactions	5, 7, Study Assignment A, 37	Ionic precipitation reactions; net ionic equations; definitions of acids and bases; strong and weak acids and bases; neutralization; acidic and basic character and the periodic table; oxidation and reduction; oxidation numbers; oxidizing and reducing agents; balancing simple redox reactions
L, M: Reaction Stoichiometry; Limiting Reactants	5, 8	Mole-to-mole and mass-to-mass calculations; volumetric analysis (titrations); reaction yield and the concept of a limiting reactant
1. Atoms: The Quantum World	17	Electromagnetic radiation; quanta and photons; spectra; wavefunctions, orbitals and energy levels; periodicity of atomic properties; physical and chemical properties of the main group elements and transition metals
2. Chemical Bonds	18, Study Assignment B	Ionic and covalent bonds; electron configuration of ions; writing Lewis structures, Lewis acids and bases; electronegativity and polarizability; bond strengths and bond lengths
3. Molecular Shape and Structure	19	The VSEPR model; valence bond theory; molecular orbital theory
4. The Properties of Gases	11, 12, 13, 44	Pressure; the gas laws; diffusion and effusion; kinetic model of gases; real gases and intermolecular forces
5. Liquids and Solids	20	Intermolecular forces; liquid structures; viscosity and surface tension; solid structures; metallic solids; metals and alloys; ionic, molecular, and network solids
6. Thermodynamics: The First Law	14, 15, 16	Energy and work; heat; state functions; PV work of expansion; calorimetry; enthalpy; Hess's law; enthalpies of formation; Born-Haber cycle; bond enthalpy; variation of reaction enthalpy with temperature

Preface

While timeless perfection is our goal each time we submit an edition of *Chemistry in the Laboratory,* each time we take it into the laboratory with our students, read the monthly *Journal of Chemical Education,* or discuss a topic with a colleague, we find ourselves making notes on improvements we'd like to make to the manual. We hope this fifth edition is a step closer to your idea of perfection, whether you are an instructor or a student of chemistry. As with past editions, *Chemistry in the Laboratory* contains forty-four experiments and two study assignments, sufficient for a one-year course in general chemistry. The manual is a companion to the general chemistry text *Chemistry: Molecules, Matter, and Change,* by Loretta Jones and Peter Atkins, or it can be used with other general chemistry texts.

We all envision an ideal educational laboratory setting where a student and an instructor explore the physical world together at an unhurried pace, discovering the important laws and theories of science by systematic observation, challenging discussion, and insightful deduction. But the realities of limited budgets, demanding schedules, and inadequate preparation, coupled with the need to "just get it done" or the desire to "make it work right," leave us all unsettled about the laboratory experience and its place in a modern education system. Some have suggested abandoning the lab altogether or simulating lab experiences facilitated by technology. We reject the former out of hand and are very cautious of the latter. Computer simulations are wonderful augmentations to the educational experience, providing models that students can manipulate and thus explore arenas of the physical world that were previously inaccessible. But even the finest models cannot connect us to natural processes and their accompanying colors, textures, and smells, nor to the startling realities of the imperfections of such models.

NEW TO THIS EDITION

Thus we have retained many of the features of previous editions of *Chemistry in the Laboratory* and have carried through the improvements of each edition. The student-tested, directive laboratory exercises that have characterized our laboratory manual (and most other successful texts) create a foundation of positive laboratory experiences and techniques to build from. We have also retained the somewhat expansive background discussions (not without some paring down) to hopefully create an interesting context for the experiment.

New to the fifth edition is a feature that we hope creates added enthusiasm for the lab experience in students and teachers alike. We've called it *Consider This.* We share the desire of many instructors to open the vistas for students and confine them less to a procedure (or recipe, as some would describe it), to allow for more exploration and discovery in the general chemistry laboratory. These discovery/exploration models of lab experience, however, are very demanding of instructor time and expertise, patience, and support systems such as stockroom help, budgetary support, and equipment availability. As we evaluated our own experiences and our experience with students, we felt that a combination of the two extremes would create a healthy, positive laboratory experience. In the *Consider This* feature, the principles of the experiment are expanded into interesting applications, open-ended experiments, or unexplored corners. Some of these activities can serve as a jumping-off point for undergraduate research projects and perhaps even publications. We are eager to hear how this feature is applied and what insights have been gleaned as these explorations are pursued. We hope you will use the W. H. Freeman and Company Web site (at www.whfreeman. com), e-mail, or traditional communication to let us know of your explorations.

EXPANDED USES OF MICROSCALE TECHNIQUES

In this edition we have continued and expanded our emphasis on reduced-scale experimentation (e.g., titrations involving 1-mL burets), but we have also expanded and added detail for macroscale experimental procedures for those who prefer to use 25- or 50-mL burets. In general we see advantages to the smaller scale manipulations, in that they:

- reduce student exposure to toxic and corrosive chemicals
- improve laboratory air quality
- reduce the amount of hazardous waste by a factor of ten and cut down on disposal costs and burdens on the environment
- significantly lower purchase costs for apparatus and chemicals
- reduce breakage and replacement costs for apparatus
- take less time than their macroscale counterparts
- expose students to microscale experimentation, which is now widely used in the second year organic course

On the minus side we note that microscale procedures also have some drawbacks:

- Working on a smaller scale may demand more attention to development of laboratory skills (although many students are now being introduced to reduced scale experimentation in their high school chemistry courses).
- The precision of volumetric titrations is generally reduced. This is partially offset by the increase in the sharpness of the endpoint as the titration volume is reduced. (These points are discussed more fully in the Introduction.)
- Balances with at least milligram sensitivity are required for decent quantitative work.

Many experiments have not been reduced in scale tenfold. For example, a chemical synthesis on a milligram scale doesn't produce a satisfying aesthetic experience; there's not enough product to see and handle. In such experiments we have generally reduced the scale, but they are still at the gram level. Experiments to measure the enthalpy of a reaction also suffer when scaled too small. The larger surface-area-to-volume ratio means that a greater proportion of energy will leak through the walls of the container. Also the thermal mass of thermometers and stirrers will become a larger proportion of the heat capacity of the system, necessitating corrections that complicate the calculation of the enthalpy change. For these reasons we have not reduced the scale of the reaction enthalpy experiments.

Although the term "microscale" is by now firmly entrenched, it is something of a misnomer. The scale of almost all such experiments in the current chemical education literature would more appropriately be called "milliscale," because sample sizes are in the milligram-to-gram range and reaction volumes, and titration volumes are in the range of 1–10 mL. In coining the term microscale to describe the reduced scale approach, modern practitioners appear to be ignoring the priority of the already well-established fields of microchemistry and microanalysis that work with truly micro samples—in the microgram and microliter range. These microchemical techniques began to be developed as early as 1910, the journal *Mikrochemie* made its first appearance in 1923, and the use of microchemical techniques was well established in the 1930s. In short, microscale is new only to undergraduate laboratories and, in the form presented in contemporary microscale lab manuals, is not truly microscale. Nonetheless, we think the trend to reduced scale experimentation in undergraduate laboratories is laudable and will prove not to be simply a fad.

Most of us who have been teaching chemistry for more than a few years are familiar with passing fads in chemical pedagogy, and we all have a tendency to stick with methods that are tried and true. So there is a natural reluctance to change, particularly when institutional inertia is a barrier or if the idea is not our own. Reduced scale experimentation is not a panacea, but it is worth trying simply because it's safer for students and because small messes are easier and less costly to clean up than big ones. So we urge you to begin by looking for ways to reduce the scale of your experiments whenever it appears feasible. This lab manual is offered in that spirit. If you use reduced scale experiments, we believe that in many instances you will come to prefer them.

NEW AND EXPANDED FEATURES

There are four new experiments in the fifth edition: "Observing the Reactions of Household Chemicals, Deductive Chemical Reasoning" (Exp. 4), "Intermolecular Forces/Slime Gel: Making and Killing Slime" (Exp. 20), "Qualitative Analysis of Unlabeled Solutions: The N-Solution Problem" (Exp. 35), and "The Atmospheric Chemistry of Nitric Oxide/Smog and Acid Rain" (Exp. 44). In addition to these we have substantially modified six experiments (2, 3, 21, 27, 28, and 40) with features that we hope you will find more interesting or easier to accomplish. The sugar solutions of Experiment 2 and the tomato paste of Experiment 3 provide nice entries into real-world situations.

We have tried to give the manual a more modern "look" with a substantial number of new or redrawn figures and by working to make the writing clearer and simpler. Where possible we have updated the references and the bibliography section of the experiments.

We believe quite strongly in the traditional format of our laboratory manual. Each exercise contains a substantial introduction that provides a context for the experiment and discussion of the theory. We believe these materials can set the stage for important student insights. We provide all relevant equations so that each experiment can stand on its own without the need to consult other texts for lab preparation. And we provide options, where appropriate, of reduced scale and macroscale directions with sufficient detail to allow success in either format.

SAFETY AND WASTE COLLECTION PROCEDURES

We have maintained from previous editions our emphasis on safety and on handling laboratory wastes properly. Based on our own experience, our reading of the chemical literature, and user comments, we have improved the experimental procedures, safety precautions, and waste collection procedures. All of these changes are designed to make the laboratory a safer place to work, to raise the safety consciousness of students, and to protect students and the environment from exposure to hazardous materials.

SAFETY

No instructor should assume that students will carefully read and practice the precautions contained in the experiments. Safety consciousness is not inborn; it must be developed by training and practice. Consequently, general chemistry laboratory instructors must take a few minutes each laboratory period to emphasize safety precautions and waste collection procedures.

Instructors must also rigorously enforce good safety practices, particularly the rule that students must wear safety goggles at all times. Here students will take their cue from the instructor. If the instructor is lax in observing good safety practices, students will be lax. A number of experiments in this manual require the use of a fume hood. Where these are not available we suggest that the experiment not be performed.

Good safety practices are in the best interests not only of students but also the instructor and the institution, which may be liable if a student is injured because of negligence or ignorance on the part of the instructor.

WASTE COLLECTION

Good waste collection and waste management procedures require that wastes not be mixed together indiscriminately. For each hazardous waste, provide a properly labelled waste receptacle. The segregated and properly identified wastes can then be combined and packed

for proper disposal by someone who is trained in waste management.

Most sizable research institutions have already put waste management programs in place, but many smaller institutions continue to grapple with the problem of maintaining proper procedures for handling and disposing of laboratory wastes. In the process, these schools have discovered that the costs of disposing of chemical wastes can easily surpass the costs of purchasing the chemicals in the first place. Thus, for economic reasons, as well as protection of the students and the environment, we have tried to devise experiments that produce a minimum amount of hazardous waste.

THE INSTRUCTOR'S MANUAL

We believe that our *Instructor's Manual* can be a most useful tool for the instructor, the person who grades laboratory results, and stockroom personnel. It includes equipment and materials lists, directions for chemical preparations, and filled-in worksheets for each of the experiments. It is our hope that this will save time and energy and facilitate the grading process.

ACKNOWLEDGMENTS

We wish to acknowledge with thanks the assistance of Jessica Fiorillo, Brian McCooey, Christopher Miragliotta, Danielle Swearengin, and Susan Wein at W. H. Freeman and Company and Gail Savage at Proof Positive/Farrowlyne Associates, Inc., in preparing the fifth edition of the manual. Special thanks are due to Fred E. Wood, Supervisor of the Freshman Laboratory Program; Kevin Ferrero, Dispensary Supervisor of the Freshman Program, Department of Chemistry, University of California, Davis; and Margaret Korte of the Chemistry Department at California State University, Chico, for their detailed critiques of each experiment and suggestions for further improvement. Our thanks also to Professors Richard E. Brown, Michigan Technological University; Chris A. Clausen, University of Central Florida; Gordon J. Ewing, New Mexico State University; Gordon Fricke, Indiana University–Purdue University at Indianapolis; David B. Green, Pepperdine University; David B. Macaulay, Harper College; Kenneth Schug, Illinois Institute of Technology; and Robert A. Wilkins, Andrews University; all reviewed various editions and made valuable comments, criticisms, and suggestions.

Welcome to the Chemistry Lab

Your laboratory work is the core of your chemistry course. You have a challenging opportunity to make many observations of chemical reactions under controlled laboratory conditions. You also will experience firsthand the method of inquiry that is the foundation of all experimental sciences. Making and interpreting observations is at the heart of the scientific method.

Here are a few brief points that will give you a good start in this course.

1. Gain self-confidence by working individually, unless the experiments call for teamwork. However, don't hesitate to ask questions of the instructor if you are uncertain about a procedure or the interpretation of a result.

2. Use your ingenuity and common sense. Although laboratory directions are quite specific, they leave ample opportunity for clear-cut, logical, original, and imaginative thinking. This attitude is a prerequisite in any scientific endeavor.

3. Don't waste time by coming unprepared to the laboratory. Prepare for each experiment by studying it (both the Pre-Lab Preparation and, briefly, the Experimental Procedure) before you come to the laboratory.

4. Note beforehand any extra equipment required from the stockroom, and obtain all of it at the same time.

5. Prepare your Laboratory Reports on each experiment with care. Use a permanently bound laboratory notebook, as prescribed by your instructor, to record your data. All data should be recorded directly into this notebook, not on loose sheets or scraps of paper. Where calculations of data are involved, show an orderly calculation for the first set of data, but do not clutter the calculation section with arithmetic details. Likewise, *as you perform the experiment*, think through and answer important questions that are intended to give you an understanding of the principles on which the experimental procedure is based.

6. Scientists learn much by discussing their work with one another. You may likewise profit by discussions with your classmates, but do not copy their work. Integrity is the keystone of scientific work. You may also find it useful to refer to your text while working in the laboratory. (Books are generally even more reliable and complete sources of information than are your classmates!)

7. For tabular data on the properties of substances, you may wish to consult handbooks such as the *Handbook of Chemistry and Physics* (CRC Press, Inc., Boca Raton, FL) or *Lange's Handbook of Chemistry* (McGraw-Hill, NY).

SAFETY RULES

Familiarize yourself with the safety rules given in the Introduction. Observing these rules, as modified or added to by your instructor, is essential for the sake of your safety and that of others in the laboratory.

Your instructor will indicate the location and show you the proper use of the fire extinguishers, fire blanket, eyewash fountain, safety shower, and first aid cabinet and supplies. The instructor will also tell you where to obtain safety glasses or goggles.

GOOD LABORATORY PRACTICES

Familiarize yourself with the good laboratory practices in the Introduction. It is essential that you carefully follow these regulations, as modified or added to by your instructor.

BASIC LABORATORY EQUIPMENT AND PROCEDURES

Check the equipment in the laboratory locker assigned to you, following the procedure described in the Introduction or as directed by your instructor. If time remains, or if your instructor so directs, read over the procedures for the handling of chemicals, use of laboratory burners, the use of microscale equipment, the care of laboratory glassware, and volumetric measurements of liquids.

REDUCED SCALE (MICROSCALE) EXPERIMENTATION

Several of the experiments in this lab manual involve reduced scale titrations, commonly called "microscale," using 1-mL burets in place of the conventional 50-mL buret. The description of several devices used as microburets and instructions for using conventional burets can be found in the Introduction. Some of the advantages of microscale experimentation are discussed in the Preface.

Introduction

The chemistry laboratory can be a place of enjoyment, discovery, and learning. It can also be a place of frustration—and danger.

Although every effort has been made in these experiments to minimize exposure to hazardous, toxic, or carcinogenic substances, there is an unavoidable hazard involved in using a variety of chemicals and glass apparatus. In experiments where a potential danger exists, you will find a Safety Precautions section at the beginning of the experimental procedure. Read this section very carefully before you proceed with the experiment.

Your skin and the delicate membranes of your eyes, mouth, and lungs are largely made of protein. We hope that you do not experience firsthand the adverse effect that concentrated solutions of acids and bases have on protein. The eyes are especially sensitive to acids, bases, and oxidizing agents and must be protected at all times by wearing safety goggles. In addition, the open flame of a Bunsen burner presents a constant hazard to loose clothing and hair.

It's likely that you will become very frustrated if you come unprepared to the laboratory, neglect to record important data, or frantically try to write up reports an hour before they are due. You can minimize these problems by reading the experiments carefully beforehand, noting the critical data that must be recorded, and thoughtfully considering the data as you collect it in order to avoid careless blunders.

We strongly advise you to learn and observe at all times the following laboratory rules and good laboratory practices. By doing so you will minimize the potential dangers and frustrations of laboratory work and maximize the enjoyment.

SAFETY RULES

These rules are designed to ensure that all work done in the laboratory will be safe for you and your fellow stu-dents. In addition to the rules listed here, your institution may have a set of rules that you will be asked to read and to sign as evidence that you have read them.

Your laboratory instructor should also point out the location of various pieces of safety equipment such as fire extinguishers, fire blankets, safety showers, eyewash fountains, and equipment for handling spills, and should demonstrate how to use these items.

Finally, you should know where first aid supplies are kept and where a telephone is located for emergencies that require paramedical, fire, or police assistance. The telephone numbers of these on- or off-campus services should be posted in a prominent place.

1. The most important rule is that *safety goggles* or *safety glasses with side panels* must be worn at all times in the laboratory. Goggles provide the best protection. Ordinary prescription glasses sometimes cover only parts of the eyes and lack side panels that protect the wearer from chemical splashes that might get into the eyes from the side. For this reason, eyeglasses should be covered by safety goggles. Contact lenses should not be worn in the laboratory, even under safety goggles. By themselves they offer no protection from splashes, and they are considered unsafe even under safety goggles because various fumes—for instance, hydrogen chloride gas—may accumulate under the lens and cause injury. If you do not own a pair of safety goggles, obtain one from the stockroom or your campus bookstore.

If any chemical comes in contact with the eye, the most effective first aid is to flush the eye immediately with a steady stream of tap water. You are seldom more than a few seconds from a faucet or eyewash fountain. Continue flushing for at least five minutes and then consult a physician as soon as possible. If your laboratory is equipped with eyewash fountains, familiarize yourself with their use and their location.

2. Fire is an ever present danger. Learn where the nearest fire extinguisher is and how to use it. Your laboratory should also be equipped with a fire blanket and safety shower. *If your hair or clothing should catch fire, smother the fire with a blanket or douse yourself in the shower.*

3. Minor burns, cuts, and scratches are fairly common injuries. However, you must report every such injury to your instructor, who will determine what first aid is appropriate. If you or any student must report to the infirmary or hospital, be certain that someone accompanies the injured person.

4. Bare feet are not allowed in the chemistry laboratory. Broken glass and spilled chemicals, such as concentrated acids, are all too common on the floors of chemistry labs. Wear shoes that protect the tops of the feet from spills; don't wear sandals or flip-flops. Because some synthetic fabrics (polyester) melt and stick to the skin when they catch fire, we also recommend that you wear old, preferably cotton, clothes or a laboratory apron or labcoat. A pair of cotton jeans affords some protection to the legs from minor spills, and cotton shirts are more fire resistant.

5. The vapors of a number of solutions are quite potent and can irritate or damage the mucous membranes of the nasal passages and the throat. Use the technique displayed in Figure I-1 when you need to note the odor of a gas or vapor.

FIGURE I-1

Chemical substances should not be inhaled or smelled unless you are instructed to do so. Even in such cases, you must exercise great care in noting the odor of a substance, using your hand to waft its vapor gently toward your face. Whenever possible, avoid breathing fumes of any kind.

6. In many experiments, it is necessary to heat solutions in test tubes. Never apply heat to the *bottom* of the tube; always apply it to the point at which the solution is highest in the tube, working downward if necessary. Be extremely careful about the direction in which you point a tube; a suddenly formed bubble of vapor may eject the contents violently (an occurrence called "bumping"). Indeed, a test tube can become a miniature cannon. (See Figure I-2.)

NEVER point a test tube that is being heated at your neighbor—it may bump and eject its contents. **SAFETY GOGGLES** worn regularly in the laboratory will protect your eyesight.

FIGURE I-2

Two important safety precautions.

7. Avoid tasting anything in the laboratory. (Poisonous substances are not always labeled as such in the laboratory.) Do not use the laboratory as an eating place, and do not eat or drink from laboratory glassware.

8. Perform no unauthorized experiments.

9. Never work in the laboratory alone.

10. Beware of hot glass tubing; it *looks* cool long before it can be handled safely.

11. For reactions involving poisonous gases, *use the hood*, which draws gases or vapors out of the lab and into the exhaust system.

12. Neutralize a spilled acid or base as follows: (a) acid on clothing: use dilute sodium bicarbonate solution; (b) base on clothing: use boric acid solution (50 g/L); (c) acid or base on the desk: use solid sodium bicarbonate for either, followed by water.

13. To insert glass tubing (including thermometers, long-stemmed funnels, thistle tubes, etc.) through a rubber stopper, first lubricate the tube and stopper with water or *glycerol*. Hold the tubing with a cloth near *the end to be inserted*, and insert with a twisting motion. (If you twist a long-stemmed funnel by the funnel end, it is easily broken. See Figure I-3.)

GOOD LABORATORY PRACTICES (DO'S AND DON'TS)

These practices are designed to guide you in developing efficient laboratory techniques and in making your laboratory a pleasant place in which to work.

1. Read each experiment thoroughly before entering the lab. If you do not, you will waste a great deal of time (both your own and your instructor's), you may expose yourself and others to unnecessary hazards, and you will probably not obtain reliable, useful data. (You will also routinely fail all pre-lab quizzes if your instructor chooses to use them.)

2. *Don't throw matches, litmus, or any insoluble solids into the sink.* Labeled waste containers should be provided to collect hazardous solid or liquid wastes.

3. Leave reagent bottles at the side shelves. Bring test tubes or beakers *to the shelf* to obtain chemicals.

4. Read the label *twice* before taking anything from a bottle.

5. Avoid using excessive amounts of reagent—1 to 3 mL is usually ample for test-tube reactions.

6. *Never* return unused chemicals to the stock bottle. You may make a mistake that later will cause other students' experiments to suffer.

7. Don't insert your own pipets or medicine droppers into the reagent bottles. Avoid contamination of the stock solution by pouring only what you need into a small, clean beaker or test tube.

8. Don't lay down the stopper of a bottle. The stopper may pick up impurities and thus contaminate the solution when the stopper is returned. (The proper way to pour liquids from a glass-stoppered bottle is shown later.)

9. Don't heat thick glassware such as volumetric flasks, graduated cylinders, or bottles; they break easily, and heating distorts the glass so that the calibrations are no longer valid. (See Figure I-4.) If test tubes are heated above the liquid level, they may break, then splash liquid over the hot glass. Evaporating dishes and crucibles may be heated red hot. Avoid heating any apparatus too suddenly; apply the flame intermittently at first.

FIGURE I-3

The procedure for inserting a glass tube into a stopper. NEVER force a thistle tube or funnel into a stopper by grasping the large end. Grasp the stem and twist as you push. ALWAYS wrap your hands in a towel when putting a glass tube into a stopper. Moisten the tube with water or glycerol, and insert with a twisting motion.

FIGURE I-4

If heat is applied to the wrong type of laboratory apparatus, the outcome can be disastrous. NEVER heat a graduated cylinder or bottle.

BASIC LABORATORY EQUIPMENT AND PROCEDURES

THE LABORATORY LOCKER

Check the equipment in the locker assigned you. Refer to Figure I-5 for the identification of any unfamiliar items. Ascertain that all items are present, and examine them carefully to be sure they are in acceptable condition. You are responsible for this equipment and may be charged for any breakage or shortage at the conclusion of the course.

THE HANDLING OF CHEMICALS

Some suggestions bear repeating: Be considerate of others by always bringing your container to the reagent shelf to obtain a chemical. Don't take the bottle to your desk. Maintain the purity of the chemicals in the reagent bottles. Don't withdraw more than you need, and never return any chemical to the bottle. Never contaminate the stopper by laying it down; hold it by your fingers. Don't insert your own transfer pipet into a reagent bottle or the transfer pipet from a reagent bottle down into your own test tube or solutions (Figure I-6). If necessary, clean the outside of the reagent bottle of accumulated dust, ammonium chloride, or other contaminant; rinse the neck and stopper with distilled water; and wipe dry before removing the stopper.

Some of these simple suggestions on the proper handling of solid and liquid chemicals are illustrated in Figures I-7 and I-8.

Observing these suggestions carefully will prevent the contamination of the stock bottles. If you do spill any chemical, clean it up completely, at once. A dirty laboratory prevents good work from being done.

LABORATORY BURNERS

The Bunsen burner, used for most laboratory heating, produces a cone-shaped flame, as illustrated in Figure I-9. Ordinary beakers, crucibles, and other objects to be heated are placed just above the hottest portion of the flame. This allows the most heat to spread about them. If placed down in the cold inner cone of the flame, which consists of unburned gas, the objects are not heated effectively.

The modern Fischer burner is designed to give a concentrated, very hot flame (Figure I-10). For the maximum temperature, have the gas on full pressure, and, with the air vents open, adjust the needle valve at the base (or the air valve if the compressed air type of burner is used) to produce a short blue flame of many short cones that are about 0.5 cm high. The object to be heated is placed about 1 cm above the grid.

OPERATIONS WITH GLASS TUBING

Glass is not a true crystalline solid and therefore does not have a sharp melting point. (In this respect, glass more nearly resembles a solid solution or an extremely viscous liquid that gradually softens when heated.) It is this property that makes glassworking possible.

Soda-lime glass, made by heating a mixture of soda (Na_2CO_3), limestone ($CaCO_3$), and silica (SiO_2), softens readily at about 300 °C to 400 °C in the burner flame. Tubing of this glass is easily bent, but because of its high temperature coefficient of expansion, it must be heated and cooled gradually to avoid undue strain or breakage. Annealing by mild reheating and uniform cooling helps reduce breakage. Such glass must not be laid on a cold surface while it is hot, since this introduces strains and causes breakage.

Borosilicate glass (such as Pyrex or Kimax) does not soften much below 700 °C to 800 °C and must be worked in an oxygen–natural-gas flame, preferably using a glassworking torch. Because it has a low temperature coefficient of expansion, objects made of it can withstand sudden temperature changes.

Figure I-11 shows the proper way to cut glass tubing and fire polish the ends. Fire polishing smooths the sharp edges so that the glass tubing can be inserted easily into a rubber stopper without cutting your fingers. Figure I-12 shows the way to make a bend or a constricted tip.

CARE OF LABORATORY GLASSWARE

Examine all glassware for cracks and chips. When heated, flasks or beakers with cracks may break and cause injury. Small chips in borosilicate glassware can sometimes be eliminated by fire polishing; otherwise, chipped glassware should be discarded because it is easy to cut oneself on sharp glass edges.

The recommended procedure for cleaning glassware is to wash the object carefully with a brush in hot water and detergent, then rinse thoroughly with tap water, and finally rinse once again with a small quantity of distilled or deionized water. Then allow the glassware to drain dry overnight in your locker. If you must use a piece of glassware while it is still wet, rinse it with the solution to be used.

In a general chemistry laboratory we don't recommend the use of "cleaning solution" (a solution of CrO_3 or $K_2Cr_2O_7$ in concentrated sulfuric acid). Such hazardous solutions should be employed only under the *direct* supervision of the instructor, because the solution is very corrosive and under certain conditions concentrated acids can produce noxious gases, endangering you and others nearby. Using such cleaning solutions also poses a considerable hazardous waste disposal problem.

FIGURE I-5
Common laboratory equipment.

FIGURE I-5

(continued)

When putting contents of a medicine dropper into a test tube,

do this

but NOT this.

FIGURE I-6
The proper use of a medicine dropper or transfer pipet.

VOLUMETRIC MEASUREMENTS OF LIQUIDS
Volumetric measurements of liquids are made with graduated cylinders, burets, volumetric transfer pipets, and volumetric flasks.

GRADUATED CYLINDERS. The graduated cylinder (Figure I-13) is usually used to measure approximate volumes of liquids. Aqueous solutions wet the glass walls, forming a concave meniscus; the bottom of the meniscus is used to indicate the volume of liquid. To avoid parallax error (a misreading caused by an improper location of the observer's eye level), your eye should always be level with the meniscus when you are making a reading. The volume is estimated to one fifth of the smallest division. Graduated cylinders may be calibrated to contain (TC) or to deliver (TD). If it is calibrated to deliver, the cylinder actually contains slightly more than the volume read, thus compensating for the thin film of liquid left on the walls when the contents are poured out.

BURETS FOR REDUCED SCALE TITRATION (MICROBURETS)
The benefits of reduced scale experimentation are firmly established, as described in greater detail in the Preface.

The generic term "microscale" has become the established term for reduced scale experimentation in undergraduate laboratories. We will conform to this usage, although it would be more appropriate to call these techniques "milliscale" because sample sizes are typically greater than 100 milligrams, and titration volumes are in the milliliter range.

In general, we prefer a 10–25-fold scaling down of macroscale titration volumes, using 1-mL burets in place of 25- or 50-mL burets. Typically, microscale titration procedures in this manual are scaled to require about 0.8 mL of titrant. (However, we continue to provide instructions for traditional macroscale titrations using 25- or 50-mL burets.) When 1-mL burets are used, students can put into a 13 × 100-mm test tube enough reagent for three or four titrations. This markedly reduces consumption and incidental spilling and waste of reagent.

We have used three different devices for microscale titrations. These are pictured in Figure I-14. All give satisfactory results when properly used. The conventional 50-mL buret is more fragile and costs 300 to 1000 times more than the devices described here. Although they can be reused and will last a semester, they are inexpensive enough to be considered disposable.

The first microburet (as shown in Fig. I-14A) uses a 1 mL × 1/100 mL Pyrex glass, disposable serological pipet.[1] (A polystyrene pipet can be substituted, although there is some sacrifice in chemical resistance and ease of reading.) Added to the bottom of the serological pipet is a fine tip, cut from a polyethylene transfer pipet (shown in Fig. I-14C). Addition of the fine tip allows 0.02 mL (1 drop) volume resolution.

At the top, to control the flow, is a 2-cm length of 1/8 inch ID × 1/32 inch wall thickness silicone rubber tubing containing a 4-mm soda-lime glass ball. The ball is located close to the top of the serological pipet to give maximum control of delivery by minimizing the space between the bead and the top of the pipet. Otherwise, there is a tendency to pinch the tubing below the bead, causing a rebound effect that can draw air into the tip when the pressure on the silicone tubing is relaxed. Silicone rubber tubing is preferred to polyvinyl chloride (Tygon) tubing because silicone rubber is more flexible and allows easier control of the flow.

The second device is a 1 mL × 0.01 mL all polypropylene tuberculin syringe,[2] shown in Figure I-14B. A fine tip, cut from a polyethylene transfer pipet, makes a snug fit over the Luer tip of the syringe and allows 0.02 mL (1 drop) volume resolution.

[1] 1 mL × 1/100 mL Pyrex disposable serological pipet (Corning 7078D-1 or equivalent, cost approx. $0.20 each in quantities of 500.)

[2] 1 mL × 0.01 mL tuberculin syringe with Luer tip (Air-Tite Norm-ject; cost approx. $0.15 each in quantities of 100; Air-Tite of Virginia, Inc., 565 Central Drive, Virginia Beach, VA 23454, tel 800-231-7762.)

First—   Roll and tilt the bottle until some of the contents
         enters the inside of the plastic cap.

Scoop out a little of the material with the spatula provided.

Second—   Carefully remove the cap so that some
          of the contents remains in it.

Tap the spatula until the desired amount falls off.

Second Method

Third—   Tap the cap with a pencil until the
         desired amount falls out.

First Method

Roll and tilt the jar until
the desired amount falls out.

Third Method

FIGURE I-7
Methods for transferring powders and crystals.

REMOVING A STOPPER

First—   Read the label twice.

Second—   Hold the stopper in and tilt the bottle until the contents wet the stopper.

Third—   Moisten the inside of the neck and the lip with the wet stopper.

Fourth—   Replace the stopper and withdraw it again with the back of the hand.

POURING A LIQUID

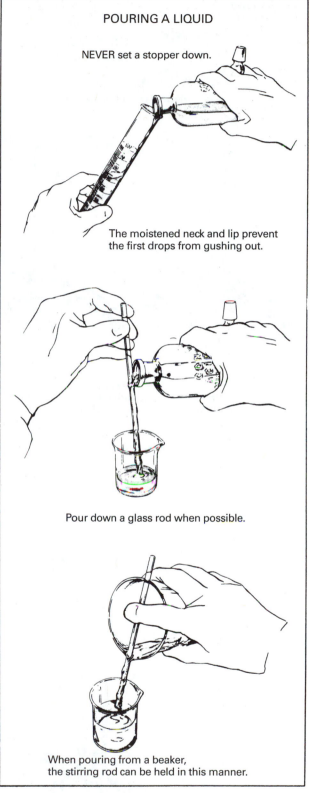

NEVER set a stopper down.

The moistened neck and lip prevent the first drops from gushing out.

Pour down a glass rod when possible.

When pouring from a beaker, the stirring rod can be held in this manner.

FIGURE I-8

Methods for transferring liquids. When handling corrosive materials, latex gloves should be worn to protect the skin.

Oxidizing region
(pale violet)

Reducing
region
(pale blue)

Hottest part of flame

Yellow flame

This gets
very hot.

Violet

Flame rises
from the burner

If the flame is like this,
turn down the gas.

Yellow tip

Violet

Blue
absent

Adjustment for
regulating the
air supply

Adjust to a uniform,
nonluminous flame.

If the burner ignites at the base,
turn off the gas, close the
air vent slightly, and relight.

If the flame is like this,
open the air regulator.

Always use a flame spreader
for bending glass tubing.

FIGURE I-9
Instructions for operating a Bunsen burner.

Oxidizing region (pale violet)

Hottest part of flame

Reducing region
(tiny blue cones)

Air

Needle valve for
adjusting gas supply

Air

Gas

Gas

FIGURE I-10
Fischer burners.

BREAKING A TUBE

First—Make a single scratch with the triangular file.

Scratch

Second—Place the thumbs together opposite the scratch.

Third—Pull and bend quickly.

FIRE POLISHING THE END OF A TUBE

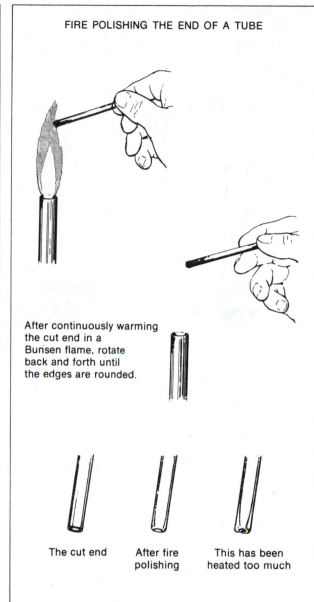

After continuously warming the cut end in a Bunsen flame, rotate back and forth until the edges are rounded.

The cut end   After fire polishing   This has been heated too much

Trim a jagged edge by stroking it with wire screen.

To flare the end of a tube, first heat it, then insert a file and rotate the tube.

FIGURE I-11

Some elementary manipulations of glass tubing.

MAKING A BEND

Roll the tube back and forth in the high part of a flat flame until it has become quite soft.

Remove from the flame and hold for a couple of seconds to let the heat become more uniform.

Bend quickly to the desired shape and hold until it hardens.

A good bend

Inadequate heating

Local overheating

MAKING A CONSTRICTED TIP

Roll the tube in a Bunsen flame until it softens. Don't use a flame spreader.

Allow the tube to become shorter as the walls thicken to about twice their original thickness.

Remove from the flame and after a moment pull until the softened region is as small as desired.

Cut to length

Fire polish, or file the tip

FIGURE I-12

Additional manipulations of glass tubing.

A

Too high

Proper line of sight

B

Too low

C

Always read the bottom of the meniscus.
This reads 87.5 mL.

FIGURE I-13
The proper way of reading a meniscus to avoid parallax error.

count method, in our view, has some drawbacks. It's relatively easy to become distracted and lose count. It's difficult to interrupt, then resume the addition of reagent, particularly when air gets introduced into the pipet tube after the pressure on the bulb is temporarily relaxed. Some who like this type of buret have improved it by constructing some device (operating like a tubing screw clamp) to replace finger pressure. This allows you to interrupt the titration without getting air into the pipet tube.

The transfer pipet is well-suited to weight titrations; one simply weighs the transfer pipet, bulb down, in a small beaker before and after the titration. The quantity of reagent delivered can be determined by the simple relation *mmols of titrant = g of titrant solution × moles of reagent/kg of solution.* This approach requires that solutions be made up by mass rather than by volume, which is a simple procedure. The concentration unit *moles/kg solution* also has the advantage of being temperature invariant. The concentration in moles/kg solution is essentially equal to the molar concentration for dilute aqueous solutions whose density is close to 1.0 g/mL.

Tuberculin syringe. We generally prefer the tuberculin syringe or serological pipet versions of the buret. The syringe is robust and compact. It can be laid down on the bench top. With a little practice, it can be operated with one hand, pressing the thumb on the plunger and applying pressure on the shaft of the plunger held between the first two fingers and the palm of the hand. This technique avoids "stick-slip" friction, which can cause inadvertent addition of more reagent than intended. (See Figure I-15.) It's easy to interrupt the titration and not lose track of the volume added. The syringe can also be used to stir the solution.

Filling the syringe and expelling air bubbles must be done with care. Reagent may dribble over the outside of the syringe and get on the fingers, or worse, squirt out the end as students work to expel the bubbles. Students with vision problems may find the graduations too closely spaced to easily read the volume scale printed on the side of the syringe. Instructors may also be concerned about the possible theft of syringes and the possibility that they could be put to illicit uses.

Serological pipet. The buret made using a serological pipet overcomes some of the disadvantages of the syringe. It is easily filled, using as a pipet filler a 3- or 5-mL disposable plastic syringe with a Luer tip that can be inserted into the tubing at the top of the buret. The graduations are three times farther apart, so they are easier to read than those on the syringe. The serological

The simplest and least expensive device is a fine-tipped polyethylene transfer pipet,[3] producing about 50 drops per mL, shown in Figure I-14C. The pipet is calibrated by counting the number of drops per mL of solution delivered, then using the drop count method to determine the volume of titrant added (volume added = no. of drops × 1 mL/50 drops).

ADVANTAGES AND DISADVANTAGES OF THE MICROBURET DEVICES DESCRIBED.
We consider that any of the three inexpensive devices described works satisfactorily; the choice of which is best is somewhat a matter of personal preference. (At the University of Redlands we use the 1-mL tuberculin syringe.)

Transfer pipet. The polyethylene transfer pipet has an appealing simplicity and is the least expensive, but the drop

[3]Disposable polyethylene transfer pipet, 50 drops/mL (Rainin catalog no. TP-232; cost approx. $0.06 each in quantities of 500; Rainin Instrument Co., Inc., tel 800-472-4646.)

TABLE I-1

Relative Precision of Different Kinds of Burets

Type of buret	Total capacity, mL	Volume resolution, mL	Relative precision, %
Conventional buret	50.0	0.05 (1 drop)	0.1
Conventional buret	25.0	0.05 (1 drop)	0.2
Gilmont syringe buret[a]	2.0	0.002 (tip immersed)	0.1
Serological pipet	1.0	0.02 (1 drop)	2.0
Tuberculin syringe	1.0	0.02 (1 drop)	2.0
Fine-tipped polyethylene transfer pipet	1.0 (3 mL bulb)	0.02 (1 drop)	2.0

[a] The Gilmont syringe buret provides 2-mL capacity with precision as good as a 50-mL buret. It has a polypropylene barrel that rotates like a micrometer, driving a Teflon plunger that displaces the titrant solution. It is much more expensive than the simple devices we have described (cost approx. $80).

pipet, like the syringe version, can be operated with one hand and can also be used to stir the solution. Although we prefer glass, because of its chemical resistance and the clarity and durability of its volume markings, polystyrene serological pipets are also available and have the advantage that they won't break if dropped. Serological pipets generally come with a cotton plug. This plug should be removed to eliminate the possibility of cross-contamination when different solutions are put in the buret, since it's almost inevitable that the cotton plug will become wet with the titrant solution.

PRECISION IN MICROSCALE TITRATION. The precision (repeatability) of a titration is limited by the smallest volume that can be delivered and accurately measured, which we will call the *volume resolution*. We show in Table I-1 the *percentage relative precision*, which we define as the *volume resolution(mL)/buret capacity(mL) × 100*.

Some authors prefer to use a 2-mL pipet as a microburet, implying that you can obtain four significant figures with a 2-mL buret.[4] The claim of four significant figures can only be justified if the uncertainty in locating the endpoint is 0.001 mL and if the leading digit can assume ten different values (0 through 9). So, although superfi-

cially it looks as if you might have four significant figures, in practice you can obtain about 2 2/10 significant figures using a 2-mL buret.

First, in student hands, the uncertainty in locating the endpoint is equal to the volume resolution, typically about 1 drop or 0.02 mL. For an absolute volume uncertainty of ±0.02 mL, the relative precision would be 2 percent for a 1-mL buret and would be improved to 1 percent for a 2-mL buret. So the relative precision improves by a factor of two, not a factor of ten. (In this context an improvement in relative precision means that the relative precision gets smaller, as shown in Table I-1).

Second, to increase the relative precision by a full significant digit implies a tenfold improvement in relative precision, but this would require you to increase the buret volume by tenfold, allowing the leading digit to be anything between 0 and 9 (up to 9.99 mL). Clearly, using a buret capacity of 2 mL does not allow ten values for the leading digit, but only two values, 0 or 1 (up to 1.99 mL).

To summarize, going from a 1-mL buret to a 2-mL buret does not really give you the benefit of another full significant figure, but does improve the relative precision by a factor of two, equivalent to about 2/10 of a significant figure.

ENDPOINT SHARPNESS AND THE TITRATION VOLUME PARADOX.[5] A smaller relative precision means that one can, in principle, achieve a more precise titration. Superficially, it would appear from Table I-1 that the relative precision achievable with a 1-mL buret is a factor of 10 or 20 times larger (larger is worse) than can be attained with a conventional 25- or 50-mL buret. In practice, we have observed that the relative precision is typically a factor of 2 to 4 times larger using 1-mL burets. Part of this has to do with the level of training, practice, skill, and motivation of general chemistry students. Using conventional macroscale burets, general chemistry students typically achieve relative precision of about 1 to 2 percent, rather than the theoretical 0.1 percent that can be achieved in very careful work.

The second factor is the increase in the sharpness of the endpoint as the volume of the titration is reduced. This effect is often overlooked because it is so counterintuitive that it has been called paradoxical.[5] In simple terms, in a potentiometric titration, where the response variable R is a logarithmic function of concentration, the rate of change of system response (R) with volume (V) of added titrant, $\Delta R/\Delta V$ (be it pH reading or a change of the

[4]Singh, M. M.; McGowan, C.; Szafran, Z.; Pike, R. M., "A Modified Microburet for Microscale Titration," *J. Chem. Educ.* **1998,** *75,* 371.

[5]Rossi, M.; Iha, M. E. V. S.; and Neves, E. A., "A Paradoxical Effect of the Titrated Volume in Potentiometric Titrations," *J. Chem. Educ.* **1990,** *67,* 65–66.

4-cm length
1/8 in. ID × 1/32 in. wall
silicone rubber tubing

4-mm glass ball

1-mL in 1/100
serological pipet

1-mL in 1/100
polypropylene
syringe

Disposable
polyethylene
transfer pipet
50 drops/mL

15 cm

Fine tip cut from
disposable
transfer pipet

A   B   C

FIGURE I-14

Three devices that can be used as simple and inexpensive
microburets.

USING A CONVENTIONAL (MACRO) BURET.

The use of a 25-mL buret is shown in Figure I-16. If it is
clean, the solution will leave an unbroken film when it
drains; if drops of solution adhere to the inside of the
buret, it should be cleaned with a brush, hot water, and
detergent until it drains properly (Figure I-17). Cleanli-
ness is important because the volume of a 25- or 50-mL
buret can ordinarily be estimated to the nearest 0.02 mL,
and the error caused by a single large drop adhering to
the inside of a buret causes an error of about 0.05 mL.
The presence of several drops would obviously cause an
error in reading the volume delivered.

When filling a freshly cleaned buret with solution, add
a 5- to 10-mL portion of the solution, being sure the stopcock is turned off, and tip and rotate the buret so that the
solution rinses the walls of the buret completely. Repeat
the procedure with at least two more fresh portions of
solution; then fill the buret above the zero mark and
clamp in a buret holder. Open the stopcock wide to flush
out any air bubbles between the stopcock and the tip of
the buret. Next, drain the buret below the zero mark, and
take the initial reading, being careful to avoid parallax
error. The smallest division on a 25- or 50-mL buret is
ordinarily 0.1 mL; estimate the volume to the nearest fifth
of the smallest division, or the nearest 0.02 mL.

Burets may have stopcocks made of glass, which must
be periodically cleaned and lubricated as shown in Figure
I-18A. Teflon stopcocks (Figure I-18B) ordinarily require
no lubrication, but they may have to be cleaned if they are
plugged or adjusted if the tension nut is too tight or too

FIGURE I-15

Using a 1-mL syringe as a microburet. The thumb applies pressure to the plunger, while the first two fingers lightly grip the
plunger, and the last two fingers hold the barrel of the syringe.
The left hand swirls the solution in the flask.

indicator color) is greater as the sample volume is made
smaller, *even when the solution concentrations (titrant and
solution titrated) remain constant.* If the sample volume is
reduced tenfold, the endpoint sharpness increases tenfold. This effect makes it easier to locate the endpoint in a
small scale titration. The sharper endpoint obtained in a
smaller sample volume partially offsets the tendency of
the relative precision to get larger (worse) as you make
the buret volume smaller.

Using a small funnel, rinse a clean buret with a few milliliters of the solution. Allow the buret to drain.

Fill the buret to above the zero mark with the solution.

Open the stopcock wide for a few seconds to remove all air from the tip.

Refill to just *below* the 0.00 mark (somewhere between 0–1 mL). Take initial reading with eye level with meniscus. Do not attempt to set initial reading at 0.00 or 1.00 or any other specific reading.

FIGURE I-16
The use of a conventional (macro) buret.

Water spreads out smoothly on CLEAN glass

but stands in droplets on SOILED glass.

FIGURE I-17
Clean and dirty glassware.

loose. A drawn-out glass tip that is connected to the buret by a short length of rubber tubing containing a round glass ball makes the simple yet effective stopcock shown in Figure I-18C.

The buret is calibrated to deliver (TD) and is capable of a precision of approximately ± 0.02 mL when carefully used. The tip should be small enough that the delivery time is not less than 90 s for a 50-mL buret. This allows adequate time for drainage, in order to enable you to obtain a proper reading. If the delivery time of your buret is faster, wait a few seconds before taking a reading, so that you allow the buret to drain.

Figure I-19 illustrates the recommended technique for manipulation of a buret stopcock. You may add the solution from the buret quite rapidly until it is close to the endpoint, but then you should reduce the flow until individual drops fall into the flask. As you add the last few drops slowly, swirl the flask to obtain thorough mixing.

THE USE OF VOLUMETRIC PIPETS. Volumetric transfer pipets are designed to deliver a single fixed volume of liquid. The graduation mark is located on the narrow part of the pipet to assure precision. These pipets come in sizes ranging from less than 1 mL to 100 mL. They are calibrated to deliver (TD) the specified volume if they are handled in the prescribed manner.

Pipets are ordinarily calibrated at 20 °C. In very careful work, temperature corrections are necessary if the solution temperature is markedly different from the calibration temperature of the pipet. Fill the pipet by placing the tip in a flask of the solution and using a suction bulb to draw the liquid up past the calibration mark (Figure I-20). Then slip off the bulb and, with the forefinger, quickly seal the top of the pipet before the solution drops below the calibration mark. Wipe the outside of the pipet with tissue or a clean towel; then allow the liquid to flow out until the bottom of the meniscus is just at the calibration ring. To pick off the last drop adhering to the

FIGURE I-18
Three varieties of stopcocks. (A) Glass stopcock: To grease a glass stopcock, remove old grease from both parts with organic solvent such as cyclohexane; then wipe dry; then apply a thin film of stopcock grease as shown. (B) Teflon stopcock: No grease is used on a Teflon stopcock. (C) Rubber-tubing, glass-bead stopcock: pinching the tube near the bead allows the solution to drain from the buret.

Adjust liquid flow by
the thumb and two fingers
held around the stopcock,
with a slight inward
pressure on the Teflon
plug to prevent leakage.

Swirl the liquid
during the titration,
and wash the walls
with distilled water
from the wash bottle.

White paper

FIGURE I-19
Recommended technique for manipulation of a buret stopcock. Most left-handed students will manipulate the stopcock with the right
hand (A), whereas most right-handed students will prefer to manipulate it with the left hand (B).

outside of the tip, touch the tip to the side of the flask. Then withdraw the pipet from the flask and hold it over the vessel into which the liquid is to be transferred. Allow the pipet to drain in a vertical position, with the tip against the side of the vessel. After it appears that most of the liquid has drained out, allow an additional 15 to 20 s for drainage. The tip of the pipet will still contain some liquid. Ordinarily, this has been accounted for in the calibration of volumetric pipets (marked TD) and should not be blown out. (Some pipets, such as calibrated serological pipets, are calibrated to be blown out. Most of these will have a ring, or double ring, at the top of the pipet.) Like the buret, the pipet must be scrupulously clean if precise results are to be obtained. If the pipet is still wet from cleaning, rinse with several portions of the solution to be pipetted.

THE USE OF VOLUMETRIC FLASKS. Volumetric flasks come in a variety of common sizes from 5 mL to 2 L. They are calibrated to contain the specified volume at a given temperature, usually 20 °C. They are used in the preparation of accurate solution concentrations by weighing a pure substance, quantitatively transferring the substance to the volumetric flask, adding water to dissolve the substance, and diluting the solution precisely to the mark inscribed on the neck of the volumetric flask. (The bottom of the curved portion of the meniscus should coincide with the mark.)

Small amounts of solid (1–2 g) can be weighed on a creased sheet of weighing paper and transferred directly into the mouth of a volumetric flask. A small spatula is used to push the solid into the mouth of the volumetric flask. Larger amounts of solid are usually first dissolved in a small amount of water in a beaker, then quantitatively transferred to a volumetric flask using a ribbed-funnel in the mouth of the volumetric flask, a glass stirring rod, and a wash bottle, as shown in Fig. I-21.

Volumetric flasks also are used to produce accurate dilutions of solutions, by pipeting a known volume of solution into the volumetric flask, then diluting the solution to the mark. (See Figure I-21.)

Suction by bulb **or aspirator**

Drain to graduation mark,
touching off adhering drop.

After discharge, allow a
20-s drainage period.

Touch the tip to the
side of the flask.

The liquid remaining in the
tip of the pipet should not
be blown out. The pipet
was calibrated for this
amount to remain.

A   B   C   D

FIGURE I-20
The procedure for using a transfer pipet.

A   B

FIGURE I-21
Using volumetric flasks. (A) Method for quantitatively transferring a dissolved solid to a volumetric flask. (B) Accurate dilution
using a volumetric pipet and a volumetric flask.

Scientific Measurements

PURPOSE

To gain an understanding of the utility of measurements in chemistry by examining how measurements are made, what the limitations of measurements are, and how the data obtained from measurements can be presented in a useful way.

PRE-LAB PREPARATION

The introduction of *quantitative* measurement brought chemistry from the magical, mystical alchemist's laboratory to the practical science it is today. In the 1600s experimenters began to measure the properties of gases in a quantitative way, and the relationship between the pressure and volume of a fixed quantity of gas was discovered and published by Robert Boyle.

It wasn't until the late 1700s, when chemists began to measure the masses of the starting materials and of the products of a reaction, that some of the fundamental laws of chemistry were recognized and stated. These include the laws of mass conservation, of definite composition, and of multiple proportions.

Such quantitative observations provided supporting evidence for the concept that matter was composed of atoms with definite masses. The ability to determine accurately the relative masses of the elements and to systematize their chemical properties led eventually to the creation of the periodic table of elements. If you look carefully at the periodic table you will note that, with few exceptions, the elements are arranged in order of increasing atomic mass.

When quantitative measurements were made of living things, it became apparent that the processes of life were themselves chemical reactions. Measurements showed that these biochemical reactions were subject to the same laws as other chemical processes.

HOW MEASUREMENTS ARE MADE

Most quantitative measurements are operations in which two quantities are compared. For example, to measure the length of an object you compare it to a ruler with calibrated marks on it. As the ruler itself was designed and manufactured, it was compared to other standards of length that conform to the standard unit of length agreed upon by international conventions.

The double-pan equal-arm analytical balance is another example of measurement by direct comparison. With this apparatus the mass of an object is determined by comparing it directly with a set of masses of known value. The single-pan analytical balance you may have in your laboratory also involves a comparison of masses, using the principle of substitution—an unknown mass on the pan substitutes for known masses on the beam above the pan. The point of balance occurs when the sum of the known masses that have been removed is exactly equal to the unknown mass. Figure 1-1 shows how unknown and known masses are compared by means of double-pan, single-pan, and electronic balances. Even a modern electronic balance involves an indirect comparison because standard masses are used to calibrate the force coil of the balance.

Time is usually measured by counting the number of periods of a system that oscillates at some natural frequency. The pendulum of a grandfather clock swings at a natural frequency. An escapement mechanism turns a toothed wheel that is connected by gears to the hands of the clock. The hands of an electric clock are driven by a synchronous electric motor supplied with alternating current of a stable frequency. A quartz digital watch

Equal-arm double-pan balance with three knife edges

A

Substitution-type single-pan balance with only 2 knife edges
Constant load = constant sensitivity
Lever arm error impossible

B

Electronic balance

C

FIGURE 1-1

(A) How masses are compared on an equal arm balance. The lever arms of the balance must have equal lengths. (B) How masses are compared on a substitution-type single-pan balance. Masses whose sum is equal to the unknown mass are removed from the hanger to keep the beam in equilibrium. (C) Calibration of an electronic balance with a known mass. A sensitive detector/DC current amplifier (not shown) keeps the balance pan precisely at the rest (or null) position by adjusting the electric current in the force coil. The current, proportional to the mass on the pan, is transformed into a digital readout of the mass.

measures time by counting (and subdividing) the oscillations of a tiny quartz crystal. The most accurate clocks, which serve as our fundamental time standard, are based on counting the frequency of an oscillator that is continuously compared with the frequency associated with a transition between two energy states of the electrons in gaseous cesium atoms. From these descriptions of various timekeeping devices, we see that the measurement of time is an operation in which the natural periods of various oscillating systems are compared.

BASIC AND DERIVED UNITS

By an international agreement reached in 1960, certain basic metric units and units derived from them are to be preferred in scientific use. The preferred units are known as the International System units (commonly called SI units, from the French, *Système Internationale*). The basic units of the SI system are given in the Appendix, Table 1.

The National Institute of Standards and Technology (NIST) Internet site (http://physics.nist.gov) provides a full discussion of the fundamental physical constants, including historical background information and the best values of the constants and their uncertainties.

Units obtained from the basic units are called *derived* units. The volume of a cube or rectangular box can be determined by measuring the lengths of its three dimensions and computing the volume as the product of those lengths. Thus volume is derived from length measurements and has the dimensions of length cubed, for example, cubic centimeters (cm^3). Cylindrical containers are easier to manufacture than rectangular containers, so volume is more often measured in containers such as graduated cylinders and burets. The volume of a cylinder is equal to the product of its cross-sectional area times its length. This formula allows us to use a length scale, often inscribed along the length of the cylinder, to determine volume.

THE ASSESSMENT OF EXPERIMENTAL ERRORS

Every measurement involves some measurement error (or measurement uncertainty). Because all generalizations or laws of science are based on experimental observations involving quantitative measurements, it is important for a scientist to take into account any limitations in the reliability of the data from which conclusions are drawn. In the following section we will discuss different kinds of errors—systematic, random, and personal errors—and show how a series of measurements can give us an estimate of the precision of a measurement.

PRECISION AND ACCURACY

The limitations of both precision and accuracy will contribute to uncertainty in the measurement. The error in a measurement or result is the difference between the true value of the quantity measured and the measured value. The smaller the error, the closer the measured value is to the true value and the more accurate is the result. *Accuracy is a measure of the correctness of a measurement.*

Unless we have precise standards against which we can test our measurement, we often do not know the true value of a measured quantity. If we do not, we can obtain only the mean, or average, value of a number of measurements, and measure the spread or dispersion of the measurements. A measure of the spread of individual values from the mean value is the *deviation*, δ—defined as the difference between the measured value, x_i, and the arithmetic mean, $\bar{x}$, of a number, n, of measurements (a bar over a quantity means the average value of the quantity):

$$\delta_i = x_i - \bar{x} \qquad (1)$$

The mean value is obtained by adding all of the individual measurements and dividing by the total number of measurements:

$$\bar{x} = \frac{\sum x_i}{n} \qquad (2)$$

where Σ represents the operation of summation.[1]

The smaller the deviations in a series of measurements, the more precise the measurement is. *Precision is a measure of the reproducibility of a measurement.*

The difference between accuracy and precision is shown in Figures 1-2 and 1-3. Figure 1-2 illustrates the distribution of a number of weighings of a sample on a less precise balance (curve A) and on a more precise balance (curve B). In this series of measurements, both balances give the same mean value, so their accuracy is the same. However, the precision of the measurement is much better for balance B; therefore, we can have more confidence in the result of that measurement.

---

[1]Therefore Σx_i means to calculate the sum $x_1 + x_2 + x_3 + x_4 + \dots$

SYSTEMATIC ERRORS.   Errors are of two general types, systematic (determinate) and random (indeterminate). A *systematic error* causes an error in the same direction in each measurement and diminishes accuracy, although the precision of the measurement may remain good. A miscalibrated scale on a ruler, for example, would cause a systematic error in the measurement of length.

FIGURE 1-2

The distribution of a number of weighings of a sample on a less precise balance (curve A) and on a more precise balance (curve B). The standard deviation for curve A is ±0.05 g and for curve B, ±0.02 g.

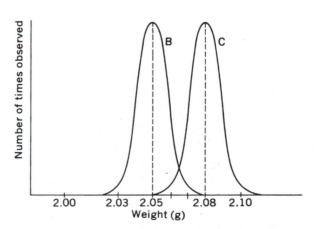

FIGURE 1-3

The distribution of a large number of weighings of a sample on two balances of equal precision (B and C). However, there is a systematic error in at least one of the balances.

Similarly, Figure 1-3 shows the result of weighings of a sample on two balances of equal precision, but at least one of the balances has a systematic error, perhaps due to a miscalibrated weight that was used in each measurement. Ordinarily we can detect an inaccurate (but precise) measurement only by use of precision standards. Such standards are produced and tested by the National Institute of Standards and Technology (NIST), formerly called the National Bureau of Standards. Even if all the systematic errors in a measurement are found by careful work with precise standards, the possibility remains that the measurement will be in error because it is impossible to make any measurement with infinite accuracy.

RANDOM ERRORS AND STANDARD DEVIATION.

If the measurement is made a large number of times, you will obtain a range of values (like the distributions shown in Figure 1-2) caused by the *random errors* inherent in any measurement. For random errors, small errors are more probable than large errors and negative deviations are as likely as positive ones. The resulting distribution of the measurements (like that shown in Figure 1-2) is called a *normal error distribution*. The mean value of the set of measurements is the most probable value, corresponding to the center of the distribution curve. The spread, or dispersion, of the results is expressed by the *standard deviation, s*:

$$s = \left(\frac{\sum (x_i - \bar{x})^2}{n - 1} \right)^{1/2} \qquad (3)$$

This formula says: Sum the squares of the deviations, divide by $n - 1$, and take the square root of the result. So the standard deviation could also be called the *root-mean-square deviation*. This formula actually gives only an estimate of the standard deviation unless the number of measurements is large (>50). We must recognize that when we repeat a measurement only two or three times,[2] we are not obtaining a very large sample of measurements, and the confidence we can place in the mean value of a small number of measurements is correspondingly reduced.

Although the formula may look forbiddingly complex, the steps are very simple. First calculate the arithmetic mean, or average, value, $\bar{x}$, of the measurements. Then subtract the mean value, $\bar{x}$, from each one of the individual values, x_i, to obtain the deviation. Square each devia-

tion, and add all of the squares. Divide the total by $n - 1$, where n is the total number of measurements. Finally take the square root of the result to obtain the estimate of the standard deviation. The procedure is illustrated for the calculation of the standard deviation in Table 1-1.

The curves like those shown in Figure 1-2 correspond to distributions that would be obtained from a very large number of measurements. Curve A corresponds to a mean value of 2.05 g with a standard deviation of 0.05 g. Curve B corresponds to a mean value of 2.05 g with a standard deviation of 0.02 g. The standard deviation can be related to the *confidence interval*, or the range about the mean value in which one of a group of measurements may be expected to fall.

If we recognize that the normal error distribution is a distribution of the probabilities of obtaining a particular measurement, we see that the probability of measurements occurring close to the mean will be greater than the probability of a measurement occurring far away from the mean. In fact, there is a 0.68, or 68 percent, probability that a given measurement will fall within plus or minus one standard deviation of the mean value. There is a 0.95 (95 percent) probability that a measurement will fall within plus or minus *two* standard deviations of the mean value. This means that for curve A of Figure 1-2, we can expect that 95 percent of the measurements will fall between 1.95 and 2.15 g. If we were to measure the total area under the curve, we would also see that this interval of 2s corresponds to 95 percent of the

TABLE 1-1

The Procedure for Calculating the Standard Deviation

Measured value[a] (x_i)	Deviation $(x_i - \bar{x})$	Square of deviation $(x_i - \bar{x})^2$
4.28	−0.01	0.0001
4.21	−0.08	0.0064
4.30	0.01	0.0001
4.36	0.07	0.0049
4.26	−0.03	0.0009
4.33	0.04	0.0016
$\sum$ = 25.74	$\sum$ = 0.00	$\sum$ = 0.0140

The mean, $\bar{x} = \dfrac{\sum x_i}{n} = \dfrac{25.74}{6} = 4.29$ g

The standard deviation, $s = \left(\dfrac{0.0140}{6 - 1} \right)^{1/2} = 0.053$ g

The best value of the measurement is written as 4.29 ± 0.05 g.

[a]The measured values are those obtained from a series of six replicate measurements of the weight of a sample on a triple-beam balance.

---

[2]If only two or three measurements are made, the standard deviation may be approximated by the average deviation, which is the mean value of the absolute values of the deviations, δ_i:

$$\bar{\delta} = \frac{\sum |x_i - \bar{x}|}{n}$$

area under the normal error curve. Note that we cannot make a definite prediction about any single measurement. We can only say that if we make the measurement a large number of times, we can expect that 95 percent of the values obtained will fall within plus or minus two standard deviations of the mean.

PERSONAL ERRORS.

PERSONAL ERRORS. To the types of errors already described—systematic and random—we might add a third category, the personal error, or blunder. Such errors are all too common in student work. Thus, if the numbers on a scale are misread, or recorded incorrectly, or if part of a solution is spilled in a titration, the result will contain an error. Careful work will not contain any blunders, and any work suspected of containing one should be repeated.

PRECISION OF LABORATORY OPERATIONS

The precision associated with various pieces of equipment you may use in the laboratory is summarized in Table 1-2. These uncertainties express limitations in the reading of the instruments and do not reflect systematic errors.

SIGNIFICANT FIGURES

The limited precision of any measurement must be considered whenever that measurement is to be used, whether that use is to report the value of the quantity, or to constitute one of a series of measurements, or to be combined in a calculation with other measured quantities. The limited precision creates uncertainty, which limits the conclusions that can be drawn from a measurement.

TABLE 1-2

Typical Instrument Uncertainties

Instrument	Typical uncertainty
Platform balance	±0.50 g
Triple-beam (centigram) balance	±0.01 g
Top-loading semimicro balance	±0.001 g
Analytical balance	±0.0001 g
100-mL graduated cylinder	±0.2 mL
10-mL graduated cylinder	±0.1 mL
50-mL buret	±0.02 mL
25-mL pipet	±0.02 mL
10-mL pipet	±0.01 mL
Thermometer (10 °C to 110 °C, graduated to 1 °C)	±0.2 °C
Barometer (mercury)	±0.5 torr

Keeping track of this uncertainty would be rather cumbersome if the uncertainty needed to be reported each time the measurement itself were reported or used in a calculation. Introducing the concept of *significant figures* allows us to *imply* the precision of a measurement without having to state explicitly the uncertainty. Significant figures also allow us to estimate easily the precision of a value that is calculated from a combination of different measured quantities.

For example, when the mass of an object is reported as 56.78 g, it is implied that the object was weighed in such a way that there is no question that the mass is at least 56.7 g and that the value of the hundredths place is thought to be 8, but that some judgment was required in deciding this last digit. We should infer that there is an uncertainty of at least 0.01 g in the value 56.78 g.

The number 56.78 has four significant figures (i.e., digits); three are completely certain (56.7), and one has some uncertainty in it (the 8), but all four digits contain useful information. When the significant figure convention is used, the last digit written is assumed to have an uncertainty of at least 1.

Note that although the use of significant figures to imply the uncertainty is practical, the convention has limitations. The values 35.4 ± 0.1 and 35.4 ± 0.4, for instance, would both be expressed as 35.4 to the correct number of significant figures. Nevertheless, the convention provides a convenient way of stating the approximate uncertainty in a number, and significant figures are especially helpful in estimating the uncertainty in a value calculated from a series of uncertain quantities.

Other examples of the number of significant figures in quantities are given in Table 1-3.

Zeros require special attention in order to determine whether or not they are significant figures in a number. The general rule is that zeros used to hold a decimal place are not significant. The implications of this statement are as follows:

1. Zeros in front of a number are not significant. In Table 1-3 the examples 0.0862 (with three significant figures) and 0.001740 (with four, the last zero being significant) illustrate this.

2. Terminal zeros are significant if they are to the right of the decimal point. Thus the terminal zeros in 0.001740, 1.6300, 14.00, and 250.00 are significant digits. This also means that if a measurement was not made precisely, the terminal zeros must not be written. For instance, 14.00 implies a measurement made to ±0.01. If the measurement is only precise to ±1, only "14" should be reported.

3. Zeros as digits between other significant digits are significant (note 12.0004 with six significant figures).

TABLE 1-3
Number of Significant Figures

Quantity	Number of significant figures
857. 29	5
435	3
6.02×10^{23}	3
0.0862	3
0.001740	4
12.0004	6
1.6300	5
62,000	2
14.00	4
1500.	4
250.00	5

4. By convention, terminal zeros are not significant if there is no decimal point, but a decimal point implies that all digits to the left of it are significant. This is the origin of the implied four significant figures in the quantity 1500., but only two in the 62,000 quantity shown in Table 1-3. To specify 62,000 as having three significant figures it would be necessary to write it in scientific notation as 6.20×10^4.

PROPAGATION OF ERROR IN CALCULATIONS

The limitations of each measurement must be recognized whenever that measurement is to be used or reported. When the value of a measurement is used to calculate another quantity, the uncertainty in the measurement limits the certainty of the calculated value. This uncertainty must be recognized, but the relationship between the uncertainty of the measurement and that of the calculated quantity is not always easy to predict. Significant figures are useful because they enable us to estimate the error in calculated quantities. Two simple rules allow us to produce this estimate for the basic mathematical operations.

1. For multiplication and division the answer is rounded to the same *number* of significant figures as that of the *least* significant factor in the calculation. Examples include:

$$1.48 \times 3.2887 = 4.87$$
$$\frac{2.62}{8.1473} = 0.322$$

$$0.023 \times 1.482 \times 13.25 = 0.45 \qquad (4)$$

$$\frac{3.457}{0.00015} = 23,000 \quad \text{or} \quad 2.3 \times 10^4$$

$$\frac{1.918 \times 0.47523 \times 81.96}{53} = 1.4$$

In each case, the least significant factor determines the significance of the answer, no matter how significant the other factors are. If you perform the foregoing calculations on a calculator, you will notice that calculators are unaware of the significant-figure rules and are more than happy to create an abundance of nonsignificant digits. You must decide which of the digits are to be retained as significant and which are to be discarded.

2. For addition or subtraction the last significant digit in the answer is the right-most common digit of all of the terms in the sum. This rule is more easily seen and rationalized by considering an example, such as adding these two numbers:

$$
\begin{array}{ll}
12.34 & \text{addend} \\
+ 5.6 & \text{addend} \\
\hline
17.94 & \text{sum} \qquad (5)\\
\uparrow & \\
\text{not significant} &
\end{array}
$$

The digit "4" in the sum is not significant because the addend 5.6 does not have a significant digit in the hundredths place. This rule can be understood by realizing that the hundredths digit of 5.6 is unknown and could have any value. When this unknown value is added to the "4" of 12.34, the result is also an unknown (nonsignificant) digit.

Other examples of significant figure determinations in addition and subtraction are:

$$
\begin{array}{lll}
73.213 & 1.00257 & 8.5672 \\
+ 14.84 & + 0.0013 & + 153 \\
\hline
88.05 & 1.0039 & 162
\end{array}
$$
(6)
$$
\begin{array}{lll}
6.02 \times 10^{23} = 6.02 \times 10^{23} & & 1.76541 \\
+ 5.2 \times 10^{22} = .52 \times 10^{23} & & -1.75893 \\
\hline
6.54 \times 10^{23} & & 0.00648
\end{array}
$$

GRAPHING

In many instances the goal of making measurements is to discover or study the relationship that exists between two variables. The pressure and the volume of a gas, the volume and the temperature of a substance, or the color of a solution and the intensity of that color are examples of sets of variables that are related. As one variable changes, so does the other.

We often use graphs to visualize the relationship between two variables. If there are two variables, the graph will be a two-dimensional plot of the points that represent pairs of values of the two variables. An example of a well-drawn graph is shown in Figure 1-4. You should notice that the graph has several features that help to make the meaning of the graph clear.

FEATURES

1. *A Title*   The title on a graph should be a brief but clear description of the relationship under study. Titles like "Lab Number 1" or "Volume and Temperature" are not acceptable because their meaning is clear only to those familiar with the experiment and because the meaning will be lost as memory fades with the passage of time.

2. *Labeled Axes*   Each axis of the graph should be clearly labeled to show the quantity it represents and the units that have been used to measure the quantity. You should recognize the distinction between the quantity measured (pressure, volume, temperature, time, etc.) and the units that have been used to measure that quantity (atmospheres, liters, degrees Celsius, seconds, etc.).

It is convenient to label each axis with the name of the measured quantity followed by the unit, separating the unit (usually abbreviated) from the quantity by a slash mark, for example, Volume/L. Then only numbers need to appear beside each axis, and the axes are not cluttered up with the units of each variable. In effect, the slash means that the variables represented by numbers with units have been divided by the units, leaving dimensionless numbers plotted along the axes.

3. *Scales*   The scale on each axis should be chosen carefully so that the entire range of values can be plotted

FIGURE 1-4

This is an example of a well-made graph. Notice the clear title, the properly labeled axes with variable names and units, the clearly marked data points, and the smooth curve showing the observed trend and the extrapolated trend.

on the graph. For practical reasons, 2, 4, 5, or 10 divisions on the graph paper should represent a decimal unit in the variable. This equivalence will make it easy to estimate values that lie between the scale divisions. For greatest accuracy and pleasing proportions, the scales selected should be chosen so that the graph nearly fills the page. Be sure, however, that no plotted points fall outside the borders of the graph.

Note that the lower left corner of the graph does not *have* to represent zero on either axis. If the range of measured values extends to zero, the latter may be a good choice; but if not, there will be much wasted space on the graph.

4. *Data Points*   It is good practice to mark the location of each data point with a very small dot and then draw a small circle around the dot to make the point more visible.

5. *The Curve*   A smooth curve should be drawn through the points. The curve should pass as close as possible to each of the points but should not be connected point-to-point with short line segments. If the relationship appears to be linear, the smooth curve should be a straight line. If the line is extended past the range of the measured values, this extension should be indicated by a dashed rather than a solid line.

LINEAR RELATIONSHIPS

Although many variables in chemical systems may be related in a complex nonlinear way, some of the relationships turn out to be direct proportions; that is, the value of one variable is a constant factor times the value of the other variable plus or minus a second constant. When this proportional relationship exists, the graph of the two variables will be a straight line. An example of a linear graph is shown in Figure 1-5, which is a graph of the volume of a sample of gas vs. the temperature of the gas.

The *slope* of the line is the constant factor relating the two variables. The slope can be determined by using any two points on the line, as shown in Figure 1-5. Note that the slope is calculated in the units of the two variables.

In Figure 1-5 the point on the line where the value of the volume is zero (which is where the line crosses the temperature axis) is called the temperature *intercept*.

If a relationship exists between two variables, it is possible to estimate what the value of one variable would be for any value of the other variable by using the graph. If the point of interest lies within the range of measured values, this estimation is called *interpolation*. For example, from Figure 1-4 we can determine by interpolation that at 125 °C the vapor pressure of water is about 1800 torr.

If the estimation is made beyond the measured range, the process is called *extrapolation*. From Figure 1-4 we can see that the vapor pressure of water at 155 °C is estimated by extrapolation to be about 4000 torr.

FIGURE 1-5

The relation between the temperature and the corresponding volume of a sample of hydrogen gas. A direct proportion, as represented by Charles's law, $V = kT = k(t\,°C + 273)$, plots as a straight line. (T is the Kelvin temperature.)

Interpolation and extrapolation are useful techniques, but both are estimates and assume that the graph is accurate or that it extends beyond the measured values. For extrapolation, especially, this assumption may lead to incorrect conclusions.

EXERCISES IN MEASUREMENT

Materials and Supplies: Sets of four pennies dated 1981 and earlier and sets of four pennies dated 1983 and later (plus some extra ones if you plan to do experiments on the pennies as described in Consider This), meter sticks, 30-cm metric rulers, milk cartons (quart or half-gallon size), beakers of five or six sizes, 60-cm lengths of string, fine-tipped marking pens.

1. MASS MEASUREMENT. Pennies minted in the United States in 1981 and earlier years are significantly different from pennies minted in 1983 or later. Is it possible to tell the difference between these two kinds of pennies by weighing them?

Work in pairs. One student will receive from the instructor four pennies dated 1981 and earlier, and the other student will receive four pennies dated 1983 and later. Carefully check the zero on the analytical balance. Weigh and record the mass of each of the four pennies to the nearest milligram as well as the year in which each penny was minted. Then put all four pennies on the pan and weigh and record the total mass of the four pennies. Finally, record the masses and dates of the four pennies weighed by your partner.

In your report calculate the mean (average) value and standard deviation for the masses of the pennies that you and your partner weighed. Can you tell from these data if there is a significant difference in the average mass of the two groups of pennies? If there is a significant difference, what hypothesis could you offer to explain the difference?

In Part 3, your instructor may also ask you to plot the data (penny mass vs. year in which the penny was minted) for the class data. This can be conveniently done using a spreadsheet program on a computer, if one is available, or it can be done by hand.

2. VOLUME MEASUREMENT. (a) Using a metric ruler or meter stick, measure the length of each of the three edges of a quart or half gallon cardboard milk carton to the nearest 0.5 mm. (Ignore the irregular portion of the carton at the top.) Record these values. In your report calculate the volume of the milk carton in units of cubic centimeters, liters, and quarts.

(b) Read the volumes of partially filled 10-mL and 50-mL graduated cylinders provided by the instructor. Take care to avoid parallax error and note the position of the bottom of each meniscus as described in the Introduction. Read the value of the smaller cylinder to the nearest 0.02 mL, and read the larger one to the nearest 0.2 mL. Record these values. Report the two values to the instructor. If they are not correct, the instructor will show you how to read the meniscus correctly.

3. GRAPHING. (a) *Circumference vs. diameter of different size beakers.* Using a length of string, measure the circumference of four different sizes of beakers. Draw the string snugly around the beaker, and mark the overlapped ends of the string with a fine-tipped marking pen. Measure the distance between the marks on the string with a meter stick to get the circumference. Record this value. Now measure the diameter of each beaker. An easy way to do this is to place the edges of two books or blocks of wood against the beaker on opposite sides. Then carefully remove the beaker and measure the distance across the gap with a metric ruler or meter stick. Record each diameter.

In your report make a graph of the circumferences of the beakers vs. their diameters. Draw a line through the plotted points. Pick two points on the line (located near the ends of the line) and determine the slope of the line, where

$$\text{slope} = \frac{\text{change in circumference}}{\text{change in diameter}} \quad (7)$$

Is the numerical value of the slope, the ratio of the circumference to the diameter of a circle, what you expected it to be?

(b) *Penny mass vs. year of minting.* If your instructor so directs, use the accumulated data for the class to make a graph of the average mass of your pennies vs. the year in which they were minted. The class data can be entered and plotted by hand or by using a computer and spreadsheet program. If you have several pennies minted in the same year, you can enter a data point for each penny or plot the average mass for that year. From looking at your graph, is it possible to see evidence of any change in the average mass of pennies over time? What year(s) does it appear that this change took place? What hypothesis would you suggest to explain your findings? Was there a change in size of the pennies? If not, how could they be the same size, but have different masses?

☞ **Consider This**

To get to the bottom of the question of how pennies could be different without looking much different to casual observation, some further experimentation is called for. Here are some experiments you can try. Take a penny from each group (1981 or earlier and 1983 or later) and use a triangular file to file four equally spaced notches on the edge of the penny. First carefully observe the notches you have filed. Are there any obvious differences? Then, working in a fume hood, put two notched pennies in a small beaker and cover them with 10 mL of 6 M HCl. Observe the pennies for a few minutes and after standing for an hour or two. Do you see any difference in the way the pennies behave in HCl?

Or try this. First, file notches on the edges of the pennies. Then, using a pair of steel tongs to hold the pennies, try heating the pennies in the flame of a Bunsen burner. Be careful to tilt the burner so that if the penny melts, the molten metal won't fall into the burner and clog it, or worse, fall on your hand. Is there a difference in the way the pennies behave in the flame of a Bunsen burner? Now try to put these observations together to explain any differences you have found in the pennies.

BIBLIOGRAPHY

The National Institute of Standards and Technology (NIST) Internet site (http://physics.nist.gov) provides a full discussion of the physical constants, including historical information and the best values of the constants and their uncertainties.

Mauldin, R. R. "Introducing Scientific Reasoning with the Penny Lab," *J. Chem. Educ.,* **1997,** *74:* 952–955.

Miller, J. A., "Analysis of 1982 Pennies," *J. Chem. Educ.* **1983,** *60,* 142.

Ricci, R. W. and Ditzler, M. A. "Discovery Chemistry: A Laboratory-Centered Approach to Teaching General Chemistry," *J. Chem. Educ.,* **1991,** *68:* 229. The authors describe a measurements experiment they call the Pennies Lab.

Sardella, D. J. "An Experiment in Thinking Scientifically: Using Pennies and Good Sense," *J. Chem. Educ.,* **1992,** *69:* 933.

Stolzberg, R. J. "Do New Pennies Lose Their Shells? Hypothesis Testing in the Sophomore Analytical Chemistry Laboratory," *J. Chem. Educ.,* **1998,** *75:* 1453–1455.

EXERCISES

1. Determine the number of significant figures in each of the following quantities.

45.8736	_____	1.70×10^{-4}	_____
0.00239	_____	0.00590	_____
48,000	_____	1.00040	_____
93.00	_____	4.0000	_____
3.982×10^6	_____	3800	_____

2. Express each of the following quantities using significant figures to imply the stated error.

2.3 ± 0.001	_____	1.77 ± 0.06	_____
1.989 ± 0.0003	_____	$22,500 \pm 10$	_____
1.77 ± 0.01	_____	0.0043 ± 0.000007	_____

3. Round each of the following to three significant figures.

1367	_____	12.85	_____
0.0037421	_____	1.6683×10^{-4}	_____
1.5587	_____	1.632257	_____

4. Perform each of the indicated operations and report the answer to the proper number of significant figures.

32.27×1.54 _____

$3.68 / 0.07925$ _____

$1,750 \times 0.0342$ _____

$0.00957 / 2.9465$ _____

$(3.2650 \times 10^{24}) \times (4.85 \times 10^3)$ _____

$7.56 + 0.153$ _____

$8.2198 - 5.32$ _____

$10.052 - 9.8742$ _____

$6.75 \times 10^{-8} + 5.43 \times 10^{-7}$ _____

$0.01953 + (7.32 \times 10^{-3})$ _____

$(8.52 + 4.1586) \times (18.73 + 153.2)$ _____

$(8.52 \times 4.1586) + (18.73 \times 153.2)$ _____

$\sqrt{8.32 \times 10^{-3}}$ _____

$(3.84 \times 10^{-2})^3$ _____

$(0.000738 - 8.3 \times 10^{-5}) / (6.298 \times 10^{-8})$ _____

5. To demonstrate that the rules for significant-figure propagation in multiplication do provide an estimate of the uncertainty in the answer, consider a block with rectangular sides whose dimensions have been measured with a millimeter ruler.

Height: 254.7 mm   Width: 136.8 mm   Depth: 25.3 mm

(a) Determine the volume of the block in cubic millimeters.

(b) Determine the volume of the block if you assume that each of the foregoing measurements is actually 0.1 mm too low.

(c) With which digit do the answers to parts a and b begin to differ? If you report the volume using all of the digits that are the same in parts a and b plus one more digit where the two values do not agree, how many digits should you report?

(d) How many digits would you report according to the rules for significant figures? Is this the same as in part c?

6. The density of water is determined by measuring the mass of a known volume of water. If you determined the density of a 25-mL sample of water, how many significant figures should you report if you used the following instruments? (Refer to Table 1-2 for the instrument uncertainties.)

(a) an analytical balance for measuring mass and a 25-mL pipet for the volume measurement?

(b) a centigram balance and a pipet?

(c) an analytical balance and a graduated cylinder?

(d) a centigram balance and a graduated cylinder?

3. Graphing

(a) Circumference vs. diameter for different size beakers

Beaker size (mL)	Circumference (cm)	Diameter (cm)

On the graph paper provided, plot the circumference along the vertical axis and the diameter along the horizontal axis. Following the examples shown in Figures 1-4 and 1-5, give the graph a title and label the variables (with units) plotted on each axis.

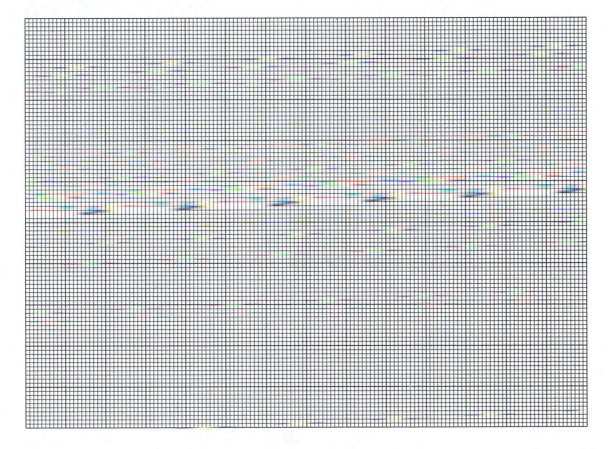

Calculate the slope of your graph, following the example shown in Figure 1-5, and discuss the significance of the numerical value of the slope.

$$\text{slope} = \frac{\Delta \text{ circumference}}{\Delta \text{ diameter}} = \underline{\hspace{3cm}}$$

(b) Penny mass vs. year of minting

Make a plot of the average mass of the pennies vs. the year in which they were minted for each year for which you have data. Expand the mass scale (a range of 2.3 to 3.3 grams is suggested) so that you can easily see any trends. The horizontal axis should range over the years from your oldest to newest penny (there may be gaps in your plot if you have no data for a particular year).

From looking at your graph is it possible to see any evidence of a change in the average mass of pennies over time?

What year(s) does it appear that this change took place?

Was there a change in the size of the pennies? If not, how could they be the same size, but have different masses?

Consider This

If you tried any of the experiments, briefly summarize your observations and what you conclude from them.

EXPERIMENT 2

Mass and Volume Relationships

PURPOSE

To become familiar with metric units of mass, length, and volume. To measure the densities of liquids and solids. To investigate the relationship between solution density and solution concentration.

PRE-LAB PREPARATION

UNITS OF MEASUREMENT

Chemistry is distinctly an *experimental* science. The establishment of the fundamental laws and theories of the nature and behavior of matter depends on the careful measurement of various quantities—mass, volume, length, temperature, time, or electrical magnitudes

The *metric system* of units is especially convenient because it is a decimal system, like our system of numbers based on powers of ten. The standard metric units were originally intended to be related to certain quantities in nature. For example, the meter was intended to be one ten-millionth of the distance from the North Pole to the equator along the meridian line through Paris. The first international standard meter was defined as the distance between two fine lines engraved on a bar of platinum-iridium alloy that is kept in a vault at the International Bureau of Weights and Measures near Paris. In 1960, a new standard was adopted by the Eleventh General Conference on Weights and Measures, and the standard meter was defined to be 1,650,763.73 wavelengths of a particular orange-red line of the gaseous element krypton-86. This standard has the advantage that it permits comparisons that are ten times more accurate than is possible with the meter bar and can be precisely reproduced, so it is not subject to accidental loss or destruction, as is the meter bar.

A cornerstone of relativity theory is the notion that the speed of light is a universal constant that is the same for all observers, regardless of the motion of the observer or of the source of light. This makes it reasonable to treat the speed of light as a constant of nature and use it in conjunction with the unit of time to provide a unit of length. Because time can be measured accurately with atomic clocks and because the speed of light is now known accurately, the 1983 General Conference on Weights and Measures was able to redefine the meter in terms of the velocity of light as "the length of the path traveled by light in a vacuum during a time interval of 1/299,792,458 of a second."[1]

Multiples and decimal fractions of the meter (represented as m), as well as their symbols, are listed below.

1 km = 1 kilometer = 1000 meters
1 dm = 1 decimeter = 0.1 meter
1 cm = 1 centimeter = 0.01 meter
1 mm = 1 millimeter = 0.001 meter
1 μm = 1 micrometer = 10^{-6} meter
1 nm = 1 nanometer = 10^{-9} meter
1 pm = 1 picometer = 10^{-12} meter

A related unit of length sometimes encountered in the older chemical literature is the *angstrom* (1 Å = 10^{-8} cm = 10^{-10} m). The size of a hydrogen atom is of the order of 1 angstrom = 100 picometers.

The kilogram was intended to be the mass of a cubic decimeter (1000 cm^3) of water. The present standard of mass is a platinum-iridium cylinder kept at the International Bureau of Weights and Measures and assigned, by international agreement, a mass of 1 kilogram. Each nation has its own metric standards, but all are based on and have been carefully compared with the originals kept in France. The U.S. copy of the international standard of

---

[1] Pipkin, F. M., and Ritter, R. C. "Precision Measurements and Fundamental Constants," *Science*, **1983**, *219*: 913.

mass, known as Prototype Kilogram No. 20, is housed in a vault at the National Institute of Science and Technology (NIST). Practically all scientists working on research and development today use metric units.

The common submultiples of the *kilogram* (kg) are the *gram* (g) and the *milligram* (mg). Chemists often use the terms *mass* and *weight* interchangeably, although they represent different concepts. Mass is defined in terms of the standard kilogram. Weight is a *force*, actually a mass times the acceleration due to the gravitational attraction of the earth. (On the moon, an object weighs less than it does on earth because the moon's mass and gravitational attraction are smaller than the earth's.) In chemistry we are concerned primarily with mass, but because virtually all weighings in the laboratory involve the comparison of an unknown mass with a standard mass, the operation we call *weighing* is actually a measurement of mass. This dual terminology seldom causes problems, and in this manual when we say "weigh an object," we mean "determine the mass of the object."

The practical laboratory unit of volume unit is the *liter* or cubic decimeter. This is a derived unit, having the units of length cubed, m^3. It is the volume enclosed by a 1-decimeter cube. The *milliliter* (mL) is the most commonly used volume unit in the chemical laboratory. The prefix *milli* tells us that it is one thousandth the volume of the liter. Recently, the milliliter was redefined slightly so that $1 \text{ mL} \equiv 1 \text{ cm}^3$. (The triple equal sign means "is *exactly* equal to.")

A condensed table of the most commonly used metric units of length, mass, and volume, and their English equivalents, is included in the Appendix as Table 1. Consult the internet site listed in the Bibliography for more detailed information.

PRECISION OF MEASUREMENT

See Experiment 1 for a more complete discussion of the treatment of experimental errors, the concept of precision, and related topics.

When measuring physical quantities, it makes sense that the measuring devices should have a precision consistent with the overall precision desired. So, for a precision of 1 percent in the weighing of a 10-g sample, a balance that is accurate to 0.1 g ($0.01 \times 10 \text{ g} = 0.1 \text{ g}$) is sufficient. For the same precision with a 1-g sample, a balance accurate to 0.01 g is necessary. In this experiment, balances that read to 0.01 g and pipets that can deliver with an accuracy of about 0.05 mL should not cause uncertainty greater than about 1 percent in the calculated density. In a determination that combines several measurements, there is no point in using a much greater precision for one measurement than for the other. Choose your measuring devices according to the precision desired.

DENSITY. Ordinary matter, the stuff of which our familiar world is made, has two important properties. We feel its weight when we hold it in our hands, and it takes up space. In the language of science we say that a particular sample of matter has a definite *mass* and *volume*. The ratio of these quantities, or the mass per unit volume, is called the density, written $d = m/V$. In the metric system, this ratio is usually expressed as grams per cubic centimeter (g/cm^3) or the equivalent grams per milliliter (g/mL). Density is a fundamental physical property of matter.

The measurement of density is necessary for a variety of important procedures in the science of chemistry, such as the calculation of Avogadro's number from unit-cell dimensions of crystals, the determination of the molecular weight of a substance from its gas density, the conversion of the height of a column of liquid to hydrostatic pressure units, the conversion from mass to volume, and the determination of the concentration of a solute from density measurements.

An understanding of density is also useful in the everyday world. For example, the service station attendant determines the charge of an automotive battery by measuring the density (and hence the concentration) of the sulfuric acid solution in the battery. Also, a wine maker (even the amateur making wine at home) measures the density of the grape juice to determine whether the sugar content is sufficient for fermentation.

This experiment focuses on the density of liquids (aqueous solutions) and solids. To give you some idea about the volume of space occupied by 0.2 g samples of different substances, Figure 2-1 shows their relative volumes. The smaller the volume the greater the density of the substance.

Note in Figure 2-1 the difference between the density of gases and that of liquids or solids. In liquids and solids the atoms or molecules are about as close together as they can get. In gases there is a lot of empty space between the molecules. Liquids and solids are not very compressible. For example, doubling the pressure on any volume of water from one to two atmospheres decreases the volume by only 0.005 percent. Doubling the pressure on a gas will reduce its volume by about a factor of 2.

Most liquids and solids expand slightly on heating, and the volume of a sample of water increases about 4 percent on heating from 4 °C to 100 °C. Water is a curious substance. Its density *increases* slightly as it is heated from 0 °C to 4 °C. At 4 °C water has its maximum density of 1.000 g/cm^3. As heating is continued above 4 °C, the density of water *decreases*, following the normal behavior of liquids. Because density changes with temperature (and pressure), it is necessary to specify the temperature when reporting the measured density of a liquid or solid. (The pressure is ordinarily assumed to be one atmosphere unless otherwise specified.)

would be numerically equal to the density of the substance at 20 °C.

If the mass of the substance is compared to the mass of an equal volume of water at some temperature other than 4 °C, the specific gravity and density are no longer equal because the density of water is generally less than 1.0000 g/cm³ at temperatures other than 4 °C. For example, if the specific gravity of a solution, D_{20}^{20}, is specified as equal to 1.0400, its density at 20 °C would be equal to

$$1.0400 \times d_{water} \text{ at } 20 \text{ °C } (0.9982 \text{ g/cm}^3) =$$
$$1.0381 \text{ g/cm}^3 \text{ at } 20 \text{ °C.}$$

THE BUOYANCY AND DENSITY OF LIQUIDS AND SOLIDS. Buoyancy is the tendency of a substance to remain afloat or to rise in a fluid. A cork floats on water. An ice cube also floats on water, but it does not ride as high out of the water as a cork. Each substance that floats on water finds a level where the mass of water displaced is equal to the mass of the substance. Based on these observations, which is denser, ice or cork?

A hydrometer, shown in Figure 2-2, operates on this principle. The greater the density of the liquid in which it

FIGURE 2-1

The relationship of density to volume. The cubes of different sizes represent the relative volumes of equal masses (about 0.2 g) of the various materials at 0 °C. Note the difference in density units of solids (g/cm³) and gases (g/L).

SPECIFIC GRAVITY. Specific gravity is a property closely related to the concept of density. It is the mass of a substance compared to the mass of an equal volume of water. This is equivalent to the dimensionless ratio of the density of a substance measured at some temperature divided by the density of water at 4 °C or some other specified temperature. (The two temperatures do not need to be the same but must be specified.) For example, the specific gravity of a substance, symbolized by D_4^{20}, is equal to the density of the substance measured at 20 °C divided by the density of water at 4 °C.

$$\text{Specific gravity } (D_4^{20}) = \frac{d_{substance} \, (20 \text{ °C})}{d_{water} \, (4 \text{ °C})}$$

Because water has its maximum density of 1.0000 g/cm³ at 4 °C, the specific gravity, D_4^{20}, of the substance at 20 °C

FIGURE 2-2

The hydrometer is used by wine makers to monitor the progress of fermentation of the grape sugars to alcohol. The dual scale reads both density and mass% sugar (grams of sugar per 100 g of solution).

floats, the higher the stem of the hydrometer rises above the liquid. The density of the liquid is read from a scale in the stem of the hydrometer. In wine making, the density of the grape juice, which can be measured by the hydrometer, depends on the amount of sugar in the juice. In the process of fermentation, the density of the fermenting juice decreases as sugar is converted to alcohol.

Cork and ice don't mix with water to form a *homogeneous* solution. (A solution is said to be *homogeneous* if every part of the solution has the same composition.) What do you think would happen if we added two homogeneous solutions having different densities to one another? If the solutions are completely miscible, would you expect them to immediately mix? (Solutions are said to be *miscible* if they can be mixed together in all proportions to form a homogenous solution.) We will be doing some experiments to discover answers to these questions.

MEASUREMENT TECHNIQUES

Your instructor will demonstrate the correct techniques to use in reading the meniscus in a graduated cylinder and in careful weighing with the balance. See also Figures 2-3 and 2-4. For every weighing, observe the following rules and precautions:

1. *Keep the balance pans clean and dry.* Clean up *immediately* any chemical that is spilled.

2. *Make certain the balance is level.* Sensitive balances usually have a bubble level. The balance is level when the

FIGURE 2-3
The proper method of reading a meniscus (curved surface of a liquid) in order to measure the volume of the liquid.

bubble is centered in the black circle on the bubble level. Error will be introduced into the reading if the balance is not level.

3. *Check the rest point (or tare) of the empty balance.* Electronic balances have no swinging beam but usually have a tare button that allows you to zero the reading.

If you have a platform or triple-beam balance, first be sure all movable beam weights are at their zero position. Then release the beam release (if the balance has one), give the pan a little push to cause it to swing gently, and note the central position on the scale about which the pointer oscillates. Use this point as the reference-zero rest point in your weighings.

Never take readings with the beam and pointer at rest. When the balance beam is stationary, it is possible for the balance beam suspension to stick in a position that is not the true equilibrium position. If the pointer reading differs by more than two to three scale divisions from the marked zero point, have your instructor adjust the balance. *Do not change the balance adjustments yourself.*

4. *Never weigh an object while it is warm.* The convection currents of warm air will affect the mass reading of an electronic balance or the rest point of a balance beam.

5. *After weighing an object, return the beam weights to the zero position, and restore the beam release to its rest position.*

EXPERIMENTAL PROCEDURE

Special Supplies: One six-pack each of diet and regular (sugared) cola drinks in 12-ounce aluminum cans; one or more buckets of 2–3 gallons capacity; fine-tipped (50 drops/mL) polyethylene transfer pipets; 10 × 75 mm and 13 × 100 mm test tubes; pipetting devices; red and yellow food coloring; aluminum metal pellets or lengths of aluminum rod (¾ in. diameter × 2 ¾ in. long); other metal samples of regular shapes (cylinders or rectangular blocks); metric rulers and/or metric vernier calipers.

Reduced scale procedure: 10-mL polystyrene serological pipets.

Macroscale procedure: 50-mL polystyrene serological pipets.

Chemicals: liquid samples: 8% and 16% by weight sucrose solutions, each colored a different color by adding a small amount of red and yellow food coloring.

1. THE BUOYANCY AND DENSITY OF DIET AND REGULAR COKE (A GROUP EXERCISE). For this first experiment you will need a six-pack of carbonated diet drinks and a six-pack of regular (sugared) drinks in 12-oz (355-mL) aluminum cans. (either Diet Coke and Coca-Cola Classic or Pepsi will work fine.) This experiment will be done as a group exercise and need not be carried out by each individual. Weigh and record the mass (to the nearest 0.1 g) of each of the cans of sugared cola. (Make sure that the outsides of the cans are dry.) Then fill a 2–3 gallon bucket with tap water, and place the previously weighed cans in the bucket. Do they sink or float? Repeat the experiment with six previously weighed

FIGURE 2-4

Laboratory balances. (A) A platform balance for weighing to the nearest 0.1 g. (B) A triple-beam balance for weighing quantities to the nearest 0.01g. (C) An electronic top-loading balance for weighing to the nearest 0.001 g (the necessary draft shield is not shown).

cans of diet cola. Do you observe a difference in the buoyancy of the cans of regular and diet drinks?

(a) Calculate the average mass for each sample of six cans. Is there a significant difference in the average mass of a can of regular and diet cola?

(b) Next, determine the approximate density of diet and regular cola. To do this you need to know the average mass and volume of the beverage contained in the can. You can get the net mass of beverage in the can by subtracting the average mass of an empty aluminum can from the average mass of a can of the beverage. If you are supplied with some clean, dry, empty aluminum cans, you can determine this yourself. (We found the average mass of an empty 12-oz aluminum can to be about 13.7 g.)

Assuming that the average volume of the drink is that specified by the manufacturer on the can's label (355 mL), calculate the density of regular and diet cola.

Compare this density to that of water at the same temperature. You can find the density of water in a handbook, such as the *CRC Handbook of Chemistry and Physics* or *Lange's Handbook of Chemistry*.

(c) Also calculate the mass percentage of sugar in the regular (sugared) cola, using the manufacturer's specified sugar content and volume (355 mL), and the density you calculated in Part 1(b). The grams of cola = 355 mL × density (g cola/mL).

$$mass\% \ sugar = \frac{g \ sugar}{g \ cola} \times 100$$

2. THE BUOYANCY OF SUGAR (SUCROSE) SOLUTIONS. We will see if two miscible solutions having different densities can be floated on one another. The two solutions will at first be identified only by their different colors, produced by adding a tiny amount of food coloring to the solution. One solution is red, and the other solution is yellow. One of the solutions contains 8.0% sucrose by mass (or weight), the other 16% sucrose by mass. (Sucrose is common table sugar.)

Your job is to try to find out which solution is the most dense by trying to float one solution on the other and later to make an accurate density measurement of the two solutions. At this point, your instructor will tell you the composition of the two solutions, and you will then be able to see if there is a correlation between the density of a solution and its composition, expressed as mass% sucrose. (An aqueous solution that is 8% sucrose by mass contains 8.0 grams of sucrose for every 100 g of solution, meaning that 100 g of solution contains 8.0 g of sucrose and 92 g of water.)

For the experiment you will need two fine-tipped (50 drops/mL) polyethylene transfer pipets, two 13 × 100 mm test tubes, and about six 10 × 75 mm test tubes. Obtain about 5 mL each of the red and yellow sucrose solutions in two clean 13 × 100 mm test tubes. Put a transfer pipet in each colored sucrose solution, and suck up some solution into the pipets. As you fill the pipets, leave the tip below the surface of the solution, and take your fingers completely off the pipet bulb, so that you do not draw air into the stem of the pipet. Put about a 1-cm deep layer of red solution in a clean, dry 10 × 75 mm test tube. Put the same amount of yellow solution in a second, and the same amount of pure water in a third 10 × 75 mm test tube.

Now observe and record what happens when you add solution of the opposite color (red to yellow and yellow to red). Begin by adding red to the yellow in the following way. Put the tip of the pipet with yellow solution about 1 millimeter below the surface of the red solution, and *gently* squeeze the bulb to add the red solution one drop at a time. You will be able to see what's happening more clearly if you view the tube against a white piece of paper. Do the red drops sink or rise as they are added? Keep adding solution drop by drop until you have added a volume of the red solution that is about equal to the volume of the yellow solution already in the tube. Repeat the process, adding yellow solution to a tube of red solution. After finishing, view both tubes against a white background. In which tube is one layer most clearly floating on top of the other? What conclusions do you draw from these observations? Which solution has the higher density, the red or yellow sucrose solution?

3. QUANTITATIVE MEASUREMENT OF THE DENSITY OF SUGAR (SUCROSE) SOLUTIONS.

Reduced scale procedure. This procedure is preferred if your balance has a sensitivity of ±0.01 or 0.001 g. You

will need a 10-mL polystyrene serological pipet and six 25- or 50-mL Erlenmeyer flasks to do parts (a), (b), and (c). (A polystyrene pipet is preferred because it is not wet by aqueous solutions, so that when the pipet drains, little or no solution will be left behind.)

Macroscale procedure. If your balance has a sensitivity of ±0.1 g, the following procedure employing a 50-mL graduated cylinder can be used. Weigh the empty, clean, and dry graduated cylinder to the nearest 0.1 g. Put slightly less than 50 mL of solution in the graduated cylinder, and reweigh the cylinder with solution to the nearest 0.1 g. Also read and record the exact volume of the solution, estimating the volume to the nearest 0.2 mL. The ratio of the mass of liquid to its volume gives the density.

(a) *Density determination of the red sucrose solution.* Number each flask; then weigh and record the masses of the clean, dry Erlenmeyer flasks. Use a pipeting device to draw up a little more than 10 mL of the red sucrose solution into the 10-mL serological pipet. Adjust it exactly to the full mark; then deliver the contents into a previously weighed Erlenmeyer flask. Blow out any liquid in the tip of the serological pipet into the flask. (The double ring at the top of the pipet means that it is designed to deliver 10 mL when all of the liquid is blown out.) Reweigh the Erlenmeyer flask to obtain the mass of sucrose solution. Calculate the density of the sucrose solution as mass/volume (g/mL). (A more accurate value for the volume of the pipet can be determined by calibration with pure water as described later in section 3[c].)

Also measure and record the temperature of the solution.

No matter whether you are using the reduced scale or macroscale procedure, repeat all of these measurements with a fresh sample of the same liquid, including temperature. The average of duplicate determinations will be more reliable and will provide you with a check on gross errors in measuring mass and reading volumes correctly.

(b) *Density determination of the yellow sucrose solution.* Now repeat the entire process described in Part 3(a) in duplicate using the yellow sucrose solution. (Be sure that you have rinsed out all of the previous liquid before measuring the volume and mass of the next liquid.)

When you have finished the measurements on both sucrose solutions your instructor will tell you which is the 8.0 mass% and which is the 16.0 mass% sucrose solution.

(c) *Density determination of water and pipet volume calibration.* As a final check on your technique, and to provide a way to determine the volume of your pipet, repeat the entire process in duplicate using pure water, if your instructor so directs. Record the temperature of the water, and compare your calculated density for water with the value (at the same temperature) found in the *CRC Handbook of Chemistry and Physics* or some other reference source. If the density you measure is significantly different (± 0.01 g/mL) from the literature value, it may indicate that the volume of the polystyrene pipet was significantly different from the nominal value, since manufacturing tolerances on the pipet volumes are on the order of 0.5–1 percent. Use the average mass of water delivered divided by the known (literature) value for the density of water at the same temperature to calculate the corrected volume of the pipet. You can use the corrected volume of the pipet in all of your density calculations.

$$\text{Corrected pipet volume } (mL) = \frac{\text{mass of water delivered } (g)}{\text{density of water } (g/mL)}$$

4. MAKING A GRAPH OF THE DATA.

Is there a simple relationship between density and mass% sucrose? To help visualize it, construct a graph of **density (g/mL)** vs. **mass% sucrose**. Include a data point for pure water (0% sucrose), so that you have a total of three data points. If you did not determine the density of pure water as described in Part 3(c), use the literature value for the density of water at the same temperature as your sucrose solutions. If your instructor so directs, plot the average values of density using the combined data of all of the members of your laboratory section. (The average of several values is statistically a more reliable measure of the true value of a quantity than any single measurement.) Draw a straight line that looks like a reasonable fit to the points. (Don't connect the points together by straight line segments). Better yet, use a spreadsheet or graphing program to plot the points and to add the best-fit line to the data points.

The graph you make can be used to estimate the sugar content in regular or diet cola or any common beverage such as apple juice or other fruit juice drinks. (In these drinks, besides water, the main ingredient is sugar.) We will use it to estimate the sugar content in the regular (sugared) drink we used in Part 1.

On the graph of density vs. mass% sucrose, draw a horizontal line corresponding to the density of regular cola calculated in Part 1. Where this line intersects the best-fit straight line (drawn through your three data points), drop a perpendicular line to the mass% sucrose axis and read off the corresponding mass% sucrose. Compare this value with the mass% sucrose you calculated in Part 1 (g sugar/g cola $\times$ 100). Do the two values agree, considering possible sources of uncertainty in the measurement of density and mass% sucrose?

5. THE DENSITY OF SOLID ALUMINUM.

(a) Use the following technique to determine the density of an irregularly shaped solid. You will be supplied with aluminum metal pellets or some other sample that is selected by your instructor. (Lengths of aluminum rod 3/4 in. diameter $\times$ 2 3/4 in. long are also suitable.) Use about 50 g of the aluminum pellets (avoiding fine or powdered material), or an amount that will give a volume increase of around 20 mL.

Weigh a small beaker to a precision of 0.1 g. Add the sample of metal pellets and weigh again. Place about 25 mL of water in a 50- or 100-mL graduated cylinder, and read the volume to the nearest 0.1 mL. Tilt the cylinder, and slide the weighed sample pieces into it carefully, to avoid loss of water by splashing, then tap the sides to dislodge any adhering air bubbles. Again read the volume. The increase is the volume of the sample. (See Figure 2-5.) Calculate the density of the sample. If time permits, make a duplicate determination with another sample, to increase the accuracy of your measurements.

Save the aluminum pellets or other material you used, since they can be dried and reused.

(b) *Optional.* If you wish to determine the density of a regular-shaped solid such as a rectangular block or a cylinder of aluminum (or some other material), the volume can be calculated from measurements of the dimensions. The mass may be obtained directly on the balance, and the density can then be calculated. In making these measurements, note how the quality and precision of the rulers and balances you use affect the number of significant figures you retain, and hence determine the degree of precision you can expect to obtain in your measurement.

The difference between these readings is the volume of these pieces of material.

FIGURE 2-5
A method of measuring the volume of an irregularly shaped solid.

Consider This

If you add ice at 0 °C to a container of water maintained at 0 °C, the ice will float on the water with the water finding a certain level in the container. As the ice gradually melts, forming liquid water at 0 °C, would you expect the water level in the container to go up, down, or remain the same? Explain the reasoning by which you arrived at your conclusion.

The density of diet cola is about the same as pure water. The cola was contained in an aluminum can, and if you did Part 5 of the experiment, you know that the density of aluminum is considerably greater than that of water. Considering all this information and your own experience drinking carbonated beverages, explain how it is that cans of diet cola can float in water while cans of regular (sugared) cola sink? Wouldn't you expect both kinds of cola cans to sink since both diet cola and regular cola have a density equal to or greater than water and they are contained in an aluminum can that has a density greater than water?

The mass of the earth is known to be about 5.98 $\times$ 10^{24} kg. Assume that it is approximately spherical, with an average diameter of 1.276×10^7 meters. From these data calculate the average density of the earth (g/cm³). Could the earth be made mostly of water

(density = 1.00 g/cm³), abundant elements like calcium silicate (Ca_2SiO_4, density = 3.27 g/cm³), or aluminum oxide (Al_2O_3, density = 4.0 g/cm³)? What hypothesis would explain why the earth has a considerably higher density than any of these materials, which are abundant at the surface of the earth?

BIBLIOGRAPHY

The National Institute of Standards and Technology (NIST) Internet site (http://physics.nist.gov) provides a full discussion of the physical constants, including historical information and the best values of the constants and their uncertainties.

Henderson, S. K.; Fenn, C. A.; Domijan, J. D. "Determination of Sugar Content in Commercial Beverages by Density: A Novel Experiment for General Chemistry Courses," *J. Chem. Educ.*, **1998,** *75*(9):1122–1123.

Pipkin, F. M.; Ritter, R. C. "Precision Measurements and Fundamental Constants," *Science,* **1983,** *219:* 913.

Stein, M.; Miller, D. "Density Explorations," *The Science Teacher,* February **1998,** 45–47.

Summerlin, L. R.; Borgford, C. L.; Ealy, J. B. "66. Sugar in a Can of Soft Drink: A Density Exercise," *Chemical Demonstrations: A Sourcebook for Teachers,* Volume 2, Second Edition, American Chemical Society, Washington, D.C., 1988, 126–127.

REPORT

2

Mass and Volume
Relationships

NAME

SECTION   LOCKER

INSTRUCTOR   DATE

DATA, CALCULATIONS, AND OBSERVATIONS

1. The buoyancy and density of diet and regular cola

    Brand name(s) of beverages _____

    Volume of cola listed on the label _____ mL

    Sugar content of regular cola listed on the label _____ g

    Describe the buoyancy behavior of cans of regular and diet cola. Which float and which sink?

    (a) Determination of the average mass of cans of diet and regular cola and empty cans. (Optionally, assume that
        the mass of an empty 12 oz aluminum can is about 13.7 g.)

Mass of cans of regular cola	Mass of cans of diet cola	Mass of empty cans
g	g	g
g	g	g
g	g	g
g	g	g
g	g	g
g	g	g
Average: g	Average: g	Average: g

    (b) After subtracting the average mass of an empty can from the average mass of cans of regular and diet colas,
        calculate the average density of regular and diet cola as the net mass of cola divided by the volume of the cola.

    Calculated density of regular cola

    Calculated density of diet cola

    (c) Calculated mass% of sugar in cola: g sugar/g cola × 100

2. The buoyancy of sugar (sucrose) solutions

    Record what you observed when you layered the red sucrose solution on top of the yellow sucrose solution and vice versa. What are your conclusions about which solution is denser, the red or yellow sucrose solution?

3. Quantitative measurement of the density of sugar (sucrose) solutions

    In calculating the density, use the nominal volume of the solution (10.00 mL) or preferably the corrected volume determined from the pipet calibration in Part 3(c): V(mL) = mass of water(g)/density of water (Literature value, g/mL):

    Volume calculated from data of Part 3(c) _____ mL

	(a) Red sucrose solution		(b) Yellow sucrose solution		(c) Water	
Temperature, ºC						
	Trial 1	Trial 2	Trial 1	Trial 2	Trial 1	Trial 2
Mass of empty flask	g	g	g	g	g	g
Mass of flask + solution	g	g	g	g	g	g
Net mass of solution	g	g	g	g	g	g
Average mass of solution		g		g		g
Volume of solution		mL		mL		mL
Calculated density		g/mL		g/mL		g/mL

4. Making a graph of density vs. mass% sucrose

Make a graph of density / (g/mL) vs. mass% sucrose. (A density scale running from 0.99 to 1.08 g/mL and a mass% sucrose scale running from 0 to 20% will be suitable.)

There should be three data points, one for pure water (0 mass% sucrose) and data points for the red and yellow sucrose solutions. If possible, make your graph using the average of the data for your entire laboratory section. Draw the best straight line through the points.

Using a spreadsheet program (such as Microsoft Excel), makes the drawing of graphs and the addition of the best-fit line (called a "trendline" in Excel) easier, once you have learned how to use the program.

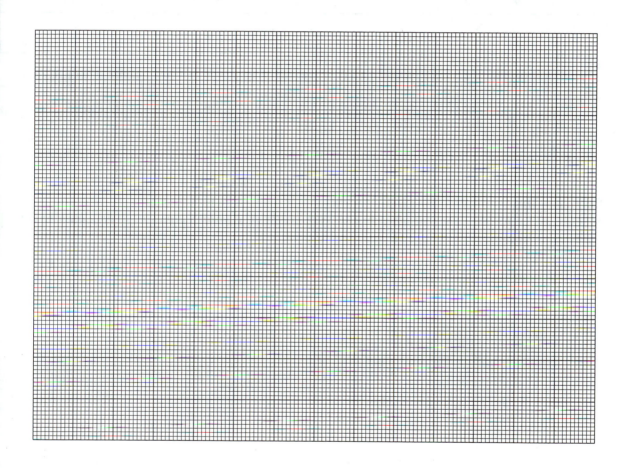

5. The density of solids

   (a) Density of aluminum metal

	Trial 1	Trial 2
Initial volume of water in the graduated cylinder	mL	mL
Final volume of water after addition of the solid metal sample	mL	mL
Net volume of the solid metal sample	mL	mL
Mass of the metal sample	g	g
Calculated density of the metal sample	g/mL	g/mL

   (b) *Optional*: Record the measurements and mass of any regular-shaped solid that you measure, and calculate its density.

☞ **Consider This**

Discuss in the space below the three questions posed at the end of the experimental procedure in the Consider This *paragraph.*

EXPERIMENT 3

Separation of Food Dyes by Paper Chromatography

Separation of Lycopene and Beta-Carotene by Thin Layer Chromatography

PURPOSE

To become acquainted with chromatographic separation techniques by using paper chromatography and thin layer chromatography to separate mixtures of food color dyes and plant pigments.

PRE-LAB PREPARATION

Separating two substances that have quite different physical properties is usually a simple and straightforward task. A mixture of sand and salt can easily be separated by adding water to dissolve the salt, filtering the mixture to remove the sand, and recovering the salt by evaporating the water. But what if the substances to be separated have similar physical and chemical properties? Separation of a mixture of several closely related substances presents a more difficult problem, but the problem has been solved by the development, over the past 75 years, of a family of powerful separation techniques that collectively are called *chromatography.*

The word *chromatography,* formed from the Greek words *khroma,* meaning color, and *graphein,* meaning to draw or to write, was coined by the Russian botanist M. S. Tswett around 1906 to describe his process of separating mixtures of plant pigments. He allowed a solution of plant pigments to percolate down a glass column of absorbent powder where they separated into colored bands, literally a "color drawing," on the powder. Even though most chromatographic methods are now used on solutes that do not produce bands of color visible to the human eye, the term *chromatography* is applied to any separation procedure employing the same principle as the method described by Tswett. Chromatography is widely used for qualitative or quantitative analysis or for isolation and recovery of the components of a mixture.

All forms of chromatography employ the same general principle: A mixture of solutes in a *moving phase* passes over a selectively absorbing medium, the *stationary phase.* Separation occurs because the solutes have different affinities for the stationary and moving phases. Solutes that have a greater affinity for the moving phase will spend more time in the moving phase; therefore, they will move along faster than solutes that spend more time in the stationary phase.

There are several ways of achieving a flow around or through a stationary phase. In *column chromatography* small particles of the stationary phase are packed in a tube, and the moving phase (a liquid or gas) flows around the confined particles. In *thin-layer chromatography* *(TLC)* the stationary phase is a thin layer of fine particles spread on a glass or plastic plate. In *paper chromatography* the paper itself forms the stationary phase, along with any solvent adsorbed to the paper. Figure 3-1 illustrates the separation of a sample mixture as a moving liquid phase flows through a stationary phase (such as paper).

In this experiment we will employ two forms of chromatography, paper chromatography and thin layer chromatography (TLC) using silica gel. The cellulose structure of paper (shown in Figure 3-2) contains a large number of hydroxyl groups ($-OH$), which can form weak bonds (called hydrogen bonds) to water molecules, so the stationary phase can be regarded as a layer of water that is hydrogen-bonded to cellulose. If the solvent is water, the moving phase is also aqueous; but if a mixture of an organic solvent is used with water, the moving phase is apt to contain a high proportion of the organic solvent.

The solvents employed in chromatography must wet the stationary phase so that the mobile phase will move through the paper fibers (or through the particles in TLC) by capillary attraction. A solute in the mobile phase

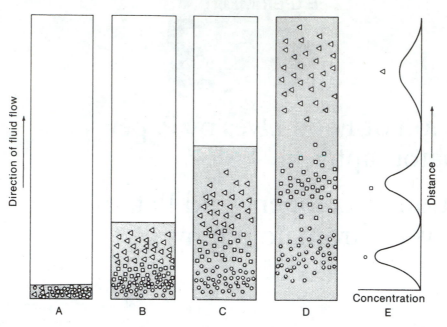

FIGURE 3-1

Schematic illustration of the separation of a three-component mixture by differential migration on a strip of paper. Parts A through D represent a time sequence showing how the components move along at different rates. (A) At the beginning of a separation, the three components are clustered together at the point of sample application. (B–D) As the sample migrates along the paper strip, the three sample components are gradually separated, the fastest-moving component spending the most time in the moving liquid phase. (D) The final distribution of substances. (E) A smoothed plot of the concentration of each substance as a function of the distance from the origin, for the distribution shown in D.

FIGURE 3-2

The structure of cellulose. Carbon atoms are located also at the intersections of the lines. The wavy lines indicate that the overall structure consists of multiple chains of these units. Note the many polar $-OH$ groups on the cellulose. Such groups have an affinity for water molecules.

moves along with the solvent during the chromatographic *development* (the term for the process that occurs as the solute moves along and is partitioned between two phases). As the solute moves, it undergoes many successive distributions between the mobile and stationary phases. The fraction of time it spends in the mobile phase determines how fast it moves along. If it spends all of its time in the moving phase, it will move along with the solvent front (the leading edge of the solvent moving through the stationary phase). If it spends nearly all of its time in the stationary phase, it will stay near the point of application.

After the chromatogram has been developed and the solutes on the paper have been located, the movement of the solute on the paper is expressed by the R_f value (called the *retention factor*), where

$$R_f = \frac{\text{distance traveled by the solute}}{\text{distance traveled by the solvent front}}$$

The distances used in calculating R_f values are measured as shown in Figure 3-3. The distance traveled by the solute is measured from the point of application to the center or densest part of the spot; the distance traveled by the solvent front is measured from the point of application to the limit of movement of the solvent front (which must be marked immediately after the paper is removed from the developing chamber because the front may become invisible after the solvent evaporates).

If all conditions could be maintained constant, R_f values would be constant. However, they can be altered by variations in temperature, the composition of the solvent phase, or changes in the paper. The R_f value is useful mainly for expressing the relative mobility of two or more solutes in a particular chromatographic system. Absolute R_f values may change from day to day, but their values in relation to each other remain nearly constant.

Paper chromatography can be performed on strips or sheets of paper, in which case the solvent is allowed to flow either upward or downward. Horizontal development may also be used, in which case the solvent is allowed to flow radially outward from the center. Two-dimensional chromatography can be used to separate complex mixtures by applying two different solvent systems. The paper is first developed in one direction, then turned by 90° and developed again in a second solvent.

Development is accomplished in the following way. First, the solution containing the solutes to be separated is spotted on the dry paper—that is, a micropipet or glass capillary is used to put very small drops (microliters) of solution on the paper. After the spots have dried, the filter paper is placed in a tightly closed receptacle containing the developing solvent, whose vapor saturates the atmosphere of the container. The paper is supported so that the developing solvent moves through the spots that have been applied. The point of application must *not* be immersed in the developing solvent; otherwise, the solute spots will diffuse into the solvent and be greatly diluted.

The solvent flow is allowed to continue for a fixed time or until the solvent front reaches a specified point on the paper. Then the paper is removed, the solvent front is marked with a pencil, and the paper is dried. This fixes the separated solute spots in their position at the end of development.

If the solutes are colored, they can be readily seen on the paper. Colorless spots are located by various means. Often the paper is exposed to the vapor of a reagent (such as iodine, which makes organic compounds visible), or it can be dipped in or sprayed with a reagent that reacts with the colorless substances to form a visible spot. When placed under an ultraviolet light, many substances can be located by their fluorescent glow. Or, if fluorescent dyes are added to TLC plates, the spots are seen as dark spots against a fluorescent background.

After the spots are located and their R_f values measured, they are compared with standard known materials run under the same conditions. If possible, known and unknown substances are run on the same paper. If the known and unknown compounds have the same R_f values in different solvent systems and give the same reactions with various chromogenic (color-forming) reagents, it is probable that they are identical.

SEPARATION OF FD&C DYES

In the past few years there has been more awareness and concern about the safety of food, drug, and cosmetic (FD&C) dyes. At present only seven of these dyes are allowed in foods: reds 3 and 40, yellows 5 and 6, blues 1 and 2, and green 3. Examples are listed in Table 3-1. The

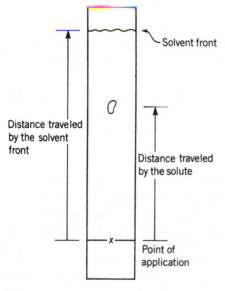

FIGURE 3-3

The movement of the solute on the paper chromatogram is expressed by the R_f value, the distance traveled by the solute divided by the distance traveled by the solvent front.

TABLE 3-1

FD&C Dyes Present in Food Products

Food	FD&C dye
Red maraschino cherry juice	Red 40
Red cough medicine	Red 40, yellow 6
Red mouthwash	Red 40
Green mouthwash	Yellow 5, blue 2
Pickle brine	Yellow 5
Orange soda	Yellows 5 and 6
Grape soda	Red 40, blue 1

FIGURE 3-4

Structures of FD&C dyes. A carbon atom is located at the corner of each hexagon and—whenever no other atom is bonded to a corner—a hydrogen atom as well.

structures of these dyes are shown in Figure 3-4, and information about them can be found in the references cited in the Bibliography. Present law does not require listing of specific dyes on labels, so there is genuine interest in dye separation and identification.

In order to separate and identify dyes in food products, it is necessary to add an extraction and concentration step like those described in the references cited in the Bibliography. In this experiment, however, we will separate and try to identify the dyes in the concentrated form usually sold in supermarkets as assorted food colors. This will allow us to omit the extraction and concentration step.

SEPARATION OF PLANT PIGMENTS

Plants contain a number of natural pigments. Green leaves contain chlorophylls (the green pigments in plants), carotenes (yellow pigments like those found in carrots), and xanthophylls (which also are yellow and may be regarded as mildly oxidized carotenes). Tomatoes

are rich in lycopene, an isomer of carotene with a deep yellow-orange color in dilute solution. The structures of typical compounds that fall in these three categories are shown in Figure 3-5.

EXPERIMENTAL PROCEDURE

Special Supplies: *Separation of food dyes:* four colors of food dyes (blue, green, red, yellow) such as those sold by McCormick under the Schilling label or by Smart & Final under the Iris label (S & F Iris "double strength" red food color should be diluted by a factor of 10 before using it); Whatman No. 1 filter paper, 46 × 57 cm sheets (pre-cut each sheet into twelve 11.5 × 19 cm pieces); centimeter rule; soft lead pencils; applicators (75 mm × 0.5 mm ID glass capillaries, such as Drummond 1-00-800); plastic food wrap (Handi-Wrap); size 33 rubber bands; scissors; stapler.

Separation of plant pigments: canned tomato paste; 15 × 125 mm and 13 × 100 mm test tubes; test-tube rack; fine-tipped (50 drops/mL) polyethylene transfer pipets; double-bladed stainless steel spatula (7 mm blade width); hot plate; centrifuge (for 15 × 125 mm test tubes); #3 and #4 size corks; plastic food wrap (Handi-Wrap); 10-mL beakers; J.T. Baker Baker-flex Precoated Flexible TLC sheets, 2.5 × 7.5 cm; Silica Gel IB-F (or IB) coating; developing chamber (Qorpak or other brand 120 mL (4 oz) wide-mouth screw cap bottles with Teflon-lined screw caps; the bottles are nominal 48 mm OD × 90 mm high); applicators (75 mm × 0.5 mm ID glass capillaries, such as Drummond 1-00-800); soft lead pencils.

Chemicals: *Separation of food dyes:* developing solvent solution is 0.2% (2 g/L) aqueous NaCl.

Separation of plant pigments: 0.1 M K_2CO_3, anhydrous sodium sulfate, $Na_2SO_4(s)$ (freshly prepared 50 mg/100 mL beta-carotene solution dissolved in toluene/acetone (50:50 by volume). (Approximately 0.5 mL of the beta-carotene solution is needed per student; beta-carotene tablets can often be purchased in the vitamin section of drug stores; cut the gelatin tablets open and remove the beta-carotene powder.) The extracting solvent is a mixture of petroleum ether (bp 30–60° C)-acetone (50:50 parts by volume; the developing solvent is petroleum ether-toluene (90:10 parts by volume). A toluene-acetone solution (50:50 by volume) is used to re-dissolve lycopene pigment if the sample solution evaporates to dryness. To avoid the hazard of mixing flammable solvents, all solvent solutions should be prepared for the students in advance.

! SAFETY PRECAUTIONS: Petroleum ether, acetone, and toluene are all flammable solvents. There must be no open flames in the laboratory while this experiment is performed. To minimize exposure to the volatile vapors of these solvents, work in a fume hood, if one is available.

WASTE COLLECTION: The solvents used for the separation of plant pigments (Part 2) should be collected in a bottle labeled "Waste petroleum ether, acetone, and toluene."

1. SEPARATION OF FD&C DYES IN FOOD COLORS.

If pre-cut sheets are not already prepared for you, cut an 11.5 × 19 cm piece of Whatman No. 1 filter paper. Using a soft lead *pencil* (ink pens contain dyes that may migrate from the point of application), draw a line parallel to the long dimension about 2 cm from the edge. Still using a pencil, put four small Xs on the line, beginning 5 cm from the edge of the paper and spacing the marks about 3 cm apart. (The Xs mark the points where you will later spot the dyes.) Put 70 mL of the developing solvent (0.2% NaCl in water) in the bottom of a clean, dry

Color	Name	Structure
Yellow	α-Carotene	
Yellow	ß-Carotene	
Yellow orange	Lycopene	
Yellow	Xanthophyll (lutein)	
Green	Chlorophyll *a* (R = CH₃)	
Green	Chlorophyll *b* (R = CHO)	

FIGURE 3-5

The structures of pigments found in green plants and the tomato. A carbon atom is located at each line intersection and—whenever no other atom is bonded—a hydrogen atom as well. For purposes of this experiment, note the great similarities of structure of the carotenoid pigments. Also note the small differences (at the ends of the molecules) in the structures of beta-carotene and lycopene. Xanthophyll (found in green leaf pigments) contains two waterlike −OH groups. The chlorophylls are green leaf pigments and are catalysts for photosynthesis.

800-mL beaker, and cover the beaker with plastic wrap held in place by a rubber band.

Get four or five applicators (0.5 mm ID glass capillaries). Using a scrap of filter paper, practice applying small (2–3 mm diameter) spots on the paper. (Use about 2 to 3 mm of dye in a 0.5 mm ID glass capillary applicator). When you are confident you have mastered the technique of applying the spots, lay the 11.5 × 19 cm piece of paper down on a clean paper towel or sheet of clean paper.

Carefully apply a small (2–3 mm) spot of the individual food colors on the four Xs, one color per X. Label in pencil near each X with the appropriate letter symbol (B, G, R, Y) to identify the color of the dye spotted there. Allow the spots to dry by gently waving the paper in the air.

After the spots have dried, form the paper into a cylinder as shown in Figure 3-6, and staple the edges together, leaving a gap so that the edges do not quite meet. Put the paper cylinder in the beaker containing the developing

FIGURE 3-6

Form the chromatographic paper into a cylinder, and staple as shown in A. Put the paper cylinder carefully into the beaker containing the solvent, and cover with plastic film held in place by a rubber band as shown in B.

solvent, placing the edge nearest the spots at the bottom and taking care that the paper does not touch the wall of the beaker. *The spots of dye must be above the surface of the solvent in the beaker.* Immediately seal the beaker with plastic wrap, and allow the solvent to rise to within 1 cm of the top of the paper. (This usually takes 30 to 40 minutes.) Note the position of the *solvent front* (the wet-looking leading edge of the solvent) every few minutes, so that you do not let the solvent front rise all the way to the top of the paper. Also note how the spots begin to separate into bands of color as they move along.

When development is finished, remove the paper from the beaker. Immediately mark the solvent front with a pencil line before the solvent evaporates. Then dry the paper thoroughly, preferably in the hood.

After the paper has dried, locate the densest part of the band for each color and draw a pencil line through it. Measure the distance (in millimeters) (a) from the pencil line at the point of application to the densest part of the band of color, and (b) from the distance from the point of application to the solvent front. Record the values in your report. Calculate and record the R_f values for every colored spot on the paper. For each original spot of food color, describe the number of components you find and their colors. Is there a green dye in the green food color? Explain.

Pencil your name and the date on the chromatogram and turn it in with your lab report if your instructor so directs.

2. EXTRACTION OF LYCOPENE FROM TOMATO PASTE AND SEPARATION OF PLANT PIGMENTS.

If possible, this procedure should be carried out in subdued lighting, with the overhead lights turned off if the lab has

outside windows. Lycopene and other carotenoid pigments are light sensitive and also react with oxygen in the air. Since we have no easy way to exclude air, it is best to carry out the extraction and separation during the same lab period. If this is impossible, the lycopene solution can be stored in a well-stoppered vial in a refrigerator.

Obtain a 100-mL beaker and 15 × 125 mm test tube. Put the empty test tube in the beaker. Tare the balance to read zero, or weigh and record the mass of the beaker + test tube to the nearest 0.1 g. Using a narrow blade spatula, place about 3 g of tomato paste in the mouth of the test tube. Use the spatula to work the paste into the tube, leaving a tunnel through the paste so that extracting solvent can be added.

Pour 10 mL of petroleum ether-acetone solution (50:50 by volume) into the test tube. Next, use the spatula to disperse the tomato paste in the extracting solvent. Work the paste down into the solution, scraping it off the sides of the test tube and twirling the spatula between your fingers to disperse the tomato paste as much as possible. Continue this process until the tomato paste is well dispersed into smaller pieces, with little or no paste adhering to the sides of the test tube. Find a cork to fit the mouth of the test tube, cover the cork with a small piece of plastic food wrap, and insert the film-covered cork into the test tube. Then vigorously shake the test tube for about 2 minutes. It's O.K. to rest your arm after a period of shaking, then resume shaking. At this point the extracting solvent should have a definite yellow-orange color from the lycopene pigment extracted from the tomato paste.

If a centrifuge is available, centrifuge the test tube to compact the tomato paste. (Be sure to counterbalance the tube with another tube filled to the same height with water.) If no centrifuge is available, filter off the tomato paste using filter paper in a filter funnel.

After centrifuging the sample, pour off the solution into a clean 15 × 125 mm test tube. Add 5 mL 0.1 M K_2CO_3, stopper the tube, and vigorously shake the tube. This removes acidic compounds. Recentrifuge to separate the phases. Use a fine-tipped transfer pipet to draw off the yellow-orange upper layer, and transfer the material to a 13 × 100 mm test tube. When you carry out this transfer, be careful to put the mouths of the two test tubes right next to each other; otherwise, the high vapor pressure of the solvent may cause the solution to squirt out of the transfer pipet, and you will lose much of the solution containing the extracted lycopene pigment.

Now add 0.5 g of anhydrous sodium sulfate to the 13 × 100 mm test tube, stopper with a cork covered with plastic wrap, and vigorously shake. The anhydrous sodium sulfate acts to dry the solvent by absorbing any dissolved water. If the solution looks cloudy, centrifuge to clarify it.

Next, pour the dried solution into a 10-mL beaker. Put the 10-mL beaker containing the yellow-orange solution on a hot plate set to its lowest heat setting. Allow the solvent to boil and evaporate until there is just a small amount of solution in the beaker (less than 1 mm). Watch the solution continuously as the last of the solvent evaporates. In the last stage, some solid may come out of solution and float on the surface of the solution. This is O.K., but be very careful not to take the solution to complete dryness. This could allow the pigment to overheat, destroying some of it.

At this point remove the beaker from the hot plate. If there is any solid present, it may be redissolved by adding a few drops of toluene-acetone solution (50:50 by volume). (The toluene-acetone mixture is preferred over the petroleum ether-acetone solution used as the extracting solvent because toluene is less volatile and is a good solvent for the lycopene pigment.) To help redissolve any solid, rapidly suck up and expel the solid-solvent mixture in a clean, fine-tipped transfer pipet until the solid has completely redissolved. The solution is now ready to be spotted on a TLC plate, along with a spot of beta-carotene solution.

Prepare the developing chamber by pouring 6 mL of petroleum ether-toluene (90:10 by volume) into a 120 mL (4 oz) wide-mouth screw capped bottle. Next, obtain a Baker-flex 2.5 × 7.5 cm thin layer chromatography sheet. (This consists of a 200 micrometer thick layer of fine silica gel particles coated on a polyester plastic backing. At the time of manufacture a binder is added to make the silica gel particles adhere to the plastic backing.) Since the sheets are rather expensive, follow instructions carefully so that they are not wasted.

First, wash and dry your hands. Get a TLC sheet, handling it by the edges. Place it on a clean towel or sheet of paper, silica gel side up. Using a soft pencil, *lightly* draw a line about one centimeter from the end of the sheet. You will spot on this line, about 10–12 mm apart, two samples—one of the tomato extract (lycopene) and one of the 10 mg/100 mL beta-carotene standard.

To spot the solutions on the plate, use a clean applicator (0.5 mm ID glass capillaries) for each solution. Dip the tip of the glass capillary into the tomato extract solution so that about 2–3 mm of the capillary is filled. Lightly touch the tip of the capillary to a point on the pencil line about 6 mm from the right edge. Put a second spot of beta-carotene solution on the line, about 6 mm from the left edge. The spots should be at least 1 mm in diameter, but ideally not larger than 3 mm. After adding the spots to the plate, allow the solvent to evaporate from the spot. You can speed up this process by holding the TLC plate by the edges and waving it gently in the air.

After the spots have dried, unscrew the cap of your developing chamber bottle, to which you have previously added the developing solvent. Place the TLC plate into the bottle, spot end down, angling the plate across the bottle as shown in Figure 3-7. The spots must be above the level of the solvent; otherwise, they will be dispersed and diluted in the solvent. You will see the solvent immediately begin to climb up the plate, and as you watch you will see the spots being carried along by the moving solvent.

Keep an eye on the solvent front. Where the solvent has traveled, the TLC sheet will appear somewhat wet and more transparent. When the solvent front has climbed to a point about 1 cm from the top, get ready to unscrew the cap. As soon as you unscrew the cap, you must *quickly* remove the TLC plate and mark the solvent front with a soft pencil line drawn on the silica gel side of the plate. So, get yourself prepared with pencil in hand, and make a last observation of the approximate location of the solvent front *before* you unscrew the cap and remove the plate. Once the plate is removed, the solvent will evaporate in seconds, leaving no visible trace of the position of the solvent front.

Identify the main spots, and mark them with a pencil on the plate (B for beta-carotene and L for lycopene contained in the tomato extract). Measure and record in your lab notebook the distances from the pencil line where you placed the spots to the solvent front and to the middle of

FIGURE 3-7

Separation of beta-carotene and lycopene by thin layer chromatography. The Baker-flex TLC plate is a 2.5 × 7.5 cm strip of plastic coated with silica gel. The developing chamber is a 4 oz (120 mL) wide-mouth screw cap bottle.

solvent front

point of application

each spot. These measurements will be used to calculate the R_f values for the spots. Minor components may be made more visible under a fluorescent lamp, if you have one, or by leaving the TLC plate for a few minutes in a bottle containing a few crystals of solid iodine.

The TLC plate can be wrapped in plastic food wrap (Handi-Wrap) and turned in with your lab notebook. An adhesive label with your name and other information can be placed on the plastic wrap, and the plate can be turned in with your lab report if your instructor so directs.

☞ Consider This

We have said that the main component of the tomato extract is lycopene. How can this assertion be further tested? First, can you definitely say that it is not beta-carotene, based on the TLC result? Next, based on the TLC result, does the tomato extract show evidence of several colored spots or mainly one component? It would be desirable to compare the tomato extract with an authentic sample of lycopene, but pure lycopene is expensive (approximately $70 for 1 mg from Sigma-Aldrich).

If you have a UV-visible spectrophotometer available, you can record the spectrum of the tomato extract over the range 700–350 nm. An authentic sample of lycopene is reported in the Merck Index, 12th edition, to have absorption bands at 505, 472, and 446 nm (with molar absorptivities, respectively, of 1.69×10^5, 1.85×10^5, and 1.21×10^5 L/mole-cm, that is, the peak intensities are in the ratio 0.91:1.0:0.65). You can also compare the tomato extract spectrum with the spectrum of the beta-carotene standard. (It will probably be necessary to dilute both the tomato extract and the beta-carotene standard further before recording their spectra.)

Carefully compare the structures of beta-carotene and lycopene shown in Figure 3-5. They are both

hydrocarbons and are isomers of one another, each having the molecular formula $C_{40}H_{56}$. How would you describe the difference in their structures?

BIBLIOGRAPHY

Epp, D. *The Chemistry of Food Dyes*, Terrific Science Press, Miami University Middletown, Middletown, Ohio, 1995.

Experiments in Liquid Chromatography—Separations of FD&C Dyes, 885 (April, 1979), University Marketing Division, Waters Associates, Inc., Milford, MA. This experiment describes the use of a Waters SEP-PAK C_{18} cartridge to provide a simple, convenient, and efficient dye extraction procedure, followed by analysis using thin layer chromatography.

Furia, T. E. (ed.) *Handbook of Food Additives*, 2nd ed., Vol. 1 (1973); Vol. 2 (1980); CRC Press, Boca Raton, FL. These handbooks provide information about the nature and regulatory status of food additives.

Goodrich, J.; Parker, C.; Phelps, R. "The Microscale Separation of Lycopene and Beta-Carotene from Tomato Paste," *J. Chem. Educ.*, **1993**, *70:* A158.

Markow, P. G. "The Ideal Solvent for Paper Chromatography of Food Dyes," *J. Chem. Educ.*, **1988**, *65:* 899.

McKone, H. T. "Identification of FD&C Dyes by Visible Spectroscopy," *J. Chem. Educ.*, **1977**, *54:* 376.

McKone, H. T., and Nelson, G. J. "Separation and Identification of Some FD&C Dyes by TLC," *J. Chem. Educ.*, **1976**, *53:* 722.

Strain, H. H. "Paper Chromatography of Chloroplast Pigments," *J. Phys. Chem.*, **1953**, *57:* 638–640.

Strain, H. H., and Sherma, J. "Michael Tswett's Contributions to Sixty Years of Chromatography," *J. Chem. Educ.*, **1967**, *44:* 235. This article is immediately followed on page 238 by a translation of Michael Tswett's article, "Adsorption Analysis and Chromatographic Methods," published in 1906 in *Berichte der deutschen botanischen Gesellschaft*.

Separation of Food Dyes and Plant Pigments by Paper and Thin Layer Chromatography

NAME

SECTION   LOCKER

INSTRUCTOR   DATE

OBSERVATIONS AND DATA

1. Separation of FD&C dyes in food colors

   (a) Describe the color and measure the R_f value of each colored spot on the chromatogram. (Begin at the top with the fastest-moving spot.)

Food Color	Blue	Green	Red	Yellow
Component 1				
Color of spot	_____	_____	_____	_____
Distance traveled by spot	_____ mm	_____ mm	_____ mm	_____ mm
Distance traveled by solvent front	_____ mm	_____ mm	_____ mm	_____ mm
R_f value	_____	_____	_____	_____
Component 2				
Color of spot	_____	_____	_____	_____
Distance traveled by spot	_____ mm	_____ mm	_____ mm	_____ mm
Distance traveled by solvent front	_____ mm	_____ mm	_____ mm	_____ mm
R_f value	_____	_____	_____	_____
Component 3				
Color of spot	_____	_____	_____	_____
Distance traveled by spot	_____ mm	_____ mm	_____ mm	_____ mm
Distance traveled by solvent front	_____ mm	_____ mm	_____ mm	_____ mm
R_f value	_____	_____	_____	_____

   (b) Describe the composition of the food colors. Do they consist of single components or a mixture of dyes? Is there a green dye in the green food color?

   Blue: _____

   Green: _____

   Red: _____

   Yellow: _____

2. Separation of lycopene and beta-carotene

   (a) What was the brand name of the tomato sauce you extracted? _____

   (b) Describe the color and measure the R_f value for each colored spot on the chromatogram. (Begin at the top with the fastest-moving spot.)

	Tomato extract (lycopene)	Beta-carotene
Component 1 Color of spot		
Distance traveled by spot	mm	mm
Distance traveled by solvent front	mm	mm
R_f value of component 1		
Component 2 Color of spot		
Distance traveled by spot	mm	mm
Distance traveled by solvent front	mm	mm
R_f value of component 2		

☞ Consider This

1. *Compare the TLC results for the tomato extract and beta-carotene. Could the main components be the same compound?*

2. *Describe the results of any further measurements such as UV-visible spectra. What do they tell you about the identity of the sample?*

Observing the Reactions of Household Chemicals
Deductive Chemical Reasoning

PURPOSE

To sharpen your powers of observation. To practice describing reactions of ordinary household chemicals with several reagents. To deduce from observed reaction patterns the identity of a household chemical presented as an unknown.

PRE-LAB PREPARATION

One of the important goals of the first course in general chemistry is to make connections or build bridges between three worlds of human experience:

- **The world of everyday life experience.** This includes the natural world around us and increasingly the products of human technology, including a vast array of chemical products not previously found in the natural world even most recently as 60 years ago.
- **The visible world of the laboratory.** The laboratory is where we perform experiments under conditions that we control, then observe what happens. Our observations are often assisted by instruments that work as extensions of our five senses (sight, hearing, smell, touch, and taste). Instruments allow us to measure things that we ordinarily can't sense at all, such as weak electric or magnetic fields, or the flux of particles from a radioactive source.

  In the general chemistry lab, touching and tasting chemicals are almost never done, because of unknown hazards. Smelling chemicals is sometimes cautiously done, with supervision of a knowledgeable instructor and when there is every reason to believe that there will be no harmful consequences.
- **The invisible microscopic world of atoms.** This is largely a world that we construct in our imaginations to provide an atomic or molecular interpretation of the visible world.

- **Experimentation is a cycle of planning, observation, and interpretation.** The planning of this experiment already has been done. You are coming in on the observation and interpretation steps of the cycle. We call it a cycle because the planning, observation, and interpretation of one experiment often naturally lead into the next experiment.

When you see something happen in a chemical reaction, try to begin thinking about how it can be understood or interpreted at the atomic level. In simple language, ask yourself the question, "What are the atoms doing, and why are they doing it?" To answer these questions you need to find out what chemical substances are present and to begin constructing mental ideas and pictures of the possible ways that they can interact. An important part of this process is learning to write chemical equations and chemical formulas and structures to explain your laboratory observations.

We are not born with any knowledge of chemical formulas and reactions. One step at a time, we must patiently build our knowledge about what atoms can do and why they do it. To make connections between the observations and our atomic and molecular interpretation of them, we must actively think about, or conceptualize, what the atoms and molecules are doing. The quicker and better we learn to do this, the greater the probability that we will be successful students of science and ultimately practicing scientists, if that is our goal. Conceptualizing chemical processes is the hardest thing to do at the beginning, when we have the least information stored in our heads and the least experience in interpreting things at the atomic and molecular level. The rewards are well worth the effort.

In this first introductory experiment, we will focus primarily upon learning to make careful observations by studying the reactions of various household chemicals with a series of reagents that test for different chemical properties. These household chemicals are available at most supermarkets.

STUDYING CHEMICAL REACTIONS

Before we begin our experiments, let's first think about the best way to carry them out. Usually we don't try to mix together two solid substances. Chemical reaction requires intimate contact between the atoms of the reacting substances. When the reactants are solids, we can achieve this only by grinding them together. This is less convenient and often is unsafe because pure solid reactants are in a very concentrated form. A reaction that evolves a lot of heat could lead to fires or even explosions. If the reactants are in a more dilute form, dissolved in water, the water can absorb any heat evolved and also can provide a clear, transparent medium for observing any reaction that takes place. Also, reactants in solution are easily mixed together. So, usually one or both of the substances that we are going to mix together will be a liquid or dissolved in water to form an aqueous solution.

OBSERVING CHEMICAL REACTIONS

Relying mainly on sight, what things should we look for when we mix two chemical substances? The most common visible changes that accompany chemical reactions are the following:

- **The formation of an insoluble solid, when two solutions are mixed.** Besides the fact that it does form, the insoluble solid, or *precipitate,* may appear to have particular characteristics. Formation may occur immediately, or it may take some time. The precipitate may be a finely divided suspension, or it may coagulate to form larger particles that rapidly settle out. It may be *flocculent* (forming cottony looking tufts called *flocs*) or *gelatinous* (looking like a thick starch suspension). So you may need to include a number of specific adjectives to give a more complete and accurate description, for example, "a brown gelatinous precipitate formed."

  The word *precipitate* comes from the Latin word meaning to "throw down suddenly," describing the tendency of most precipitates to be formed as soon as the substances are mixed and for the solid material to eventually settle to the bottom. Note: Be careful not to mispronounce and misspell "precipitate" as "percipitate."

- **A visible color change.** The color change may take place in solution or may be associated with the formation of a precipitate.

- **The evolution of a gas.** Generally, you will see bubbles of the gas forming in solution or on the surface of a solid. If the evolution of the gas is sufficiently vigor-

ous, you may see the foam as it forms and hear a fizzing sound. We must be careful about incorrect assumptions, however. If you heat a solution to speed up the reaction, do not be fooled into thinking a gaseous reaction product is being formed, when in actuality you may only be observing the boiling of water to form bubbles of water vapor (or steam).

- **Evolution or absorption of heat.** If the reaction is carried out in a test tube, the tube may get warm or cold, depending on whether the reaction evolves or absorbs energy (called *heat* in this context). Temperature changes involving liberation or absorption of small amounts of heat may be detectable by using a thermometer, an example of using a simple instrument as an extension of the senses.

- **You may detect no changes.** This result is ambiguous. It may mean that no chemical reaction is taking place, or it could mean that although a chemical reaction is taking place, it is invisible to the eye. Many chemical substances, when dissolved in water, do not absorb visible light and therefore appear colorless.

PATTERN RECOGNITION

The human eye and brain are very good at recognizing subtle features and discriminating between different patterns. For instance, babies very soon learn to recognize the characteristic features of their parents' faces and to recognize the differences between them and the faces of strangers. We can make use of this inborn human skill by thinking of our collected observations in terms of patterns. By comparing chemical reaction patterns of various substances with several test reagents, we can learn to distinguish one substance from another. Or, if your instructor presents to you as an unknown one of the substances whose reaction pattern you have studied, you can identify the substance from its characteristic reactions. When applied to chemical reactions, this form of pattern recognition is called *qualitative analysis.*

WHAT YOU WILL DO

In this experiment you will be provided with a set of household chemicals purchased at the supermarket. Each is composed predominantly of one active ingredient that is specified on the label. You will mix each of these substances with a series of reagents whose composition is specified. (If the household chemical is a solid or concentrated liquid, you will first make an approximately 5% solution of the substance in water.) You will carefully observe and record what happens when you add each reagent to a fresh sample of the household chemical, working on a reduced scale in a 24-well reaction plate, shown in Figure 4-1. Using the observed reaction patterns you will be asked to identify two unknowns. Each unknown will be one of the household chemicals whose reactions you observed. All of the unknowns will be aqueous solutions, so if the unknown is a solid, it will be presented to you as a 5% solution of the unknown in water. If you are keeping a separate lab notebook, read through the experimental procedure and prepare a table to record your observations.

EXPERIMENTAL PROCEDURE

Special Supplies: 24-well reaction plates, ten polyethylene transfer pipets (fine-tipped, 50 drops/mL), Handi-Wrap plastic film, pre-printed labels and Scotch tape (see *Note to Instructor*), ten 13 × 100 mm test tubes.

Household chemicals: Clorox or other brand of liquid bleach (5% sodium hypochlorite); Lysol Toilet Bowl Cleaner (9.5% hydrochloric acid); Arm & Hammer Baking Soda (100% sodium bicarbonate); Vanish Crystal Bowl Cleaner (62% NaHSO$_4$); S & F Iris or other brand of blue food color (FD & C Blue #1, listed in the *Merck Index* as Brilliant Blue FCF).

Chemicals: 0.1 M AgNO$_3$, 1.0 M CaCl$_2$, 6 M HCl, 0.1% Thymol Blue indicator (1.0 g/L Thymol Blue, sodium salt).

NOTE TO INSTRUCTOR: It's helpful to provide students with preprinted labels for the four household chemicals and five reagent solutions. Students can tape the labels to 13 × 100 mm test tubes that contain the household chemicals and reagents. Fine-tipped polyethylene transfer pipets can be stored in the test tubes when they are not being used. Alternatively, the polyethylene transfer pipets can be labeled by taping a label to the bulb with a circle of Scotch tape. If time is short, the pipets can be pre-filled with the household chemicals and reagents by the stockroom assistant.

! SAFETY PRECAUTIONS! Several of the household chemicals are concentrated corrosive solutions, as is the 6 M HCl reagent. WEAR EYE PROTECTION AT ALL TIMES. If you spill any of the household chemicals or reagents on your skin, immediately rinse the affected area with water under the tap. Neutralize any acid or base spills with sodium bicarbonate, and sponge them up immediately.

WASTE COLLECTION AND CLEAN-UP: Your instructor will tell you if you are going to be given two unknowns to identify. If so, you need to save the solutions and reagents that you will use in Part 1.

The contents of the reaction plate wells can be disposed of by emptying them into a large beaker, neutralizing the solutions with about 2 grams of sodium bicarbonate (baking soda), adding enough water to dissolve the sodium bicarbonate, then flushing the solution down the drain with plenty of water. Then rinse out the reaction plate with tap water, using a test-tube brush or paper towel and detergent to clean out the wells in which a precipitate was formed, and follow with a final rinse with deionized water. The reaction

FIGURE 4-1
Four household chemicals and five reagents are mixed in a 24-well reaction plate.

plate can be used for Part 2, identifying two unknowns, if your instructor so directs.

At the end of the experiment, the unused household chemicals can be collected together in a beaker, along with the unused 6 M HCl, thymol blue indicator, blue food color, and 1 M CaCl$_2$. Neutralize the mixture with 2–3 g of sodium bicarbonate, and flush it down the drain. The unused 0.1 M AgNO$_3$ (silver nitrate) should be collected in a waste container labeled "silver nitrate waste."

1. OBSERVING REACTION PATTERNS FOR HOUSEHOLD CHEMICALS. Obtain a 24-well reaction plate, nine clean 13 × 100 mm test tubes in a test-tube rack, a 30-mL beaker, and ten fine-tipped polyethylene transfer pipets. Now obtain samples of the household chemicals you will study: 2–3 mL of Clorox Liquid Bleach, approximately 0.5 mL Lysol Toilet Bowl Cleaner, and 0.5 g baking soda. Also place about 0.5 g of Vanish Crystal Bowl Cleaner in a 30-mL beaker. (We're using a beaker for the Vanish rather than a test tube because it foams a lot when water is added to it.) The Clorox will be used as received (without dilution).

Read the labels of the four household chemicals, and record in your lab notebook the following: the full commercial name of the product, the intended purpose of the product, the active ingredient(s) listed on the label and their amounts or concentrations, and any special warnings about the proper use of the product.

Next, prepare solutions of the four household chemicals you will study. Add about 8 mL of deionized water to each of the two test tubes containing the Lysol Toilet Bowl Cleaner and baking soda and to the beaker containing the Vanish Crystal Bowl Cleaner. Cover the mouths of the test tubes with a small piece of plastic film (Handi-Wrap) and, putting your thumb over the mouth of the tube, shake them until the solution is thoroughly mixed and any solids have dissolved. Stir the Vanish Crystal Bowl Cleaner in the beaker with a glass stirring rod until the solids have dissolved.

Next obtain samples of each of the five reagents in five clean 13 × 100 mm test tubes. About 3 mL of each solution (a test tube approximately 1/3 full) will be sufficient. The five reagents are: blue food color (FD & C Blue No. 1 dye), 0.1 M $AgNO_3$ (silver nitrate), 0.1% Thymol Blue indicator, 1.0 M $CaCl_2$ (calcium chloride), and 6 M HCl (hydrochloric acid).

Now obtain a 24-well reaction plate. Think of it as having four columns of six wells each. Each column will correspond to one of the four household chemicals. To work as efficiently as possible, we will put a household chemical solution in each of five wells in a column (the last well in each column will be empty). Put enough of each solution in a well to cover the bottom to a depth of 2–3 mm (about 0.5 mL).

Now think of the plate as having five rows of four wells each. Each row will correspond to one of the five reagents, and the wells are filled with the four household chemicals.

Add about 3 drops of each reagent to the four household chemicals contained in a row. *Caution!* A noxious substance (chlorine) is produced when 6 M HCl is added to Clorox Liquid Bleach. Do the addition of 6 M HCl in the fume hood. It may help to view the reactions against a white sheet of paper, or to lift up the plate and view the wells through transmitted light reflected from the white sheet of paper. The formation of white precipitates is often best seen by viewing against a dark background. As you add the test reagents, note what happens when you first add the reagent, then watch for 30 seconds or so to see if any further reaction takes place; finally, come back and look at each well after five or ten minutes have passed and note any further changes. When you finish, you will have observed 20 possible reactions.

If you were well-prepared, you would already have read this and prepared a data table in your lab notebook before you came to lab, with columns and rows to provide an organized space to record your observations. Leave plenty of space—a full page in your lab notebook—for recording the observations.

When you have finished recording all of your observations of the 20 possible reactions, look carefully at your data table. Is there a unique pattern of observations for each household chemical that would serve to distinguish that chemical from the others? For example, how is the pattern for baking soda different from the pattern for Lysol Toilet Bowl Cleaner? Could you tell the difference between them with just one test reagent? If so, which reagent would you use, and what differences would you look for?

Save your solutions of household chemicals and reagents if you are going to do Part 2, identifying two unknown household chemicals.

2. IDENTIFYING AN UNKNOWN BY ITS REACTION PATTERN.

Obtain a clean 24-well reaction plate, or clean and use the same plate used in Part 1. Put samples of each unknown in each of the wells, and react them with the same five reagents used in Part 1. Compare the reaction patterns used for the unknowns with those previously determined for the four household chemicals in Part 1. Record your observations, your deductions based on the reaction patterns, and your conclusions about the identity of the unknown(s) in your lab notebook. Also include them in your final laboratory report.

☞ Consider This

The label on Clorox Liquid Bleach has several warnings. One says, "Do not mix with other household chemicals, such as toilet bowl cleaners, rust removers, acids, or products containing ammonia. To do so will release hazardous gases."

*Likewise, the label on Lysol Toilet Bowl Cleaner contains the following warning: "**Danger: Corrosive—** produces chemical burns. Contains hydrochloric acid . . . Do not use with chlorine bleach or any other chemical products."*

What did you observe when you added 6 M HCl to Clorox Liquid Bleach? Let's begin in a small way to try to answer the question, "What are the atoms doing in the reaction of Clorox with hydrochloric acid?"

First, let's provide a little background information. The label says Clorox contains sodium hypochlorite (NaOCl). Clorox can be produced by bubbling chlorine gas into an aqueous solution of sodium hydroxide. In the process, some NaCl is also produced. A chemical equation describing the process is as follows:

$$Cl_2(g) + 2NaOH(aq) \rightarrow$$
$$NaCl(aq) + NaOCl(aq) + H_2O \text{ (Clorox bleach)}$$

The sodium hypochlorite gives Clorox its bleaching action. Recall what happened when you added the blue food color and the thymol blue indicator to Clorox. Did these colored materials get bleached?

The process can be reversed, forming noxious chlorine gas, when hydrochloric acid (or any acid) is added to Clorox:

$$NaOCl(aq) + 2HCl(aq) \rightarrow Cl_2(g) + H_2O + NaCl(aq)$$

Now we have a clearer picture, at the atomic level, of what the atoms are doing, and why it is a bad idea to mix chlorine bleach with acid-containing products to clean toilet bowls.

Now you try one: What happens when you add a few drops of 0.1 M AgNO_3 (silver nitrate salt) to Lysol Toilet Bowl Cleaner, in which the active ingredient is HCl (hydrochloric acid)? What did you see happen?

What combinations of positive ions (Ag^1 and H^1) with negative ions (NO_3^2 and Cl^2) could lead to formation of a precipitate of an insoluble salt? **Hint:** *The possible combinations we need to consider are: AgCl and HNO_3. We already know that the other two possible combinations, AgNO_3 and HCl, are soluble in water. (How do we know that?) So, a good guess might be that the precipitate is AgCl (silver chlo-ride) or HNO_3 (nitric acid). Try looking up the properties of AgCl and HNO_3 in your chemistry text, or in the CRC Handbook of Chemistry and Physics. Which one is least soluble? Now try to write a plausible set of products for the reaction:*

$$Ag^+ + NO_3^- + H^+ + Cl^- \rightarrow ?$$

This is an example of a very common type of reaction in aqueous solution: the combination of a positive ion (called a cation) with a negative ion (called an anion) to form an insoluble salt (or precipitate).

BIBLIOGRAPHY

Bosma, W. B. "Using Chemistry and Color to Analyze Household Products: A 10–12 Hour Laboratory Project at the General Chemistry Level," *J. Chem. Educ.*, **1998,** *75*: 214–215.

Myers, R. L. "Identifying Bottled Water: A Problem-Solving Exercise in Chemical Identification," *J. Chem., Educ.* **1998,** *75*: 1585–1587.

Ricketts, J. A. "A Laboratory Exercise Emphasizing Deductive Chemical Reasoning," *J. Chem. Educ.*, **1960,** *37*: 311–312.

OBSERVATIONS

1. Reactions of Four Household Chemicals with Five Reagents

Record your observations in the table below.

Reagent	Household Chemical			
	Lysol Toilet Bowl Cleaner	Baking Soda	Vanish Crystal Bowl Cleaner	Clorox Liquid Bleach
Blue food color				
0.1 M $AgNO_3$				
0.1% Thymol blue				
1.0 M $CaCl_2$				
6 M HCl				

67

2. Identifying Unknown Household Chemicals

Record in the table the identifying number(s) of your unknown(s).
Record your observations in the table below.

Reagent	Unknown No. _____	Unknown No. _____
Blue food color		
0.1 M AgNO$_3$		
0.1% Thymol blue		
1.0 M CaCl$_2$		
6 M HCl		

Summarize your observations and conclusions about the identity of your unknown(s).

☞ **Consider This**

1. *Explain concisely, using clear sentences and chemical equations, why it is dangerous to use an acid containing toilet bowl cleaner with a chlorine liquid bleach.*

2. *Write a plausible chemical equation to interpret, at the atomic level, what you observed in the reaction of Lysol Toilet Bowl Cleaner with 0.1 M AgNO$_3$.*

EXPERIMENT 5

A Cycle of Copper Reactions

PURPOSE

To observe a sequence of reactions of copper that form a cycle, along with the color and physical property changes that indicate those reactions. To gain skill in recording observations and interpreting them in terms of chemical equations. To use a simple classification scheme for grouping chemical reactions by reaction type. To practice quantitative laboratory techniques by determining the percentage of the initial sample of copper that is recovered.

PRE-LAB PREPARATION

To a beginning student of chemistry, one of the most fascinating aspects of the laboratory is the dazzling array of sights, sounds, odors, and textures that are encountered there. Among other things, we believe that this experiment will provide an interesting aesthetic experience. You will be asked to carry out a series of reactions involving the element copper and to carefully observe and record your observations. The sequence begins and ends with copper metal, so it is called a *cycle of copper reactions*. Because no copper is added or removed between the initial and final steps, and because each reaction goes to completion, you should be able to quantitatively recover all of the copper you started with if you are careful and skillful. This diagram shows in an abbreviated form the reactions of the cycle of copper:

$$Cu \xrightarrow[\text{(1)}]{HNO_3} Cu(NO_3)_2 \xrightarrow[\text{(2)}]{NaOH} Cu(OH)_2$$
$$Cu \xleftarrow[\text{(5)}]{Zn, HCl} CuSO_4 \xleftarrow[\text{(4)}]{H_2SO_4} CuO \xleftarrow[\text{(3)}]{heat} $$

Like any good chemist, you will probably be curious to know the identity of each reaction product and the stoichiometry of the chemical reactions for each step of the cycle. Here they are, numbered to correspond to the steps shown in the chemical equation.[1]

$$8HNO_3(aq) + 3Cu(s) + O_2(g) \rightarrow$$
$$3Cu(NO_3)_2(aq) + 4H_2O(l) + 2NO_2(g) \quad (1)$$

$$Cu(NO_3)_2(aq) + 2NaOH(aq) \rightarrow$$
$$Cu(OH)_2(s) + 2NaNO_3(aq) \quad (2)$$

$$Cu(OH)_2(s) \rightarrow CuO(s) + H_2O(l) \quad (3)$$

$$CuO(s) + H_2SO_4(aq) \rightarrow CuSO_4(aq) + H_2O(l) \quad (4)$$

$$CuSO_4(aq) + Zn(s) \rightarrow ZnSO_4(aq) + Cu(s) \quad (5)$$

These equations summarize the results of a large number of experiments, but it's easy to lose sight of this if you just look at equations written on paper. You can easily be overwhelmed by the vast amount of information found in this lab manual and in chemistry textbooks. It is in fact a formidable task to attempt to learn or memorize isolated bits of information that are not reinforced by your personal experience. This is one reason why it is important to have a laboratory experience. Chemistry is preeminently an experimental science. As you perform this—and any other—experiment, watch closely and record what you see. Each observation should be a little hook in your mind on which you can hang a more abstract bit of information, such as the chemical formula for the compound you are observing.

---

[1]Labels specify the states of the reactants and products: (*s*) means a solid, (*l*) means a liquid, (*g*) means a gas, and (*aq*) means an aqueous (water) solution.

It is also easier to remember information that is organized by some conceptual framework. Observations and facts that have not been assimilated into some coherent scheme of interpretation are relatively useless. It would be like memorizing the daily weather reports when you have no knowledge of or interest in meteorology.

Chemists look for relationships, trends, or patterns of regularity when they organize their observations of chemical reactions. The periodic table, which groups the elements into chemical families, is a product of this kind of thinking. Each element bears a strong resemblance to other members of the same chemical family but also has its own unique identity and chemistry.

In a similar fashion, it is useful to classify reactions into different types. Because no one scheme is able to accommodate all known reactions, several different kinds of classification schemes exist. A simple classification scheme we will use at the beginning is one based on ideas of precipitation (ion combination), acid-base (proton transfer), and redox (electron transfer). Here we present an outline and some examples of this kind of classification:

A SIMPLE SCHEME FOR CLASSIFYING CHEMICAL REACTIONS

1. PRECIPITATION REACTIONS (THE COMBINATION OF POSITIVELY CHARGED IONS WITH NEGATIVELY CHARGED IONS TO FORM AN INSOLUBLE NEUTRAL COMPOUND THAT PRECIPITATES FROM SOLUTION).

If we add a solution of sodium chloride (NaCl) to a solution containing silver nitrate ($AgNO_3$), an insoluble white solid forms. The solid, called a *precipitate,* is silver chloride (AgCl), and we may write the following chemical equation to describe the reaction, using symbols to represent the substances in solution.

$$Na^+(aq) + Cl^-(aq) + Ag^+(aq) + NO_3^-(aq) \rightarrow$$
$$AgCl(s) + Na^+(aq) + NO_3^-(aq)$$

First we eliminate the ions called *spectator ions,* which appear on both sides of the equation but do not participate in the precipitation reaction. What is left is the *net ionic equation:*

$$Ag^+(aq) + Cl^-(aq) \rightarrow AgCl(s)$$

The net ionic equation concisely summarizes the net result of mixing the two solutions: the formation of an insoluble precipitate when a positively charged silver ion is combined with a negatively charged chloride ion. The

solid AgCl precipitate is easily separated from the solution containing the soluble sodium nitrate salt. If we desired, we could also recover the sodium nitrate by evaporating the water from the solution.

2. ACID-BASE (PROTON TRANSFER REACTIONS).

An acid is a substance that reacts with water to form hydronium ions (H_3O^+) by transferring a proton to a water molecule.

(a strong acid) $HCl(g) + H_2O \rightarrow H_3O^+(aq) + Cl^-(aq)$

(a weak acid) $CH_3COOH(aq) + H_2O \leftrightarrow H_3O^+(aq)$
$$+ CH_3COO^-(aq)$$

HCl is a strong acid, completely dissociated in aqueous solution, while acetic acid (CH_3COOH) is a weak acid that only partially dissociates into hydronium ion and acetate ion.

A base is a substance that forms hydroxide ions when dissolved in water:

(a strong base) $NaOH(s) \rightarrow Na^+(aq) + OH^-(aq)$

(a weak base) $NH_3(g) + H_2O \leftrightarrow NH_4^+(aq) + OH^-(aq)$

NaOH is a strong base, completely dissociated in aqueous solution, while NH_3 is a weak base, only partially dissociating into ammonium ion and hydroxide ion.

Acids react with bases to form salts and (usually) water. Both are neutral compounds, being neither strongly acidic nor strongly basic. So acid-base reactions are also called *neutralization reactions.* Two examples follow:

(a strong acid + a strong base)
$$H_3O^+(aq) + Cl^-(aq) + Na^+(aq) + OH^-(aq) \rightarrow$$
$$2\,H_2O + Na^+(aq) + Cl^-(aq)$$

(the net ionic equation)
$$H_3O^+(aq) + OH^-(aq) \rightarrow 2\,H_2O$$

(a weak acid + a weak base)
$$CH_3COOH(aq) + NH_3(aq) \rightarrow NH_4^+(aq) + CH_3COO^-(aq)$$

Note that in these two examples of acid-base reactions a proton is transferred from the acid to the base.

3. REDOX REACTIONS (ELECTRON TRANSFER REACTIONS).

Oxidation-reduction reactions, called *redox reactions,* are reactions that involve the shift or transfer of electrons from one kind of atom to another. In some reactions, the transfer is obvious, as in this reaction:

$$Mg(s) + 2H_3O^+(aq) + 2Cl^-(aq) \rightarrow$$
$$H_2(g) + Mg^{2+}(aq) + 2Cl^-(aq)$$

Here, each magnesium atom is giving up two electrons to two hydrogen ions, forming magnesium ion and hydrogen gas.

Sometimes, the transfer of electrons between atoms is less obvious, as in the following reaction:

$$2SO_2(g) + O_2(g) \rightarrow 2SO_3(g)$$

Here, the reactants and products are all gases, and no ions are formed. In classifying this reaction as an oxidation-reduction reaction, we use the concept of assigning oxidation numbers (also called oxidation states) to each atom in the compound. A simple set of rules defines the procedure for assigning the oxidation number. For a simple binary compound (a compound composed of two different elements), we imagine that all of the electrons in the chemical bonds are assigned to the atoms with the greatest affinity for electrons. The ability of an atom to attract electrons to itself is called *electronegativity,* and the atoms having the greatest electronegativity are found in the upper right hand corner of the periodic table, with fluorine having the most.

In sulfur dioxide, SO_2, and sulfur trioxide, SO_3, we imagine that all of the electrons in the S—O bonds are assigned to the O atoms, giving each oxygen atom a full valence shell; this formally gives each oxygen atom a net charge of -2. So if oxygen is assigned oxidation number -2, the sulfur in SO_2 must have oxidation number $+4$ and the oxidation number of S in SO_3 would be $+6$, since the sum of the oxidation numbers on all the atoms must add up to the net charge on the molecule (zero, in this case). The oxidation number of atoms in their elemental form is always assigned zero. This seems reasonable for O_2 because the oxygen atoms are equivalent so that there would be no tendency for one oxygen atom to transfer electrons to its partner in the O_2 molecule.

Once we have assigned oxidation numbers to each element in the chemical reaction, we will see that in this particular reaction the oxidation number of the sulfur atoms increases from $+4$ to $+6$ while the oxidation number of the oxygen atoms decreases from zero (in O_2) to -2 (in SO_3). From this viewpoint, the change in oxidation number is formally equivalent to transferring electrons from sulfur to oxygen. We say that the sulfur atoms have been oxidized (because their oxidation number increases), while the oxygen atoms in O_2 have been reduced (because the oxidation number of oxygen atoms decreases from zero in O_2 to -2 in SO_3).

We must be careful to note, however, that the oxidation numbers we assign do not necessarily represent the real distribution of electronic charge in the molecule. By assigning the oxidation number according to fixed rules, we have artificially assigned integer changes in oxidation number to particular atoms (sulfur and oxygen in this case), but the changes in the electron density on the sulfur and oxygen atoms may not be as large as implied by the assigned oxidation numbers. Nevertheless, it is reasonable to suppose that the sulfur atom in SO_3 is more positive than the sulfur atom in SO_2 because the added oxygen atom would tend to draw electrons away from the sulfur atom.

4. DECOMPOSITION REACTIONS (A SUBSTANCE BREAKING DOWN INTO SIMPLER SUBSTANCES UNDER THE INFLUENCE OF HEAT).
Although many chemical reactions can be classified into one of the three reaction types we described earlier, it is possible to find examples of chemical reactions that do not neatly fit into this scheme. For example, when calcium carbonate is heated, it breaks down into simpler substances:

$$CaCO_3(s) \rightarrow CaO(s) + CO_2(g)$$

This reaction might be called a decomposition or dissociation reaction. Any substance heated to a sufficiently high temperature will begin to decompose into simpler substances, so this kind of reaction is common.

DOES A COMPREHENSIVE CLASSIFICATION SCHEME EXIST?
If we searched we would find examples of other reactions that do not fit into these four categories. Indeed, it is probably fair to say that there is no completely comprehensive classification scheme that would accommodate all known chemical reactions. However, many, if not most, of the chemical reactions described in your general chemistry text will fit into this simple scheme. As you carry out each step of the cycle of copper reactions, think about what is happening in each reaction, and try to fit it into one of the four categories we described.

EXPERIMENTAL PROCEDURE

Special Supplies: Infrared lamps or steam baths, porcelain evaporating dish.

Chemicals: 18- to 20-gauge copper wire, concentrated (16 M) HNO_3, 3 M NaOH, 6 M H_2SO_4, 30-mesh zinc metal, 6 M HCl, methanol.

! SAFETY PRECAUTIONS: Concentrated nitric acid, HNO_3, is hazardous. It produces severe burns on the skin, and the vapor is a lung irritant. When you handle it, you should use a fume hood while wearing safety glasses (as always) and rubber or polyvinyl chloride gloves. A polyethylene

FIGURE 5-1
If a fume hood is not available, substitute this apparatus.

squeeze pipet can be useful for transferring the HNO₃ from a small beaker to your 10-mL graduated cylinder. Rinse your hands with tap water after handling HNO₃. The dissolution of the copper wire with concentrated HNO₃ should be carried out in a fume hood. If no hood is available, construct the apparatus shown in Figure 5-1 to substitute for the fume hood. The brown NO₂ gas that is evolved is toxic and must be avoided.

NaOH solutions are corrosive to the skin and especially dangerous if splashed into the eyes—*wear your safety glasses.*

Methanol and acetone are flammable and their vapors are toxic. Use them in the hood to avoid breathing the vapor, and *keep them away from all open flames.*

WASTE COLLECTION: The supernatant solution that is decanted in Step 5 contains zinc sulfate and zinc chloride and should be collected.

A waste container should also be provided for the methanol used to dry the product in Step 5.

1. Cu TO Cu(NO₃)₂. Cut a length of pure copper wire that weighs about 0.5 g (about a 10-cm length of 20-gauge copper wire). If it is not bright and shiny, clean it with steel wool, rinse it with water, and dry it with a tissue. Weigh it to the nearest milligram, recording the weight in your laboratory book. Coil the wire into a flat spiral, place it in the bottom of a 250-mL beaker, and—in the fume hood—add 4.0 mL of concentrated (16 M) nitric acid, HNO₃. (If a fume hood is not available, use the apparatus shown in Figure 5-1.)

Record in your notebook a description of what you see. Swirl the solution around in the beaker until the copper has completely dissolved. What is in the solution when the reaction is complete? After the copper has dissolved, add deionized water until the beaker is about half full. Steps 2 through 4 can be conducted at your lab bench.

2. Cu(NO₃)₂ TO Cu(OH)₂. While stirring the solution with a glass rod, add 30 mL of 3.0 M NaOH to precipitate Cu(OH)₂. What is formed in the solution besides Cu(OH)₂? Record your observations in your lab book.

3. Cu(OH)₂ TO CuO. Stirring gently with a glass rod to prevent "bumping" (a phenomenon caused by the formation of a large steam bubble in a locally overheated area), heat the solution just barely to the boiling point over a burner, using the apparatus shown in Figure 5-2. If the solution bumps you may lose some CuO, so don't neglect the stirring. Record your observations. When the transformation is complete, remove the burner, continue

FIGURE 5-2
Setup for heating Cu(OH)₂ to convert it to CuO.

stirring for a minute or so, then allow the CuO to settle. Then decant (pour off) the supernatant liquid, being careful not to lose any CuO. Add about 200 mL of hot deionized water, allow to settle again, and decant once more. What is removed by this washing and decantation process?

4. CuO TO CuSO$_4$. Add 15 mL of 6.0 M H$_2$SO$_4$, while stirring. Record your observations. What is in solution now? Now transfer operations back to the fume hood.

5. CuSO$_4$ TO Cu. In the fume hood, add all at once 2.0 g of 30-mesh zinc metal, stirring until the supernatant liquid is colorless. What happens? What is the gas produced? When the evolution of gas has become very slow, decant the supernatant liquid, and pour it into the waste container provided. If you can see any silvery grains of unreacted zinc, add 10 mL of 6 M HCl and warm, but do not boil, the solution. When no hydrogen evolution can be detected by eye, decant the supernatant liquid, and transfer the copper to a porcelain dish. A spatula or rubber policeman is helpful for making the transfer. Wash the product with about 5 mL of deionized water, allow it to settle, and decant the wash water. Repeat the washing and decantation at least two more times. *Move to the hood, away from all flames.* Wash with about 5 mL of methanol, allow to settle, and decant. *Dispose of the methanol in the proper receptacle.* Place the porcelain dish under an infrared lamp or on a steam bath or hot plate, and dry the copper metal. What color is it? Using a spatula, transfer the dried copper metal to a preweighed 100-mL beaker and weigh to the nearest milligram. Calculate the mass of copper you recovered by subtracting the weight of the empty beaker from the weight of the beaker plus the copper metal.

CALCULATION OF PERCENTAGE RECOVERY.

Express the percentage of copper recovered as

$$\text{percentage recovery} = \frac{\text{mass of recovered copper}}{\text{initial mass of copper wire}} \times 100\%$$

If you are careful at every step, you will recover nearly 100 percent of the copper you started with.

☞ Consider This

If you used a penny as the source of your original copper in this experiment, would it matter if you used a pre-1982 penny (essentially pure copper) or a post-1982 penny (copper cladding over a zinc core)? Describe what would happen in each step if you used a post-1982 penny. Test your predictions.

Why would it be hard to perform a cycle of oxygen or a cycle of hydrogen experiment similar to this cycle of copper exercise? Can you design an experimental apparatus for a cycle of oxygen lab?

BIBLIOGRAPHY

Bailar, Jr., J.C. "A Further Improvement on the Copper Cycle Experiment," *J. Chem. Educ.,* **1983,** *60:* 583.

Condike, G. F. "Near 100% Yields With the 'Cycle of Copper Reactions' Experiment," *J. Chem. Educ.,* **1975,** *52:* 615.

Todd, D. and Hobey, W.D. "An Improvement in the Classical Copper Cycle Experiment," *J. Chem. Educ.,* **1985,** *62:* 177.

Umans, T. and de Vos, W. "An Improved Copper Cycle Experiment," *J. Chem. Educ.,* **1982,** *59:* 52.

A Cycle of Copper Reactions

NAME

SECTION   LOCKER

INSTRUCTOR   DATE

DATA AND CALCULATIONS

1. Initial mass of copper wire		g
2. Mass of beaker plus dry copper		g
3. Mass of empty beaker		g
4. Mass of recovered copper		g

5. Percentage recovery

$$\text{percentage recovery} = \frac{\text{mass of recovered copper}}{\text{initial mass of copper wire}} \times 100\% = \underline{\hspace{2cm}} \%$$

Show your calculation.

EQUATIONS AND OBSERVATIONS

For each step of the cycle, write the products of the reaction and balance the chemical equation(s). Using the classification scheme presented in the Pre-Lab Preparation, write the *reaction type* (combination, decomposition, or single/double replacement). Also record your observations of what happens at each step, and answer the questions posed earlier in the Experimental Procedure.

1. $Cu + HNO_3 + O_2 \rightarrow$

Reaction type: _____

What is in the solution after reaction is complete?

Observations (Be sure to include color and texture changes that occur in each step.)

2. $Cu(NO_3)_2 + NaOH \rightarrow$

Reaction type: _____

$HNO_3 + NaOH \rightarrow$

Reaction type: _____

What is formed in the solution besides $Cu(OH)_2$?

Observations:

3. $Cu(OH)_2 \xrightarrow{\text{heat}}$

Reaction type: _____

What is removed by the washing and decantation process at the end of Step 3? (Consider the products of the reaction as well as reagents from previous steps.)

Observations:

4. $CuO + H_2SO_4 \rightarrow$

Reaction type: _____

What is in the solution at the end of Step 4?

Observations:

5. $Zn + CuSO_4 \rightarrow$

Reaction type: _____

$Zn + H_2SO_4 \rightarrow$

Reaction type: _____

$Zn + HCl \rightarrow$

Reaction type: _____

What happens when the zinc is added?

What is the gas produced in the reaction?

What is removed by the washing and decantation near the end of Step 5?

What color is the recovered copper?

Observations:

QUESTIONS

1. Describe whether the error introduced by each of the following problems would result in a high or a low value for the Cu recovery, or would not affect the results.

  (a) Some of the copper nitrate solution is splashed out of the beaker in Step 1.

  (b) Insufficient NaOH is added in Step 2.

  (c) The solution bumps in Step 3, splashing out some CuO.

  (d) Some solid is lost in the decanting process.

  (e) An excess of H_2SO_4 is added in Step 4.

  (f) Some unreacted zinc remains with the product at the end of the experiment.

(g) The washings in Step 5 are insufficient to remove all of the solution residues from the copper.

(h) The copper is not completely dried.

2. How many milliliters of 3.0 M NaOH are required to react with 4.0 mL of 16 M HNO_3?

3. How many milliliters of 3.0 M NaOH are required to react with 0.5 g of Cu^{2+} to form $Cu(OH)_2$?

. (a) Add together the results you calculated for Questions 2 and 3 and compare the sum with the milliliters of 3.0 M NaOH added in Step 2.

(b) Is there an excess of NaOH above that required to react with the HNO_3 and Cu^{2+}?

(c) What would be the effect on the percentage recovery if not enough NaOH were added in Step 2 to react with both the unreacted HNO_3 and the Cu^{2+} present at the end of Step 1?

Consider This

If you used a penny as the source of your original copper in this experiment, would it matter if you used a pre-1982 penny (essentially pure copper) or a post-1982 penny (copper cladding over a zinc core)? Describe what would happen in each step if you used a post-1982 penny. Test your predictions.

Why would it be hard to perform a cycle of oxygen or a cycle of hydrogen experiment similar to this cycle of copper exercise? Can you design an experimental apparatus for a cycle of oxygen lab?

6

Synthesis of a Chemical Compound: Making Alum from Aluminum

PURPOSE

To illustrate the principles of synthetic chemistry, the process by which one pure substance is transformed into another.

PRE-LAB PREPARATION

Anyone who lives in a modern city surrounded by the necessities that make urban life possible and the luxuries that make it enjoyable should ponder this question: Where does all this stuff come from? The answer, of course, is that it comes from the elements found on earth. The same elements that make up the mountains and valleys, the seas and the air, and all living and nonliving matter can be transformed into new substances that are used in a variety of ways in our technological society.

The process of making new substances from other, usually simpler, substances is called *chemical synthesis*. The word *synthesis* implies the notion of putting things together. The opposite of synthesis is *analysis*, which implies taking things apart, that is, determining the chemical composition of substances.

In this experiment we will transform the aluminum from a beverage can into a substance known as *alum*, more formally designated "aluminum potassium sulfate dodecahydrate," which has the chemical composition $AlK(SO_4)_2 \cdot 12H_2O$. Alum is a versatile substance used in dyeing fabrics, tanning leather, and clarifying water in municipal water treatment plants. Medicinally, it has applications as an astringent and a styptic; for example, if you nicked yourself while shaving, you might use a stick of alum to coagulate the blood and stop the bleeding. Derivatives of alum are also widely used in formulating antiperspirants, thus exploiting the astringent properties of Al^{3+} ions, which bind water strongly.

Since aluminum cans have some intrinsic value, empty cans are normally recycled, that is, collected, melted down, and used to make more cans or other aluminum metal products. So this experiment represents another way of recycling scrap aluminum into something potentially useful. (To be truthful, we would have to say that if you wanted to make tons of alum, it would be cheaper to start with natural ore already containing Al^{3+} ions. This is because a lot of chemical reagents and electrical energy go into making aluminum metal from natural ores. On a large scale, making alum from aluminum metal would be uneconomical.)

Now let's look at the chemistry involved in transforming an aluminum can into alum. Aluminum cans are made of fairly pure aluminum, except for the decorative paint on the outside and the thin coating of plastic on the inside, which prevents acidic chemicals in the beverage from reacting with the aluminum can. The first step in this experiment is to remove these coatings before the chemical reactions with aluminum can begin.

The next step is to react the clean aluminum with potassium hydroxide solution:

$$2Al(s) + 2K^+(aq) + 2OH^-(aq) + 6H_2O(l) \rightarrow$$
$$2K^+(aq) + 2Al(OH)_4^-(aq) + 3H_2(g) \qquad (1)$$

This reaction can be classified as an oxidation-reduction reaction, in which aluminum is oxidized and water is reduced, producing hydrogen gas and hydroxide ion by transfer of electrons from aluminum to water. The KOH helps to dissolve the thin oxide coating on the aluminum that normally protects it from further oxidation. The hydroxide ion produced when aluminum replaces hydrogen forms a complex ion, the tetrahydroxyaluminate ion, $Al(OH)_4^-$.

Equation (1) also explains why alkaline products like liquid detergents and bleaches are never stored in aluminum containers. The aluminum would slowly dissolve, producing hydrogen gas. In a sealed container, a high pressure of hydrogen gas might build up, possibly causing the container to rupture or explode if the hydrogen ignited.

Addition of a slight excess of sulfuric acid to the products produced in Equation (1) neutralizes the excess KOH from the previous step and neutralizes the four OH^- groups on the tetrahydroxyaluminate complex ion:

$$K^+(aq) + Al(OH)_4^-(aq) + 2H_2SO_4(aq) \rightarrow$$
$$K^+(aq) + Al^{3+}(aq) + 2SO_4^{2-}(aq) + 4H_2O(l) \quad (2)$$

This is an acid-base (or neutralization) reaction in which the H^+ ions in H_2SO_4 react with the OH^- ions in $Al(OH)_4^-$ to form water, leaving behind $Al_2(SO_4)_3$. The reaction is strongly driven to the right by the formation of the very stable H_2O molecule.

The resulting solution contains K^+, Al^{3+}, and SO_4^{2-} ions in the proportions in which they are found in alum: $AlK(SO_4)_2$. After sufficient cooling, crystals of hydrated potassium aluminum sulfate, or alum, will form.

There are 12 molecules of water associated with each formula unit of the salt, so the formula of the crystalline alum that forms is $AlK(SO_4)_2 \cdot 12H_2O$. This is typically true of compounds that contain metal ions, but in a beginning chemistry course we often ignore these water molecules as we write chemical formulas—unless they are critical to the results of an experiment, as they are here. For example, what you usually see written as $FeCl_3$ is actually $FeCl_3 \cdot 6H_2O$ (or $Fe[H_2O]_6Cl_3$ to show that the waters surround the metal). X-ray crystallographic studies show that each K^+ ion and each Al^{3+} ion is surrounded by six water molecules. In solution each is also surrounded by a cluster of approximately six water molecules. As you will study later in greater detail, water is a polar molecule, with the oxygen atom having a slightly negative charge while the hydrogen atoms are slightly positively charged. So the negative oxygen atoms in the water molecules are attracted to the positively charged K^+ and Al^{3+} ions, and there is just enough room for about six water molecules to crowd around each positive ion.

The solution containing the dissolved alum is cooled with ice because alum is quite soluble in water, but its solubility decreases as the temperature is lowered. So cooling the alum solution increases the amount of solid alum that can be recovered.

To get pure alum it is necessary only to filter off the crystals and wash them to remove the excess H_2SO_4 used in the second step. The washing is done with methanol, which does not dissolve the alum or remove any of the water of crystallization. Methanol also has the advantage of evaporating quickly.

EXPERIMENTAL PROCEDURE

Chemicals: 1.4 M KOH, 9 M H_2SO_4, methanol, $NaHCO_3$.

Special Supplies: Aluminum beverage can, sturdy scissors or snips, coarse steel wool (or metal pot scrubber), smooth board or other material on which the aluminum metal can be scrubbed, crushed ice, glass filter funnel, Whatman No. 4 filter paper, filter flask, Büchner filter funnel, filter paper to fit Büchner funnel, aspirator or house vacuum, hot plate or steam bath, rubber policeman, container for the alum produced.

! SAFETY PRECAUTIONS: WEAR EYE PROTECTION AT ALL TIMES. Potassium hydroxide, KOH, and sulfuric acid, H_2SO_4, are very corrosive. Both can cause burns if left in contact with the skin. If either chemical gets on your skin, immediately rinse with plenty of cool water for several minutes. Your instructor will discuss emergency measures to follow if your eyes are affected. If spills occur, notify the instructor, who will know how to neutralize the spill with sodium bicarbonate and can advise you on proper cleanup procedures. Wear gloves if there is any chance of contact with strong acids or bases.

Methanol vapors are toxic, and liquid methanol is toxic if swallowed. Use methanol in the hood, and avoid getting it on your skin. If it does contact your skin, rinse it off immediately with water. Because methanol is flammable, no open flames should be allowed in the laboratory.

THE SYNTHESIS OF ALUM. To begin the synthesis of alum, cut a piece of aluminum about 5 × 5 cm from an aluminum beverage can. Be careful with the sharp-edged piece of metal. Place it on a smooth board surface, and scour off the coatings on both sides with coarse steel wool or a metal pot scrubber. When the piece of aluminum is clean and bright, rinse and wipe it with a paper towel. Using scissors, cut the clean aluminum into small squares about 0.5 × 0.5 cm.

Tare,[1] or weigh, an empty 150-mL beaker to the nearest milligram, and record the mass. Add between 0.5 and 0.6 g of the small pieces of aluminum, and record the mass to the nearest milligram.

Obtain 25 mL of 1.4 M KOH solution.

CAUTION: KOH will dissolve aluminum and your skin! Make certain you are wearing your eye protection, and immediately rinse off with water any KOH that gets on your skin.

---

[1]To "tare" a container is to weigh it on a balance while adjusting the balance to read "zero" while the container is on the balance. You can then place an object in the container, and the weight of the object can be read directly from the balance display. If you are to use this feature on your balance, your instructor will need to instruct you in the taring process. The alternative is to weigh the container, then weigh the container with the object in it, and then subtract one weight from the other to determine the weight of the object. This method is referred to as "weighing by difference." Under no circumstances should reagents be weighed by placing them directly on the balance pan.

Working in the hood, carefully add the 25 mL of 1.4 M KOH to the beaker containing the small pieces of aluminum. Use a stirring rod to guide the solution stream as you pour it.

Place the beaker on a hot plate set at low heat or on a steam bath. Warm the solution, but do not allow it to boil. With occasional stirring, all the aluminum should react in about 20 min. While the aluminum is dissolving, set up a glass funnel using a ring stand and iron ring or funnel support. See Figure 6-1 for an example of the equipment setup. Place the stem of the funnel in a 150-mL beaker.

Fold and crease lightly

Fold again

Tear off the corner unequally (The torn corner prevents air from leaking down the fold.) Then open the filter out like this

Place the filter paper in the funnel and seal the edge against the funnel wall by moistening the paper

The filtrate should run down the walls of the beaker

Fill the funnel with water and let it run until the air is washed out of the stem. When the water level drops to the top of the stem, add the mixture to be filtered

The weight of this column of water hastens filtration

Stem

FIGURE 6-1
The process of filtration.

Fold a piece of coarse filter paper (Whatman No. 4 or equivalent) into quarters, and place it in the funnel. Pour a little water into the filter paper so that it will stay in place. Discard any water that runs through the filter into the beaker.

When there are no visible pieces of aluminum or bubbles of hydrogen gas, you will know that all of the aluminum has reacted. Remove the beaker from the hot plate temporarily. Obtain 10 mL of 9 M H_2SO_4, and carefully but rapidly pour all of the 9 M H_2SO_4 into the beaker, stirring steadily. If the H_2SO_4 is added too slowly, you may see the transient formation of insoluble $Al(OH)_3$ that will precipitate when just enough acid has been added to react with one of the OH^- groups of the $Al(OH)_4^-$ complex ion. Because you are adding a strong acid to a solution of a strong base, there is an exothermic neutralization reaction, making the solution hotter.

Stir the solution thoroughly. If white crystals are visible, return the beaker to the hot plate or steam bath, and stir until all such material dissolves. Any tiny black specks may be disregarded.

When the beaker can be handled safely, place your stirring rod against the lip of the beaker to guide the hot solution into the previously prepared glass filter funnel. Do not touch the filter paper with the stirring rod or you may puncture the paper. Don't overfill the filter funnel. The level should not be closer than 1 cm to the top of the filter paper. The filtrate should be clear and colorless.

Allow the 150-mL beaker containing the filtrate to cool. Fill a 600-mL beaker about half full of crushed ice. Place the 150-mL beaker in the ice bath for about 20 min, stirring frequently as crystals of alum form.

While the solution is cooling, set up a filter flask and Büchner filter funnel attached to an aspirator or house vacuum (see Figure 6-2).

When the crystallization is complete, stir *carefully* with a thermometer, and record the temperature when the reading stabilizes. The solution temperature should be 6 °C or less.

Place a circle of filter paper (preferably Whatman No. 4, although No. 1 will do) of the appropriate size in the Büchner funnel. Moisten the filter paper to hold it in place. Turn on the aspirator or house vacuum. Remove the 150-mL beaker from the ice bath, and stir the contents of the beaker with a stirring rod to loosen all of the alum crystals. Transfer all of the suspended crystals to the Büchner funnel. A rubber policeman on the end of your stirring rod is helpful in scraping out as many crystals as practical.

Add two 10-mL portions of methanol in succession to wash the excess H_2SO_4 out of the crystals. Continue to draw air through the alum for another 10 to 15 minutes to dry the crystals.

Moist filter paper

Filter pump

FIGURE 6-2

Vacuum filtration technique using a Büchner funnel. To prevent aspirator water from entering the flask, disconnect the rubber tube before turning off the aspirator.

Weigh a clean, dry 150-mL beaker to the nearest milligram. Record the mass. Shut off the source of vacuum, and remove the Büchner funnel. Carefully transfer the dried alum crystals and filter paper circle to the weighed 150-mL beaker. Use a stirring rod or metal spatula to scrape off any crystals adhering to the filter paper and remove the filter paper. Reweigh the beaker containing the alum crystals, and record this mass.

Follow the instructor's directions for disposing of the sample of alum that you prepared.

Weigh 10 g of sodium bicarbonate, $NaHCO_3$. Add about 80 percent of the $NaHCO_3$ to the solution left in the filter flask. Swirl the flask until the bubbling ceases and the $NaHCO_3$ has dissolved. Now add the rest of the $NaHCO_3$, and note whether any additional bubbles of CO_2 are produced. If additional bubbles form, add 2 g more of $NaHCO_3$. When there is no more bubbling upon further addition of a small amount of $NaHCO_3$, rinse the solution in the filter flask down the drain.

☞ Consider This

What does the CRC Handbook of Chemistry and Physics, the Merck Index, or another reference tell you about what it takes to remove the 12 water molecules from alum? Why do we often "forget" to include the "$\cdot 12H_2O$" when we write the formula of alum or similar hydrated salts? Would it be better to write the formula as $Al(H_2O)_6K(H_2O)_6(SO_4)_2$? Comment on other ways the formula could be written, such as: $AlH_{24}KO_{20}S_2$ or $AlK(SO_4)_2(H_2O)_{12}$.

When we combine salt (NaCl) and water, we don't describe the process as a synthesis or the product, "salt water," as a compound. What is the difference between the synthesis of a compound such as alum and the production of a mixture or solution such as salt water? As you write your description, suppose you are an early chemist and cannot use the concepts of an element or a formula in your explanation.

Clean, dry, and weigh an aluminum can. Find the price of aluminum in a newspaper. (It's usually found on the stocks page in the business section.) What is the value of the aluminum in a beverage can? Then determine how much a local recycling center is paying for aluminum cans. How does the value of aluminum compare to its "worth" in your community? Why are the two prices so different?

Synthesis of a Chemical Compound: Making Alum from Aluminum

NAME

SECTION   LOCKER

INSTRUCTOR   DATE

DATA

1. Weighing the aluminum

(a) Mass of empty 150-mL beaker	g
(b) Mass of 150-mL beaker + aluminum	g
(c) Temperature of cooled alum crystals	°C

2. Weighing the alum

(a) Mass of empty 150-mL beaker	g
(b) Mass of 150-mL beaker + alum	g

CALCULATIONS

Mass of aluminum metal used	g
Mass of alum obtained	g

Calculate the number of moles of aluminum that you used. Show your calculations:

_____ moles Al

Calculate (or look up in a reference such as the *CRC Handbook of Chemistry and Physics*) the formula mass (molar mass) of alum, $KAl(SO_4)_2 \cdot 12H_2O$, including its water of crystallization. (You might find it under the name potassium aluminum sulfate.)

_____ g/mole

Noting that one aluminum atom is in each formula unit of alum, we know that each mole of aluminum will give rise to one mole of alum. Based on the number of moles of aluminum you used, calculate the maximum number of grams of alum that you could produce, assuming that aluminum is the limiting reagent:

_____ g alum

Using the number of grams of alum that you actually obtained, calculate the percentage yield of alum for your synthesis:

$$\% \text{ yield} = \frac{\text{g alum from experiment}}{\text{maximum g alum from theory}} \times 100\% = \underline{\hspace{3cm}} \%$$

Show your calculation.

QUESTION

1. Experiments show that the solubility of alum in 25 mL of 1.4 M KOH plus 10 mL of 9 M H_2SO_4 is about 1.0 g at 1.0 °C and 1.7 g at 6.0 °C. Using your measured solution temperature, estimate the amount of alum left in your chilled solution.

   Alum left in solution: _____ g

2. Does the amount of alum left in solution account for most of the difference between the maximum grams of alum and your experimental grams of alum? Explain.

☞ **Consider This**

What does the CRC Handbook of Chemistry and Physics, the Merck Index, or another reference tell you about what it takes to remove the 12 water molecules from alum? Why do we often "forget" to include the "· 12H$_2$O" when we write the formula of alum or similar hydrated salts? Would it be better to write the formula as $Al(H_2O)_6K(H_2O)_6(SO_4)_2$? *Comment on other ways the formula could be written, such as:* $AlH_{24}KO_{20}S_2$ *or* $AlK(SO_4)_2(H_2O)_{12}$.

When we combine salt (NaCl) and water, we don't describe the process as a synthesis or the product, "salt water," as a compound. What is the difference between the synthesis of a compound such as alum and the production of a mixture or solution such as salt water? As you write your description, suppose you are an early chemist and cannot use the concepts of an element or a formula in your explanation.

Clean, dry, and weigh an aluminum can. Find the price of aluminum in a newspaper. (It's usually found on the stocks page in the business section.) What is the value of the aluminum in a beverage can? Then determine how much a local recycling center is paying for aluminum cans. How does the value of aluminum compare to its "worth" in your community? Why are the two prices so different?

The Chemistry of Oxygen: Basic and Acidic Oxides and the Periodic Table

PURPOSE

To learn a method for the laboratory preparation of dioxygen, O_2. To observe the reactions of O_2 with several metallic and nonmetallic elements. To determine whether aqueous solutions of these oxides are acidic or basic. To see the relationship between acidic or basic properties of the oxides and the position of the element in the periodic table.

PRE-LAB PREPARATION

OXYGEN AND LIFE

Oxygen is one of the most abundant elements on earth. Large amounts of it are found in the molten mantle, in the crust that forms the great land masses, in the water of the vast oceans that cover most of the earth's surface, and in the gaseous atmosphere that surrounds the earth. Oxygen is found in elemental form, as molecular O_2, mainly in the atmosphere, with smaller amounts of dissolved oxygen found in the oceans, lakes, rivers, and streams, where it is vital to all water-dwelling creatures. Elsewhere, oxygen is found in combination with other elements.

Dioxygen, O_2, reacts so avidly with both metals and nonmetals that the presence of a large amount of free O_2 in the atmosphere raises the question, where did all of that O_2 come from? Studies over the past 200 years have provided the general outline of the answer, although some details are still being actively investigated. Most of the dioxygen on earth has been produced by plants, from the smallest single-celled algae to the majestic redwoods. Plants use water, carbon dioxide (CO_2), and sunlight to form carbohydrates and oxygen in a complex process called *photosynthesis*. Animals reverse the process. They react carbohydrates with oxygen inside their cells in a process called *respiration*, forming CO_2 and water. So, plants and animals exist together in a grand symbiotic cycle, each supplying the other's needs. The energy involved in this biological cycle is about 30 times the amount of energy expended each year by all of the machines on earth.

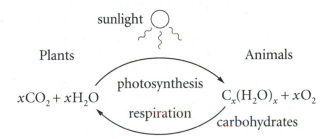

THE CHEMICAL PROPERTIES OF OXYGEN

To observe and understand some of the chemistry of oxygen, we will first prepare some pure oxygen by catalytic decomposition of hydrogen peroxide, then react the O_2 with a variety of elements. The oxides formed will be dissolved in water, and the solutions tested to see if they are acidic or basic.

Ask yourself these questions while doing the experiment:

1. Is there a difference in the vigor of reaction (rate, light and heat evolved, etc.) of the different elements with oxygen and if so, do those differences relate to the positions of these elements in the periodic table?

2. Is there a relation between the acidity or basicity of the water solution of these oxides and the positions of the elements in the periodic table?

The elements whose oxides (or hydroxides) you will study are circled in the abbreviated periodic table shown in Table 7-1. The heavy zigzag line in the figure approximately divides the metals from the nonmetals, so you will study samples from each group.

TABLE 7-1

Main Groups of the Periodic Table

I	II					III	IV	V	VI	VII	VIII
H											He
Li	Be		Metals ← → Nonmetals			B	C	N	O	F	Ne
		(Transition metals)									
Na	Mg					Al	Si	P	S	Cl	Ar
K	Ca	Sc	... Fe ...	Zn	Ga	Ge	As	Se	Br	Kr
Rb	Sr	Y	... Ru ...	Cd	In	Sn	Sb	Te	I	Xe

The heavy zigzag line divides the metals from the nonmetals. The oxides (or hydroxides) of the elements that are encircled will be studied in the experiment.

ACIDS AND BASES FROM OXIDES: A CONCEPTUAL FRAMEWORK

Let's begin with a microscopic view of what the atoms are doing and why they are doing it. Take a look at Figure 7-1. It's a picture of how we might imagine a solid element E reacting with gaseous dioxygen molecules, O_2.

For reaction to take place, the O_2 molecules have to collide with the atoms of E in the surface of a piece of the solid element. What happens next depends on the nature of the atoms of E and of oxygen. We're going to construct two scenarios that lead to two different outcomes depending on a property of atoms called their *electronegativity*. This is defined qualitatively as *the ability of atoms to attract electrons to themselves.*

Elements on the left side of the periodic table (the metals) have less ability to attract electrons to themselves and therefore exhibit smaller electronegativity than elements on the right hand side of the periodic table. Why is this? The increase in electronegativity, moving from left to right in a period (or row) of the periodic table, is largely the result of an increasing nuclear charge. The greater the nuclear charge, the more attraction there will be to draw in valence electrons.

The polarity of the bonding between an element and oxygen correlates with the difference in electronegativity between the element and oxygen. If the difference is large, the E—O bond will have a more ionic character. If the electronegativity difference is relatively small, the E—O bond will be less polar, so there will be more electron sharing, which we call *covalent bonding*. Thus, reaction products involve two kinds of bonding, ionic and covalent, as shown in Figure 7-1.

In the first scenario, let's suppose that we have an element oxide, EO, in which there is mostly covalent character in the E—O bond. Such compounds tend to be molecular rather than ionic solids. Oxide ions, as such, don't exist in the covalent oxide (EO), and when the covalently bonded oxides are dissolved in water, they too form a covalently bonded hydrated form of the oxide, $E(OH)_2$:

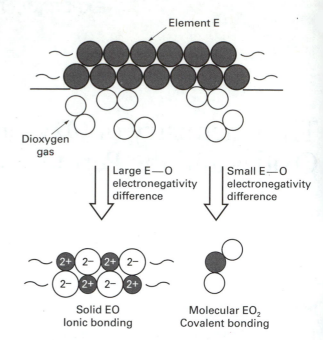

FIGURE 7-1

A piece of solid element E reacts on its surface with gaseous dioxygen, O_2. If the electronegativity of O is much larger than that of E, the reaction product will be an ionic solid. If the electronegativity difference is smaller, E and O_2 react to form covalently bonded molecules, $EO_2(g)$ in this example. E atoms shrink when they lose electrons to form positive ions while O atoms expand when they add electrons to form negative ions.

$$EO + H_2O \rightarrow E(OH)_2$$

This reaction could be viewed as a near simultaneous transfer of a proton from a water molecule to the oxygen atom and the formation of a second E—OH bond to give $E(OH)_2$.

A covalent bond implies more electron density concentrated in the bond between the E and O atoms, drawing electron density from the O—H bond and tending to make the H atom more positive. This makes it easier for the $E(OH)_2$ molecule to give up (or donate) a proton to a second water molecule:

$$E(OH)_2 + H_2O \rightarrow H_3O^+ + E(O)OH^-$$

and by further dissociation

$$E(O)OH^- + H_2O \rightarrow H_3O^+ + EO_2^{2-}$$

So, because it forms hydronium ions (H_3O^+), $E(OH)_2$ is an *acid*.[1] Acids are called *proton donors* in the Brønsted-Lowry definition of acids and bases.

---

[1] A number of important acids are not derived from oxides, such as the binary hydrides of the nonmetals: HF, HCl, HBr, and H_2S.

In the second scenario, let's imagine that the element E transfers all of its valence electrons to oxygen atoms forming an ionic oxide, EO, which we schematically represent by formulating the oxide as $E^{2+}O^{2-}$. What might we expect to happen if we put this oxide in water? A plausible possibility is described by the reaction

$$EO + H_2O \rightarrow E^{2+} + 2OH^-$$

This reaction can be viewed as a water molecule donating a proton (H^+) to the oxide ion, forming two OH^- ions:

$$O^{2-} + \overset{+}{H} - \overset{-}{OH} \rightarrow 2OH^-$$

where we have deliberately exaggerated the polarity of the H—O bond in water.

The net result is that the compound EO reacts with water to form hydroxide ions. Compounds that form hydroxide ions in water are called *bases*. Bases are called *proton acceptors* in the Brønsted-Lowry definition of acids and bases.

To summarize, the striking difference in the outcome of the two scenarios, one producing an acid and the other a base, is determined by the nature of the element E and the polarity of the E—O bond. That polarity is in turn related to the electronegativity difference between element E and O. The polarity of the E—O bond, ranging over a continuum from ionic to covalent, correlates well with the position of E in the periodic table.

ACID AND BASE ANHYDRIDES

If the oxide of an element forms an acid in water, it is termed an *acidic oxide* or *acid anhydride*. If the oxide in water forms a base, we speak of a *basic oxide* or *base anhydride*. To determine the formula of the anhydride of an oxy acid or base, simply subtract water to eliminate all hydrogen atoms. For example,

$$2NaOH(s) \text{ minus } H_2O \rightarrow Na_2O(s)$$

$$Mg(OH)_2(s) \text{ minus } H_2O \rightarrow MgO(s)$$

$$2B(OH)_3(s) \text{ minus } 3H_2O \rightarrow B_2O_3(s)$$

There are some borderline elements whose oxides are not very soluble in water but that can display both weakly acidic and basic properties by dissolving in both strong bases and acids. Examples of these *amphoteric oxides* include SnO_2, ZnO, Al_2O_3, and PbO.

PREDICTING CHEMICAL FORMULAS FOR OXIDES OF THE ELEMENTS

When an element combines with oxygen, there is a tendency for electrons to be transferred from atoms of the element E to atoms of oxygen. A molecule of dioxygen has the capacity to take up a maximum of four electrons to form two oxide ions:

$$O_2(g) + 4e^- \rightarrow 2O^{2-}$$

An oxide ion, O^{2-}, corresponds to an oxygen atom with a filled valence shell. Assuming that the element E gives up all its valence electrons to oxygen, the formula of the resulting compound will depend on the number of electrons in the valence shell of the element E.

When a metal atom forms an ion by transferring its valence electrons to a nonmetal such as oxygen, the charge of the metal cation can be determined by its position in the periodic table for main-group elements. It is simply equal to the group number: +1 for Group I, +2 for Group II, and so on. The charge corresponds to the loss of the valence electrons from the metal atom.

A second important principle is that anything stable enough to be collected and put in a bottle must be electrically neutral, which means that the formula for the compound must show zero net charge. (This is sometimes called the *electroneutrality principle*.) In practice, this means that we must determine the least common multiple of the cation and anion charges to find the formula of the compound.

AN EXAMPLE. Cesium oxide contains cesium ions (cesium cations). Cesium is in Group I, so it forms Cs^{1+} cations; the oxide ion is O^{2-}, and the requirement that every stable substance be electrically neutral requires that two Cs^+ ions combine with one O^{2-} ion to form neutral cesium oxide, Cs_2O. Aluminum is from Group III, so the neutral product between Al^{3+} and O^{2-} ions would be Al_2O_3.

In contrast, the bonding in nonmetal oxides is best described by the *covalent bonding model*, which involves the sharing of electron pairs rather than the complete transfer of electrons implied by the ionic bonding model. The covalent model is compatible with the ability to form more than one oxide of an element. For example, carbon can form the oxides CO and CO_2. Sulfur can form the oxides SO_2 and SO_3. This obviously creates some difficulties when we are trying to predict the formulas of the oxides of the elements. How do we know which oxide will form? In truth, for a particular set of reaction conditions, we have to do the experiment to find the answer to this question.

Some of the nonmetal oxides appear to follow the rules that we used to determine the formulas of the metal oxides. From their formulas, the molecules CO_2, SO_3, and N_2O_5 all look as if they could be ionic (but they are not). They correspond to the oxides we would predict if C, S, and N each gave up all of their valence electrons to oxygen.

However, the molecules CO, SO_2, NO, and NO_2 also exist. If they were ionic compounds (they are not), we would be tempted to say that these compounds resulted from incomplete transfer of electrons from the element to oxygen (CO resulting from a C^{2+} ion combining with an O^{2-} ion, SO_2 resulting from S^{4+} combining with two O^{2-} ions, etc.). It's as if there were formally only a partial transfer of the valence electrons from the element to oxygen. In a sense, this is what we mean by covalent sharing of electrons. Covalent bonding implies a tendency for electron density to partially shift from C, S, and N atoms to O atoms, without complete transfer of electrons to form ions. This allows more than one way for C, S, and N to share their valence electrons with oxygen atoms.

For purposes of this exercise, you will be given the formulas of the nonmetal oxides that are already prepared: B_2O_3, CO_2, P_2O_5, SO_2, and Cl_2O_7. In this group only SO_2 does not follow the pattern of the other members of the group. If we pretend that the oxygen atoms in SO_2 have full valence shells, formally equivalent to oxide ions, O^{2-}, then by the electroneutrality principle we would have to assign a charge of +4 to the sulfur atom. Later on, we will see that this is exactly what we do when assigning oxidation numbers to each element in a binary compound (a binary compound is one composed of two different elements).

If a sulfur atom gave up all of its six valence electrons to oxygen atoms, the predicted formula for the oxide would be SO_3. This is a known oxide of sulfur, but it is a molecular and not an ionic compound. *A good general rule is that the nonmetal oxides are better described by a covalent bonding model than by an ionic bonding model.*

HYDRATION OF THE OXIDES— SUMMARY CONCEPTS

We've now established a conceptual framework for understanding what happens when oxides come in contact with water. The outcome depends on the polarity of the E—O bond. If it is strongly ionic, the oxide ions react with water to form a base. If the bonding is largely covalent, the oxide will form a molecular hydrate with a stronger E—O bond, leaving the H atom more positive, so that the hydrate acts as an acid by donating a proton to a water molecule. Now we illustrate these concepts with examples of the reaction of representative elements.

IONIC OXIDES. For an ionic metal oxide, a plausible prediction is that one water molecule is added for each oxide ion present in the original oxide compound. Thus calcium oxide, CaO, would react with a water molecule to form calcium hydroxide.

$$CaO + H_2O \rightarrow Ca(OH)_2$$

Lithium oxide, Li_2O, reacts to give lithium hydroxide, LiOH, and a plausible equation describing the reaction would be:

$$Li_2O + H_2O \rightarrow 2LiOH$$

We do not write $Li_2O_2H_2$ because we always prefer the simplest empirical formula for ionic compounds.

For Fe_2O_3 the reaction would be:

$$Fe_2O_3 + 3H_2O \rightarrow Fe(OH)_3$$

COVALENT OXIDES. Because the bonding in the nonmetal oxides is much less polar than the bonding in ionic compounds, the behavior of the hydrated oxides is different. Not all of the nonmetal oxides react with one water molecule per oxygen atom, as the metal oxides do, and the actual structure of the final compound must be determined by experiment. Therefore, for this part of the exercise you will be given the formulas of the hydrated forms of the nonmetal oxides.

For example, when the acidic oxide N_2O_5 is added to water, we might expect the following reaction to occur:

$$N_2O_5 + 5H_2O \rightarrow 2N(OH)_5$$

But we find that molecules with a lot of —OH groups bonded to the same atom are often unstable and tend to dehydrate. So, the partially hydrated oxy acid $HONO_2$ proves to be the most stable configuration, which you may recognize as nitric acid when written with the more familiar formula HNO_3.

Although acidic oxides seldom form fully hydrated oxy acids, it is still easy to determine the formula of the anhydride by subtracting water so as to leave no hydrogen atoms. For example, the formula of the anhydride of perchloric acid is obtained by subtracting one mole of water from two moles of perchloric acid.

$$2HClO_4 \text{ minus } H_2O \rightarrow Cl_2O_7$$

A WARNING ABOUT THE STRUCTURES OF THE OXY-ACIDS

Don't let the conventional way of writing the formulas for nitric acid and perchloric acid as HNO_3 and $HClO_4$, respectively, mislead you into thinking that the hydrogen atoms in the acids are directly bonded to the nitrogen or chlorine atoms. The structural formula for nitric acid might more accurately be written as $HONO_2$ or

It is a good general rule for oxy acids that the hydrogen atoms that are acidic (or dissociable as protons) are bonded to oxygen atoms.

CONVENTIONS FOR WRITING THE FORMULAS OF ACIDS AND BASES

If a hydroxide compound is basic in water, it is because it releases hydroxide ions into the water. For example,

$$LiOH \rightarrow Li^+ + OH^-$$

and

$$Ca(OH)_2 \rightarrow Ca^{2+} + 2OH^-$$

If, instead, the hydrated oxide were acidic, it would react with water by donating a proton to a water molecule to form hydronium ions, H_3O^+.

$$LiOH + H_2O \rightarrow H_3O^+ + LiO^-$$

or

$$Ca(OH)_2 + H_2O \rightarrow H_3O^+ + CaO_2H^-$$

Which set of reactions is most plausible? Do LiOH and $Ca(OH)_2$ behave as bases or as acids? You will discover the answer by making the oxides of lithium and calcium and testing solutions of them in water.

If the compounds are acids, there is a convention that we write the acidic hydrogen atoms first in the formula. So if LiOH were an acid, it would be written HLiO, and $Ca(OH)_2$ would be written H_2CaO_2. Here are some examples:

BASE FORMULA	CONVENTIONAL FORMULA FOR AN ACID
KOH	HKO
$Sr(OH)_2$	H_2SrO_2
$B(OH)_3$	H_3BO_3
$SO(OH)_2$	H_2SO_3
$ClO_2(OH)$	$HClO_3$

REMINDER: This convention of writing the formulas for acids can be misleading. The hydrogen atoms written first are not bonded to the element that immediately follows. As we have already pointed out, you need to remember that, *in oxy acids, the hydrogen atoms dissociable as protons are bonded to the oxygen atoms.*

EXPERIMENTAL PROCEDURE

Special Supplies: Deflagrating spoon, porcelain crucible and clay triangle, crucible tongs, six glass squares, plastic basin, six wide-mouthed 250-mL (8 oz) bottles, 250-mL Erlenmeyer flask with 1-hole rubber stopper, pH meter and/or universal indicator, cylinder of oxygen (if students do not generate O_2 by decomposition of hydrogen peroxide).

Chemicals: 3% (0.9 M) H_2O_2, 1 M $FeCl_3$, $MnO_2(s)$ (powder), active dry yeast (Fleischmann's or other brands used in baking are satisfactory), Ca metal (turnings), Li metal (shot, oil coated), Mg metal (ribbon), phosphorus (red), sulfur (powder), carbon (small pieces of charcoal), universal indicator,[2] $H_3BO_3(s)$ (labeled "H_3BO_3 or $B(OH)_3$"), 0.1 M $HClO_4$ (labeled "0.1 M $HClO_4$ or $ClO_3(OH)$").

NOTE TO INSTRUCTOR: If $O_2(g)$ is available in cylinders, then Part 1 of the experiment can be done to study O_2 production, but the bottles of O_2 for Part 2 can be filled from the O_2 cylinder.

! SAFETY PRECAUTIONS: Although we expect you to ALWAYS WEAR EYE PROTECTION in the laboratory, it is especially important while doing this experiment.

Magnesium ribbon burns so brightly that the light can damage your eyes. *Do not stare directly at the burning ribbon.* Instead, glance out of the corner of your eye to note the flame.

Do not use white (or yellow) phosphorus in Part 2e of this experiment. It spontaneously ignites in air, is highly irritating to the skin, and is acutely toxic. The red form of phosphorus

[2]Universal indicator can be purchased from a laboratory supply house (Fisher). To prepare 200 mL of Yamada's universal indicator, dissolve 0.005 g thymol blue, 0.012 g methyl red, 0.060 g bromthymol blue, and 0.100 g phenolphthalein in 100 mL ethanol. Neutralize the solution to green with 0.01 M NaOH and dilute to 200 mL with distilled water. From L. S. Foster and I. J. Gruntfest, *J. Chem. Educ.* **1937,** *14,* 274.

used in this experiment is not spontaneously flammable but can be ignited by friction. Do not handle it with your bare hands because it is irritating to the skin.

The combustion of red phosphorus produces a choking, irritating smoke of P_2O_5. (The actual molecular species produced is P_4O_{10}.) Use a fume hood for the combustion. If a fume hood is not available, the laboratory instructor will perform the experiment as a demonstration.

The combustion of sulfur in Part 2f produces SO_2, a choking, irritating gas. Use a fume hood for the combustion. If a fume hood is not available, the laboratory instructor will perform the experiment as a demonstration.

WASTE COLLECTION: The solutions containing $FeCl_3$ and MnO_2 in Part 1 should be collected in a bottle marked "$FeCl_3$ and MnO_2 waste." All solutions produced in Part 2 can be collected in a beaker, neutralized with sodium bicarbonate, $NaHCO_3$, and then flushed down the drain.

1. PREPARATION OF OXYGEN. To observe the effect of *catalysts* (what is a catalyst?) on the rate of decomposition of hydrogen peroxide,

$$H_2O_2(aq) \rightarrow H_2O(l) + O_2(g)$$

place four 13×100 mm test tubes in a beaker or rack and put 3 mL of 3% H_2O_2 in each test tube. To the first one add 0.5 mL (10 drops) of 1 M $FeCl_3$. To the second add a pinch of solid MnO_2 powder. To the third add several grains (10–20) of active dry yeast. Add nothing to the fourth tube; it will be your control. Shake the tubes to mix the contents, and observe the rate of evolution of oxygen. Are there any visible changes in the colors of the solutions? Does the MnO_2 dissolve or appear to react in any way? Record your observations.

Now assemble the apparatus shown in Figure 7-2.[3] Place about 0.2 g of active dry yeast in the 250-mL Erlenmeyer flask. Fill six 250-mL wide-mouthed bottles with water, and invert these in the basin of water as needed. Add 200 mL of 3% H_2O_2 to the flask and replace the one-hole stopper. After about 15 s, when air has been expelled from the generator, fill the six bottles with oxygen by displacement of water. Leave about 10 mL of water in each bottle. As soon as each bottle is full, cover it with a glass square and place it right side up on the bench top.

2. PREPARATION OF OXIDES. Prepare oxides of the following elements (except calcium) by burning them in the bottles containing oxygen gas, being sure to keep

the bottles covered as much as possible (see Figure 7-3). Number or label each bottle to avoid confusion. Just before each combustion, slide the glass square far enough aside to quickly add 30 to 50 mL of deionized water. Then replace the glass square, perform the combustion, shake the bottle to dissolve the oxide formed, and set it aside for later use.

(a) **Lithium.** Take three or four pieces of the lithium metal shot, and place them in the deflagrating spoon. Heat the lithium in the spoon over a Bunsen burner. You may note an initial wisp of smoke as the mineral oil coating evaporates. When the lithium starts to burn, as indicated by a violet glow, stick the spoon into the bottle until the reaction stops. Leave the spoon and its contents in the bottle to cool, and then use water from your wash bottle to rinse the contents of the spoon into the solution at the bottom of the bottle.

(b) **Magnesium.** Grip a 10-cm length of magnesium ribbon with crucible tongs, ignite it using a match or Bunsen burner, and at once thrust it into a bottle of oxygen. (*Do not look directly at the brilliant light*, because it can injure your eyes.)

(c) **Calcium.** Calcium metal is difficult to ignite but burns brilliantly. Place a shaving of calcium metal in a crucible, heat it, cautiously at first and then very strongly, in the air over a Bunsen burner for 15 min. After the crucible has cooled, wash out the product with water into a beaker.

(d) **Carbon.** Ignite a small piece of charcoal in a Bunsen burner, holding it with tongs or in a clean deflagrating spoon, and thrust the charcoal into a bottle of oxygen.

(e) **Phosphorus.** **Caution:** *Carry out the ignition in a fume hood!* If a fume hood is not available, your instructor will demonstrate the combustion. Put a bit of red phosphorus (the yellow form is too toxic) no larger in volume than half a pea in a clean deflagrating spoon. Ignite it over the burner, then thrust it into a bottle of oxygen. After the combustion dies down, reheat the deflagrating spoon (in the fume hood) to burn out all the remaining phosphorus. *Avoid breathing the residual combustion products.*

(f) **Sulfur.** *Caution: Carry out the ignition in a fume hood!* Prepare the oxide of sulfur by the same procedure that you used for phosphorus and take the same precautions. Your instructor will demonstrate this combustion if a fume hood is not available.

3. ACIDS AND BASES FROM OXIDES. For each bottle in which the oxide of an element has been formed by the foregoing procedures, and to which water has been added to form a solution, either use your pH meter to

---

[3]Alternatively, the bottles of oxygen can be filled from an oxygen cylinder if one is available.

FIGURE 7-2

Collection of O_2 prepared by catalytic decomposition of hydrogen peroxide.

determine the acidity of the solution or add a few drops of universal indicator solution to the bottle. The results of adding the universal indicator are interpreted as follows:

INDICATOR COLOR	PH	PH VALUE
red	acidic	≤ 4
orange	slightly acidic	5
yellow	acidic	6
green	neutral	7
blue	slightly basic	8
indigo	basic	9
violet	basic	≥ 10

Is there a relationship between the chemical character of the reaction product and the position of the element in the periodic table? Is there any difference between the behavior of the metal and the nonmetal oxides?

The preceding tests include representative elements from Groups I, II, IV, V, and VI in the periodic table. To complete the series, let us examine an oxide or hydroxide from each of the other principal groups, III and VII. (Why not include an element from Group VIII?)

The *oxide of boron*, B_2O_3, is not readily available and in some forms is very insoluble. Dissolve a small amount (0.1 g) of "boron hydroxide," labeled "$B(OH)_3$ or H_3BO_3,"

FIGURE 7-3

The use of a deflagrating spoon to burn a substance in oxygen gas.

in 15 mL of hot water. Cool the solution and test with universal indicator or a pH meter. Is this substance an acid or a base? How is the substance usually named? How is its formula usually written, $B(OH)_3$ or H_3BO_3?

The *oxides of chlorine*—$Cl_2O(g)$, $ClO_2(g)$, and $Cl_2O_7(g)$—are quite unstable. They react with water. A solution of the hydrated form of Cl_2O_7, labeled 0.1 M "$ClO_3(OH)$ or $HClO_4$," will be available to you. Test some of this solution with universal indicator or a pH meter, and in your report write the formula as an acid, $HClO_4$, or base, $ClO_3(OH)$, according to your observations of its properties.

Consider This

Let's try out the concepts we have been learning in this experiment on a real molecule, sulfur trioxide (SO_3). First, locate sulfur in the periodic table. Is it among the metals or the nonmetals? Would you expect sulfur trioxide to form an acidic or a basic solution when it reacts with water?

Based on your answer to the previous question, what would be the preferred conventional way of writing the formula of a solution of sulfur trioxide in water: $SO_2(OH)_2$ or H_2SO_4?

If you reacted SO_3 with one molecule of water, how would you write the formula for the resulting hydroxide?

$$SO_3 + H_2O \rightarrow ?$$

If we think of the reaction of a Group I metal with oxygen as involving transfer of electrons from the metal atom to an oxygen molecule, what would happen if the addition of electrons to the oxygen molecule occurred in a step-by-step fashion? Each of the ions named below is known to exist. What would be the formulas for the resulting compounds if these ions were combined with sodium ions (Na^+)?

$$O_2(g) + e^- \rightarrow O_2^- \text{ (superoxide ion)}$$

$$O_2(g) + 2e^- \rightarrow O_2^{2-} \text{ (peroxide ion)}$$

$$O_2(g) + 4e^- \rightarrow 2O^{2-} \text{ (two oxide ions)}$$

BIBLIOGRAPHY

Emsley, J., 3rd edition, *The Elements*, Oxford University Press, 1998. A most useful handbook summarizing properties of all known elements.

Mingos, D. M. P., *Essential Trends in Inorganic Chemistry*, Oxford University Press, 1998.

The Chemistry of Oxygen: Basic and Acidic Oxides and the Periodic Table

NAME		
SECTION		LOCKER
INSTRUCTOR		DATE

OBSERVATIONS AND DATA

1. Preparation of Oxygen

(a) Write the equation for the reaction by which you prepared oxygen.

(b) Hydrogen peroxide slowly decomposes, even when no $FeCl_3$ is present. Compare the decomposition of H_2O_2 in the presence and absence of $FeCl_3$, $MnO_2(s)$, or yeast.

　　(i) Does the amount of oxygen that may be obtained when reaction is complete depend on whether $FeCl_3$ is added?

　　(ii) What is the purpose of adding $FeCl_3$, $MnO_2(s)$, or yeast to the H_2O_2 solution?

　　(iii) What is the name applied to a substance used as $FeCl_3$ in this reaction?

2. Preparation of the Oxides

Describe the changes that occur during the reaction with oxygen of each of the following elements, and note any distinctive characteristics of the products formed. Write a chemical equation for each reaction. (For carbon, phosphorus, and sulfur, the formula of the oxide cannot be determined from charge considerations. The formulas of these oxides are CO_2, P_2O_5, and SO_2, respectively.)

(a) Lithium 　　＿＿＿＿＿＿＿＿＿＿＿＿＿＿＿＿＿＿＿＿＿＿＿＿＿＿＿

　　　Equation 　＿＿＿＿＿＿＿＿＿＿＿＿＿＿＿＿＿＿＿＿＿＿＿＿＿＿＿

(b) Magnesium 　＿＿＿＿＿＿＿＿＿＿＿＿＿＿＿＿＿＿＿＿＿＿＿＿＿＿＿

　　　Equation 　＿＿＿＿＿＿＿＿＿＿＿＿＿＿＿＿＿＿＿＿＿＿＿＿＿＿＿

(c) Calcium 　　＿＿＿＿＿＿＿＿＿＿＿＿＿＿＿＿＿＿＿＿＿＿＿＿＿＿＿

　　　Equation 　＿＿＿＿＿＿＿＿＿＿＿＿＿＿＿＿＿＿＿＿＿＿＿＿＿＿＿

(e) Carbon 　　＿＿＿＿＿＿＿＿＿＿＿＿＿＿＿＿＿＿＿＿＿＿＿＿＿＿＿

　　　Equation 　＿＿＿＿＿＿＿＿＿＿＿＿＿＿＿＿＿＿＿＿＿＿＿＿＿＿＿

(e) Phosphorus 　＿＿＿＿＿＿＿＿＿＿＿＿＿＿＿＿＿＿＿＿＿＿＿＿＿＿

　　　Equation 　＿＿＿＿＿＿＿＿＿＿＿＿＿＿＿＿＿＿＿＿＿＿＿＿＿＿＿

(f) Sulfur 　　　＿＿＿＿＿＿＿＿＿＿＿＿＿＿＿＿＿＿＿＿＿＿＿＿＿＿＿

　　　Equation 　＿＿＿＿＿＿＿＿＿＿＿＿＿＿＿＿＿＿＿＿＿＿＿＿＿＿＿

3. Acids and Bases from Oxides

(a) *Prediction of the formulas of the hydrated oxides*

Write balanced chemical equations to describe the formation of the hydrated oxides (hydroxides) from the respective oxide, for example, oxide + water → hydroxide compound. For those elements which are classified as nonmetals whose oxide and hydroxide reaction product formulas are more difficult to predict, two formulas are given, the first written as a base formula, and the second written as the conventional acid formula. You must balance these equations.

(a) Lithium _____ _____ _____ _____

(b) Magnesium _____ _____ _____ _____

(c) Calcium _____ _____ _____ _____

(d) Carbon $CO_2(g)$ + _____ H_2O → _____ $CO(OH)_2$ _____ or H_2CO_3 _____

(e) Phosphorus $P_2O_5(s)$ + _____ H_2O → _____ $PO(OH)_3$ _____ or H_3PO_4 _____

(f) Sulfur $SO_2(g)$ + _____ H_2O → _____ $SO(OH)_2$ _____ or H_2SO_3 _____

(g) Boron $B_2O_3(s)$ + _____ H_2O → _____ $B(OH)_3$ _____ or H_3BO_3 _____

(h) Chlorine $Cl_2O_7(l)$ + _____ H_2O → _____ $ClO_3(OH)$ _____ or $HClO_4$ _____

(b) *Summary of experimental results*

(i)   Write in the first column the periodic table group to which the element belongs.

(ii) In the next two columns write the formulas of the oxides and hydrated oxides (see Part 3a above for the formulas). If the hydrated oxide is acidic, write its formula in such a way as to indicate this fact, using the conventional acid formula.

(iii) From your pH measurements or the indicator color, describe the acidic or basic properties of the solutions formed from each oxide compound (or hydroxide, if the oxide was not available and you did not make it yourself).

(iv) Write a chemical equation showing how the hydrated oxide forms H_3O^+ or OH^- in the solution.

Element symbol	Periodic table group	Formula of oxide	Formula of hydrated oxide	Acidic or basic properties	Equation
Li					
Mg					
Ca					
C					
P					
S					
B					
Cl					

APPLICATION OF CONCEPTS

1. Comment on the acidic or basic character of the oxide of phosphorus, P_2O_5 (which actually exists as molecular P_4O_{10}), as predicted by its position in the periodic table and its classification as a metal or nonmetal.

2. Considering the positions of the elements in the periodic table, write the formulas of three other acidic oxides and three other basic oxides. On the longer line write the equation for the reaction of the oxide with water.

   Acidic oxides: _____ _____

   _____ _____

   _____ _____

   Basic oxides: _____ _____

   _____ _____

   _____ _____

3. What does the term *anhydride* mean?

4. Deduce and write the formulas of the anhydrides of the following.

   H_2SO_3: _____   HClO: _____

   H_2SO_4: _____ $HClO_3$: _____

   $La(OH)_3$: _____ HNO_2: _____

   $Sn(OH)_4$: _____ HNO_3: _____

   RbOH: _____ H_2CO_3: _____

   $Ba(OH)_2$: _____ H_4SiO_4: _____

👉 **Consider This**

1. *Locate sulfur in the periodic table. Is it among the metals or nonmetals? Would you expect sulfur trioxide, SO_3, to form an acidic or a basic solution when it reacts with water?*

   Based on your answer to the previous question, what would be the preferred conventional way of writing the formula of a solution of sulfur trioxide in water: $SO_2(OH)_2$ or H_2SO_4?

$$SO_3 + H_2O \rightarrow \underline{\hspace{3cm}} ?$$

2. *If we think of the reaction of a Group I metal with oxygen as involving transfer of electrons from the metal atom to an oxygen molecule, what would happen if the addition of electrons to the oxygen molecule occurred in a step-by-step fashion? Each of the ions named below is known to exist in a compound with sodium ion. What would be the formulas for the resulting compounds if these ions were combined with sodium ions (Na^+)?*

Formula of sodium salt

$O_2(g) + e^- \rightarrow O_2^-$ *(superoxide ion)* _____

$O_2(g) + 2e^- \rightarrow O_2^{2-}$ *(peroxide ion)* _____

$O_2(g) + 4e^- \rightarrow 2O^{2-}$ *(two oxide ions)* _____

Determination of a Chemical Formula
The Reaction of Iodine with Zinc

PURPOSE

To determine the iodine/zinc mass ratio for zinc iodide, which is prepared by reacting zinc with iodine in solution. To use the mass ratio and the iodine/zinc atom mass ratio to determine the simplest chemical formula for zinc iodide. To write a balanced chemical equation for the reaction of zinc with iodine. To show that mass is conserved in a chemical reaction.

PRE-LAB PREPARATION

When you react two elements to form a binary compound, it is necessary to keep track of the masses of the reactants and products in order to write a balanced chemical equation. This atom bookkeeping is called chemical *stoichiometry*; a word derived from the Greek words *stoikheion* (element) and *metron* (to measure).

In this experiment you will make zinc iodide by heating granular zinc metal with iodine that is dissolved in methanol. We want to know two things about this compound: (1) the mass of iodine that reacts with a given mass of zinc, that is, the *mass ratio*, g I/g Zn, and (2) the *atom ratio*, atom I/atom Zn, which defines the simplest chemical formula for the compound. The *mole ratio*, mol I/mol Zn, is exactly the same as the atom ratio, atom I/atom Zn, because a mole is just an Avogadro's number of the particles defined by the simplest chemical formula. Taking an Avogadro's number of particles doesn't change the ratio of iodine atoms to zinc atoms. We will represent the simplest chemical formula as Zn_aI_b or ZnI_x, where $x = b/a$ = atom I/atom Zn = mol I/mol Zn.

There are at least two approaches we can use to determine the stoichiometry of the reaction. We can react zinc and iodine, then analyze a sample of pure zinc iodide to determine the masses of zinc and iodine that are present.

In the second approach (the one used in this experiment), we can synthesize zinc iodide by reacting a known mass of iodine with an excess of zinc. Because zinc is a granular solid and the reaction product is soluble in the solvent (methanol), the reaction product can be easily separated from the unreacted zinc. If we know the mass of iodine reacted and the mass of zinc before and after the reaction, we can determine the amount of each that reacted. By evaporating the volatile solvent, we can recover the zinc iodide as a solid and weigh it, too. This gives us a way to see if the mass of the zinc iodide formed equals the mass of zinc and iodine that reacted.

From our data we will be able to determine the mass of zinc that reacts with a known mass of iodine and to calculate the mass ratio, g I/g Zn, for the reaction. We will also be able to answer this question: Is mass conserved in a chemical reaction?

The second ratio we want to determine is the atom ratio, atom I/atom Zn. As we pointed out, this ratio is equal to the mole ratio, mol I/mol Zn. Determining the atom or mole ratio is more difficult than determining the mass ratio because it requires knowledge about the relative masses of Zn and I atoms.

Pretend for the moment that you don't know the relative mass of a zinc atom and an iodine atom. (Put away your periodic table or table of atomic masses.) This, of course, was exactly the situation that existed during the time of John Dalton (1766–1844) and his contemporaries: In the early 1800s no one knew the relative masses of atoms of different elements.

Dalton found it useful to think in terms of a model, and we will do the same. Imagine that atoms are small indivisible spheres that can link together in various combinations to form compounds with a definite atom ratio that is equal to the ratio of small integers. To be specific, let's assume that zinc and iodine form a compound with the formula Zn_2I, so that $x = b/a = ½ = 0.50$. The formula $ZnI_{0.50}$ would be an equally correct way of writing the chemical formula, but this format doesn't lend itself to drawing a simple picture as Zn_2I does. We will picture the simplest particle of Zn_2I in the following way, using circles to represent the atoms:

Atoms are too small to see and count, so we can't directly determine the atom ratio by counting, but we can perform the experiment we described to determine the mass ratio, g I/g Zn. How are the two ratios related? To see the connection, let's write an equation, making use of the dimensions of each quantity to verify that the units come out correctly:

$$\text{mol ratio} \times \text{atomic mass ratio} = \text{mass ratio}$$

$$\frac{\text{mol I}}{\text{mol Zn}} \times \frac{\text{g I/mol I}}{\text{g Zn/mol Zn}} = \frac{\text{g I}}{\text{g Zn}} \qquad (1)$$

If we multiply the mole (or atom) ratio by the atom mass ratio (*mass of a mole of I*)/(*mass of a mole of Zn*), we get the mass ratio for the compound, g I/g Zn. The mass ratio can be directly determined by a chemical experiment.

Dalton and his contemporaries could measure the mass ratio, g I/g Zn, as you can, but they didn't know either the mole ratio (1 mol I/2 mol Zn in our hypothetical example) or the atomic mass ratio [(g I/mol I)/(g Zn/mol Zn)]. But if two of the three ratios shown in Equation (1) were unknown, how could they be determined? To be perfectly truthful there is no simple way to do this, and in the early days scientists sometimes made mistakes. For example, Dalton knew of only one compound formed between hydrogen and oxygen, namely, water. When he thought that only one compound existed, he usually made the simplest possible assumption: One atom of one element reacted with one atom of the other. So he assumed that water had the formula HO. He knew that the mass ratio, g O/g H, in water was about 8, so in his first table of atomic weights he assigned 1 to the lightest element, hydrogen, and 8 to oxygen. It was only after many compounds had been synthesized and analyzed that inconsistencies began to show up, and these inconsistencies could only be resolved by assigning the formula H_2O to water. As a result, the proper mass ratio for the mass of an oxygen atom relative to the mass of a hydrogen atom was found to be 16.

Now look at your periodic table or table of atomic masses, and find the values for the relative masses of zinc and iodine. Calculate for Zn_2I the ratio g I/g Zn from Equation (1). You should get the value 0.97 g I/g Zn. Now do a second calculation, in reverse: Suppose there is an iodide of zinc that has the composition 1.94 g I/g Zn. By rearranging Equation (1) you can calculate the mole ratio, $x = b/a$, for this compound. You should get a value of 1.0, so we could write the formula for this compound as ZnI. What is the true formula for zinc iodide? Let's find out by doing the experiment.

EXPERIMENTAL PROCEDURE

Special Supplies: Variable-temperature electric hot plate, aluminum foil, glazed weighing paper, desiccator (optional).

Chemicals: 20-mesh granular zinc metal, $I_2(s)$, methanol.

!　SAFETY PRECAUTIONS: Handle solid iodine with care. The solid and its concentrated solutions can cause skin burns. Wash thoroughly with soap and water if spilled on flesh. Iodine is also corrosive to stainless steel balance pans, so clean up spills quickly. The action of iodine on skin and clothing is similar to chlorine bleach, so wash it off with water if it spills. Methanol is flammable and toxic. Avoid breathing methanol vapor or spilling the liquid on your hands. There must be no open flames in the laboratory. Working in a fume hood, conduct the reaction of zinc with iodine dissolved in methanol and the subsequent evaporation of the methanol to recover the zinc iodide.

1. THE REACTION OF IODINE WITH ZINC.

Weigh a clean, dry 125-mL Erlenmeyer flask as precisely as possible on an analytical balance (at least milligram sensitivity). Record the mass of the empty flask. Preweigh 2.0 ± 0.1 g of 20-mesh granular zinc metal on a glazed weighing paper using a lab balance (0.1-g sensitivity). Add the zinc metal grains to the Erlenmeyer flask, and reweigh the flask, using the more accurate analytical balance. Record the mass of the flask plus added zinc. Using a lab balance, preweigh 2.0 ± 0.1 g of iodine crystals on a glazed weighing paper, taking care to avoid spilling any iodine crystals. If any iodine does spill, clean it up immediately as directed by your instructor. Iodine is quite corrosive to balance pans if left in prolonged contact. Add the iodine crystals to the Erlenmeyer flask containing the zinc metal, and reweigh as accurately as possible on an analytical balance. Record the mass of the flask plus added zinc and iodine.

Working in the fume hood, obtain 25 mL of methanol, and add it to the Erlenmeyer flask. (**Caution:** Methanol vapors are flammable and toxic.) Put an aluminum foil cap on the flask, crimping it around the edges to hold it in place. Place the flask on a hot plate in the fume hood. Adjust the temperature setting of the hot plate to a low level, sufficient to heat the methanol to boiling. Swirl the flask contents occasionally to dissolve the iodine. After 15 to 20 minutes, the reaction between the zinc and iodine will be complete, as shown by the disappearance of the dark brown color of the dissolved iodine. When the reaction is complete, the methanol solution should be pale yellow and nearly colorless, and the unreacted zinc granules should be easily visible.

Accurately weigh a clean, dry 250-mL beaker on the analytical balance while waiting for the zinc-iodine reaction to go to completion. Record the mass of the empty beaker. (The weighed beaker will be used in Part 2 of the experiment.)

When the reaction is complete, remove the Erlenmeyer flask from the hot plate. (A folded strip of paper towel wrapped around the neck of the flask will protect your fingers from the heat.) Remove the aluminum foil cap. Working in a hood, and using a glass stirring rod to guide the solution as shown in Figure 8-1, pour the warm liquid contents of the flask into the previously weighed 250-mL beaker. Be careful not to let any of the solid zinc particles escape from the Erlenmeyer flask.

Add 5 mL of methanol to the Erlenmeyer flask, and swirl for 15 s. Allow the solid zinc to settle; then pour the methanol into the beaker. Repeat with two additional methanol rinsings of 5 mL each. The three rinses ensure that all of the zinc iodide reaction product gets transferred to the beaker.

Place the Erlenmeyer flask containing the unreacted zinc back on the hot plate to dry. Leave the foil cap off Adjust the hot plate to a medium setting.

Also, put the beaker containing the zinc iodide in methanol on the hot plate, following the procedure described in Part 2.

Swirl the Erlenmeyer flask containing the zinc granules occasionally to redistribute them and to speed up the drying of the zinc. When all the methanol has evaporated from the flask and the zinc particles look dry and move freely on the bottom of the flask, remove the flask from the hot plate. Allow it to cool to room temperature (10 to 15 min). Does the dry product look like zinc granules, or does it contain a white residue of zinc iodide? If white residue is present, a few more rinses are in order. Be sure to transfer these rinses to the collection beaker as well, then redry the zinc granules. Reweigh the flask containing the unreacted zinc on the analytical balance. Record the mass.

Calculate the exact masses of iodine and zinc that reacted. From these masses, calculate the ratio g I/g Zn in the zinc iodide product. Using the atomic masses of zinc and iodine, calculate the ratio mol I/mol Zn and the simplest formula for zinc iodide.

FIGURE 8-1

Use a glass stirring rod to guide the methanol solution into the beaker so that the solution does not run down the side of the Erlenmeyer flask.

2. ISOLATION OF THE ZINC-IODINE REACTION PRODUCT.

Carry out the following procedure to isolate and weigh the zinc iodide produced in Part 1 when zinc and iodine reacted. Place on the hot plate the 250-mL beaker (whose mass you have previously recorded) containing the methanol solution of zinc iodide and washings. Adjust the hot plate to medium heat, and allow the methanol to evaporate. (**Caution:** This must be done in the fume hood.) If it shows a tendency to boil irregularly or spatter, stir it occasionally with a glass stirring rod until the methanol has nearly evaporated. It normally takes about 25 min to evaporate all of the methanol. When the last traces of methanol have completely disappeared, allow the beaker to heat 2 min more; then remove it from the hot plate. Allow the beaker to cool to room temperature for 15 min. Reweigh the beaker containing the dry zinc iodide (but not the stirring rod!) on the analytical balance. Record the mass.

NOTE: Zinc iodide is hygroscopic (that is, it readily absorbs moisture from the air), which means its mass may increase by 1 to 2 mg/min, depending on the relative humidity. If a desiccator is available, put the beaker in the desiccator while it is cooling. Otherwise, cover the beaker with a watch glass and then weigh the sample after it has cooled for 15 min.

Subtract the mass of the empty beaker to get the mass of the zinc iodide reaction product. Compare this mass with the sum of the masses of zinc and iodine that reacted as determined from your data for Part 1. Do your results confirm the conservation of mass principle? Record the appearance of the zinc iodide product. Is the appearance distinctly different from that of the zinc and iodine reactants?

☞ **Consider This**

Develop a procedure to decompose the zinc iodide product back into zinc metal and solid iodine. (A 9-volt battery can serve as the power supply to electrolyze a solution of zinc iodide.) Can you recapture the zinc and iodide? Are the masses consistent with the Laws of Constant Composition and Conservation of Mass? (See the DeMeo article listed in the Bibliography for hints and a possible procedure.)

Other binary compounds that can be prepared from the elements include the sulfides of copper, iron, nickel, and lead, as well as the oxides of magnesium, tin, and copper. Antimony iodide can also be prepared by a procedure similar to that used in this experiment. Prepare some of these other compounds using procedures from the sources discussed in the Bibliography or other lab books, and compare your results to those obtained for zinc iodide.

BIBLIOGRAPHY

DeMeo, S. "Synthesis and Decomposition of Zinc Iodide: Model Reactions for Investigating Chemical Change in the Introductory Laboratory," *J. Chem. Educ.*, **1995,** *72:* 836–839.

Walker, N. "Tested Demonstrations, Synthesis and Decomposition of ZnI$_2$," *J. Chem. Educ.*, **1980,** *57:* 738.

Wells, N., and Boschmann, E. "A Different Experiment on Chemical Composition," *J. Chem. Educ.*, **1977,** *54:* 586.

QUESTIONS

1. Iodine is known to exist in the solid state as a diatomic molecule, I_2. Using the simplest formula you found for zinc iodide, write a balanced chemical equation for the reaction of Zn with I_2.

2. What observations support the assumption that all of the iodine reacted with the zinc? Explain.

3. Comment on the statement, "The properties of a chemical compound are the averages of the elements that make up the compound." What evidence do you have from this experiment about the truth (or falsity) of this statement?

4. Do your observations support the principle that mass is conserved in a chemical reaction, that is, that the mass of the products equals the mass of the reactants? Explain.

5. If at the end of the reaction some of the unreacted zinc escaped from the Erlenmeyer flask, the mass of the flask plus the unreacted zinc would be lower than it should be. Would this cause the calculated mass ratio, g I/g Zn, to be higher or lower than the value predicted?

Consider This

Develop a procedure to decompose the zinc iodide product back into zinc metal and solid iodine. (A 9-volt battery can serve as the power supply to electrolyze a solution of zinc iodide.) Can you recapture the zinc and iodide? Are the masses consistent with the Laws of Constant Composition and Conservation of Mass? (See the DeMeo article listed in the Bibliography for hints and a possible procedure.)

Other binary compounds that can be prepared from the elements include the sulfides of copper, iron, nickel, and lead, as well as the oxides of magnesium, tin, and copper. Antimony iodide can also be prepared by a procedure similar to that used in this experiment. Prepare some of these other compounds using procedures from the sources discussed in the Bibliography or other lab books, and compare your results to those obtained for zinc iodide.

Determination of a Chemical Formula by Titration

The Reaction of Calcium with Water

PURPOSE

To learn what happens when active Group I or Group II metals react with water. To introduce you to the use of an acid-base titration to determine the chemical formula of the product formed when calcium reacts with water.

RE-LAB PREPARATION

The reaction of an active metal with an acid produces hydrogen gas. For example, zinc reacts with hydrochloric acid to produce hydrogen gas and zinc chloride:

$$Zn(s) + 2HCl(aq) \rightarrow H_2(g) + ZnCl_2(aq) \qquad (1)$$

Hydrogen is a lighter-than-air flammable gas. At one time it was used in rigid, hydrogen-filled luxury airships that made regular Atlantic crossings. One of them, the German dirigible *Hindenburg,* caught fire and burned as it was landing in New Jersey in 1937. World War II ended the era of the rigid airship.

In the chemical reaction shown in Equation (1), the element hydrogen has been replaced by zinc, in the sense that hydrogen chloride has been converted to zinc chloride.

What if we put zinc in water instead of in an HCl solution? We would see that over the course of a few minutes or even an hour, very little reaction would take place. But if we put a more active metal, like sodium, in water, we would observe a violent reaction that produces hydrogen gas and a strong base, sodium hydroxide:

$$2Na(s) + H_2O(l) \rightarrow H_2(g) + 2NaOH(aq) \qquad (2)$$

By boiling the solution to evaporate the water, we could recover the sodium hydroxide as a white, translu-

cent solid that is used in the manufacture of paper and soaps and in petroleum refining. It is extremely corrosive to the skin and other tissues and is sometimes called *caustic soda* or *lye.*

The heavier Group II metals, such as calcium, strontium, and barium, also react readily with water at room temperature to form hydrogen and the metal hydroxide. (Magnesium does not react rapidly with water at room temperature but will react with high-temperature steam.)

The reaction of calcium with water is classified as an oxidation-reduction reaction (sometimes also called a single replacement reaction) in which calcium replaces the hydrogen in water by transferring its two valence electrons to water molecules:

$$\text{Calcium} + \text{water} \rightarrow \text{hydrogen gas} + \text{calcium hydroxide}$$
$$(3)$$

We did not write the chemical formulas for this reaction because the point of this experiment is to determine the chemical formula for the calcium hydroxide produced when calcium reacts with water. We don't want to give away the end of the story, *but your first clue has already been revealed in Equation (2).*

Like sodium hydroxide, calcium hydroxide is corrosive to the skin. When calcium hydroxide is heated to drive off water, calcium oxide (often called *lime*) is produced. Lime is an important ingredient in the mortar used in plastering walls.

Let's look at the reactions of active metals with water in greater detail. The reactions are classified as oxidation-reduction reactions because they involve changes in the charges on the metal atom and on the oxygen (and hydrogen) atoms in water.

In the reaction of sodium with water, a sodium atom loses an electron to a water molecule. In the process the sodium becomes a positively charged sodium ion with an electron configuration like that of neon, a noble gas.

Having received the electron from the sodium atom, the water molecule becomes unstable. It splits up into a neutral hydrogen atom and a hydroxy group with a negative charge, called a *hydroxide ion*. That means that the product of the reaction is ionic, and solid NaOH is best characterized as containing sodium ions, Na^+, and hydroxide ions, OH^-.

The neutral hydrogen atoms, each with one electron, wander around until they find each other. (This random wandering around is called *diffusion*.) When two hydrogen atoms bump into each other, they combine to form molecular hydrogen gas, H_2. (The "wander and combine" process occurs in a tiny fraction of a second.) The two hydrogen atoms each contribute an electron to form a shared electron pair bond, called a *covalent* bond. The net result is that when an active Group I or Group II metal reacts with water, an ionic metal hydroxide is formed along with molecular hydrogen.

Here is your second clue: If a calcium atom tends to give up electrons until it has the same number of electrons as the noble gas nearest to it in the periodic table, what would be the charge on the calcium ion that is left after the electrons have been removed?

Calcium ion charge: _____

Now for the third and last clue: If every substance that is stable enough to be collected and stored in a bottle must be electrically neutral, what would you predict for the formula of calcium hydroxide, keeping in mind that each hydroxide ion has a −1 charge?

Expected formula: _____

Now let's put your predictions to the test. If we react a known amount of metal with water, we will obtain the metal hydroxide, a base. If we then react the base with a known amount of acid, we can determine how many moles of the base were present. Knowing the moles of base produced and the mass of the metal and its atomic mass, we can easily calculate how many moles of base were produced for every mole of metal. This allows us to write the correct formula for the metal hydroxide produced in the reaction.

To determine the amount of hydroxide ion produced from a given amount of metal, we will use a procedure called an *acid-base titration*. The base is contained in a reaction flask. To the base we add a known concentration of acid dispensed from a buret. Knowing the volume and concentration of the acid, we can calculate the moles of acid added, which will be equal to the moles of hydroxide because they react one-to-one, as illustrated by the reaction of sodium hydroxide with hydrochloric acid:

$$NaOH(aq) + HCl(aq) \rightarrow NaCl(aq) + H_2O(l) \quad (4)$$

This is an acid-base reaction, driven to completion by the formation of water, a very stable compound.

The reactants and products shown in Equation (4) are all colorless, so we cannot see anything happening as the reaction proceeds. We need something else to indicate when we have added exactly enough hydrochloric acid to react with the hydroxide. This is called the *end point* of the titration. To detect the end point we use a substance called an *acid-base indicator*. It has the property of changing color when the solution goes from slightly basic to slightly acidic (or vice versa).

EXPERIMENTAL PROCEDURE

Special Supplies: Glazed weighing paper, forceps for handling calcium turnings,[1] sturdy side-cutting pliers to cut calcium turnings into smaller pieces.
Microscale: 1-mL buret
Macroscale: 25–50-mL buret

Chemicals: Mossy zinc (see footnote 1), fresh calcium turnings, 6 M HCl, 0.02 M HCl, 1% phenolphthalein indicator, 0.1% thymol blue indicator, $NaHCO_3(s)$.
Microscale: standardized 6.00 M HCl for titration
Macroscale: standardized 0.300 M HCl for titration

NOTE TO INSTRUCTOR: Test the calcium turnings beforehand by dropping a turning into water. Fresh calcium turnings react vigorously, and the reaction is complete within 30 seconds. Turnings that have been oxidized by contact with air or water vapor may take 20–30 minutes to react, even when the water is heated, and are unsuitable for this experiment.

If you prefer to use conventional 25- or 50-mL burets, use standardized 0.300 M HCl titrant in place of the 1-mL buret and standardized 6.00 M HCl titrant.

! SAFETY PRECAUTIONS: HCl solutions are corrosive and irritating to the eyes and skin. WEAR EYE PROTECTION, and wash your skin immediately with cool water if contact with HCl solution should occur. Spills should be neutralized with bicarbonate solution and cleaned up according to your instructor's directions (wear protective gloves).

Calcium metal is corrosive and irritating to the skin. Use forceps or tongs to handle the calcium turnings.

1. THE REACTION OF ZINC AND CALCIUM WITH HCl AND WATER. Place 5 mL of 6 M HCl in a 15 × 125 mm test tube. Fold a piece of glazed weighing paper to make a crease in it. Weigh out one or two pieces of mossy zinc (approximately 0.5 g) on the creased

---

[1] The term "mossy zinc" is descriptive of the pieces of zinc that are formed b dropping molten zinc metal into water. Calcium "turnings" are the curly chips that form when the metal is turned on a lathe.

weighing paper; then slide the zinc into the test tube containing the 6 M HCl. Observe what happens, and record your observations.

While the reaction is proceeding briskly, invert over the reaction tube a second 15 × 125 mm test tube, holding the two tubes so they are mouth-to-mouth with no gap between them. After about 30 s, ask a neighbor to light a match for you. Quickly move the mouth of the inverted test tube close to the flame of the match as shown in Figure 9-1. Be prepared to hear a "woof" or sharp "bark" if the hydrogen ignites. Describe what happens. If you neither see nor hear anything, try the procedure again.

Next put 5 mL 6 M HCl into another 15 × 125 mm test tube. Using forceps to handle the calcium, put one or two calcium metal turnings (approximately 0.1–0.2 g) on a creased sheet of glazed weighing paper. Add the calcium turnings to the 6 M HCl in the test tube. Observe what happens, and record your observations. Repeat the procedure for collecting the hydrogen in an inverted test tube and igniting it. Again, record your observations.

Next, on sheets of creased weighing paper weigh out fresh samples of zinc and calcium of the same size as before. Record the weight, then add each metal by itself to separate 15 × 125 mm test tubes half filled with deionized water. Observe what happens, and record your observations. Did both the zinc and calcium react with water? Do your observations confirm the notion that active metals react more rapidly with acid solutions than with water? What is the insoluble white compound that forms when calcium reacts with water?

Add two drops of phenolphthalein indicator to the water in the test tube containing the calcium metal.

FIGURE 9-1
Testing for hydrogen gas.

Describe the color of the indicator. Phenolphthalein is colorless in acidic solutions and red or pink in basic solutions. Does the color indicate that the solution containing calcium and water is acidic or basic?

Transfer the solution containing the calcium-water reaction products and any unreacted calcium to a 250-mL beaker. Add 6 M HCl one drop at a time while stirring until all of the calcium has dissolved and the solution has just become colorless. At this point the solution has been neutralized and may now be discarded into the sink drain.

2. TITRATION OF THE CALCIUM–WATER REACTION PRODUCT. (Read the instructions for the use of a micro- or macroburet found in the Introduction of this manual.)

Fold a crease in a piece of glazed weighing paper, and weigh it on a balance to the nearest milligram. Record the mass of the paper. (If you have an electronic balance with a taring feature, it is convenient to tare the paper to zero before putting the calcium on the paper. Then the balance reading gives the mass of the calcium directly.) Using forceps, get a piece of calcium metal weighing between 0.08 and 0.10 g and put it on the paper, again weighing to the nearest milligram. Record the mass of the paper + calcium metal. If your instructor has directed you to carry out two titrations, weigh a second sample of calcium metal, recording the masses as before.

In the procedure that follows we will assume that you are making a single determination. If you are making two determinations, it will save time if you weigh out both your samples and carry out the reactions at the same time, as described in the following paragraph.

Add the previously weighed calcium metal sample to a clean 250-mL Erlenmeyer flask. Mark the flask with your initials. Add about 150 mL of deionized water to the flask. Swirling the flask gently every few minutes may speed the reaction. If the calcium turnings are fresh, the calcium should react vigorously and completely with the water. If they are not fresh, the reaction may be slow, and it may be necessary to heat the flask containing the calcium and water until the calcium has completely reacted.

Microscale procedure: Obtain about 5 mL of standardized 6.00 M HCl in a clean, dry 13 × 100 mm test tube. Fill a 1-mL buret with the solution, following the directions given to you by the instructor. Read the initial buret reading, estimating to the nearest 0.01 mL. Record this volume.

Macroscale procedure: Obtain about 75 mL of standardized 0.300 M HCl in a clean flask. Fill a 25- or 50-mL buret with the solution, rinsing the buret with two or three 5 mL portions before finally filling it. Read the initial buret reading, estimating to the nearest 0.02 mL. Record this volume.

When the calcium metal previously added to the Erlenmeyer flask has completely reacted, add 10 drops of 0.1% thymol blue indicator. With the tip of the buret slightly below the mouth of the flask, begin adding the 6.00 M HCl one drop at a time, swirling the flask as you add the HCl solution. (If you are using 0.300 M HCl in a 25- or 50-mL buret, begin by adding about 1 mL at a time.)

As you near the end point of the titration, you will notice that the blue color of the indicator turns yellow at the point where the acid enters the solution. If you approach the end point carefully, the solution will change within 1 to 2 drops from a blue color to a yellow color that remains for at least a full minute when the solution is swirled (swirling is very important here). Read and record the final volume. Also record the exact molarity of the HCl titrant used, which may be different from the nominal 6.00 M (or 0.300 M) value.

If you add the HCl solution too rapidly near the end point, you may overrun it and spoil the measurement. As you near the endpoint, take care to add the titrant drop by drop because if you add a few drops in excess, the solution will have the same yellow color as if the correct volume had been added. A pink or red color indicates that you have grossly overrun the end point. If you overrun the end point by more than 2 to 3 drops, you should repeat the entire procedure with a fresh sample of calcium metal.

If you are carrying out two titrations, repeat the procedure with the second sample.

When you have finished all of your titrations, drain the HCl from the buret into a 250-mL beaker. Add to the beaker the unused HCl titrant contained in the test tube or flask. Neutralize the HCl by adding solid sodium bicarbonate, $NaHCO_3$, until the solution stops foaming (about 2 g is sufficient). The neutralized solution may now be discarded down the drain. The contents of the Erlenmeyer flask(s) that you have titrated may also be discarded down the drain. Rinse the buret with water. Use deionized water for the final rinse.

From your titration data, you will calculate the moles of HCl used to titrate the hydroxide ion produced in the reaction and the moles of hydroxide per mole of calcium for each sample.

Consider This

Look up the first and second ionization energies of zinc and calcium in your text or in the CRC Handbook of Chemistry and Physics. *If we make the hypothesis that the most active metals will have the smallest ionization energies, which metal might we expect to be most active, that is, which metal atoms give up their electrons more easily, zinc or calcium?*

Write a plausible balanced chemical reaction for the reaction of calcium with water. In doing this, think about what happens when a calcium atom gives up its two electrons to two water molecules. Which atoms in a water molecule are the more positive and therefore the most likely to accept the electrons donated by a calcium atom? What could be formed when you add an electron to a water molecule? What form of the calcium atom is left when it gives up two electrons?

Then write a second reaction in which the products of the titration reaction (the calcium-water reaction, i.e., calcium hydroxide) react with the HCl titrant. The two reactions may be combined (added together) to get the net chemical reaction that shows the initial reactants (calcium and water) and the final reaction products after the titration is completed (calcium chloride, $CaCl_2$). What is the net chemical reaction?

NAME		
SECTION		LOCKER
INSTRUCTOR		DATE

OBSERVATIONS AND DATA

1. The Reaction of Zinc and Calcium with HCl and Water

(a) Describe what happened when you added zinc to 6 M HCl in the test tube.

(b) What happened when you brought the hydrogen gas in the inverted test tube near a flame?

(c) What happened when you added calcium metal to 6 M HCl in the test tube?

(d) Compare the reaction of the gas collected from the calcium-HCl reaction with the reaction of hydrogen collected from the zinc-HCl reaction when each is brought near a flame.

(e) Describe what happened when you added the zinc and calcium metal to test tubes containing only water. Which metal is more reactive with water?

(f) What gas is produced by the reaction of Group I and Group II metals with water? (Consult your text or ask your instructor if you are uncertain.)

(g) What color results when phenolphthalein indicator is added to the solution in which calcium is reacting?

(h) The base produced in the reaction of Group I and Group II metals with water is called hydroxide ion. Write the formula for hydroxide ion. (Consult your text or ask your instructor if you are uncertain.)

2. Titration of the Calcium-Water Reaction Product

Weighing data	Trial 1	Trial 2
Mass of calcium + glazed weighing paper:	_____ g	_____ g
Mass of glazed weighing paper:	_____ g	_____ g
Mass of calcium metal:	_____ g	_____ g

Titration data

Final buret reading:	_____ g	_____ g
Initial buret reading:	_____ g	_____ g
Net volume of HCl titrant:	_____ g	_____ g

Record the exact molarity of the HCl titrant used _____ mol/L
(the nominal value is 6.00 M or 0.300 M):

CALCULATIONS

Look up the molar mass and calculate the moles of calcium in each sample. Show your calculations below:

_____ mol Ca   _____ mol Ca

Calculate the moles of HCl that were required to titrate each sample. Show your calculations below:

_____ mol HCl   _____ mol HCl

Calculate the ratio mol HCl/mol Ca for each sample. Show your calculations below:

_____   _____

Round the ratio, mol HCl/mol Ca, to the nearest integer: _____

If one mole of HCl reacts with one mole of hydroxide (recall Equation [2]), how many moles of hydroxide must have been produced for every mole of calcium that reacted?

From your experimental result, what must be the formula for the calcium hydroxide produced in the reaction of calcium with water?

☞ Consider This

1. Look up in your text or the *CRC Handbook of Chemistry and Physics* the first two ionization energies for the elements calcium and zinc. Record them below. Based on the values of the ionization energies, which metal would you expect to be the most active, calcium or zinc?

	Ca	Zn
IE(1)	_____	_____
IE(2)	_____	_____

2. Complete and balance the following sequence of chemical reactions that takes place when the calcium reacts with water and when the calcium reaction products are titrated with HCl solution:

$$Ca(s) + H_2O \rightarrow$$

$$Ca(OH)_2 + HCl \rightarrow$$

The two reactions may be combined (added together) to get the net chemical reaction that shows the initial reactants and the final reaction products after the titration is completed. Write the net chemical reaction:

3. If some of the calcium metal remains unreacted when the titration is begun, the following reaction would occur as HCl is added:

$$Ca(s) + 2HCl(aq) \rightarrow CaCl_2(aq) + H_2(g)$$

Would this cause an error in the titration? If so, would more or less HCl be required?

A

The Language of Chemistry:
Chemical Nomenclature

INTRODUCTION

This study assignment provides you with a basic introduction to inorganic chemical nomenclature, and shows you how to write simple chemical formulas, using the periodic table to correlate the ionic charges or oxidation states of the elements. Rules for naming compounds are essential to avoid the massive confusion that would result if each person invented his or her own names and symbols for the elements and their compounds. The importance of this task is underscored by noting that in January 1965, *Chemical Abstracts* (a periodical that provides short summaries of all articles appearing in chemistry journals) started a compound registry index. By 1985 this index contained about 7 million different chemical substances, and since then the list has been growing at the rate of 300,000 new substances each year. It is estimated that about 10 million chemical substances are known, the great majority of which are organic compounds (those containing carbon). Mastering the details necessary to name all of these substances would require a great deal of study. Fortunately, most of the time we work with a limited number of chemical substances, so you will not need to learn more than a few hundred names. This task is made even easier because (1) most of the names are established according to simple rules and (2) the elements are grouped in chemical families of the periodic table, and members of each family bear strong resemblances to one another. In addition, you do not have to learn all of the names at once. Just as you are able to learn the names of new friends a few at a time, you will find it easy to learn the names of new chemical substances a few at a time, as you encounter them.

However, one problem cannot be avoided. The grand traditions of chemistry go back several centuries, and some compounds are commonly referred to by names coined long ago, as well as by their more systematic names. Consequently, for some substances it is necessary to learn both the *common,* or *trivial,* name and the systematic name. Fortunately, the use of older common names is diminishing: For example, in industrial commerce it is common to refer to sodium carbonate (Na_2CO_3) as "soda ash," but in today's chemistry lab you are not likely to find a bottle of sodium carbonate labeled as soda ash.

CHEMICAL NOMENCLATURE

In many respects the best way to represent a chemical substance is to make a three-dimensional model of it, showing the arrangement of all the atoms in space and their relative sizes. With the advent of molecular modeling programs and fast computers, this has now become a possibility. Most of the time, however, this process too cumbersome, requiring us to have a computer always at hand, so we usually represent atoms and chemical substances in a more compact way, using symbols (such as Na) and formulas (such as Na_2CO_3) and using names (such as sodium and sodium carbonate) for these symbols and formulas. We call the system of naming substances *chemical nomenclature.* In rational nomenclature, the name of a compound would indicate (1) the elements of which it was composed and (2) the relative proportions of each element. Recent changes in chemical nomenclature have moved in this direction but still fall considerably short of this goal. Nomenclature, like language, is not static. It continually changes as needs and current usage dictate.

The standards of nomenclature are established by the International Union of Pure and Applied Chemistry (IUPAC). Since this international body must coordinate different languages and conflicting views, it is not surprising that the rules reflect a compromise. As a result, American chemists occasionally employ usages not officially sanctioned by the IUPAC rules. In this study assignment we have followed the IUPAC rules[1] except those conflicting with current American usage. Most of the conflicts are minor, and the beginning student of chemistry is not likely to encounter any serious conflicts in applying the rules listed here.

CHEMICAL SYMBOLS AND FORMULAS

You may have already used chemical symbols and the formulas of substances and may have written equations for a number of chemical reactions. Now we must pause to emphasize the exact meaning and correct usage of these and other terms that constitute the unique language of chemistry.

Each element is represented by a *chemical symbol.* The symbol consists of either one or two letters, such as C for carbon and Ba for barium. Several of the elements have symbols derived from their ancient Latin names: Cu for copper from *cuprum,* Fe for iron from *ferrum,* Au for gold for *aurum,* Pb for lead from *plumbum,* and Ag for silver from *argentum.* Several elements discovered after 1780 have names derived from Latin or Germanic stems: Na for sodium from *natrium,* K for potassium from *kalium,* and W for tungsten from *wolfram.* We recommend that you train yourself to recognize an element name when you see its symbol and vice versa, recognize the symbols for the elements, at least those up to radon, element 86.

A *chemical formula* represents the composition of a given *substance,* which may be either an element or a compound. Thus H and O are the symbols for the elements hydrogen and oxygen, and they can also represent the atomic state of the elements, whereas H_2 (hydrogen gas) and O_2 (oxygen gas) represent the more stable molecular forms of the elements hydrogen and oxygen. When we speak of the chemical properties of oxygen, it is usually the stable molecular form of oxygen, O_2, that is meant. In this manual we will always try to specify the chemical formula of a substance in order to avoid any confusion or misinterpretation.

A *chemical compound* is a substance formed from two or more elements, such as H_2O (water), H_2O_2 (hydrogen peroxide), or NaCl (sodium chloride).

---

[1]For the most recent IUPAC rules on the nomenclature of inorganic chemistry, see Leigh, G. J., ed., *International Union of Pure and Applied Chemistry—Nomenclature of Inorganic Chemistry—Recommendations 1990,* Blackwell Scientific Publications, Cambridge, MA, 1990. Other IUPAC publications on nomenclature have been listed in the *Journal of Chemical Education,* **1973,** *50:* 341; **1975,** *52:* 482.

THE DIVISIONS OF CHEMICAL ELEMENTS, COMPOUNDS, AND NAMES

There are three distinct classes of chemical compounds: *covalent compounds, ionic compounds,* and *acids.* This classification is based upon generalizations regarding the physical and chemical properties of compounds. Each of these classes has its own naming system, so the first order of business is to train yourself to identify the compound type. This classification system requires that you know the distinctions between the metallic elements and the nonmetals. The bold stairstep-shaped line that you see on most periodic tables, including the one on the inside front cover of your lab book and in Figure A-1, serves as the divider between the elements that are metals and the nonmetals. The metallic elements are those found below and to the left of this divider; the nonmetals are above it and to the right.

The distinction between metals and nonmetals is based on shared characteristics. If you saw samples of the metals you would quickly notice their typical shiny surfaces (even when they are colored, such as gold and copper.) They also have the typical properties of metals: they conduct electricity and heat, and they are malleable and ductile. The nonmetals cannot be so simply described; some are gases when pure (e.g., hydrogen, oxygen, nitrogen), others are solids (e.g., sulfur, iodine, phosphorus), and bromine is a liquid at room temperature. In contrast to the metals, the nonmetals are electrical insulators (i.e., nonconductors), none of them have shiny surfaces, and the solids are not readily formed into shapes.

The elements shown in dark gray in the periodic table have properties that lie between those of the metals and nonmetals, so they often are referred to as the semimetals or the metalloids. For purposes of our naming scheme, however, we will treat them as either metals or nonmetals based on which side of the stairstep line they fall. Hydrogen plays a unique role in chemistry, mimicking the behavior of metals in some contexts, but for naming purposes you should classify it as a nonmetal.

I. COVALENT COMPOUNDS

Covalent compounds are those compounds that are composed of nonmetals bonded to nonmetals. They are easily recognized from their name or formula because they contain no metal elements.

More than 90 percent of the known compounds are covalent compounds, and their naming system gets quite complicated because of the variety of forms and structures of these compounds. We will limit this discussion to the *binary covalent compounds,* i.e., those containing only two nonmetal elements. The system for naming these

TABLE A-1

The periodic table showing the categories of metals (in white) and nonmetals (gray).

Element	Element	Element	Element	Element	Element	Element	Element
1 H 1.008							2 He 4.003
3 Li 6.941	4 Be 9.012						
11 Na 22.99	12 Mg 24.31						
19 K 39.10	20 Ca 40.08	21 Sc 44.96	22 Ti 47.88	23 V 50.94	24 Cr 52.00	25 Mn 54.94	26 Fe 55.85
37 Rb 85.47	38 Sr 87.62	39 Y 88.91	40 Zr 91.22	41 Nb 92.91	42 Mo 95.94	43 Tc (98)	44 Ru 101.1
55 Cs 132.9	56 Ba 137.3	57 La 138.9	72 Hf 178.5	73 Ta 180.9	74 W 183.9	75 Re 186.2	76 Os 190.2
87 Fr (223)	88 Ra (226)	89 Ac (227)	104 Unq (257)	105 Unp (260)	106 Unh (263)	107 Uns (262)	108 Uno (265)

27 Co 58.93	28 Ni 58.69	29 Cu 63.55	30 Zn 65.39	5 B 10.81	6 C 12.01	7 N 14.01	8 O 16.00	9 F 19.00	10 Ne 20.18
45 Rh 102.9	46 Pd 106.4	47 Ag 107.9	48 Cd 112.4	13 Al 26.98	14 Si 28.09	15 P 30.97	16 S 32.07	17 Cl 35.45	18 Ar 39.95
77 Ir 192.2	78 Pt 195.1	79 Au 197.0	80 Hg 200.6	31 Ga 69.72	32 Ge 72.59	33 As 74.92	34 Se 78.96	35 Br 79.90	36 Kr 83.80
109 Une (266)				49 In 114.8	50 Sn 118.7	51 Sb 121.8	52 Te 127.6	53 I 126.9	54 Xe 131.3
				81 Tl 204.4	82 Pb 207.2	83 Bi 209.0	84 Po (210)	85 At (210)	86 Rn (222)

58 Ce 140.1	59 Pr 140.9	60 Nd 144.2	61 Pm (147)	62 Sm 150.4	63 Eu 152.0	64 Gd 157.3	65 Tb 158.9	66 Dy 162.5	67 Ho 164.9	68 Er 167.3	69 Tm 168.9	70 Yb 73.0	71 Lu 175.0
90 Th 232.0	91 Pa (231)	92 U 238.0	93 Np (237)	94 Pu (242)	95 Am (243)	96 Cm (247)	97 Bk (247)	98 Cf (249)	99 Es (254)	100 Fm (253)	101 Md (256)	102 No (254)	103 Lr (257)

binary covalent compounds is simple; the formula is transliterated into a name. The numbers in the formula are converted to the Greek prefixes, such as the following:

$1 \rightarrow$ *mono*	$6 \rightarrow$ *hexa*
$2 \rightarrow$ *di*	$7 \rightarrow$ *hepta*
$3 \rightarrow$ *tri*	$8 \rightarrow$ *octa*
$4 \rightarrow$ *tetra*	$9 \rightarrow$ *nona*
$5 \rightarrow$ *penta*	$10 \rightarrow$ *deca*

The elements are listed in the same order as in the formula (which is from the least to the greatest *electronegativity*, a concept that will be discussed as an important part of chemical bonding ideas), and the suffix of the last element is changed to *-ide*.

Examples: N_2O_5: dinitrogen pentaoxide

P_4O_{10}: tetraphosphorous decaoxide

ClO_2: monochlorine dioxide or chlorine dioxide

Na_2O: NOT A COVALENT COMPOUND!

• •

NOTE: The prefix "mono-" is usually *not* used (although it is not incorrect) except in cases where there are similar and more common compounds of the same element. Thus CO is carbon monoxide because CO_2, carbon dioxide, is more common. NO, however, is usually called nitrogen oxide.

It is critical to identify that the compound is a covalent compound before this naming system is used; Na_2O is *not* disodium oxide because it is an ionic compound, not a covalent one. (It contains a metal!)

• •

TRIVIAL NAMES. Some common binary compounds are designated by trivial names that have been assigned historically. Examples are

H_2O	water
NH_3	ammonia
CH_4	methane
PH_3	phosphane (or phosphine)
AsH_3	arsane (or arsine)

II. IONIC COMPOUNDS

Ionic compounds are those that form between a metal and one or more nonmetals. Ionic compounds contain a *metal cation* and a *nonmetal anion*. The name of the compound is simply the cation name followed by the anion name. Ionic compounds do *not* use Greek prefixes (except in rare cases where the ion itself contains one of these prefixes in its name, such as dichromate). The formula can be determined from the fact that all compounds are electrically neutral, so once the charge on the ions is known, the formula can be readily deduced.

Learning the names of ionic compounds then consists of learning the cation and anion names. To determine the formula from the name of a compound requires knowing the individual ion charges as well.

A. CATIONS. Cations are metal atoms with electrons removed. Because it is easy to recognize when a metal is in a compound or not (in a compound other elements are bonded to the metal and named with it), the metal name can be used for both the element and the ion. In some cases the name needs to be augmented to specify the charge on the metal cation.

1. The cations from periodic table groups IA, IIA, and groups III (A and B) have charges of $+1$, $+2$, and $+3$ respectively. Because of this observation, the unmodified metal name is used for the ions. For example: Na^+, sodium; Ba^{2+}, barium; Sc^{3+}, scandium; and Al^{3+}, aluminum.[2]

2. All other metal cations are named by augmenting the metal name with a roman numeral in parentheses to indicate the charge on the ion. This is necessary because these metals have different ion charges in different compounds. For example: Fe^{2+}, Iron(II); Fe^{3+}, Iron(III); Pb^{2+}, Lead(II); Pb^{4+}, Lead(IV).

• •

NOTE: In speaking, the roman numeral is pronounced as the number, such as "iron two," "iron three," "lead two," and "lead four" in the above examples.

There is an older, nonsystematic naming process for cations which uses suffixes to indicate the charge, such as ferrous for Fe^{2+}, and ferric for Fe^{3+}, but this is a cumbersome system because the suffixes *do not* correspond to a specific charge but rather to the possible range of charges; e.g., stannous is Sn^{2+}, but stannic is Sn^{4+}, whereas cupric is Cu^{2+} and cuprous is Cu^+. You will not be responsible for knowing this naming system, but you should be aware of its existence because you will run across such names from time to time. If so, look them up! Here are a few of the most common examples.

• •

---

[2]There are a few exceptions to the "rule" that groups III A and B always have a $+3$ charge, such as thallium which forms a few Tl^+ compounds.

-ous: Fe^{2+} Co^{2+} Cu^+ Au^+ Hg_2^{2+} Sn^{2+}

-ic: Fe^{3+} Co^{3+} Cu^{2+} Au^{3+} Hg^{2+} Sn^{4+}

3. The metal ions Ag^+, Zn^{2+}, and Cd^{2+} are the only ions formed by the respective metals so the roman numeral designations are not necessary. They are known as silver, zinc, and cadmium respectively as ions in their compounds. It is all right to use the roman numerals, but it is not necessary.

4. The mercury(I) ion, which you would expect to be Hg^+, is always found paired as Hg_2^{2+} in ionic compounds.

5. The polyatomic cation NH_4^+, known as the ammonium ion, mimics the behavior of the Group IA metal ions and is thus treated as such in ionic compounds. This means it is possible to have ionic compounds that do not contain any metals such as NH_4Cl or $(NH_4)_2S$, but this should not pose a problem because the distinctive formula of ammonium in a formula should immediately remind you of this intriguing ion.

B. ANIONS. The anions can be grouped by the suffixes (and prefixes) used in their names and can be learned by description once a few basic members are memorized.

1. The -ide Anions. The monatomic anions, formed from the nonmetal elements and sufficient extra electrons to give them a *noble gas configuration,* have names that end in *-ide.* They do *not* need to be memorized, but can be deduced from the periodic table by noting that the (negative) charge is determined, in practice, by counting the groups to get to the right-hand noble gas column. The list is as follows:

		Hydride, H^-
Nitride, N^{3-}	Oxide, O^{2-}	Fluoride, F^-
Phosphide, P^{3-}	Sulfide, S^{2-}	Chloride, Cl^-
Arsenide, As^{3-}	Selenide, Se^{2-}	Bromide, Br^-
	Telluride, Te^{2-}	Iodide, I^-

In addition to the above list, there are a few other anions which use the *-ide* suffix. They are OH^-, hydroxide; CN^-, cyanide; O_2^{2-}, peroxide; I_3^-, triiodide; N_3^-, azide; NH_2^-, amide. These polyatomic anions actually fit the above description in an indirect way, but for now we will have to memorize them, because they are important ions. The O_2^{2-} (hydrogen peroxide), Na_2O_2 (sodium peroxide), and normal oxides, such as MnO_2 (manganese[IV] oxide), TiO_2

(titanium[IV] oxide), and SiO_2 (silicon dioxide). Note that the peroxides can only be peroxides, for instance, if Na_2O_2 were an oxide, the sodium ion would have to be a +2 ion.

The hydrogen atom is an exception to many of the patterns in chemistry. Note that here you find it forming a −1 anion as if it belongs in Group VII, while in many contexts it behaves like a +1 ion. To avoid confusion, note that if hydrogen is combined with a metal in a compound, hydrogen is acting as the nonmetal anion component. When it is combined with other nonmetals in covalent compounds, its +1 character emerges. This behavior is covered under the concept of electronegativity.

The carbide (and silicide ions), C^{4-} (and Si^{4-}), follow this noble gas pattern, but there are many examples where the nonsystematic name of carbide is used for the C_2^{2-} ion instead, such as calcium carbide, CaC_2. We will use the name carbide to refer to the C^{4-} anion.

2. The -ate Anions. These anions and the remainder of the list are all polyatomic anions containing one or more oxygen atoms in addition to another element. It is useful to think of *-ate* anions as covalent compounds that have an excess of negative charge and that serve as the anion when combined with a metal cation. Once this list of the *-ate* ions is committed to memory, the remainder of the anions' names and formulas can be elucidated. Note that the *-ate* suffix is not a description of the number of oxygen atoms nor the ion charge, so this list *must* be memorized.

CO_3^{2-}, Carbonate	NO_3^-, Nitrate
PO_4^{3-}, Phosphate	SO_4^{2-}, Sulfate
ClO_3^-, Chlorate	$C_2H_3O_2^-$, Acetate
CrO_4^{2-}, Chromate	$Cr_2O_7^{2-}$, Dichromate
MnO_4^-, Permanganate	SCN^-, Thiocyanate[3]

There are other ions in this *-ate* family, but many of their formulas and names can be determined by using the above list and the groups of the periodic table. For instance, if ClO_3^- is chlorate, then BrO_3^- must be bromate, and IO_3^- is iodate. If PO_4^{3-} is phosphate, then AsO_4^{3-} would be arsenate. (While nitrogen is also in Group V along with phosphorus, the chemistry of arsenic should more closely resemble that of phosphorus than of nitrogen.) The list of *-ate* anions is quite extensive, but the list above and its derivatives cover most of the common ions that you will encounter in chemistry.

---

[3]The prefix "thio-" in chemical nomenclature has the meaning of "Trade a sulfur atom for an oxygen atom." The cyanate ion has the formula OCN^- but is less common than thiocyanate in the general chemistry lab context, so we give its formula here.

3. The -ite Anions. Each anion in the *-ite* family is related to a corresponding *-ate* anion. The *-ite* ions have one *less* oxygen atom than the corresponding *-ate* partner. Thus: NO_3^- is nitrate, and NO_2^- is nitrite; PO_4^{3-} is phosphate, and PO_3^{3-} is phosphite, and so on. Not all *-ate* anions have a corresponding *-ite* partner. You are *not* responsible for knowing which ones do exist, but if they do exist you should be able to name them.

4. The Per–ate Anions. The *per–ate* anions are related to the corresponding *-ate* anion by having one *more* oxygen atom. Thus: ClO_3^- is chlorate, ClO_4^- is perchlorate, IO_4^- is periodate.

5. The Hypo–ite Anions. The *hypo–ite* anions have one *less* oxygen than the corresponding *-ite* anion. The full set then looks like

ClO_4^-	ClO_3^-	ClO_2^-	ClO^-	Cl^-
Perchlorate	Chlorate	Chlorite	Hypochlorite	Chloride

Only the halogens display the entire array of anions as shown above for chlorine. Again, you do not need to know which anions exist, but you must be able to name those that are presented.

6. The Hydrogen-Anion. For any of the anions that have a negative charge that is more than -1, additional anions can be formed from the combination of these anions with an H^+ unit. These form a new anion and are simply named "*hydrogen-*," after their parent anion. If more than one hydrogen is attached, prefixes such as *di-* and *tri-* are used; e.g., HS^-, hydrogensulfide; HPO_4^{2-}, hydrogenphosphate; or $H_2PO_4^-$, dihydrogenphosphate.[4] Note that if the parent anion has a -1 charge, the addition of an H^+ does *not* result in another ion, but in a covalent compound, an acid, such as HCl or HNO_3.

HYDRATED IONIC COMPOUNDS. Many ionic compounds can be found in two or more forms that incorporate water into their structures and formulas. These water molecules are not simply absorbed into pores or cracks in the solids or adsorbed onto the surface but are incorporated into specific positions in the solid lattice and are chemically bonded to the metal cation. The key feature of these *hydrates* is that they have a specific number of water molecules associated with each cation and thus are true compounds. Because a simple heating process can remove these water molecules and because we think of water as a neutral or inert

substance in many contexts, we tend to think of these not as new compounds but as modified versions of the unhydrated compounds. Given this context, a system for writing formulas and naming these compounds has arisen that treats them as slightly modified versions of the unhydrated compounds. Thus $CuSO_4$, which is found as a colorless (white) solid, turns distinctly blue when it is exposed to water vapor or when, by drying, it forms an aqueous solution of copper(II) sulfate. The blue compound has the formula $Cu(H_2O)_5SO_4$, which shows that the water molecules are bonded to the Cu^{2+} ion. However, most often you will see it written as $CuSO_4 \cdot 5\ H_2O$ and named as copper(II) sulfate pentahydrate. Upon heating to 150 °C, the blue pentahydrate returns to the colorless $CuSO_4$ form. Other examples of hydrated compounds include the alum of Experiment 6, which is aluminum potassium sulfate dodecahydrate, $AlK(SO_4)_2 \cdot 12\ H_2O$, and calcium chloride, $CaCl_2$, which is often used as a desiccant (drying agent) because it forms a dihydrate, $CaCl_2 \cdot 2\ H_2O$ and a hexahydrate, $CaCl_2 \cdot 6\ H_2O$.

III. THE ACIDS

The set of *covalent compounds* (i.e., neutral molecules) formed nominally from H^+ cations in combination with any of the anions are known as acids. They are covalent compounds (they contain no metal cations and have covalent bonding properties), but they also have many properties related closely to ionic compounds. Because of this uniqueness, they are in a special class with a special naming system. The acids are easily identified by formula, because their formulas, by tradition, are written with the *H* first. As you will see, their names always contain the word *acid* so they are simple to spot.[5]

Acid names are derivations of the anions from which they derive.

The *-ide* anions form acids with the naming structure *hydro–ic acid.*

The *-ate* anions (including the per–ates) form acids with the naming structure *-ic acid.*

The *-ite* anions (including the hypo–ites) form acids with the naming structure *-ous acid.*

---

[4]IUPAC rules recommend combining hydrogen and the name of the anion into one word, e.g., sodium dihydrogenphosphate. In American usage the single word is often separated into two words: dihydrogen phosphate.

[5]This tradition of writing the *H* first in the formula of an acid conflicts with another tradition of using the formula to convey information about structure. For example, in the case of the oxyacids (oxygen-containing acids) such as sulfuric or nitrous acids, the formulas H_2SO_4 and HNO_2 would lead you to believe that the hydrogen atoms are bonded to the sulfur or nitrogen atoms, whereas the hydrogens are invariably bonded to an oxygen atom. In this light it would be better to write these formulas as $HOSO_3$ and $HONO$, although this arrangement is not usually seen.

Examples: HCl (from Cl⁻, chloride); hydrochloric acid

H_2S: hydrosulfuric acid

HNO_3: nitric acid

HNO_2: nitrous acid

H_3PO_4: phosphoric acid

$HClO_4$: perchloric acid

$HBrO_3$: bromic acid

HIO_2:   iodous acid

HClO: hypochlorous acid

The property of *acidity* is really only displayed when a solvent like water is present. A few of the acids exist as molecules outside of solution and could be named as covalent molecules, such as hydrogen chloride, HCl, and dihydrogen sulfide, H_2S. In most cases, their acid names are used even in this context.

IV. OTHER CONTEXTS

Two large arenas of chemistry do not fit under the naming systems described here. Because of carbon's ability to bond to itself in long chains, the compounds of carbon display a tremendous variety of formulas and structures. The naming system for these molecules is an elaborate system beyond the scope of general chemistry; thus we have not described any of the rules for naming these compounds, which are known as organic molecules.

Transition metal ions have the ability to form complex structures that themselves can be anions and/or cations in compounds. We have left the naming of these transition metal complexes to a later time as well.

The Language of Chemistry:
Chemical Nomenclature

NAME

SECTION   LOCKER

INSTRUCTOR   DATE

EXERCISES ON FORMULAS AND NOMENCLATURE

Note: These exercises will help you learn how to write correct formulas and name compounds. Check your answers, if necessary, with your instructor.

1. Name the following.

FeI_2	
$FeBr_3$	
I_2	
$AlCl_3 \cdot 6\,H_2O$	
$Fe_2(SO_4)_3$	
FeS	
NCl_3	
H_2CO_3	
$CaCO_3$	
Be_2C	
$Sn(NO_3)_2$	
$(NH_4)_2S$	
N_2O_4	
MgO_2	

2. Write the correct chemical formula.

Barium chloride		Stannous chloride	
Ammonium sulfate		Barium carbonate	
Stannic chloride		Aluminum carbide	
Sodium carbonate		Sodium hydrogen carbonate	
Magnesium phosphate		Nitrogen dioxide	
Calcium hydrogen phosphate		Disulfur dichloride	

3. Complete the following table. (Note that "salt" is a synonym for "ionic compound.")

Formula	Name as acid	Formula for sodium salt	Name of salt
HF			
HNO_2			
$HBrO_2$			
$HBrO_4$			
HNO_3			
HBrO			
$HBrO_3$			
H_3AsO_4		Na_3AsO_4	
H_3AsO_3		NaH_2AsO_3	

4. Name the following compounds.

NaIO	
Mg_3N_2	
K_2HPO_4	
$Fe(NO_3)_3$	
P_4O_6	
NH_4ClO_4	
K_2CrO_4	
Na_2SO_3	
$KMnO_4$	
$BaSO_3$	
$Ca(ClO_2)_2$	
$FeSO_4$	
P_4O_{10}	
$HClO_3$	
$K_2Cr_2O_7$	
Na_2SO_4	

5. The spaces below represent portions of some of the main groups and periods of the periodic table. In the proper squares, write the correct formulas for the chlorides, oxides, and sulfates of the elements of groups I, II, and III, respectively. Likewise, write the formulas of the compounds of sodium, calcium, and aluminum with the elements of groups VI and VII. Two of the squares have been completed as examples.

Period	Group 1	Group 2	Group 3	Group 6	Group 7
2	$LiCl$ Li_2O Li_2SO_4		(omit sulfate)		
3				Na_2S CaS Al_2S_3	
4					
5					

6. What is the systematic name of water? _____

What is the name of water when viewed as an acid (which it is)? _____

7. If sulfate has the formula SO_4^{2-}, then what is the formula of the thiosulfate ion? (See footnote 3 for the meaning of

the prefix *thio-*) _____

8. If glutamic acid has the formula $HC_5H_8NO_4$, then what is the formula of sodium glutamate, commonly known as

"MSG," or monosodium glutamate? _____

The Estimation of Avogadro's Number

PURPOSE

To make an order-of-magnitude estimate of the size of a carbon atom and of the number of atoms in a mole of carbon based on simple assumptions about the spreading of a thin film of stearic acid on a water surface.

PRE-LAB PREPARATION

John Dalton (1766–1844) was so taken with the notion of atomism that he never quite grasped the distinction between an atom and a molecule of an element. The most stable form of many elements, such as oxygen, hydrogen, and the halogens, is **not a** single atom but a diatomic molecule. Because Dalton believed that the smallest unit of an element must be an atom, he had a hard time accepting Gay-Lussac's data about the combining volumes of gases.

The reconciliation between Dalton's theory and Gay-Lussac's data was brought about by Amedeo Carlo Avogadro (1776–1856). He accomplished this by making the distinction between an atom and a molecule of an element and by making the hypothesis that equal volumes of gases contain equal numbers of molecules. For example, if you believe that hydrogen and chlorine are monoatomic and that equal volumes of gases (at the same temperature and pressure) contain equal numbers of atoms, then one volume of hydrogen should react with one volume of chlorine to form one volume of HCl:

$$H(g) + Cl(g) \rightarrow HCl(g)$$

But when you do the experiment you will find that two volumes of HCl are produced. Avogadro interpreted this result by assuming that the smallest unit of hydrogen and chlorine is a diatomic molecule, not an atom, and by writing the equation as follows:

$$H_2(g) + Cl_2(g) \rightarrow 2HCl(g)$$

Many subsequent experiments have proved that Avogadro's explanation was correct.

The hypothesis, published in 1811, was perhaps ahead of its time and went virtually unnoticed. More than half a century passed before Cannizzaro demonstrated the general applicability of the hypothesis in an article published in 1858. When he distributed it in pamphlet form at the first international chemical congress at Karlsruhe, Germany, in 1860, the pamphlet so clearly and completely discussed atoms, molecules, atomic weights, and molecular weights that chemists were convinced of his views, and quietly incorporated them into chemical thinking. Fifty years of pondering Dalton's atomic theory had created the right moment for the acceptance of Avogadro's hypothesis. Cannizzaro gave credit to Avogadro as well as to Ampère and Dumas, and no doubt it was Cannizzaro who saved Avogadro's hypothesis from oblivion. As a result, today we call the number of particles (atoms or molecules) in a mole *Avogadro's number*.

How many particles are there in a mole? Avogadro did not invent or use the concept of a mole and never knew the number of particles in a mole. In 1865, a few years after Avogadro's death, the German chemist Joseph Loschmidt published an article estimating the size of a molecule and, although Loschmidt didn't present the calculation in his paper, it is possible from the calculations he did present to estimate the number of molecules in a cubic centimeter of gas. This is only one simple step away from determining the number of molecules in a mole. For this reason, German chemists often called the constant *Loschmidt's number* rather than *Avogadro's number,* and the International Union of Pure and Applied Chemistry (IUPAC) allows the use of either the symbol N_A or L for Avogadro's number, to honor the contributions of both men.

A French scientist, Jean Perrin, is often given credit for determining the first value of Avogadro's number in 1908. Perrin measured the vertical distribution in the earth's gravitational field of gamboge (a natural resin) particles suspended in water and obtained values in the range 5.4 to 6.0×10^{23}.

After the American physicist Robert Millikan determined the charge of an electron around 1915, a more accurate value was obtained by dividing the charge of a mole of electrons (the Faraday constant) by the charge of a single electron. More refined values have been obtained by accurate measurements of crystals of silicon by X-ray diffraction: dividing the volume of one mole of silicon by the effective volume of a silicon atom yields Avogadro's number.

All of the refined measurements require sophisticated and expensive equipment and great care in experimental technique and treatment of data. The payoff is the most accurate value of Avogadro's number, N_A, that we have to date: $N_A = 6.02214 \times 10^{23}$ particles/mol. This is one of the numbers you will encounter in introductory chemistry that is worth remembering.

The mole is a fundamental unit in the International System (SI) of units. A mole of carbon is precisely 12.000000 g of carbon-12, and Avogadro's number is defined as the number of atoms in a mole of carbon-12.

In this experiment you will make an approximate (order of magnitude) estimate of Avogadro's number by determining the amount of stearic acid that it takes to form a single layer (called a *monolayer*) on the surface of water. By making simple assumptions about the way the stearic acid molecules pack together to form the monolayer, we can determine its thickness, and from that thickness we can estimate the size of a carbon atom. Knowing the size of a carbon atom, we can compute its volume; and if we know the volume occupied by a mole of carbon (in the form of diamond), we can divide the volume of a mole of carbon by the volume of an atom of carbon to get an estimate of Avogadro's number. The number won't be accurate to within 10 percent, or even a factor of 2, but it will enable you to estimate to within about a power of 10 the number of particles in a mole. That's better than chemists could do only 100 years ago.

CONCEPTS OF THE EXPERIMENT

Matter exists in three states. The fact that a gas can be condensed to a liquid and a liquid frozen to a solid indicates that there are attractive forces between all mole-

cules. We can schematically represent these forces at the surface of a liquid by the arrows in Figure 10-1. In the interior of the liquid, the forces exerted on a given molecule are uniform in all directions. At the surface, however, it is clear that there is a net force attracting each surface molecule inward. These molecules have higher energies than interior molecules, thus giving rise to the force known as *surface tension*. It is because of this force that liquid droplets are spherical. A spherical shape presents the smallest surface area for a given volume.

If the liquid is water, the surface tension is especially strong because particularly strong intermolecular forces, called *hydrogen bonds,* exist between the water molecules. (While strong for intermolecular bonds, these bonds are still considerably weaker than the intramolecular covalent bonds that bind atoms together in a molecule.) Hydrogen bonding arises whenever a hydrogen atom that is attached to a highly electronegative atom, such as oxygen, has access to an unbonded pair of electrons, such as those of another oxygen atom.

Another property displayed by water is polarity. Polar molecules posses a separation of charge. In ionic compounds, such as NaCl, there is a separation of a full unit charge, $Na^+ \ldots Cl^-$. Polar covalent molecules display a partial separation of charge, denoted by a delta, δ. An arrow is used to denote this charge separation. This polarity, referred to as the *dipole moment,* is equal to the partial charge times the distance separating the charges, and it is represented by an arrow pointing toward the negative end of the molecule. Figure 10-2 displays the polar nature of the water molecule.

Polar molecules attract each other. The negative end of the dipole of one molecule is attracted to the positive end of the dipole of another molecule. For this reason, water dissolves formic acid, H—COOH, which has a dipole moment, but it does not dissolve butane, $CH_3CH_2CH_2CH_3$, which has a nearly uniform charge

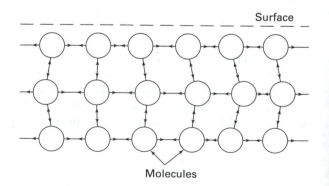

FIGURE 10-1

The action of molecular forces at a liquid surface creates a surface tension that tends to minimize the surface area of the liquid.

FIGURE 10-2

The bond polarity and angular structure of a water molecule create a dipole moment.

distribution. If a molecule possessing the properties of both of these molecules is brought up to the surface of water, the polar part of the molecule will be attracted to the surface and the nonpolar portion will be repelled. If the nonpolar part is much larger than the polar portion, the molecule will not dissolve in water but will simply stick to its surface. Consequently, it will lower the energy of the surface water molecules and of the adhering molecules.

The molecule we will use in this experiment, stearic acid, behaves in just this way. Stearic acid has a polar end consisting of a carboxyl group, —COOH, and a large nonpolar "tail" consisting of 16 methylene groups, —CH_2—, terminating in a methyl group, —CH_3. Figure 10-3 is a reasonably accurate representation of this molecule.

Adding a limited number of stearic acid molecules to a water surface results in the formation of a monolayer, as illustrated in Figure 10-4. However, after the surface is covered with a monolayer of stearic acid molecules, the addition of more molecules causes the stearic acid molecules to cluster in globular aggregates. The polar heads are attracted to the polar water molecules and form a lens-shaped surface, with the hydrocarbon tails pointing inward to form an "oily" interior.

The properties of the water surface and the stearic acid molecule permit us to perform what amounts to a titration of the water surface. We can add stearic acid molecules to the water surface until a monolayer covers the entire surface. Further addition will cause a convex lens to form on the liquid surface. At this point, the stearic acid molecules are stacked on the water surface like a layer of cordwood turned on end. If we know the area of the surface of water and have a way of measuring the volume of substance added to form the monolayer, we can calculate the thickness, t, of the layer. This thickness equals approximately the length of the stearic acid molecule.

As we saw in Figure 10-3, this molecule consists of 18 carbon atoms linked together. If we make the simple assumption that the atoms are like little cubes linked together (Figure 10-5, p. 122), the edge length of a cube of carbon is given by $t/18$. If we cube the edge length, we

FIGURE 10-3

The stearic acid molecule. (A) space-filling model; (B) structural formula; (C) schematic "matchstick" representation.

Surface area *A*

Water

FIGURE 10-4

Magnified view of a stearic acid film (not to scale). The stearic acid molecules orient themselves vertically because of the attraction of their polar ends (the —COOH groups) for the water molecules at the water surface and the weak attractive forces between their hydrocarbon tails. The volume of the stearic acid monolayer is the product of surface *Area* times *thickness* ($A \times t$).

have estimated the volume of a carbon atom. Now we have half the data needed to calculate Avogadro's number.

The other necessary data requires no experimental work on your part. What we need is the volume of a mole of carbon. Diamond is pure carbon, and the density of diamond is known to be 3.51 g/cm³. You may recall that you can calculate the molar volume of an element by dividing the molar mass (g/mol) by its density (g/cm³). Verify this idea by looking at the dimensions of this quotient.

If we assume that diamond consists of little cubes of carbon atoms stacked together, an Avogadro's number of them would equal the volume of one mole. Avogadro's number results from one final step:

$$N_A = \frac{\text{molar volume (cm}^3/\text{mole)}}{\text{atomic volume (cm}^3/\text{atom)}}$$

$$= \text{number of atoms/mol}$$

EXPERIMENTAL PROCEDURE

Special Supplies: 14-cm watch glass; cm rule or meterstick; fine-tipped (50 drops/mL) polyethylene transfer pipets; 1-mL all-polypropylene syringes (Norm-ject 1 mL × 0.01 mL tuberculin syringes are available from Air-Tite of Virginia, Inc., P.O. Box 9354, Virginia Beach, VA 23450, tel 800-232-7762); pure Type I water, free of surface active materials (preferably distilled or passed through a water purifier); disposable rubber gloves (if students will be cleaning their own watch glasses in 0.1 M NaOH in 50:50 methanol/water); 13 × 100 mm test tubes with corks to fit.

Chemicals: Hexane, 0.10 g/L stearic acid (purified grade) solution in hexane. The exact weight/volume concentration of this solution will be provided by your instructor. 0.1 M NaOH in 50:50 methanol/water (volume:volume, or v:v), used for washing the watch glasses.

NOTES TO INSTRUCTOR A purified (98%) grade of stearic acid gives the best results and is available commercially (Sigma-Aldrich).

Instead of using a dropper and counting drops, we prefer to use a 1-mL all-polypropylene syringe so that the volume of stearic acid solution in hexane can be read directly. Using the syringe also minimizes evaporation of the hexane solvent. The Luer tip of the syringe is fitted with a tip cut from a fine-tipped polyethylene transfer pipet, as shown in Figure I-14 of the Introduction. Another device for dispensing the stearic acid solution is described by Lane et al. (1984). (See the Bibliography.)

The watch glasses should be soaked in the methanolic NaOH solution in a polyethylene basin overnight or at least for an hour or two before the beginning of the laboratory period. The watch glasses should then be rinsed in distilled water before they are distributed to the students. Distilled water (or Type I water produced by a Millipore Milli-Q or Barnstead Nanopure purifier) is better for this experiment than deionized water, because deionized water often contains organic substances leached from the organic resins used to deionize the water.

FIGURE 10-5

For purposes of our estimate, each stearic acid molecule can be approximated as a vertical stack of 18 carbon atoms with a total height equal to the thickness, t, of the monolayer film. Each carbon atom then has a diameter, s, equal to $1/18$ of the film thickness. When the empty space between the atoms is included, the gross volume of each atom, $V_{\text{atom}} = s^3$.

! SAFETY PRECAUTIONS: Hexane is flammable! There must be no open flames in the laboratory while hexane is being used.

If you clean your own watch glasses by soaking them in methanolic NaOH, you must wear rubber gloves to protect your hands. As always, wear eye protection in the laboratory.

WASTE COLLECTION: At the end of the experiment, unused hexane solvent and stearic acid in hexane solution should be placed in a waste container marked "Waste hexane/stearic acid solution in hexane."

PREPARATION OF EQUIPMENT. Stearic acid is a solid. It is conveniently measured and applied to the water surface by dropping a solution of stearic acid in hexane onto the water. The hexane is insoluble in water. Because it has a high vapor pressure, it rapidly evaporates, leaving a layer of stearic acid spread on the water surface one molecule thick, called a *monolayer*.

Because stearic acid is one of the fatty acids present in soaps, you must not clean your watch glass with soap. Any soap, grease, or dirt present can form a film on the water that prevents the stearic acid from spreading on the surface, leading to meaningless results because not even the first drop will spread.

Obtain a 14-cm watch glass. If it has not been soaked in the methanolic NaOH solution described in the Chemicals section, wash it thoroughly with detergent. Rinse the detergent off completely under a full stream of cold tap water for a minute, then rinse thoroughly with distilled water. Repeat the washing after each experiment, or obtain another watch glass that has been soaked in methanolic NaOH. Handle the watch glass only by the edges, being especially careful not to let anything touch the inside of it. Be particularly careful to keep your fingers (which are usually slightly greasy) off the glass.

1. MEASUREMENT OF THE VOLUME OF STEARIC ACID SOLUTION REQUIRED TO COVER THE WATER SURFACE.

Obtain about 3–4 mL 0.10 g/L stearic acid solution in hexane in a clean, dry 13 × 100 mm test tube. Keep the tube corked when not in use. Fill the clean watch glass to the brim with distilled water. Carefully measure the diameter of the water surface with a centimeter rule or meterstick. Cut off the tip of a fine-tipped polyethylene transfer pipet about 25 mm from the tip end and push the cut-off tip onto the Luer fitting on the end of the syringe. Next, rinse and fill your 1-mL syringe with stearic acid solution, taking care to eliminate bubbles in the solution inside the syringe.

Read and record the initial volume reading of the syringe. Then add the stearic acid solution drop by drop to the water surface. Initially, the solution will spread across the entire surface, and it will continue to do so until a complete monolayer of stearic acid has been produced. As this point is approached, the spreading will become slower and slower, until finally a drop will not spread out but will instead sit on the surface of the water (looking like a little contact lens). If this "lens" persists for about 30 s, you can safely conclude that you have added 1 drop more than is required to form a complete monolayer. Now, read and record the final volume reading of the syringe.

Thoroughly clean the watch glass (or get another one), and repeat the experiment. Repeat until the results agree to within 2 or 3 drops (0.04 mL).

When you have completed all of your measurements, rinse your syringe with pure hexane, and dispose of all the hexane-containing solutions in the waste collection bottle provided.

2. CALCULATION OF AVOGADRO'S NUMBER.

The calculation proceeds in several steps. First, we calculate the volume of stearic acid solution in hexane required to deliver enough stearic acid to form a monolayer. All of the hexane evaporates, leaving only the thin monolayer film of stearic acid, so we next calculate the actual mass of pure stearic acid in the monolayer.

Then we calculate the thickness of the stearic acid monolayer, using the known density of stearic acid and the area of the monolayer. Assuming the stearic acid molecules are stacked on end and are tightly packed, and knowing that there are 18 carbon atoms linked together in the stearic acid molecule, we calculate the diameter and volume of a carbon atom.

Finally we calculate the volume of a mole of carbon atoms in diamond; dividing the molar volume of carbon (diamond) by the volume of a single carbon atom, we obtain an estimate of Avogadro's number.

☞ Consider This

The method we employed in this experiment is not especially accurate. There are other experimental ways to get values for Avogadro's number. For example, if you divide the electric charge on a mole of electrons (the Faraday constant) by the electric charge on a single electron, you will obtain the number of electrons in a mole of electrons, which is Avogadro's number. Try it, using values for the constants found in the Appendix, Table 2.

Because atoms are so small, the magnitude of Avogadro's number is so large that it is almost impossible to comprehend it. Try this exercise to get a grip on it. First, imagine that you have 64 toy alphabet blocks arranged in a cube, where each block is itself a cube, 2 inches on an edge. The cube of 64 blocks would be 8 inches in three dimensions, with 4 blocks along each edge. Then, flatten the cube into a square containing 64 blocks. How large would the square be, and how many blocks will be along its edge? Finally, arrange the 64 blocks in a line. How long would the line be?

Now that you have the idea, imagine an Avogadro's number of tiny cubes, the edge length of each tiny cube being about the size of a hydrogen atom (1.00 × 10^{-10} meter). If you had an Avogadro's number of hydrogen atom cubes arranged into a large cube (like the alphabet blocks), what would the edge length of the cube be? If you then flattened the cube into a square, what would the edge length of the square be? Finally, if you arranged the Avogadro's number of tiny H-atom cubes into a line, how long would the line be? Compare it to the earth-to-sun distance of 93 million miles.

In the method we used, we made a number of simplifying assumptions. Some of these are discussed in the Consider This section of the Report form, and their effects are quantitatively assessed. The cumulative effect of the simplifying assumptions is to cause the final result for Avogadro's number to be a factor of about three times too large.

BIBLIOGRAPHY

Boorse, H. A., and Motz, L. (eds.), "The Determination of Avogadro's Number," *The World of the Atom,* Basic Books, New York, 1966, Vol. 1, pp. 625–640. Describes Perrin's method.

Diemente, D. "Demonstrations of the Enormity of Avogadro's Number," *J. Chem. Educ.,* **1998,** *75:* 1565–1566.

Feinstein, H. I., and Sisson, R. F., III. "The Estimation of Avogadro's Number Using Cetyl Alcohol as the Monolayer," *J. Chem. Educ.* **1982,** *59:* 751.

Hawthorne, Jr., R.M. "Avogadro's Number: Early Values by Loschmidt and Others," *J. Chem. Educ.* **1970,** *47:* 751–755.

Henry, P. S. "Evaluation of Avogadro's Number" (by the method of Perrin), *J. Chem. Educ.* **1966,** *43:* 251.

King, L. C., and Neilsen, E. K. "Estimation of Avogadro's Number," *J. Chem. Educ.* **1958,** *35:* 198.

Lane, C. A., Burton, D. E., and Crabb, C. C. "Accurate Molecular Dimensions from Stearic Acid Monolayers," *J. Chem. Educ.* **1984,** *61:* 815.

Porterfield, W. W., and Kruse, W. "Loschmidt and the Discovery of the Small," *J. Chem. Educ.* **1995,** *72:* 870–875.

Robinson, A. L. "Metrology: A More Accurate Value for Avogadro's Number," *Science* **1974,** *185:* 1037.

Slabaugh, W. H. "Determination of Avogadro's Number by Perrin's Law," *J. Chem. Educ.* **1965,** *42:* 471.

Slabaugh, W. H. "Avogadro's Number by Four Methods," *J. Chem. Educ.* **1969,** *46:* 40.

The Estimation of Avogadro's Number

NAME

SECTION   LOCKER

INSTRUCTOR   DATE

DATA AND CALCULATIONS

NOTE: A number of calculations are required in this report. Show the calculations for only one of your trials, but give the results for both trials, those for trial 1 in the left blank and those for trial 2 in the right.

1. **Measurement of the Volume of Stearic Acid Solution Required to Cover the Water Surface**

	Trial 1	Trial 2
(a) Record the diameter of the water surface:	_____ cm	_____ cm
(b) Record the volume of stearic acid solution required to cover the surface:	_____ mL	_____ mL
(c) Record the concentration of the stearic acid solution:	_____ g/L	_____ g/L

2. **Calculation of Avogadro's Number**

(a) *Calculation of the thickness of a monolayer of stearic acid*

   (i) From your data of Part 1, what was the volume of stearic acid solution required to form a monolayer?

   _____ mL   _____ mL

   (ii) Calculate the mass of stearic acid contained in that volume of stearic acid solution (the concentration in grams per liter will be given to you):

   _____ g   _____ g

   (iii) Calculate the volume, V, of pure stearic acid in the monolayer on the water surface. You will need the density of solid stearic acid, which is 0.85 g/mL (or g/cm^3):

   _____ mL (cm^3)   _____ mL (cm^3)

   (iv) Calculate the area of the monolayer ($A = \pi r^2$; r is the radius of the water surface):

   _____ cm^2 _____ cm^2

   (v) Calculate the thickness of the monolayer ($t = $ *Volume/Area*):

   _____ cm   _____ cm

(b) *Estimation of the size and volume of a carbon atom*

   (i) A stearic acid molecule consists of 18 carbon atoms linked together. Assuming that the thickness, t, of a monolayer is equal to the length of the stearic acid molecule, calculate the size of a carbon atom, $s = t/18$:

   _____ cm   _____ cm

   (ii) Assuming that a carbon atom is a little cube, calculate the volume of a carbon atom, volume $= s^3$.

   V_{atom}: _____ cm^3/atom   _____ cm^3/atom

(c) *Calculation of the volume of a mole of carbon atoms*

    (i)   Calculate the molar volume of carbon (diamond) by using the density of diamond (3.51 g/cm^3) and the atomic mass of a mole of carbon:

$$V_{mol}: \underline{\hspace{4cm}} cm^3/mol \quad \underline{\hspace{4cm}} cm^3/mol$$

    (ii) Is the volume of a mole of diamond the same as the actual volume of a mole of carbon atoms?

(d) *Calculation of the volume of a mole of carbon (diamond)/volume of a single carbon atom (Avogadro's number)*

    (i)   Calculate Avogadro's number from the appropriate ratio of volumes:

$$\underline{\hspace{4cm}} atoms/mol \quad \underline{\hspace{4cm}} atoms/mol$$

    (ii) Calculate the average value of N_A from your results:

$$Average: \underline{\hspace{4cm}} atoms/mol$$

(e) *Express your results as a number $\times 10^{23}$. Are you within a power of 10 of the accepted value of 6.02×10^{23}?*

☞ Consider This

Calculate an accurate value of Avogadro's number as the ratio of the electric charge on a mole of electrons (Faraday's constant) divided by the charge on a single electron (see Table 2 of the Appendix).

Calculate the edge length of a cube containing an Avogadro's number (6.02×10^{23}) of hydrogen atoms stacked to form a cube. Assume a hydrogen atom is equivalent to a tiny cube with edge length 1×10^{-10} m.

Now imagine an Avogadro's number of hydrogen atoms formed into a square one atom layer thick. What would be the edge length of the square?

Finally, if an Avogadro's number of hydrogen atoms were arranged in a line, how long would the line be? Compare the length of this line to the distance from the earth to the sun, about 93 million miles.

In the method of estimating Avogadro's number that we used, there are a number of implicit assumptions:

1. *We assume that the stearic acid molecules form a close-packed monolayer on the water surface.*

2. *We assume that the packing of stearic acid molecules is the same in the monolayer as in solid stearic acid.*

3. *We assume that carbon atoms are like little cubes stacked up in a linear fashion. A more accurate picture is one in which the C atoms are more nearly spherical, having a zigzag linkage with tetrahedral 109.5° C—C—C bond angles as shown in Figure 10-3A. If we assume this structure instead of the linear one, show that the size (diameter, s) of a carbon atom is linked to the thickness, t, of the monolayer by the expression.*

$$s = \frac{1}{\sin 54.75°} \times \frac{t}{18}, \text{ where } \sin 54.75° = 0.8166, \text{ increasing the value of } s \text{ by a factor of } \frac{1}{0.8166} = 1.225$$

When this sine factor is cubed, the calculated volume of a carbon atom is increased by a factor of $(1.225)^3 = 1.84$.

4. *We assumed that the volume of a carbon atom was equivalent to a cube of edge length s, giving a volume of s^3. Show that if the volume of a sphere is given by the expression $V = \frac{4}{3} \pi r^3$, where r is the radius of the sphere, the volume of a sphere will be equal to $\frac{\pi}{6} \times s^3 = 0.524 \, s^3$, where s is the diameter of the sphere inscribed in a cube of edge length s.*

When we apply the two correction factors calculated for assumptions 3 and 4 to the calculation of the volume of a carbon atom, we see that their effect is to approximately cancel one another, that is, $1.84 \times 0.524 = 0.96$.

5. *Finally, we assume that diamond consists of close-packed carbon atom cubes whereas it actually has a more open structure in which each carbon atom is surrounded by four other carbon atoms with tetrahedral bond angles. In this structure, only about 34 percent of the volume of the unit cell is occupied by approximately spherical carbon atoms. Thus, the actual volume of a mole of spherical carbon atoms is about 0.34 times the volume of a mole of diamond.*

If we incorporate this factor in the calculation of Avogadro's number, we see that the calculated value of Avogadro's number, based on the simple assumptions, would be reduced by about a factor of 3.

6. *What do we mean by the size of an atom? There is good evidence that the electron clouds surrounding isolated atoms do not have sharp, well-defined boundaries. That is, there is good reason to believe that atoms aren't hard spheres like billiard balls. We can rather accurately measure the internuclear distance between two atoms in a chemical bond. If the two atoms are the same, carbon atoms for example, then we could define the radius of a carbon atom as one-half the internuclear distance (the C—C distance), meaning that the diameter of an atom would be equal to the internuclear C—C distance. Defined this way, the size (diameter) of a carbon atom is 1.54×10^{-10} meters. Compare this value to the value of s that you calculated.*

The Molar Volume of Dioxygen and Other Gases

PURPOSE

To determine the molar volume of a gas. To gain practical experience in making quantitative measurements of gas volumes, using Dalton's law of partial pressures and the ideal gas law to calculate the volume of a gas at the standard temperature and pressure.

PRE-LAB PREPARATION

AVOGADRO'S PRINCIPLE AND THE MOLAR VOLUME

This experiment deals with one of the most useful and important quantitative relationships involving gases—a relationship that depends on the following two ideas:

1. Under identical conditions of temperature and pressure, equal volumes of all gases contain equal numbers of molecules. This is called Avogadro's principle or Avogadro's hypothesis.

2. If we define a mole as a fixed number of any particle (atoms, molecules, or any particle of defined composition), then one mole of any substance contains the same number of particles as one mole of any other substance.

So if a mole contains a fixed number of particles, then according to Avogadro's principle, a mole of *any* gas will occupy the same fixed volume, called the *molar volume of a gas,* measured under identical conditions of temperature and pressure.

THE KINETIC THEORY OF GASES AND ITS CONSEQUENCES

We say that matter can exist in three states—as a solid, liquid, or gas. Solids and liquids are often called "condensed states of matter," and the fact that solids and liquids exist implies the existence of intermolecular forces of attraction between atoms and molecules in the solid and liquid state. If there were no intermolecular forces, every particle (atom or molecule) would be free to wander about constrained only by the walls of their container or, in the case of the molecules forming the earth's atmosphere, being bound only by the gravitational attraction of the earth.

So, if we grant that, in contrast to liquids and solids, gas molecules can occupy the full volume of their container, why should each volume element of the container contain the same number of particles, no matter if the gas molecules are large or small?

An explanation offered in the seventeenth century proposed that gas molecules were like same-sized balls of wool or cotton that could expand or be compressed depending on how tightly they were squeezed. An equivalent, but slightly more sophisticated, explanation proposed that gas molecules were held through opposing forces of attraction and repulsion at locations equal distances apart, about which they oscillated in a sort of stable equilibrium.

A better model, and one that has survived all tests to date, is called the *kinetic theory of gases.* It assumes that there are, on average, large distances between molecules, that the forces between molecules are negligible, and that the molecules move in straight lines without interactions with other molecules, except at moments of collision. When molecules collide with a wall, they rebound or change direction of motion just as in the collision of elastic spheres, like the collision of two balls on a pool table. In fact, if you could imagine all of the balls on a pool table in constant motion, the image would be a reasonable two-dimensional analogy for a gas.

A consequence of the frequent collisions that occur in the gas is that the molecules acquire an average kinetic energy of translation (linear motion from point to point), which is dependent only on temperature and which is the same for all molecules, large and small. This means that smaller molecules travel, on average, faster than larger molecules.

The pressure exerted by the gas is the cumulative result of the impacts on the walls of the container, smaller molecules making impacts of smaller magnitude but making more of them because they move faster. As a result, the total force exerted is the same for all molecules, no matter if they have a mass that is large or small.

In more quantitative terms, when a molecule collides with the wall, the force exerted on the wall is proportional to the molecule's momentum (mass $\times$ velocity $= mv$) multiplied by the number of impacts per unit time (proportional to the velocity, v), so the force is proportional to mv^2. But $\frac{1}{2} mv^2$ is just equal to the kinetic energy of translation. And we know from experiment that the pressure (force/unit area) is proportional to the kelvin temperature, so the average kinetic energy must be proportional to the (kelvin) temperature. The relation given by the kinetic molecular theory is

$$\text{Average kinetic energy of a molecule} = \frac{1}{2}mv^2 = \frac{3}{2}kT$$

where k is called the Boltzmann constant and is equal to the gas constant, R, divided by Avogadro's number.

To summarize, the mass and average velocity of a gas molecule are linked in such a way that as the mass, m, decreases, the square of the velocity, v^2, increases just enough to keep the product mv^2 constant (at constant temperature). An important physical insight arising from the kinetic theory of gases is that, for a fixed pressure, *the average number of molecules per unit volume (or the average volume per molecule) is determined only by the average kinetic energy (or the temperature) of the gas and does not depend on the nature of the gas.*

For a pressure of one atmosphere at room temperature, the average distance between molecules of a gas is about ten times the size of a molecule, so a gas is mostly empty space.

THE MOLAR VOLUME AT STANDARD TEMPERATURE AND PRESSURE

In this experiment we will measure the mass and volume of a sample of dioxygen gas, O_2, and from these data and the ideal gas law calculate the volume at standard temperature and pressure (STP) conditions (0 °C, 760 torr). Then we will calculate the volume of 32.0 g (one mol) of dioxygen gas.

An *ideal* gas is one in which (a) the attractive forces between molecules are vanishingly small and (b) the molecules are equivalent to "point masses" occupying a negligible volume compared with that of the container. All real gases deviate more or less from the ideal behavior described by the general ideal gas law because their molecules have some slight attraction for one another, and they do occupy some small volume of the container in which they are confined.

From measurements on real gases, the molar volume of an ideal gas has been determined to be 22.414 L STP (T = 273.15 K; P = 1 atm). This was done by calculating the molar volume (STP) at a series of progressively lower pressures and extrapolating to zero pressure, where the attractive forces between molecules and the volume occupied by the molecules are negligible. Most common gases, unless they have a high molecular weight or are measured quite near their condensation point (boiling point of the liquid), have molar volumes that do not deviate more than about 1 percent from the ideal volume at pressures around 1 atm.

CALCULATION OF GAS VOLUMES

The quantity of a gas sample can be measured more conveniently by volume than by weight. In measuring the volume of a gas, it is also necessary to measure its temperature and pressure. Why? The separate laws relating pressure to volume and relating either pressure or volume to absolute temperature are

$$PV = k_1 \quad \text{and} \quad P = k_2T \quad \text{or} \quad V = k_3T$$

For a given gas sample, these three laws may be combined into one equation that shows the way in which all three variables—pressure, volume, and absolute temperature—are interdependent:

$$PV = kT \quad \text{or} \quad \frac{PV}{T} = k \tag{1}$$

Since any two corresponding sets of PV/T measurements will be equal to k and to each other for a given amount of gas, we can write

$$\frac{P_1V_1}{T_1} = \frac{P_2V_2}{T_2} \tag{2}$$

This equation can be transposed to give

$$V_1 = V_2 \times \frac{T_1}{T_2} \times \frac{P_2}{P_1} \tag{3}$$

Note in Equations (2) and (3) that if the temperature is constant ($T_1 = T_2$), the inverse proportionality of pressure and volume (Boyle's law) is expressed. Likewise, for nstant pressures ($P_1 = P_2$), the direct proportionality of volume and absolute temperature (Charles's law) is expressed. If any five of the quantities in Equation (2) are known, the sixth can be calculated by simple algebra.

THE GENERAL IDEAL GAS LAW EQUATION

For a specific amount of any gas, Equation (1) can be restated in its most general form:

$$PV = nRT \qquad (4)$$

Here, n is the number of moles of gas, and R is a proportionality constant, called the *gas constant*, that has the same value for all gases under all conditions, namely, $0.08206 \text{ L} \times \text{atm} \times \text{mol}^{-1} \times \text{K}^{-1}$. In all calculations in which this constant is employed, pressure must be expressed in atmospheres, volume in liters, and temperature in kelvins (represented by the symbol K).

As an example of the application of the general gas law where both volume and weight of a gas sample are involved, consider the following. What weight of chlorine gas, Cl_2, would be contained in a 5.00-L flask at 21 °C and at 600 torr ($^{600}/_{760}$ atm) pressure? Substituting in Equation (4), transposed to give n, the number of moles, we have

$$PV = nRT \quad \text{or} \quad n = \frac{PV}{RT}$$

$$n = \frac{\dfrac{600}{760} \text{ atm} \times 5.00 \text{ L}}{0.08206 \dfrac{\text{L atm}}{\text{mol K}} \times 294 \text{ K}}$$

$$= 0.164 \text{ mol}$$

and, in grams,

$$0.164 \text{ mol} \times 70.9 \frac{\text{g}}{\text{mol}} Cl_2 = 11.6 \text{ g } Cl_2$$

AQUEOUS VAPOR PRESSURE AND DALTON'S LAW OF PARTIAL PRESSURES

When any gas in a closed container is collected over liquid water or is exposed to it, the water evaporates until a saturated vapor results—that is, until opposing rates of evaporation and condensation of water molecules at the liquid surface reach a balance. These gaseous water molecules contribute to the total gas pressure against the walls of the container. Thus, of all the gas molecules, if 3 percent are water molecules and 97 percent are oxygen molecules, then 3 percent of the total pressure is due to water vapor and 97 percent of the total pressure is due to oxygen. *Each gas exerts its own pressure regardless of the*

presence of other gases. This is Dalton's law of partial pressures. Stated as an equation,

$$P_{\text{total}} = P_{H_2O} + P_{O_2}$$

or, if transposed,

$$P_{O_2} = P_{\text{total}} - P_{H_2O}$$

To illustrate Dalton's law, Figure 11-1 shows a mixture of oxygen molecules and water vapor molecules. In part B, the water molecules have been removed, but all the oxy-

Pressure of oxygen and water vapor equals atmospheric pressure.

Mercury manometer

O_2

H_2O

A

The water vapor has been absorbed by the $CaCl_2$.

Differential height equals the original pressure of the water vapor.

$CaCl_2$

B

FIGURE 11-1

When water vapor is removed from a gas mixture, the pressure is reduced. This is an illustration of Dalton's law of partial pressures. Note that in this figure the gas molecules are greatly exaggerated in size.

gen molecules are still present, in the same volume. The pressure has been reduced by an amount equal to the vapor pressure of the water.

EXPERIMENTAL PROCEDURE

Special Supplies: 250-mL Erlenmeyer flask with a No. 5 one-hole rubber stopper, 500-mL Erlenmeyer flask with a No. 5 two-hole rubber stopper, 6- to 8-cm lengths of 6-mm-OD glass tubing, 30-cm length of 6-mm-OD glass tubing with a 90° bend 8 cm from one end, 10 × 75 mm test tube, 600-mL beaker, ³⁄₁₆ in.-ID rubber tubing, pinch clamp, thermometer, 500-mL graduated cylinder, barometer, analytical balance with a 150 g capacity, pipet bulb.

Chemicals: 3% (approx. 0.9 M) hydrogen peroxide, active dry yeast (Fleischmann's or other brands used in baking bread are suitable). Assemble the apparatus shown in Figure 11-2, taking care to lubricate each piece of glass tubing with a drop of glycerol and to wrap it in a towel while inserting it in the rubber stopper. (At this stage the 250-mL Erlenmeyer flask—flask A—will be empty. The solution and test tube will be added later.)

The principle of the method is simple. We determine the mass of oxygen produced in a chemical reaction by weighing the reaction flask before and after the chemical reaction. The weight loss corresponds (to a good approximation) to the mass of oxygen produced. The oxygen that is produced flows into the middle flask (flask D of Figure 11-2), displacing an equal volume of water into the

beaker. By measuring the volume of oxygen liberated at a known temperature and pressure, we can determine the molar volume of dioxygen. The oxygen is produced by the catalyzed decomposition of hydrogen peroxide (H_2O_2) solution, the same reaction we used to prepare oxygen in Experiment 7.

$$H_2O_2(aq) \xrightarrow[\text{catalyst}]{\text{yeast}} H_2O(l) + \tfrac{1}{2}O_2(g)$$

Before placing any solution in the 250-mL Erlenmeyer flask, slide an empty 10 × 75 mm test tube into the flask, and check that the test tube will stand at an angle inside the flask, without tipping over and lying flat. (Some older style 250-mL flasks will allow the test tube to lie flat; if this happens, it will be necessary to go to the next smaller size, 200 mL.) Now remove the test tube from the empty flask.

Measure 30 mL of 3% H_2O_2 solution in a graduated cylinder, and pour it into the 250-mL Erlenmeyer flask (flask A of Figure 11-2).

Put 50 to 100 grains (0.050 g) of active dry yeast into a clean, dry 10 × 75 mm test tube, being careful not to get any on the outside of the tube. Holding the 250-mL Erlenmeyer flask at an angle of about 45°, slide the test

FIGURE 11-2

Apparatus for determining the molar volume of O_2. Flask A contains 3% aqueous H_2O_2. The inner test tube in flask A contains baker's yeast to catalyze the decomposition of H_2O_2 when the two are mixed.

ube into the flask or lower the test tube with a pair of forceps. Next, place the one-hole stopper, with a short length of glass tubing already inserted, in the flask, and seat the stopper firmly. During this step take care not to allow any solution to run into the test tube—yeast must not contact the hydrogen peroxide solution before the stopper is inserted. If it does, the reaction will start prematurely, and some oxygen will be lost. If this happens, you must start over with a clean flask and test tube and a fresh portion of hydrogen peroxide solution.

Weigh the flask, with contents, stopper, and glass tube, as precisely as possible on the analytical balance (nearest 0.001 g). (The balance must have at least 150-g capacity.) Record the mass.

Fill the 500-mL Erlenmeyer flask D (in Figure 11-2) to the neck with tap water. Put in place the two-hole rubber stopper with the short glass inlet tube and the long glass siphon tube. Connect the rubber tubing C to the short glass inlet tube and the rubber tube E to the long siphon tube, as shown in Figure 11-2. (Do not yet connect the free end of tube C to the reaction flask containing the hydrogen peroxide.)

Fill beaker F about one-third full with tap water. Make sure the pinch clamp is open. Then, using a pipet bulb, blow into the open end of rubber tube C (whose free end is not yet connected at point B) to start water siphoning into the beaker. (A pipet bulb is preferred to blowing into tube C with your mouth; you want to avoid getting saliva into tube C because it would catalyze the decomposition of hydrogen peroxide.)

Raise and lower beaker F to rapidly siphon water back and forth through tube E to sweep out all air bubbles. Fill the 500-mL flask D with water until the water level is halfway up the narrow neck of flask D. Holding the beaker F so that its water level is the same as the water level in the neck of flask D, connect the free end of tube C to the reaction flask at point B.

Test the apparatus for leaks by raising beaker F as high as possible without removing the tubing from the beaker. The water level in flask D should move a little and then remain fixed. If the water level continues to change, a leak is present. Do not proceed until the leak is located and fixed.

Equalize the pressures inside and outside flask D by raising the beaker until both water levels are the same. While you continue to maintain the same levels in flask D and beaker F, have another student close the pinch clamp. Pour out all of the water in beaker F, but do not dry it. Replace the tube in the beaker, and *open the pinch clamp*. A little water will flow out, and this should be retained in the beaker. During the generation of O_2 that follows, see that the end of tube E is below the surface of the water in the beaker.

Check to make sure the pinch clamp is open. Then tip flask A carefully while tapping it so that the yeast grains fall into the solution, or solution runs into the test tube. (Avoid getting any solution on the rubber stopper.) Swirl the contents of the flask to mix the reactants. Note and record any changes you see as the reaction progresses. Frequently swirl or gently shake the flask to speed the release of oxygen from the solution. After a few minutes, put your hand on the bottom of the flask. Is heat released or absorbed in the reaction?

After a few minutes, the reaction will be nearing completion. Note very carefully the water level in the beaker (F). Continue frequent agitation of the reaction flask (A). When the water level remains unchanged for at least 5 min, you can assume that the decomposition of H_2O_2 is complete.

Adjust the water levels in flask D and beaker F until they are equal. Ask another student to close the pinch clamp on tube E.

Loosen the stopper, and insert your thermometer into the gas in flask A. After a minute or two, note and record the temperature. Measure and record the temperature of the gas in flask D in the same way. (In the calculations that you will make later, you will take the average of these two temperatures as the temperature of the oxygen gas.)

Disconnect flask A at point B, and reweigh it with its contents, rubber stopper, and glass tube. Record the mass.

Measure the volume of oxygen produced by carefully pouring the water in beaker F into a 500-mL graduated cylinder. (Alternatively, you can weigh the water by pouring it into a tared beaker on a 0.1 g balance. The volume of water in mL is equal to its mass in grams, within 0.3%, at room temperature.) Record the volume of water displaced by the oxygen gas.

Measure and record the barometer reading and the temperature of the barometer. (The corrected barometric pressure is obtained by subtracting the approximately 3 mm Hg [torr] barometer temperature correction found in the *CRC Handbook of Chemistry and Physics;* this will correct for the difference in the coefficients of expansion of the brass scale of the barometer and the column of mercury.) You will also need the vapor pressure of water, which can be found in the Appendix, Table 3.

Repeat the experiment if time permits. From your data, calculate the volume at standard temperature and pressure of 32.0 g (one mol) of dioxygen gas, O_2. All solutions may be safely disposed of down the drain.

Consider This

Do other gases have the same molar volume as dioxygen? *The method we employed to determine the molar volume of dioxygen can also be applied to other gases. For example, by placing 50 mL of 1 M HCl in the reaction flask and 1.0 ± 0.1 g of sodium bicarbonate*

($NaHCO_3$), in the 10×75 mm test tube, you can generate $CO_2(g)$ by the reaction

$$NaHCO_3(s) + HCl(aq) \rightarrow NaCl(aq) + H_2O(l) + CO_2(g)$$

Does a mole of CO_2 gas measured at standard temperature and pressure occupy the same volume as a mole of oxygen (STP)? Try it and see. (The solutions may be disposed of down the drain after neutralization of all of the HCl with sodium bicarbonate.)

Estimating molecular size and the average distance between molecules in a gas. The density of solid nitrogen is about 1.0 g/cm^3 at its freezing point. If we assume that in solid nitrogen the molecules are about as close together as they can get, we can estimate the approximate size of a molecule from the average volume occupied by one molecule. To do this, first determine the volume of a mole of solid nitrogen. Then divide this volume by Avogadro's number to get the volume occupied by one nitrogen molecule. If we assume this volume is a tiny cube containing one molecule of nitrogen, the edge length of the cube will be the approximate size of a nitrogen mol-

ecule. (If you've done the calculation correctly, you should get a value of about 0.36 nanometers.)

At room temperature and pressure, one mole of a gas occupies about 25 L (0.025 m^3). Knowing that there is an Avogadro's number of molecules in one mole, calculate the average volume per molecule. If we imagine that this volume is a little cube with, on average, one molecule located in the center of each cube, we see that the average distance between molecules will be equal to the edge length of the cube. Comparing the results of the calculation for liquid and gaseous nitrogen, show that the average distance between molecules in the gas is about ten times the size of a molecule.

BIBLIOGRAPHY

Bedenburgh, J. H., Bedenburgh, A. O., and Heard, T. S. "Oxygen from Hydrogen Peroxide: A Safe Molar Volume—Molar Mass Experiment," *J. Chem. Educ.* **1988,** *65:* 455–456.

The Molar Volume of Dioxygen

NAME		
SECTION		LOCKER
INSTRUCTOR		DATE

DATA

Data	Trial 1	Trial 2
Mass of flask A, contents, stopper, and glass tube before reaction	g	g
Mass of flask A, contents, stopper, and glass tube after reaction	g	g
Temperature of the gas in flask A	°C	°C
Temperature of the gas in flask D	°C	°C
Volume of oxygen collected	mL	mL
Barometer reading (Temperature of barometer _____ °C)	torr	torr
Aqueous vapor pressure at temperature of gas (average the two temperatures)	torr	torr

OBSERVATIONS

1. Describe the visible changes that take place during the reaction.

2. As the reaction proceeds, does flask A get warmer or cooler to the touch? Does the reaction absorb or evolve heat (energy)?

CALCULATIONS

Data	Calculations	Trial 1	Trial 2
Mass of oxygen		g	g
Temperature, absolute[a]		K	K
Corrected barometric pressure[b]		torr	torr
Pressure of oxygen alone, in flask A		torr	torr
Volume of oxygen at standard conditions		mL	mL
Moles of dioxygen, O_2		mol	mol
Molar volume of O_2 at standard conditions		L/mol at STP	L/mol at STP

[a]Average the two temperatures.
[b]Look in the *CRC Handbook of Chemistry and Physics* for the barometer temperature correction.

☞ Consider This

Do other gases have the same molar volume as dioxygen? *Using the same apparatus and procedure as used for dioxygen, carry out measurements and calculations for determining the molar volume of carbon dioxide, as briefly described in the first paragraph of the Consider This section of the Experimental Procedure. Keep a record of your measurements and calculations in your laboratory notebook. Report the final result of your measurements here.*

Estimating the size of a molecule and the average distance between molecules in a gas. *Carry out the calculations as described in the Consider This section of the Experimental Procedure. Why are gases springy (easily compressible), while liquids like water are nearly incompressible?*

There are two sources of error in this experiment, which fortuitously almost cancel one another. Circle the correct answer in the following statements.

(a) *Initially the flask that you weigh is filled with air. At the end of the reaction it is filled with oxygen. Because oxygen is denser than air, at the end of the reaction the flask is (heavier, lighter) than it would be if filled with air.*

(b) *If the oxygen that is evolved is saturated with water vapor, some water will be carried over as well as evolved oxygen. At the end of the reaction this effect would cause the flask to be (heavier, lighter) than if no water were carried over.*

Calculate the molarity of a solution that is 3.0% hydrogen peroxide by weight (3 g H_2O_2/100 g solution), assuming that the density of the solution is 1.00 g/mL. For one of your trials, calculate the moles of hydrogen peroxide that you put in the flask, the moles of oxygen evolved, and the ratio mol O_2/mol H_2O_2. Is the value you calculate for the ratio consistent with the stoichiometry of the decomposition reaction?

$$H_2O_2(aq) \rightarrow H_2O(l) + \tfrac{1}{2} O_2(g)$$

All real gases deviate to some extent from the behavior of perfect gases. At standard conditions, the density of O_2(g) is 0.0014290 g/mL, that of H_2(g) is 0.00008988 g/mL, and that of CO_2(g) is 0.0019769 g/mL.

(a) Using these values and the exact atomic weights, calculate the molar volume of each of these gases, in milliliters, to five significant figures.

O_2 _____

H_2 _____

CO_2 _____

(b) Correlate the molar mass, molecular size, and bond polarity of these three gases with the molar volume of a perfect gas.

Which gas deviates least? _____

Which gas deviates most? _____

Can you suggest reasons for this behavior?

The Molar Mass of a Gas

PURPOSE

To determine the molar mass of a gas by measuring the mass of a known volume of gas at the ambient laboratory temperature and known pressure, as determined by a thermometer and barometer. (If ideal gas behavior is assumed, knowing these four quantities—mass, volume, pressure, and temperature—allows the calculation of the molar mass from the ideal gas law.) To understand the relationship between gas density and buoyancy.

PRE-LAB PREPARATION

In prior experiments we estimated Avogadro's number and determined the molar volume of oxygen. To refresh your memory, let's restate Avogadro's law: *Equal volumes of gases contain equal numbers of molecules.* This means that any time we take an Avogadro's number of molecules of a particular gas (which is the definition of 1 mol) we can expect that the volume of gas will be the same as the volume of 1 mol of any other gas. (This expectation is, of course, subject to the usual restrictions that [a] the gases must behave as ideal gases and [b] the volumes must be measured at the same pressure and temperature.) We determined the molar volume of oxygen at the standard temperature and pressure and found it to be about 22.4 L. From this result and Avogadro's law, we can draw the important general conclusion that *1 mol of any gas will occupy 22.4 L STP, and the mass of the gas will be the molar mass of the gas in grams.*

Of course we do not have to measure the volume of the gas at 0 °C and 1 atm. We can measure the volume at any temperature and pressure, then correct to the standard conditions by using the ideal gas law, $P_1V_1/T_1 = P_2V_2/T_2$, or its equivalent form, $PV = nRT$. Thus, if we can determine the mass of a known volume of the gas at a known temperature and pressure, we can determine the molar mass of any gas, whether it is a permanent gas or the vapor of a volatile liquid or solid.

In this experiment we will measure the mass and volume of a sample of gas at the laboratory temperature and pressure. In principle this can be done by taking a bulb (or some other container), evacuating it so that we can get the mass of the completely empty bulb, then filling the bulb with the unknown gas, and weighing it again. The difference in masses would give us the mass of the gas, and once we measure the volume of the bulb and know the pressure and temperature of the gas, we can calculate its molar mass.

However, an evacuable glass bulb and a vacuum pump are not always available, so we are going to determine the mass of the empty container by weighing it filled with air and subtracting the mass of the air, which we can calculate if we know the volume of the container and the density of air (given in various reference handbooks). Subtracting the calculated mass of air gives the mass of the empty container without the hassle of actually emptying it, and the difference between this mass and the mass of the container filled with an unknown gas will give the mass of the unknown gas.

Various gases, such as helium (He), nitrogen (N_2), nitrous oxide (N_2O), carbon dioxide (CO_2), sulfur hexafluoride (SF_6), and freons (chlorofluoromethanes), can be used in this experiment. (The use of flammable or toxic gases is not advised.)

When tanks of compressed gases are available, they are convenient, but by preparing your own sample of gas, you will have an opportunity to learn some additional chemistry. In the Experimental Procedure, directions will be given for the preparation of carbon dioxide.

151

GAS DENSITY AND BUOYANCY

The density of a gas is equal to its mass divided by the volume it occupies at a particular temperature and pressure. The density is directly proportional to the molecular weight, as can be seen by deriving a modified form of the ideal gas law. Let's start with the ideal gas law in the form

$$PV = nRT \tag{1}$$

where P is the pressure, V is the volume, n is the number of moles of gas, R is the gas constant, and T is the Kelvin, or absolute, temperature (K). The number of moles, n, can also be expressed as the mass of the gas, m, divided by the molar mass of the gas, M (take a moment to verify that the units are correct):

$$n = \frac{m}{M} \tag{2}$$

We can substitute m/M for n in Equation (1) and rearrange the resulting equation to obtain the density of the gas, m/V:

$$\frac{m}{V} = \frac{PM}{RT} \tag{3}$$

From Equation (3) we see that the density, m/V, is proportional to the molar mass of the gas, M.

If we fill a lightweight container, such as a balloon, with a gas that has a density much less than the density of air, we may find that when we try to weigh the balloon of gas on a balance, the balloon floats. Why is this?

The atmosphere is like an ocean of air, and when an object is immersed in air, the object displaces air, just as your body displaces water when you jump into a swimming pool. And just as you can float in water, the balloon will float and rise in the air if the combined mass of the balloon and the gas in it is less than the mass of the air that is displaced. Since the pressure (and density) of air decreases with altitude (note the effect on density of decreasing P in Equation [3]), the balloon will expand as it rises. Ultimately, the balloon will either explode or will stop rising when it reaches an altitude where the mass of air displaced and the mass of the balloon plus the gas are equal. You can easily observe this buoyancy effect by using a lighter-than-air gas (such as natural gas, which is largely methane, CH_4) to blow bubbles in a soap solution, then repeating the experiment with a heavier-than-air gas (such as carbon dioxide, CO_2).

Hot-air and lighter-than-air balloonists are able to enjoy their sport by filling their balloons with a gas that is less dense than air. Hot-air balloons use propane heaters at the base of the balloon to heat air. Hot air is less dense

than cold air. (Note the effect on the density of a gas of increasing T in Equation [3].) Lighter-than-air balloonists use a gas with a low molecular weight, such as helium, to obtain the necessary lower density. (Note the effect of decreasing M in Equation [3].)

The effect of buoyancy must also be considered when very accurate weighings are made, particularly for substances such as water, which has a density less than the density of the standard weights in the balance. Standard weights, which are made of stainless steel for the most commonly encountered single-pan balances, have the density of stainless steel, which is 7.76 g/mL. A sample placed on the balance pan displaces a certain amount of air. A similar effect applies to the removable weights of the balance. There will be a net effect of *buoyancy* (apparent mass less than the true mass) whenever the density of the object being weighed is less than the density of the weights. The magnitude of the buoyancy factor is illustrated by a few examples. When water is weighed (density of 1.00 g/mL), the true mass is 1.001 g when the balance reads 1.000 g. In this case, the error is −0.1%. For NaCl (which has a density of 2.16 g/mL), the error would be −0.04%. For $AgNO_3$ (with a density of 4.35 g/mL), the error is only −0.01%.

When obtaining a mass by difference (as we did in determining the molar volume of oxygen by weighing, decomposing, and reweighing a sample of dilute H_2O_2 solution), the buoyancy correction for the water and container cancels out. In most of the experiments you will do in this laboratory, the buoyancy corrections will be negligible, but if you are weighing samples of gases, as in this experiment, the corrections are significant and can be larger than the sample weights.

EXPERIMENTAL PROCEDURE

NOTE TO INSTRUCTOR: *Tell the students whether they will use compressed gases (Method A) or the gas generator shown in Figure 12-1 to generate CO_2 (Method B). If compressed gases are used, you may want to wrap the cylinders or lecture bottles with paper to conceal the identity of the gas, and have the students identify it from a list of gases posted in the laboratory.*

The experiment calls for first filling the flask with clean, dry air. If in-house compressed air is not available, the flasks can be weighed filled with moist ambient air. Students will then calculate the density of the moist ambient air from the formula given as Equation (4). Tell the students which procedure they should use.

Special Supplies: *Thermometer; barometer; calcium chloride drying tube with one-hole stoppers; No. 4 one-hole rubber stopper bored out to 8-mm-diameter hole with cork or rubber stopper to plug the hole; 6-mm-OD glass tubing; cotton; ³⁄₆ in.-ID rubber tubing; ring stand and clamp; source of clean, dry compressed air; nonflammable, nontoxic compressed*

FIGURE 12-1

Apparatus for determining the molar mass of a gas. Flask D is filled with dry air, weighed, then filled with
an unknown gas (or CO_2), and weighed again.

gases in cylinders or lecture bottles if Method A is to be used. (Suitable gases include He, Ar, N_2, O_2, N_2O, CO_2, SF_6, and freons.) If the students are to generate CO_2 by Method B, a thistle tube, a 200-mL Erlenmeyer flask with a two-hole rubber stopper, and a 6-mm-OD glass tube with a right-angle bend will be required (see Figure 12-1).

Chemicals: 4-mesh anhydrous calcium chloride, $CaCl_2$, for the drying tube; marble chips (calcium carbonate, $CaCO_3$) and 6 M HCl for CO_2 generation if Method B is to be used.

! SAFETY PRECAUTIONS: It is recommended that no toxic or flammable gases be used in this experiment. Even with this precaution, it is recommended that you fill your flasks in a fume hood.

All cylinders should be chained or securely fastened so that they cannot topple, and you should be instructed in the proper way to use the gas regulators or valves.

Carefully adjust the flow rates of compressed air or unknown gas *before* connecting the tubing to the drying tube to avoid accidents resulting from sudden pressure surges.

If you assemble your own apparatus to generate CO_2, take care to lubricate the glass tubing and wrap it in a towel before inserting it in the rubber stopper.

DETERMINING THE MASS OF THE FLASK FILLED WITH AIR

The setup shown in Figure 12-1 is used to fill a flask with clean, dry air. (If in-house compressed air is not available, the flask is weighed filled with the moist ambient air of the laboratory.) Obtain a clean and thoroughly dry 125-mL Erlenmeyer flask, preferably one that has been oven-dried and allowed to cool just before it is to be used. The flask should be fitted with a one-hole rubber stopper, bored out to have a diameter about 2 mm larger than the glass tube, so that gas passing through the tube can flow out at the top through the annular space between the tube and stopper. The glass tube is connected by a section of rubber tubing to a drying tube (C) filled with 4-mesh calcium chloride; each end of the tube is protected by a loose plug of cotton. Adjust the compressed-air valve to give a very gentle flow of air, then connect it to the other end of the drying tube with a short section of rubber tubing. Pass a gentle stream of dried air through flask D for at least 5 min.

Gently withdraw the glass tubing through which you have been passing clean, dry air, quickly plug the hole with a small cork or rubber stopper, and weigh flask D, including the one-hole rubber stopper and plug, to an accuracy of at least 0.001 g. Record the mass on your report sheet. (*You must use the same cork or rubber plug throughout the entire experiment, so put it in a safe place whenever it is necessary to remove it.*) After the weighing, do not handle the flask any more than is absolutely necessary. If you must handle the flask, grasp it by the rim, or wrap a small piece of paper around the neck to keep greasy fingerprints off the flask. Place the flask on a clean, dry square of paper to avoid contamination from the bench top.

Repeat the filling and weighing process. The two masses should agree to within 1 to 3 mg.

If compressed air is not available, use a flask that has been oven-dried and allowed to cool in the ambient laboratory air. Fit it with a one-hole stopper and plug, as described above, and weigh this assembly. Then remove the stopper and plug, allow the flask to stand open to the ambient air for 10 min, put the stopper and plug back in, and reweigh.

DETERMINING THE MASS OF THE FLASK FILLED WITH AN UNKNOWN GAS BY USING COMPRESSED GASES (METHOD A)

Connect a rubber tube to the outlet valve of a cylinder or lecture bottle of an unknown gas. Have your instructor adjust the flow rate through the tube to about 100 to 200 mL/min by displacement of water in an inverted graduated cylinder. Then connect the tube to the inlet of the drying tube, insert the glass tube on the outlet of the drying tube into flask D, and allow at least 1 L of gas to flow through it. (This will require 5 to 10 min at the recommended flow rate.) Remove the glass tube from flask D, quickly insert the plug, and weigh the flask as before. Repeat the filling and weighing operation with the same gas. The two masses should agree to within 1 to 3 mg.

DETERMINING THE MASS OF THE FLASK FILLED BY MEANS OF THE CARBON DIOXIDE GENERATOR (METHOD B)

Prepare the carbon dioxide generator shown in Figure 12-1. Use a 200- or 250-mL Erlenmeyer flask that has a thistle tube[1] reaching almost to the bottom and that has a short right-angle exit tube. This apparatus will be connected to the calcium chloride drying tube when you are ready to generate the gas. Place about 30 g of marble chips ($CaCO_3$) in the generator, and add about 20 mL of water to cover the end of the thistle tube. Do not add acid until you are ready to generate carbon dioxide. Put flask D in position, with the glass tube passing through the one-hole rubber stopper. When ready, add 6 M HCl through the thistle tube, a little at a time as needed, so as to maintain a gentle but fairly rapid evolution of CO_2 gas. (You may momentarily pinch the tubing between C and D and note the rate at which liquid backs up in the thistle tube, in order to estimate the rate at which gas is being generated.)

Let the generator run for at least 20 min to displace all the air in flask D by CO_2 gas. Finally, withdraw the glass tube from the flask, and quickly insert the plug in the stopper. Without undue handling, weigh the flask. Repeat the filling operation again by passing CO_2 through flask D for another 10 to 15 min. Reweigh the flask. The two weights should agree to within 1 to 3 mg.

DETERMINING THE TEMPERATURE, PRESSURE, AND VOLUME OF THE GAS

Take the temperature of the gas by placing a thermometer in it shortly after the last weighing. Record the barometric pressure. To measure the exact volume of the flask to the bottom of the stopper, fill the flask with water to the brim, replace the stopper, insert the plug in the stopper, and wipe off any excess water. Weigh the full flask on the platform balance to the nearest 0.5 g. (*Do not weigh the flask full of water on the analytical balance.* Use a more rugged platform balance. Note also that milligram accuracy is not required for this weight: Why not?) How does the mass of water enable you to calculate the volume? Look up in a reference source[2] the density of dry air at the laboratory temperature and pressure. Enter all data and calculations on your report sheet.

If you filled flask D with the moist ambient air of the laboratory rather than dry air, calculate the density of the moist air from the expression shown on the next page.

---

[1]If thistle tubes are not available, you may substitute a long-stemmed funnel; or you may simply use a one-hole stopper for the right-angle tubing, removing the stopper to add the acid, then quickly re-stoppering the flask.

[2]See, for example, *Lange's Handbook of Chemistry,* 15th ed., McGraw-Hill, New York, 1998; or the *CRC Handbook of Chemistry and Physics,* CRC Press, Boca Raton, FL.

$$d(\text{g/ml}) = 0.001293 \times \frac{273}{T} \times \frac{[P - 0.378(\text{RH})\, P_{\text{H}_2\text{O}}]}{760} \tag{4}$$

where d is the density in grams per milliliter of the air; T is the absolute temperature (°C + 273); P is the barometric pressure in torr (mmHg); and RH is the relative humidity. (If you don't know the relative humidity, it will be sufficiently accurate to assume that RH = 0.5, which corresponds to a 50% relative humidity.) $P_{\text{H}_2\text{O}}$ is the vapor pressure of water (Appendix, Table 3) at the laboratory temperature.

(Equation [4] may also be used to calculate the density of dry air. If $P_{\text{H}_2\text{O}} = 0$, the third term of the equation reduces to $P/760$.)

Equation (3) can be rearranged to yield an expression for M, the molar mass, as a function of the mass (m) of the gas, the volume (V), the pressure (P), the Kelvin temperature (T), and the gas constant, R. Calculate the molar mass of the unknown gas (or carbon dioxide), taking care that the units of R are consistent with the units of m, V, P, and T.

☞ **Consider This**

Weigh an inflated basketball, volleyball, or soccer ball to at least the nearest milligram. Deflate but do not collapse the ball and reweigh. Collapse the ball as much as possible and reweigh. Discuss your results.

The Molar Mass of a Gas

NAME		
SECTION		LOCKER
INSTRUCTOR		DATE

DATA AND CALCULATIONS

Data: Unknown no. _____	Trial 1	Trial 2
(a) Mass of flask, stopper, and plug, filled with air	g	g
(b) Mass of flask, stopper, and plug, filled with unknown gas (or CO_2)	g	g
(c) Mass of flask, stopper, and plug, filled with water	g	g
(d) Temperature of flask	°C	°C
(e) Barometric pressure	torr	torr
(f) Density of dry (or moist) air at the flask temperature and pressure Reference:	g/mL	g/mL

Calculations	Trial 1	Trial 2
(g) Temperature, absolute	K	K
(h) Pressure	atm	atm
(i) Volume of flask[a]	L	L
(j) Mass of air in flask at the start	g	g
(k) Mass of empty flask, stopper, and plug	g	g
(l) Mass of unknown gas (or CO_2)	g	g
(m) Molar mass of gas (CO_2). Calculate from Equation (3), assuming $R = 0.08206$ L $\cdot$ atm $\cdot$ mol^{-1} $\cdot$ K^{-1}	g/mol	g/mol
(n) True molar mass of gas (CO_2) Reference:	g/mol	g/mol
(o) Percentage of relative error	%	%

[a]Why can we neglect the mass of air in the flask when obtaining the mass of the water (from which we calculate the volume of the flask) but not when calculating the mass of the carbon dioxide? Explain. If you are unable to answer this question, discuss it with your instructor before leaving the laboratory.

PROBLEMS

1. Write the equation for the reaction taking place in the preparation of carbon dioxide.

2. How many moles are there in the 30.0 g of $CaCO_3$ used?

3. How many liters of carbon dioxide could be generated at standard conditions by this mass of $CaCO_3$?

4. What volume would this amount of carbon dioxide occupy at ordinary laboratory conditions? (Assume 25 °C and 710 torr.)

5. What is the mass of a liter of each of the following gases at standard conditions? (Use no other data than atomic masses and the molar volume.)

   Fluorine, F_2 (g):

   Propane, $C_3H_8(g)$:

   Ammonia, $NH_3(g)$:

6. A 1.60-g sample of impure sodium acetate was mixed with excess $NaOH(s)$ and heated in a test tube to decompose it completely according to the equation

$$NaCH_3COO(s) + NaOH(s) \rightarrow Na_2CO_3(s) + CH_4(g)$$

The evolved methane gas, $CH_4(g)$, was collected over water at 27 °C and 747 torr, resulting in a volume of 320 mL of gas collected.

   (a) Calculate the number of moles of methane gas that was collected. (Be sure to correct for the vapor pressure of the water.) You might save time by using $PV = nRT$:

   (b) Calculate the mass of $NaCH_3COO(s)$ that would be required to produce this amount of $CH_4(g)$:

   (c) Calculate the percentage of pure anhydrous sodium acetate, $NaCH_3COO$, in the sample:

☞ **Consider This**

Weigh an inflated basketball, volleyball, or soccer ball to at least the nearest milligram. Deflate but do not collapse the ball and reweigh. Collapse the ball as much as possible and reweigh. Discuss your results.

The Reactivity of Metals with Hydrochloric Acid

PURPOSE

To illustrate the concept of equivalent mass by determining the mass of a metal that reacts with hydrochloric acid to form 0.5 mol of hydrogen gas, H_2.

PRE-LAB PREPARATION

THE CONCEPT OF EQUIVALENT MASS

Equivalent mass, as the adjective "equivalent" implies, designates the relative amounts of substances that are chemically equivalent—that is, the masses of those substances that react with or replace one another in chemical reactions. When a metal reacts with oxygen to form an oxide, the metal's equivalent mass is defined as the portion of the overall mass that reacts with 0.25 mol (8.000 g) of oxygen (O_2) gas. Here we will consider the equivalent mass of a metal to be the mass of metal that reacts with 1 mol of hydrogen ions (H^+) to produce 1 mol of hydrogen atoms, or 0.5 mol of hydrogen (H_2) gas (1.008 g). The reaction is a replacement reaction: the metal reacts with hydrochloric acid, HCl, to replace hydrogen gas. It can also be classified as an *oxidation-reduction* reaction because the metal is *oxidized* to a positively charged metal ion (*increase* in oxidation state) and the H^+ is *reduced* to H_2 (*decrease* in oxidation state). We will measure the volume of hydrogen gas produced when a weighed sample of metal is added to an excess of hydrochloric acid. For example,

$$Ca(s) + 2HCl(aq) \rightarrow CaCl_2(aq) + H_2(g)$$

Note that for each mole of calcium that reacts to form Ca^{2+} ions (ionic charge +2), 1 mol of hydrogen gas is formed. Therefore, here the equivalent mass of calcium is half its atomic mass. Equivalent mass is always the atomic mass (for elements that form ions with ± 1 charge) or a fraction of it, depending on the charge of the ion produced. So, a general definition of equivalent mass is

$$\frac{\text{equivalent}}{\text{mass}} = \frac{\text{atomic mass}}{\text{change in charge on atom}} \quad (1)$$

Some elements, particularly transition metals like iron, can have more than one ionic charge in their compounds and therefore can have *more than one possible equivalent mass.* Consider the following possible reactions:

$$Fe(s) + 2HCl(aq) \rightarrow FeCl_2(aq) + H_2(g) \quad (2)$$

$$2Fe(s) + 3Cl_2(g) \rightarrow 2FeCl_3(aq) \quad (3)$$

$$2FeCl_2(aq) + Cl_2(g) \rightarrow 2FeCl_3(aq) \quad (4)$$

In reaction (2) the change in charge is +2, in reaction (3) it is +3, and in reaction (4) it is +1. The corresponding equivalent masses of iron are, respectively, 27.92 (half the atomic mass), 18.62 (one third the atomic mass), and 55.85 (the atomic mass). *Thus, for elements that can have multiple ionic charges (or oxidation states), you must always specify the reaction before determining equivalent mass.* An alternative definition of equivalent mass is that it is *that mass of substance that gives up or accepts 1 mol of electrons.* The defining equation then becomes

$$\frac{\text{equivalent}}{\text{mass}} = \frac{\text{atomic mass}}{\text{moles of electrons transferred}} \quad (5)$$

Thus, in reaction (2) iron changes from an oxidation state of 0 to oxidation state +2. It loses 2 mol of electrons per mol of Fe in doing so. The equivalent mass of Fe *in this reaction* therefore is

$$\frac{55.85}{2} = 27.92$$

This is the same result obtained by means of Equation (1). Verify that the equivalent masses for reactions (3) and (4) are 18.62 and 55.85, using Equation (5).

Equation (5) emphasizes that the replacement reactions studied in this experiment can also be classified as oxidation-reduction reactions, reactions in which a transfer of electrons has caused changes in oxidation states. These concepts will be studied in greater detail in later experiments.

You should keep the three (related) definitions of equivalent mass in mind as you proceed:

- The mass of a metal that reacts with acid to produce ½ mole of $H_2(g)$
- The mass of a metal that reacts with acid to produce 11,200 mL of $H_2(g)$ at standard temperature (0 °C) and pressure (760 torr)
- The atomic mass of a metal divided by the change in charge that occurs in the reaction

EXPERIMENTAL PROCEDURE

Special Supplies: Thermometer, pieces of metals (cut to size or issued as unknowns), 50-mL buret (for Method A only), fine copper wire, 25 × 250 mm test tube, two 500-mL Erlenmeyer flasks (for Method B).

Chemicals: Concentrated (12 M) hydrochloric acid, HCl

! SAFETY PRECAUTIONS: Concentrated hydrochloric acid is a lung irritant and causes skin and eye burns. Handle with care. Dispense it in a well-ventilated fume hood. Clean up any spills immediately. Protect your hands with plastic gloves.

. . . .

NOTE: The analysis samples for this experiment may be issued (1) as unknown metals, for you to calculate the equivalent mass; (2) as preweighed samples, for you to calculate and report the sample mass from the known equivalent mass; or (3) as Al-Zn alloys of different composition, for you to calculate the percentage composition from the known equivalent masses. Prepare the units in your report accordingly, *before beginning the experiment.*

Your instructor will designate which of the preceding analyses and calculations you will do and which of the following alternate procedures you will follow. Method A is faster, but it limits you to smaller samples, to a fixed acid concentration, and to room temperature. Using Method B enables you to use larger samples and to control both temperature and acid concentration during the reaction.

. . . .

NOTE TO THE INSTRUCTOR: *Samples may be preweighed by a stock assistant on a rapid single-pan balance or, if in wire or ribbon form, cut to exact length to give a known mass, and individually coded. The student may then report the corrected volume of hydrogen as a preliminary check on his or her work and finally report the calculated equivalent mass, or the mass of the sample from the known equivalent mass.*

Obtain two samples of the metal to be used. Weigh these precisely on the analytical balance, at the same time taking care that the masses do not exceed the maximum permitted, so you do not generate more hydrogen than your apparatus can accommodate in either Method A or Method B. (For all laboratories except those at high elevation, maximum masses are as follows: for Method A, with a 50-mL buret, 0.12 g Zn, 0.032 g Al, 0.042 g Mg, 0.10 g Mn; for Method B, with a 500-mL flask, 1.10 g Zn, 0.40 g Mg, 0.30 g Al, 0.90 g Mn, 0.90 g Fe, 1.90 g Sn, 1.90 g Cd.)

METHOD A

Compress the weighed samples into compact bundles, and wrap each sample in all directions with about 20 cm of fine copper wire, forming a small basket or cage, leaving 5 cm of the wire straight as a handle. This confines the particles as the metal dissolves, and also speeds the reaction.[1]

Obtain and clean a 50-mL buret. Next measure the uncalibrated volume of the buret between the stopcock and the 50-mL graduation. Measure by filling the buret with water and draining the buret through the stopcock until the liquid level falls exactly to the 50-mL mark. Then use a 10-mL graduated cylinder to measure the volume delivered when the water level is lowered to the top of the stopcock. Using a funnel, pour into the buret the required amount of concentrated hydrochloric acid. Be careful not to allow the acid to touch your skin or clothing. Because of differences in the activity of the metals used, it is necessary to vary the amount of acid. For magnesium, use about 3 mL; for aluminum or zinc, use about 20 mL; for manganese, use about 7 mL. Fill the buret completely with water slowly and carefully, to avoid undue mixing of the acid. Insert the metal sample about 4 cm into the buret, and clamp it there by the copper wire handle, using a one- or two-hole rubber stopper. Make certain no air is entrapped in the buret. Cover the stopper hole(s) with your finger,[2]

---

[1] As the more active metal dissolves, it gives up electrons that move easily to the less reactive copper, where they react with hydrogen ions of the acid to form hydrogen gas. Note that the bubbles of gas form on the copper wire, thus keeping a larger surface of active metal exposed to the acid.

[2] Use a plastic glove for this step to ensure that no acid comes in contact with your skin.

and invert the buret (Figure 13-1) in a 400-mL beaker partly filled with water. The acid, being more dense, quickly sinks, diffuses down the buret, and reacts with the metal. As the H$_2$ is generated, it collects at the top of the buret, expelling the HCl and water solution out the hole in the stopper at the bottom. (**Caution:** If the reaction is too rapid and the metal too close to the end of the buret, as the acid solution is expelled, small bubbles of hydrogen may also escape from the buret. If so, repeat the experiment using less acid.)

After complete solution of the metal, let the apparatus cool to room temperature, since heat is generated by the reaction. Tap the apparatus to free any hydrogen bubbles adhering to the sides of the vessel or the copper wire. Measure the volume of gas liberated,[3] and—without changing the position of the buret—measure the difference in height of the two water levels with a metric rule, and calculate the equivalent pressure in millimeters of mercury (torr; mmHg = mm solution × density of solution/density of mercury). Take the temperature of the gas by holding a thermometer in contact with the side of the buret. Raise the buret up out of the HCl solution in the beaker, and allow the remainder of the solution to drain out of the buret into a large beaker. Slowly add solid sodium carbonate, Na$_2$CO$_3$, to the HCl solution until there is no further fizzing. Then flush the HCl solution down the drain with plenty of water, and discard the copper wire in the wastebasket.

Obtain the barometer reading for the day. Repeat the determination with your second sample.

METHOD B

Set up the apparatus as sketched in Figure 13-2, utilizing a 25 × 250 mm test tube and a 500-mL flask to contain the evolved hydrogen. The exit tube C from the test tube, connected to the flask E, must not extend below either rubber stopper (so that gas will not be trapped). The longer glass tube in the test tube should extend nearly to the bottom and should be constricted to a small capillary and bent as illustrated. Completely fill flask E with water, and put a little water into flask F. Fill the siphon tube D, which extends to the bottoms of both flasks E and F, by blowing into the tube C.

Place the first carefully weighed sample of metal in the 25 × 250 mm test tube, as indicated. A fine copper wire, wrapped about the sample in all directions like a cage, as in Method A, may be of some help in increasing the rate of reaction and in confining small bits of metal as the sample dissolves, although this is not absolutely necessary. With the stopper in flask E loosened to permit air to escape, pour

FIGURE 13-1

The appropriate volume of concentrated (12 M) HCl is added to the buret, and water is then layered on top of it until the buret is completely filled, as in A. Inversion of the buret in a beaker partly filled with water begins the reaction (B), which continues until all of the metal is gone and the buret is nearly full of hydrogen gas, as shown in C.

water into the funnel to completely fill the test tube and tubes B and C, and close clamp B. If flask E is not completely filled with water, raise flask F so that water siphons back into it, then push in the stopper in E tightly.

When all air bubbles have been thus removed from the apparatus, release clamps B and D just enough to permit the water level in the funnel to fall just to the stem top *but*

FIGURE 13-2

An alternate apparatus for the determination of the equivalent mass of a metal.

no farther; then close clamp B. Now empty and drain flask F, but do not dry it, leaving tube D filled with water. Clamp D open. When all is ready, pour exactly 25.0 mL of concentrated hydrochloric acid into the funnel. Release clamp B momentarily to permit a little acid to flow into the test tube and react with the metal at a moderate rate.

A volume of water equal to the volume of hydrogen generated will siphon from E to F. You can control the rate of reaction by regulating the amount of acid you add and by immersing the tube (for less reactive metals) in a beaker of warm water as needed. When all the metal has dissolved and the reaction mixture has cooled to room temperature, release clamp B to permit the acid level to fall to the funnel stem, *but no farther.* Carefully measure

25.0 mL, of water, add this to the funnel, and again release the clamp to let the level fall exactly to the funnel stem; then close the clamp tightly. All gas should now be displaced from the test tube and connecting tubes into flask E. If it is not, repeat the addition of a measured amount of water. Adjust the levels in flasks E and F by raising or lowering one of them until they are even (avoid warming the gas in E with your hands on the flask); then close clamp D tightly. Obtain the temperature of the hydrogen by removing the stopper enough to place a thermometer in flask E. Measure the volume of water in flask F by pouring it into a 500-mL, graduated cylinder. This volume, minus the volumes of acid and water added to the funnel, will be the volume of the hydrogen generated. Record the barometric pressure. Pour the HCl solution from the test tube and flask E down the drain with plenty of water. Repeat the determination with your second sample.

CALCULATION OF THE EXPERIMENT

Begin by recalling that the standard molar volume of any gas is 22.4 L. The volume of 0.5 mol of hydrogen gas (1.008 g) at standard conditions is thus 11.2 L, or 11,200 mL. *The equivalent mass of your metal sample therefore is that mass which will produce 11,200 mL of hydrogen gas at standard conditions from an acid.* Also note that you must consider the difference in liquid levels (Method A) and the vapor pressure of water in arriving at the correct hydrogen gas pressure.

Calculate and report your result according to the type of sample you have been given: (1) the equivalent mass of a metal (g metal/equiv), (2) the mass of a preweighed sample, or (3) the percentage composition of an Al-Zn alloy. The percentage of Al in the alloy is given by

$$\text{percentage of Al} = \frac{\text{g Al}}{\text{g sample}} \times 100\%$$

To do this, you must know *both* the mass of aluminum (g Al) and the mass of zinc (g Zn), and you will need two independent equations to solve for the two unknown masses. These two equations may be written as

$$\text{g sample} = \text{g Al} + \text{g Zn}$$

and

$$\text{vol } H_2 \text{ (STP)} = 11,200 \frac{\text{mL } H_2\text{(STP)}}{\text{equiv}} \times$$

$$\left(\frac{\text{g Al}}{\text{g Al/equiv}} + \frac{\text{g Zn}}{\text{g Zn/equiv}} \right)$$

where g Al/equiv and g Zn/equiv are the equivalent masses of Al and Zn, respectively.

For your reported result, calculate the percentage of relative error given by

$$\text{percentage relative error} =$$

$$\frac{\text{your value} - \text{accepted value}}{\text{accepted value}} \times 100\%$$

Consider This

Which metals would work in this experiment—in other words, which metals react with 12 M hydrochloric acid? Collect as many different samples of metals as you can and test. Do you note different speeds of reaction? What happens if you use acetic acid (vinegar) instead?

Does the concentration of the acid make much difference in the reaction? Make up 6M, 3M, 1.5 M, 0.75 M, and other concentrations of hydrochloric acid solutions, and place a piece of zinc in each. Note any correlation. Is there a threshold concentration below which the reaction doesn't proceed?

Can you determine the amount of zinc in a post-1982 U.S. penny by measuring the volume of hydrogen gas produced in an experimental setup like the one used here? Does this technique work with a pre-1982 penny?

BIBLIOGRAPHY

Masterton, W. L. "Analysis of an Aluminum-Zinc Alloy," *J. Chem. Educ.,* 1961, 35: 558.

NAME		
SECTION	LOCKER	
INSTRUCTOR	DATE	

DATA AND CALCULATIONS

Data: Unknown sample identifier _____	Trial 1	Trial 2
Mass of sample (except for preweighed sample)	g	g
Volume of uncalibrated portion of buret (Method A)	mL	mL
Final buret reading (Method A)	mL	mL
Volume of hydrogen	mL	mL
Temperature of hydrogen	°C	°C
Barometric pressure	torr	torr
Difference in water levels inside and outside tube (Method A)	mm H_2O	mm H_2O
Aqueous vapor pressure at temperature of hydrogen	torr	torr

Data	Calculations	Trial 1	Trial 2
mm Hg (torr) equivalent of difference of water levels (Method A)		torr	torr
Pressure of H_2 after correction for difference in H_2O levels and for vapor pressure		torr	torr
Temperature, absolute		K	K
Volume of dry H_2 at standard conditions, i.e., corrected for vapor pressure, pressure, and temperature		mL	mL
Equivalent mass[a]		g/equiv	g/equiv
Percentage of relative error = $\dfrac{\text{your value} - \text{accepted value}}{\text{accepted value}} \times 100\%$		%	%

[a]NOTE: The units in the boxes provided for recording your calculated values should be changed as necessary, depending on whether you are reporting (1) the mass of a preweighed sample, or (2) the percentage composition of an Al-Zn alloy.

PROBLEMS

1. The atomic mass of tin is 118.69. What is its equivalent mass:

    (a) when it reacts with chlorine to form $SnCl_4$?

    (b) when it reacts with hydrochloric acid to form $SnCl_4$?

2. Write the equation for the reaction of the following three metals with H_2SO_4. What volume of hydrogen gas, at standard conditions, will be liberated by the reaction of sulfuric acid with them? Relate your results to the position of these elements in the periodic table and to their oxidation numbers in compounds.

    (a) 0.0230 g Na: Equation: _____

    (b) 0.0243 g Mg: Equation: _____

    (c) 0.0270 g Al: Equation: _____

3. A 1.70-g sample of a certain metal reacts with acid to liberate 380 mL of hydrogen gas when collected over water at 20 °C and 720 torr. What is the equivalent mass of the metal?

4. Calculate the maximum mass of calcium metal that should be used in an equivalent mass determination with a 100-mL gas measuring tube for collection. For the purposes of calculation, assume 80.0 mL of hydrogen gas at standard conditions.

☞ **Consider This**

Which metals would work in this experiment—in other words, which metals react with 12 M hydrochloric acid? Collect as many different samples of metals as you can and test. Do you note different speeds of reaction? What happens if you use acetic acid (vinegar) instead?

Does the concentration of the acid make much difference in the reaction? Make up 6M, 3M, 1.5 M, 0.75 M, and other concentrations of hydrochloric acid solutions, and place a piece of zinc in each. Note any correlation. Is there a threshold concentration below which the reaction doesn't proceed?

Can you determine the amount of zinc in a post-1982 U.S. penny by measuring the volume of hydrogen gas produced in an experimental setup like the one used here? Does this technique work with a pre-1982 penny?

The Heat Capacity of Metals

PURPOSE

To gain an understanding of the terms *heat, energy, work,* and *heat capacity.* To gain an understanding of how heat is related to molecular motion. To use a simple Styrofoam cup calorimeter to measure the heat capacity of a metal and confirm the law of Dulong and Petit.

PRE-LAB PREPARATION

THE NATURE OF HEAT, ENERGY, AND WORK

What is heat? Naming something is not the same thing as understanding it, and in science a familiar name often stands for a concept that, upon close examination, proves elusive, difficult to define in a precise way.

Heat is an effect produced by the transfer of energy. We might even call it "energy in transit." Energy naturally flows from warmer to cooler bodies. This is equivalent to saying that heat spontaneously passes from a hot to a cold body.

In describing heat we used another familiar word—*energy*—that stands for an elusive concept. What is energy? We will define it as the *capacity to produce useful work.* Work is defined as the product of *force × distance,* such as the work done in lifting a weight. The word *energy* was first employed in this sense by Rudolf Clausius (1822–1888), but its use has been traced back to the ancient Greeks (probably Aristotle), who coined the word by combining the Greek words *en* and *ergon,* meaning "at" and "work."

Energy manifests itself in many forms. Part of the energy in coal and oxygen is released when they combine chemically to form carbon dioxide. This is called *chemical energy,* and it produces heat that under high pressure can be used to turn water into steam. The energy stored in the steam is converted into kinetic energy by allowing the steam to expand through a turbine, causing it to rotate. The energy of motion in the rotating turbine is converted into *electrical energy* when the turbine is coupled to a generator. The electrical energy distributed through wires can produce *mechanical work* (another form of energy), such as the shaft rotation of an electric motor, or it can be converted into heat by passing electric current through a resistor.

The rotation of the shaft of an electric motor can be used to turn a drum, winding up a cord to which a weight is attached. In raising the weight, work is done. The work done in raising the weight increases its *potential energy.* If the motor is uncoupled from the drum and the weight allowed to descend, the potential energy stored in the raised weight is converted into kinetic energy, rotating the drum as the cord unwinds. If the rotating drum is now connected by gears to a paddle wheel churning a container of water, the temperature of the water increases. The potential energy stored in the weight is converted into work, which is in turn converted into heat.

James Prescott Joule (1818–1889) used a falling weight to churn water in exactly this way to determine that there is a fixed relationship between the amount of work done and the heat effect it produces. Joule called this quantity the "mechanical equivalent of heat." In modern units, 4.184 joules (J) = 1 calorie (cal) of thermal energy, where 1 calorie is the quantity of energy required to increase the temperature of 1 g of water by 1 Celsius degree.

MODES OF ENERGY TRANSFER

There are several ways to heat an object by energy transfer. Energy can be transferred through empty space (a vacuum) by *radiation.* You feel this effect when you turn your face to the sun on a warm spring day. Light is a form of electromagnetic radiation, and the carriers of light energy are called *photons*. Although they have no rest mass, when they are absorbed by the skin, they set the atoms and molecules in the skin into motion, producing the sensation of warmth.

Heat can pass from one body to another by *convection* (movement of a fluid). If you hold the back of your hand near your mouth and exhale gently on it, you feel the sensation of warmth caused by the flow of warm air from your lungs. Finally, energy can be transferred by *conduction* through both fluids and solids. If you grasp the end of a copper rod and stick the other end in a flame, the end you are holding soon gets uncomfortably hot. The same thing happens if you spill hot coffee on yourself. The hot water heats your skin, perhaps enough to produce a painful burn.

HEAT AND MOLECULAR MOTION

The accumulated knowledge about heat's effects provides a rather detailed picture of what happens when you heat (or transfer energy to) a substance. On an atomic or molecular scale, heating increases the motion of the atoms. If the atoms are linked by chemical bonds to form a gaseous molecule, we can focus our attention not on individual atoms but on the molecule as a whole. If we heat a gas, there are three effects:

1.   The individual molecules move about faster from place to place. Their kinetic energy of *translation* is increased.

2.   Individual atoms in molecules are linked by chemical bonds that are "springy," constraining atoms that are bonded together to move like masses connected by springs. As the molecules are heated, the amplitudes of their *vibrations* increase.

3.   Finally, molecules can rotate about their centers of mass, and as a gas is heated, the rate of *rotation* of the molecules increases.

To summarize, gaseous molecules can store energy in *translational, vibrational,* and *rotational* motions.

What about solids such as metals? When you heat a solid metal, which consists of atoms closely packed together, the atoms, on average, are not going anywhere. An atom occupies a fixed position in the solid (unless the temperature gets too near the melting point), and when heated it just increases the amplitude of its vibration about this fixed position. Because atoms have nearly all of their mass located in a very small nucleus, they have a very small moment of inertia. Therefore, individual atoms have an insignificant amount of rotational energy.

LATENT HEAT

When you heat a substance (thus putting energy into it), will a thermometer immersed in the substance always indicate a temperature increase? Not necessarily. In some circumstances, the temperature may stay constant. For example, if you drop a small piece of hot metal into a mixture of ice and water at 0 °C, some ice will melt, but the temperature will remain constant until all of the ice is melted. Heat that produces a phase change such as the melting of ice (or the boiling of water) without producing a temperature change is called *latent* heat. A thermometer, therefore, can be used to measure *temperature* changes but cannot always be used to measure *energy* changes. This observation shows the important but sometimes subtle distinction between heat and temperature.

HEAT CAPACITY

When you heat a substance and its temperature increases, the input of energy divided by the temperature increase is called the *heat capacity*. It could just as well be called the energy capacity, since the quantity has the dimensions of *energy per unit temperature change*.

THE HEAT CAPACITY OF METALS AND THE LAW OF DULONG AND PETIT

In 1819, as part of a systematic study of the heat capacity of metals, Pierre Dulong (1785–1838) and his collaborator Alexis Petit (1791–1820) published their discovery that the product of the molar mass (grams per mole) of a metal times its heat capacity (calories per gram per Celsius degree) is a constant. (This product could be called the *molar heat capacity,* and it is approximately equal to 6.0 cal $\cdot$ mol^{-1} $\cdot$ K^{-1}, or 25 J $\cdot$ mol^{-1} $\cdot$ K^{-1}.) At the time there was no explanation of why this should be true, but Berzelius (and later Cannizzaro) used this empirical result to arrive at correct atomic masses and empirical formulas for compounds of metals.

It was not until 1907 that Albert Einstein published a theory of the heat capacity of metals. He treated atoms as equivalent to little oscillators and showed that the heat capacity decreases to zero as you approach zero on the absolute temperature scale, but approaches a value of $3R$ if the temperature is sufficiently high. R is the universal gas constant. This constant has the value 8.314 J $\cdot$ mol^{-1} $\cdot$ K^{-1}. Therefore, it produces the predicted molar heat capacity of $24.9 \cdot$ J $\cdot$ mol^{-1} $\cdot$ K^{-1}.[1] (In measuring heat capacities, you are considering only temperature changes, so it makes no difference whether you measure these changes on the Kelvin or Celsius scale; the degree intervals on both scales are the same size.)

---

[1]You have been using a value for R, the gas constant of 0.08206 L $\cdot$ atm $\cdot$ mol^{-1} $\cdot$ K^{-1}. The value used here is the same constant but in different energy units, joules $\cdot$ mol^{-1} $\cdot$ K^{-1}, rather than L $\cdot$ atm which also has the units of energy.

The law of Dulong and Petit became one of the first ways to estimate the molar mass of a metallic element, and we will use it to try to determine the identity of our unknown. If the law gives us the **molar** heat capacity, C_{PM}, as 25 J $\cdot$ mol^{-1} $\cdot$ K^{-1} and we can measure the **gram** heat capacity (also known as the *specific heat*) in the laboratory with its units, J $\cdot$ g^{-1} $\cdot$ K^{-1}, then the molar mass is the ratio of the two. For example, the measured heat capacity of iron is 0.473 J $\cdot$ g^{-1} $\cdot$ K^{-1}, and the law gives us C_{PM} as 25 J $\cdot$ mol^{-1} $\cdot$ K^{-1}, thus

$$\frac{(25 \text{ J} \cdot \text{mol}^{-1} \cdot \text{K}^{-1})}{(0.473 \text{ J} \cdot \text{g}^{-1} \cdot \text{K}^{-1})} = 53 \text{ g/mol} \approx 55.847$$

In its day this measurement was as accurate as most laboratory determinations.

THE HEAT CAPACITY OF METAL SALTS AND NONMETAL COMPOUNDS

When you measure the heat capacity of something like limestone ($CaCO_3$), an organic polymer such as Teflon ($-CF_2-)_x$, or any substance composed of light atoms, you find that the heat capacity per mole of atoms is typically about 60–80 percent that of a mole of heavy metal atoms. (If you multiply the specific heat of any substance [J/g $\cdot$ K] by the molar mass [g/mol] you will get the molar heat capacity [J/mol $\cdot$ K]. Now if you divide this result by the number of atoms in the empirical formula you used for the compound, you will get the heat capacity per mole of atoms.) In general, atoms with smaller masses vibrate with higher frequencies in the solid and have lower heat capacities per mole of atoms, compared to heavy atoms. This result is in general agreement with Einstein's theory of the heat capacity of solids.

HEAT AND ENTHALPY

The amount of heat it takes to increase the temperature of a substance by a given amount depends on whether the heating is done under conditions of constant volume or constant pressure. If volume is constant, all of the energy goes into increasing the internal energy of the substance (increasing the motion of the atoms of the substance). If the volume is allowed to change and the pressure is constant (e.g., equal to the atmospheric pressure), some work will be expended in pushing back the atmosphere, and this work will require a greater amount of heat for a given temperature increase. If we represent the quantity of heat by the symbol q, we can in general say that q_p will be larger than q_V.

The heat added to a system at constant pressure is called the *enthalpy* increase (from the Greek *enthalpein*,

to heat in), and changes in enthalpy are directly related to changes in temperature by the relation

$$\Delta H = q_p = m \, C_p \, \Delta T \qquad (1)$$

where m is the mass of the substance heated (in grams), C_p is the heat capacity at constant pressure (in joules per gram per kelvin), and ΔT is the temperature change. We will consistently define ΔT as the final temperature minus the initial temperature: $\Delta T = T_f - T_i$ (The temperature change ΔT will be negative if T_f is less than T_i. Also note that the temperature *change* in kelvins is equal to the *change* in Celsius degrees.)

MEASURING THE HEAT CAPACITY OF A METAL WITH A STYROFOAM CUP CALORIMETER

In this experiment we will measure the heat capacity of an unknown metal and try to identify it from the measured heat capacity. We will first measure the effective heat capacity of a simple calorimeter containing cool (room temperature) water by adding a known mass of hot water to it and measuring the temperature increase. Then, in a second experiment we will add a known mass of hot metal to the calorimeter under similar conditions.

Employing the principle of the conservation of energy, we will write an equation that states that the sum of the enthalpy change of the calorimeter plus the enthalpy change of the added hot water must be zero. (The change in enthalpy of the calorimeter is equal in magnitude to the change in enthalpy of the added hot water, but they have opposite signs.) The change in enthalpy of the calorimeter will be divided into two parts: the Styrofoam cups with the thermometer that make up the calorimeter, and the cool water originally present in the calorimeter. Thus,

$$\underset{\substack{\text{Styrofoam cups} \\ \text{and thermometer}}}{\Delta H_{cal}} \quad + \quad \underset{\substack{\text{cool water in} \\ \text{calorimeter}}}{\Delta H} \quad + \quad \underset{\substack{\text{added hot} \\ \text{water}}}{\Delta H} \quad = 0 \qquad (2)$$

In view of the definition of the enthalpy change given in Equation (1), we can rewrite Equation (2) as

$$\underset{\substack{\text{Styrofoam cups} \\ \text{and thermometer}}}{(m \, C_p) \, \Delta T} \quad + \quad \underset{\substack{\text{cool water in} \\ \text{calorimeter}}}{m \, C_p \Delta T} \quad + \quad \underset{\substack{\text{added hot} \\ \text{water}}}{m \, C_p \Delta T = 0} \qquad (3)$$

It should be recognized that not all of the Styrofoam cups and thermometers come into contact with and are heated by the added hot water. Therefore, we can determine only the *effective* product (mC_p) for the Styrofoam cups and thermometer, which we will call the *calorimeter constant*, B (joules per kelvin). So we will rewrite Equation (3) as

$$B\,\Delta T_{CW} + m_{CW}\,C_P\,\Delta T_{CW} + m_{HW}\,C_P\,\Delta T_{HW} = 0 \quad (4)$$

where m_{CW} is the mass of cool water originally in the calorimeter, ΔT_{CW} is the temperature increase of the calorimeter and water, m_{HW} is the mass of added hot water, and ΔT_{HW} is the temperature decrease of the added hot water. These four quantities are all measured in the experiment. If we also know the heat capacity of water, C_P, we can solve Equation (4) to obtain the calorimeter constant, B.

A second experiment is then carried out, adding heated metal instead of hot water to the water in the calorimeter. We can write an equation similar to Equation (4) to describe the enthalpy changes in this experiment.

$$B\,\Delta T_{CW} \;+\; m_{CW}\,C_P\,\Delta T_{CW} + m_{HW}\,C_{P,M}\,\Delta T_{HM} = 0 \quad (5)$$

| Styrofoam cups and thermometer | cool water in calorimeter | added hot metal |

In this second experiment we again measure the four quantities (m_{CW}, ΔT_{CW}, m_{HM}, and ΔT_{HM}). We substitute these values and the previously determined value of B into Equation (5) and solve for $C_{P,M}$, the heat capacity of the unknown metal.

If you perform the experiments carefully, the measured value of the heat capacity will enable you to identify the unknown metal from the list of metals shown in Table 14-1.

EXPERIMENTAL PROCEDURE

Special Supplies: Hot plate, large iron ring or safety clamp, and ring stand (or burner, iron ring, metal gauze, safety clamp [or large iron ring], and ring stand), 1000-mL beaker, two 0 to 110 °C thermometers (if available, use a 0 to 50 °C thermometer graduated in tenths of a degree in place of one of the 0 to 110 °C thermometers),[2] fine copper wire to suspend the thermometer, two 12- to 16-oz Styrofoam cups per student, 25 × 200 mm test tubes (two to contain water and one for each unknown-metal heat capacity determination), corks or rubber stoppers to fit test tubes, beaker tongs or pliers with padded jaws, 100-mL graduated cylinder.

Chemicals: Metals (in the form of shot or pellets so that they will pour easily from a test tube); for example, aluminum, magnesium, nickel, zinc, antimony, lead.

! SAFETY PRECAUTIONS: Boiling water can cause painful and serious burns. Even if you are using a hot plate, put a large iron ring or safety clamp around the beaker, and securely fasten the ring or clamp to a ring stand,

***

[2]Moderately priced electronic digital thermometers (cost approximately $30) with a rugged stainless steel temperature sensing probe and 1-meter cable, a range of −50 to +150 °C, and temperature resolution of 0.1 °C, are available from Hanna Instruments Inc., Highland Industrial Park, 584 Park East Drive, P.O. Box 849, Woonsocket, RI 02895, tel (401) 765-7500 (Model Checktemp 1C).

TABLE 14-1

The Heat Capacity of Some Metals

Metal	Heat capacity ($J \cdot g^{-1} \cdot K^{-1}$)
Magnesium	1.04
Aluminum	0.904
Iron	0.473
Nickel	0.444
Copper	0.387
Zinc	0.386
Silver	0.236
Antimony	0.207
Gold	0.129
Lead	0.128

as shown in Figure 14-1A. The safety clamp is even more important if you use a gas burner and ring stand to heat the water. Keep the hot plate or burner and ring stand well back from the edge of the laboratory bench.

A 25 × 200 mm test tube containing 80 g of metal is too heavy to handle with an ordinary test tube holder. Use beaker tongs or ordinary pliers whose jaws have been padded by having large diameter plastic tubing slipped over the jaws.

The thermometer used to measure the temperature in the Styrofoam cups should be suspended by copper wire from a ring stand support. (See Figure 14-1B.) If not suspended, the thermometer may tip over the cups, spoiling the measurement and possibly breaking the thermometer. If the thermometer does not have a suspension ring at the top, wrap several turns of tape around the wire near the top of the thermometer.

1. DETERMINATION OF THE CALORIMETER CONSTANT. Put about 700 mL of water in a 1000-mL beaker, clamp it securely in place on the hot plate or over the burner, and begin heating it to boiling. (See Figure 14-1A.) Put 30 mL of water in each of two 25 × 200 mm test tubes, and place them in the beaker so that they can be heated to the temperature of boiling water.

To save time, also prepare the weighed and stoppered tube(s) containing the unknown metal(s) as described in Part 2, and put them in the beaker of boiling water.

Obtain two 12- or 16-oz Styrofoam cups, put one inside the other so that they are nested, and weigh the nested cups to the nearest 0.1 g on a platform balance. Record the mass of the empty cups. Add 70 mL of cool water to the inner cup, reweigh, and record the mass.

FIGURE 14-1

Apparatus for measuring the heat capacity of a metal. (A) Samples of metal pellets are heated in the test tubes. (B) The heated metal pellets are then poured into the Styrofoam cup calorimeter.

When the water in the 1000-mL beaker has reached the boiling point, wait 15 min longer, then stir the water in the nested cups with a thermometer (0 to 50 °C graduated in tenths of a degree, if available) until a constant temperature is reached. Leave the thermometer in the cup suspended by a copper wire attached to a ring stand support. (The wire suspension will keep the thermometer from tipping over the cups and possibly breaking the thermometer. If your thermometer has no ring at the top, use tape to fasten the wire.) Record the temperature.

Determine the temperature of the boiling water with a 0 to 110 °C thermometer. (**Caution:** Do not place a 0 to 50 °C thermometer in boiling water or it will break.) The bulb of the thermometer should be held 2 to 3 cm above the bottom of the beaker. Record the temperature of the boiling water. Be sure to record all temperature readings to the nearest 0.1 °C.

Using beaker tongs or pliers with padded jaws, grasp the test tube containing the 30 mL of water near the top and quickly but carefully pour the hot water into the Styrofoam cups. Be careful that no hot water on the outside of the test tube gets into the cups. Stir the water steadily, noting the highest temperature attained. Record this temperature.

Remove the thermometer from the cups. Reweigh, and record the mass of the cups and water.

Pour out the water in the cups. Dry the inside of the cups gently with a towel, and carry out a duplicate set of measurements by using the remaining test tube containing 30 mL of water. Then proceed to Part 2.

2. DETERMINATION OF THE HEAT CAPACITY OF AN UNKNOWN METAL.
Weigh a 25×200 mm test tube, fitted with a cork or rubber stopper, to the nearest 0.1 g on a platform balance. (If the platform balance has a flat pan, use a cork ring with V-grooves cut in the top to support the tube. Use the same cork ring in all subsequent weighings of the test tube.) Record the mass. Place about 80 g of a *thoroughly dry* unknown metal in the test tube, stopper it, reweigh it, and record the mass. (If you are to determine the heat capacity of more than one sample of metal, weigh these metals in similar test tubes now, so that all of the samples can be heated adequately before use. Put a label on each test tube so you can identify it.)

Loosely stopper the test tube, and place it in the beaker of boiling water prepared as described in Part 1. (The purpose of the cork or rubber stopper is to prevent condensation of water inside the tube while it is being heated. Any hot water in the metal would cause a serious error in the measurement of heat capacity.) Each test tube must be immersed in boiling water at least 15 min before the metal is transferred to the calorimeter. Determine the

temperature of the boiling water with a 0 to 110 °C thermometer, as you did in Part 1. Record the temperature.

Using a graduated cylinder, measure out exactly 100 ± 1 mL of cool water, and pour it into the nested Styrofoam cups. Stir the water gently with a thermometer (0.1 °C if available) until a constant temperature is reached. Leave the thermometer in the cup, suspended by a wire or string as described in Part 1. Record the temperature.

Using beaker tongs or pliers with padded jaws, grasp the test tube containing the heated metal near the top, remove the cork, and quickly but carefully pour the hot metal into the Styrofoam cups. Be careful that no hot water on the outside of the tube gets into the cups. Stir the water steadily, noting the highest temperature attained. Record this temperature.

Pour out the water, saving the metal as directed by your instructor. (The metal must be *thoroughly dried* before it can be reused.)

If your instructor so directs, repeat the determination with either the same metal or a different one.

From the data of Part 1, calculate the calorimeter constant. Using this value and the data of Part 2, calculate the heat capacity of the metal(s). Compare the heat capacity value(s) you determined with those in Table 14-1 to try to identify the unknown metal(s).

☞ Consider This

Search the Web for sites that have periodic tables. Find one that gives thermodynamic data for the elements. Use the data given for the specific heat (the heat capacity in units of $J \cdot g^{-1} \cdot K^{-1}$) to test the law of Du Long and Petit, i.e., find the molar heat capacity in units of $J \cdot mol^{-1} \cdot K^{-1}$. Does the law hold up? Can you distinguish a metallic element from a nonmetal this way? What about the metalloids?

The Heat Capacity of Metals

NAME	
SECTION	LOCKER
INSTRUCTOR	DATE

DATA AND CALCULATIONS

1. Determination of the Calorimeter Constant

Data	Trial 1	Trial 2
(a) Mass of empty Styrofoam cups	g	g
(b) Mass of cups + 70 mL water	g	g
(c) Mass of cups + 70 mL water + 30 mL hot water	g	g
(d) Initial temperature of water in the calorimeter (cups)	°C	°C
(e) Temperature of the boiling water bath	°C	°C
(f) Final temperature of calorimeter + added hot water	°C	°C

Calculations		
(g) Mass of cool water in cups, m_{CW}	g	g
(h) Mass of added hot water, m_{HW}	g	g
(i) Temperature change of cool water in the calorimeter ($\Delta T_{CW} = T_{lf} - T_{ld}$)	+ K	+ K
(j) Temperature change of added hot water ($\Delta T_{HW} = T_{lf} - T_{le}$)	− K	− K
(k) Calculate the heat *lost* by the hot water $\Delta H_{HW} = m_{HW} C_P \Delta T_{HW}$	− J	− J
(l) Calculate the heat *gained* by the cold water $\Delta H_{CW} = m_{CW} C_P \Delta T_{CW}$	+ J	+ J
(m) Calculate the heat *gained* by the calorimeter $\Delta H_{cal} = -\Delta H_{HW} - \Delta H_{CW}$	+ J	+ J
(n) Calculate the calorimeter constant, B $B = \Delta H_{cal}/\Delta T_{CW}$	J/K	J/K

2. Determination of the Heat Capacity of an Unknown Metal

Data Unknown identifier(s)	Trial 1	Trial 2
(a) Milliliters of water placed in cups	mL	mL
(b) Mass of empty stoppered test tube	g	g
(c) Mass of stoppered test tube + metal	g	g
(d) Initial temperature of water in the calorimeter (cups)	°C	°C
(e) Temperature of the boiling water bath	°C	°C
(f) Final temperature of water + added hot metal	°C	°C

Calculations		
(g) Mass of water in cups, m_{CW} (assume that the density of water is 1.00 g/mL)	g	g
(h) Mass of added hot metal, m_{HM}	g	g
(i) Temperature change of the water in the calorimeter $(\Delta T_{CW} = T_{2f} - T_{2d})$	K	K
(j) Temperature change of the added hot metal $(\Delta T_{HM} = T_{2f} - T_{2e})$	K	K
(k) Calculate the heat *gained* by the cool water $\Delta H_{CW} = m_{CW} \times C_P \times \Delta T_{CW}$	J	J
(l) Calculate the heat *gained* by the calorimeter $\Delta H_{cal} = B \times \Delta T_{CW}$	J	J
(m) Calculate the *total* heat gained $\Delta H_{gained} = \Delta H_{CW} + \Delta H_{cal}$	J	J
(n) Since $\Delta H_{gained} = -\Delta H_{lost}$ calculate the heat *lost* by the hot metal $\Delta H_{HM} = -\Delta H_{gained}$	J	J
(o) Calculate the heat capacity of the metal from $\Delta H_{HM} = m_{HM} \times C_{PM} \times \Delta T_{HM}$	$J \cdot g^{-1} \cdot K^{-1}$	$J \cdot g^{-1} \cdot K^{-1}$

The law of Dulong and Petit states that the heat capacity of a metal in joules per gram per kelvin multiplied by the molar mass of the metal in grams per mole is a constant: $C_P \times M = 25\ J \cdot g^{-1} \cdot K^{-1}$. In step (o) you determined the heat capacity in joules per gram per kelvin. Solve for the molar mass (grams per mole) for each of the metals studied.

——————— g/mol   ——————— g/mol

QUESTIONS

1. If a student used a sample of metal that was wet, several percent of the apparent mass of the metal would be water. Would the resulting measured value of the heat capacity be too high or too low? Explain.

2. A particular metallic element, M, has a heat capacity of $0.36 \text{ J} \cdot \text{g}^{-1} \cdot \text{K}^{-1}$, and it forms an oxide that contains 1.72 grams of M per gram of oxygen.

   (a) Using the law of Dulong and Petit ($C_p \times M = 25 \text{ J} \cdot \text{g}^{-1} \cdot \text{K}^{-1}$), estimate the molar mass of the metal M.

   _____ g/mol

   (b) From the composition of the oxide (1.72 g M/g O) and the molar mass of oxygen (16.0 g/mol), determine the mass of M that combines with each mole of oxygen.

   _____ g M/mol O

   (c) Use your estimate of the atomic mass from (a) and the information from (b) to determine the empirical formula of the oxide (the moles of M that combine with each mole of O).

   _____

   (d) Now that you know the composition of the oxide *and* its formula, what is the accurate value of the atomic mass of the metal M? What is the identity of M?

   _____ g/mol

☞ **Consider This**

Search the Web for sites that have periodic tables. Find one that gives thermodynamic data for the elements. Use the data given for the specific heat (the heat capacity in units of $J \cdot g^{-1} \cdot K^{-1}$) to test the law of Du Long and Petit, i.e., find the molar heat capacity in units of $J \cdot mol^{-1} \cdot K^{-1}$. Does the law hold up? Can you distinguish a metallic element from a nonmetal this way? What about the metalloids?

EXPERIMENT

15

Enthalpy Changes in Chemical Reactions
Hess's Law

PURPOSE

To test Hess's law (which says that the heat of reaction is the same whether the reaction is carried out directly or in several steps) by measuring the enthalpy changes for the same net chemical reaction carried out by two different paths. To make an approximate estimate, by an indirect measurement, of the enthalpy change for dissociation of water into hydronium and hydroxide ions.

PRE-LAB PREPARATION

What accounts for the energy change in a chemical reaction? The simplest explanation is that in a chemical reaction the positively charged nuclei and the negatively charged electrons of the reacting atoms rearrange themselves. It is the energy change associated with the rearrangement (or reconfiguration) of nuclei and electrons that produces the energy change of a chemical reaction.

Thermochemistry deals with the thermal energy changes that accompany chemical reactions. These energy changes usually are called *heats of reaction*. When the reaction is carried out at constant pressure, the heat of reaction is called the *enthalpy* change, ΔH. (The word *enthalpy* comes from a Greek root word *thalpein* meaning "to heat"). A reaction that produces heat is said to be *exothermic*. In an exothermic reaction the enthalpy of the products is less than the enthalpy of the reactants. If we define the change in enthalpy as $\Delta H = H_{products} - H_{reactants}$ we see that ΔH will be negative for an exothermic reaction. Conversely, if energy is absorbed from the surroundings when a reaction proceeds, the enthalpy change for the reaction will be positive, and we say the reaction is endothermic.

Although the enthalpy change for a chemical reaction tells us whether heat will be released or absorbed, this change cannot be used to determine if a reaction can proceed spontaneously. There are examples of spontaneous reactions that have either negative or positive enthalpy changes. For example, both solid NaOH and KNO_3 dissolve spontaneously in water, but with NaOH the solution warms (ΔH is negative) and with KNO_3 the solution cools (ΔH is positive). As you study thermodynamics further, you will be introduced to *entropy* and *free energy* functions, which can be used to determine if a particular reaction is possible.

A chemical reaction may also involve energy changes other than a flow of heat to or from the surroundings. For example, an electrochemical cell can produce *electrical energy* equivalent to the product of the cell voltage multiplied by the quantity of electric charge transferred to an external circuit. A chemical reaction that produces a gas must push back the atmosphere, and this mechanical *work* is another form of energy. (If the reaction is carried out in containers open to the atmosphere, the work of pushing back the atmosphere is included in the enthalpy change, ΔH.)

Enthalpy changes may be classified into more specific categories: (1) The *heat of formation* is the amount of heat involved in the formation of one mole of the substance directly from its constituent elements; (2) the *heat of combustion* is the amount of heat produced when a mole of a combustible substance reacts with excess oxygen; (3) the *heat of solution* of a substance is the thermal energy change that accompanies the dissolving of a substance in a solvent; (4) the *heats of vaporization, fusion, and sublimation* are related to the thermal energy changes that accompany changes in state; (5) the *heat of neutralization* is the enthalpy change associated with the reaction of an acid and a base.

In this experiment you will measure the enthalpy changes for the following three reactions:

177

$$H_2SO_4 \ (10 \ M) + NaOH \ (1 \ M) \rightarrow$$
$$NaHSO_4 \ (0.5 \ M) + H_2O \qquad (1)$$

$$H_2SO_4 \ (10 \ M) + solvent \rightarrow H_2SO_4 \ (1 \ M) \qquad (2)$$

$$H_2SO_4 \ (1 \ M) + NaOH \ (1 \ M) \rightarrow$$
$$NaHSO_4 \ (0.5 \ M) + H_2O \qquad (3)$$

In these reactions the numbers in parentheses specify the molar concentrations of the reactants and products. Water is both the solvent and a reaction product. In order to distinguish these two roles, we write H_2O when we mean a reaction product and write *solvent* when water is used to dissolve the reactants and serve as the reaction medium. You will also see how the measurement of the enthalpy change of reaction (3) leads to an approximate estimate of the enthalpy of dissociation of water into hydronium ions and hydroxide ions.

Reactions (1) and (3) can be regarded as neutralization reactions. Reaction (2) could be called a solution (or dilution) reaction, and we will call the enthalpy change for this process the *heat of solution*. Note also that if we add reaction (2) and reaction (3) we get reaction (1). Thus, reaction (1) and the sum of reactions (2) and (3) represent two different pathways by which we can get from the same initial state to the same final state.

HESS'S LAW

Germain Henri Hess (1802–1850) discovered the principle we now call *Hess's law:* The enthalpy change of a reaction is the same whether the reaction is carried out directly or in a number of steps. This means that the enthalpy changes for chemical reactions are additive, just like the chemical reactions themselves. If

Reaction (1) = Reaction (2) + Reaction (3)

then

$$\Delta H_1 = \Delta H_2 + \Delta H_3 \qquad (4)$$

To state Hess's law another way, if you start in enthalpy state A and proceed to enthalpy state B by two different paths, the sum of the enthalpy changes along the first path will equal the sum of the enthalpy changes by the second path. Figure 15-1 shows these relationships in a diagram.

The data you obtain for the enthalpy changes for reactions (1), (2), and (3) will be used to test the relation shown in Equation (4), which is a mathematical statement of Hess's law. (Reaction [2], the heat of solution of H_2SO_4, was described by Hess in his first thermochemical publication in 1840.)

FIGURE 15-1

Hess's law says that if you go from State A to State B by any path (any sequence of chemical reactions), the net enthalpy change will be the same. For the two paths shown in the figure, this means that $\Delta H_1 = \Delta H_2 + \Delta H_3$.

Enthalpy changes are usually measured in a calorimeter, a simple version of which is shown in Figure 15-2. The purpose of the calorimeter is to isolate thermally the solution under study. All of the heat liberated (or absorbed) by the chemical reaction goes into heating (or cooling) the calorimeter and its contents.

The amount of energy (heat) required to raise the temperature of 1 g of substance by 1 K (or 1 °C) is called the *heat capacity* of that material. (The term *specific heat* is often used for this quantity, the gram *heat capacity*.) The units of heat capacity can be reported in joules per gram per kelvin, $J \cdot g^{-1} \cdot K^{-1}$, or joules per gram per Celsius degree, $J \cdot g^{-1} \cdot °C^{-1}$, since a kelvin and a Celsius degree are the same size but are referenced to different points on the temperature scale. It requires 4.184 joules to increase the temperature of 1 gram water by 1 °C. Heat capacities are generally measured experimentally. Knowledge of the volume and density (the product $V \cdot d$ is the mass) of a substance, its heat capacity, and the temperature change allows us to calculate the enthalpy change of the substance according to the equation

$$\Delta H = V \cdot dC_P\Delta T \qquad (5)$$

Dimensional analysis of Equation (5) shows that it gives the number of joules released or absorbed:

FIGURE 15-2

A simple calorimeter for measuring the heat of a reaction.

Thermometer

Corrugated cardboard

Two 250-mL foam plastic cups, nested together

Beaker to provide extra support

enthalpy change (ΔH, J)

= volume of solution (mL)

$\times$ density of solution (g/mL)

$\times$ heat capacity (J $\cdot$ g^{-1} $\cdot$ K^{-1})

$\times$ temperature change (ΔT, K)

In principle, a correction should be made for the heat absorbed or evolved by the calorimeter itself and for the energy lost or gained from the surroundings. In the Styrofoam cup calorimeter, portions of the cups and the thermometer are in contact with the solutions contained in the calorimeter. Through this contact, energy can leak into or out of the solution in the calorimeter, depending on whether the temperature of the solution is lower than or higher than the temperature of the surroundings.

Because the energy losses or gains are small (typically about 3 percent) and are about the same order of magnitude as the uncertainty in the measurement of the temperature increase, we will omit these corrections so that you can focus your attention on the main event—the enthalpy changes of the reactant solutions that are mixed in the calorimeter. So, in effect, we will treat the Styrofoam cups and thermometer as if they were perfect insulators, allowing no energy flow into or out of the solutions contained in the calorimeter.

MEASURING HEATS OF REACTION WITH A STYROFOAM CUP CALORIMETER. We will carry out three experiments (each in duplicate) to measure the enthalpy changes for reactions (1), (2), and (3). In these experiments we mix aqueous solutions of the reactants, so water is the reaction medium, and any temperature changes result from the chemical reactions that take place in solution.

Employing the principle of the conservation of energy (which is a statement equivalent to the first law of thermodynamics), we will write an equation that states that the sum of the enthalpy changes of the solutions mixed in the calorimeter and the enthalpy change of the chemical reaction must be zero. Thus

$$\Delta H \quad + \quad \Delta H \quad + \quad \Delta H \quad = 0 \quad (6)$$

Styrofoam cups and thermometer	cool water in the calorimeter	added warm water

Substituting in Equation (6) the definition of the enthalpy change given in Equation (5), we can rewrite Equation (6) as

$$V dC_p \Delta T \quad + \quad V dC_p \Delta T \quad + \quad V dC_p \Delta T = 0 \quad (7)$$

Styrofoam cups and thermometer	cool water in the calorimeter	added warm water

In these experiments we measure V_C (the volume of reagent in the calorimeter), ΔT_C (the temperature change of the solution in the calorimeter), V_R (the volume of added reactant solution), and ΔT_R (the temperature change of the added reactant solution). We will take the product of the density and heat capacity of the solutions, dC_p, to be 4.10 J $\cdot$ mL^{-1} $\cdot$ K^{-1}, nearly the same as the

[1]The density of the solutions is about 4 to 6 percent greater than that of pure water, but the heat capacity is about 4 to 8 percent less, so the product of density $\times$ heat capacity (dC_p) is nearly the same as the product for pure water (within 2 percent).

product for pure water.[1] Substituting these values into Equation (7), we can solve for the enthalpy change for the chemical reaction. Knowing the number of moles of reactants mixed together, we can express the enthalpy change per mole of reaction, $\Delta H_{reaction}$ in kJ/mol.

EXPERIMENTAL PROCEDURE

Special Supplies: Thermometer (1 °C divisions are suitable; if available, 0.1 °C divisions will give better accuracy; relatively inexpensive electronic thermometers with a temperature resolution of 0.1 °C are available),[2] four 7-oz Styrofoam cups per student for use as calorimeters (available from a supermarket), corrugated cardboard for calorimeter covers, polyethylene transfer pipets.

Chemicals: 10.0 M H_2SO_4, 1.00 M H_2SO_4, 1.00 M NaOH, $NaHCO_3(s)$.

NOTE TO INSTRUCTOR: Prepare the solutions in advance and store them in the laboratory so they will be at the lab's ambient temperature.

! SAFETY PRECAUTIONS: Handle sulfuric acid (H_2SO_4) and sodium hydroxide (NaOH) with care. They are corrosive and can cause skin and eye burns.

WASTE COLLECTION: The solutions produced by each of the reactions are first neutralized with 5–10 g of sodium bicarbonate ($NaHCO_3$), then discarded down the drain.

PREPARATION OF THE CALORIMETERS. Prepare two calorimeters, each like that shown in Figure 15-2. (Two nested cups are used to provide greater insulation.) Using a pen, label the calorimeters as calorimeters 1 and 2.

In each experiment you will be measuring initial temperatures of two solutions and the final temperature of the mixed solutions. You must be careful to use the same thermometer to measure the initial and final temperatures in an experiment because inexpensive lab thermometers often give readings that do not perfectly agree. If you are using a thermometer with 1 °C intervals, estimate the temperature reading to the nearest $\frac{1}{5}$ of a degree (± 0.2 °C).

1. THE HEAT OF SOLUTION AND NEUTRALIZATION OF 10 M H_2SO_4 AND 1.00 M NaOH.

Place 50.0 mL of 1.00 M NaOH and 45.0 mL of deionized or distilled water in calorimeter 1. Measure out exactly 5.00 mL of 10.0 M H_2SO_4 in a 5- or 10-mL graduated cylinder. Measure the temperature of the 10.0 M H_2SO_4, and record the reading. Remove the thermometer and readjust the volume to exactly 5.00 mL, if necessary, using a polyethylene transfer pipet (or a medicine dropper). Then rinse and dry the thermometer, and insert it into calorimeter 1, which contains the NaOH solution. Stir the solution gently; read and record the temperature. Then add the 5.00 mL of 10.0 M H_2SO_4. Stir for 30 seconds (or until the reading is steady or only slowly decreasing), and record the temperature (nearest 0.2 °C if you are using a thermometer with 1 °C intervals). Neutralize the solution with 5 g of sodium bicarbonate ($NaHCO_3$), and discard the solution down the drain. Rinse and dry the cups, and carry out a duplicate set of measurements. Then rinse and dry the cups in preparation for Part 2.

2. THE HEAT OF SOLUTION OF 10 M H_2SO_4.

Measure 90.0 mL of deionized or distilled water in a graduated cylinder, and put it into calorimeter 1. Measure out 10.0 mL of 10.0 M H_2SO_4 in a 10-mL graduated cylinder. Measure and record the temperature of the 10.0 M H_2SO_4. Remove the thermometer, and readjust the volume to exactly 10.0 mL, if necessary. Then rinse and dry the thermometer and place it in calorimeter 1 (which contains the water) and gently stir. Read and record the initial temperature of the water as soon as the reading is steady. Then pour the 10.0 M H_2SO_4 into calorimeter 1. Continue to stir for 30 sec (or until the reading is steady), then read and record the final temperature of the solution. Neutralize the solution with 10 g of sodium bicarbonate ($NaHCO_3$), and discard the solution down the drain. Rinse and dry the calorimeter cups, and carry out a set of duplicate measurements. Rinse and dry the cups to prepare for Part 3.

3. THE HEAT OF NEUTRALIZATION OF 1.00 M H_2SO_4 AND 1.00 M NaOH.

Place 50.0 mL of 1.00 M NaOH in calorimeter 1 and 50.0 mL of 1.00 M H_2SO_4 in calorimeter 2. Put the thermometer in calorimeter 1, and stir gently until the temperature reading is steady; then read and record the initial temperature of the NaOH solution. Remove the thermometer, rinse and dry it, and put it in calorimeter 2. Stir the solution gently, and when the temperature reading is steady, read and record the temperature of the sulfuric acid (H_2SO_4) solution. Then pour the H_2SO_4 solution in calorimeter 2 quickly and completely into calorimeter 1. Stir, and record the final temperature after stirring for 30 seconds or until the reading is steady. Neutralize the solution with 5 g of sodium bicarbonate, and discard the solution down the drain. Rinse and dry the cups. Carry out a duplicate set of measurements.

---

[2]Moderately priced electronic digital thermometers (cost approximately $30) with a rugged stainless steel temperature sensing probe and 1-meter cable, a range of −50 to +150 °C, and temperature resolution of 0.1 °C, are available from Hanna Instruments Inc., Highland Industrial Park, 584 Park East Drive, P.O. Box 849, Woonsocket, RI 02895, tel (401) 765-7500 (Model Checktemp 1C).

CALCULATIONS. For each of the reactions in parts 1, 2, and 3, determine the temperature change for the solution in calorimeter 1 (ΔT_C) and the added reactant solution (ΔT_R) by subtracting the initial temperature from the final temperature ($\Delta T = T_{final} - T_{initial}$). Calculate the heats of reaction for parts 1, 2, and 3 by rearranging Equation (7) to solve for $\Delta H_{reaction}$. Assume that the product of the *density × heat capacity (dC_P)* for each solution is 4.10 J • mL^{-1} • K^{-1} (see footnote 1). Then calculate the enthalpy of reaction per mole for each reaction by dividing each $\Delta H_{reaction}$ by the number of moles of H_2SO_4 added in the reaction. (The number of moles is calculated from the volume and molarity of the added H_2SO_4 [mol $= V_R$ (Liter) × M_R (mol/Liter)]. Express the final calculated enthalpy changes in units of kJ/mol. Because most of the ΔT values are known only to two significant figures, the final average calculated enthalpy changes should be rounded to two significant figures.

Consider This

The enthalpies of several common reactions or physical processes can be measured quite accurately with the Styrofoam cup calorimeter. Try measuring:

(a) *Mg(s), Zn(s) or Al(s) + HCl(aq)*

(b) *dissolving NaCl(s) in water*

(c) *dissolving KNO$_3$(s) in water*

(d) *melting ice*

(e) *condensing steam (as in making a latte)*

(f) *vinegar + baking soda*

Compare your results to those found from Hess's law and tabulated enthalpies of formation.

Does stirring the water in your calorimeter change the temperature? Try to duplicate Joule's original experiment to determine the mechanical equivalent of heat.

REPORT	Enthalpy Changes in Chemical Reactions	NAME	
15	Hess's Law	SECTION	LOCKER
		INSTRUCTOR	DATE

DATA AND CALCULATIONS

1. The Heat of Solution and Neutralization of 10 M H_2SO_4 and 1.00 M NaOH

Data	Trial 1	Trial 2
Initial temperature of 10 M H_2SO_4 (5.0 mL in a graduated cylinder)	°C	°C
Initial temperature of calorimeter 1 (50.0 mL of 1.00 M NaOH + 45.0 mL water)	°C	°C
Final temperature of the mixed solutions in calorimeter 1 (50 mL 1.00 M NaOH + 5.0 mL 10 M H_2SO_4)	°C	°C

Calculations	Trial 1	Trial 2
(a) Temperature change of NaOH solution in calorimeter 1 $(\Delta T_C = T_{final} - T_{initial})$	°C	°C
(b) Temperature change of added H_2SO_4 solution $(\Delta T_R = T_{final} - T_{initial})$	°C	°C
(c) Use Equation (5) to calculate the heat gained by the NaOH solution. Assume that the product of *density × heat capacity* (dC_P) is 4.10 J $\cdot$ mL^{-1} $\cdot$ K^{-1}. $\Delta H_C = V_C dC_P \Delta T_C$	J	J
(d) Use Equation (5) to calculate the heat gained by the H_2SO_4 solution. Assume that the product of *density x heat capacity* (dC_P) is 4.10 J $\cdot$ mL^{-1} $\cdot$ K^{-1}. $\Delta H_R = V_R dC_P \Delta T_R$	J	J
(e) Calculate $\Delta H_{reaction}$ by using the principle of conservation of energy expressed by Equation (6), rearranged to solve for the enthalpy of reaction. $\Delta H_{reaction} = -\Delta H_C - \Delta H_R$	J	J
(f) Calculate the moles of added H_2SO_4 mol $H_2SO_4 = V_R(L) \times M$ (mol/L)	mol	mol
(g) Calculate $\Delta H_{reaction}$ per mole of reactants by dividing the results of Part 1(e) by the number of moles of H_2SO_4 calculated in Part 1(f). Express the final result in units of kJ/mol. Be sure to keep the proper sign of the calculated reaction enthalpy.	kJ/mol	kJ/mol
Average		kJ/mol

2. The Heat of Solution of 10 M H_2SO_4

Data	Trial 1	Trial 2
Initial temperature of 10 M H_2SO_4 (10 mL in a graduated cylinder)	°C	°C
Initial temperature of water in calorimeter 1 (90 mL deionized water)	°C	°C
Final temperature of calorimeter 1 (90 mL water + added 10 mL H_2SO_4)	°C	°C

Calculations	Trial 1	Trial 2
(a) Temperature change of water in calorimeter 1 ($\Delta T_C = T_{final} - T_{initial}$)	°C	°C
(b) Temperature change of added H_2SO_4 solution ($\Delta T_R = T_{final} - T_{initial}$)	°C	°C
(c) Use Equation (5) to calculate the heat gained by the NaOH solution. Use the same assumptions as in Part 1. $\Delta H_C = V_C d C_P \Delta T_C$	J	J
(d) Calculate the heat gained by the H_2SO_4 solution. $\Delta H_R = V_R d C_P \Delta T_R$	J	J
(e) Calculate $\Delta H_{reaction}$ by using the principle of conservation of energy expressed by Equation (6), rearranged to solve for the enthalpy of reaction. $\Delta H_{reaction} = -\Delta H_C - \Delta H_R$	J	J
(f) Calculate the moles of added H_2SO_4. mol $H_2SO_4 = V_R(L) \times M$ (mol/L)	mol	mol
(g) Calculate $\Delta H_{reaction}$ per mole of reactants by dividing the results of Part 2(e) by the number of moles of H_2SO_4 calculated in Part 2(f). Express the results in units of kJ/mol. Be sure to keep the proper sign of the reaction enthalpy.	kJ/mol	kJ/mol
Average		kJ/mol

3. The Heat of Neutralization of 1.00 M H_2SO_4 and 1.00 M NaOH

Data	Trial 1	Trial 2
Initial temperature of NaOH solution in calorimeter 1 (50 mL of 1.00 M NaOH)	°C	°C
Initial temperature of H_2SO_4 solution in calorimeter 2 (50 mL of 1.00 M H_2SO_4)	°C	°C
Final temperature of mixture in calorimeter 1 (50 mL 1.00 M NaOH = 50 mL 1.00 M H_2SO_4)	°C	°C

Calculations	Trial 1	Trial 2
(a) Temperature change of NaOH solution in calorimeter 1 $(\Delta T_C = T_{final} - T_{initial})$	°C	°C
(b) Temperature change of added H_2SO_4 solution. $(\Delta T_R = T_{final} - T_{initial})$	°C	°C
(c) Use Equation (5) to calculate the heat gained by the NaOH solution. Use the same assumptions as in Part 2. $\Delta H_C = V_C dC_P \Delta T_C$	J	J
(d) Calculate the heat gained by the H_2SO_4 solution. $\Delta H_R = V_R dC_P \Delta T_R$	J	J
(e) Calculate $\Delta H_{reaction}$ by using the principle of conservation of energy expressed by Equation (6), rearranged to solve for the enthalpy of reaction. $\Delta H_{reaction} = -\Delta H_C - \Delta H_R$	J	J
(f) Calculate the moles of added H_2SO_4. mol $H_2SO_4 = V_R(L) \times M$ (mol/L)	mol	mol
(g) Calculate $\Delta H_{reaction}$ per mole of reactants by dividing the results of Part 3(e) by the number of moles of H_2SO_4 calculated in Part 3(f). Express the results in units of kJ/mol. (Be sure to keep the proper sign of the reaction enthalpy.)	kJ/mol	kJ/mol
Average		kJ/mol

SUMMARY OF RESULTS

Write the chemical equations corresponding to the reactions of parts 1, 2, and 3, and summarize the calculated heats of reaction, $\Delta H_{reaction}$, in units of kJ/mol.

	Chemical equations	Average $\Delta H_{reaction}$(kJ/mol)
Part 1		
Part 2		
Part 3		

Add the chemical reactions for parts 2 and 3. Do they give the same net chemical reaction as the reaction for Part 1? Compare the average $\Delta H_{reaction}$ for Part 1 with the sum of the results for parts 2 and 3 (pay attention to the number of significant figures).

$\Delta H_{reaction}$ for Part 1: _____ kJ/mol

Sum of the $\Delta H_{reaction}$ for parts 2 and 3: _____ kJ/mol

Difference: _____ kJ/mol

Calculate the percentage relative error for the difference between the two pathways:

$$\text{percentage relative error} = \frac{\text{difference}}{\Delta H_{reaction} \text{ for Part 2}} \times 100\% = \underline{\qquad} \%$$

Are your results in agreement with Hess's law? Explain.

(See Question 1 that follows for a discussion in the uncertainty in the measured values of the reaction enthalpies.)

QUESTIONS

1. To get some idea of the uncertainty in your measured reaction enthalpies, repeat the calculation of the ΔH_c (in joules) for Part 1(c), Trial 1, assuming that ΔT_c is 0.4 °C larger than the original one you measured. (We are assuming that the worst case uncertainty in ΔT_c would be 0.4 °C, because ΔT is the result of subtracting two temperature measurements, each having an uncertainty of 0.2 °C when measured with a thermometer with 1 °C intervals.) We are using only the ΔH_c term because it is the largest and most important term. The uncertainty in this term is essentially the same as the overall uncertainty in $\Delta H_{reaction}$ calculated in Part 1(g).

Calculated values of ΔH_C (joules):

original value calculated for ΔT_C new value calculated for $\Delta T_C + 0.4\,°C$

_____ J   _____ J

Now compare the new value to the old (first) value by calculating the relative percentage uncertainty in ΔH_C:

$$\text{Relative percentage uncertainty} = \frac{\Delta H_C(\text{new}) - \Delta H_C(\text{original})}{\Delta H_C(\text{original})} \times 100 =$$

Typically this relative percentage uncertainty is about 3 percent. Also keep in mind that our neglect of the heat losses to the calorimeter biases our results to values that are about 3 percent too low.

2. **The enthalpy of dissociation of water.** The H_2SO_4 and NaOH reactants and the $NaHSO_4$ product in Part 3 actually exist as dissociated ions, so the equation

$$H_3O + HSO_4^- + Na^+ + OH^- \rightarrow Na^+ + HSO_4^- + 2H_2O$$

better represents the reaction (the *total* ionic equation).

Write the equation that represents the net ionic equation by canceling out the spectator ions (those that appear on both sides of the total ionic equation)

Net ionic equation: _____

This net ionic equation represents the reaction of a strong acid and a strong base in water.

Based on this result, and the measured $\Delta H_{reaction}$/mol from Part 3, what is the approximate enthalpy change for the dissociation of water into H_3O^+ and OH^- ions? (Hint: how is the net ionic reaction written above related to the reaction written below?)

$$2\,H_2O \rightarrow H_3O^+(aq) + OH^-(aq)$$ Estimated $\Delta H_{reaction}$: _____ kJ/mol

(*Note:* This is only an approximate estimate because HSO_4^- ion is nearly undissociated in 1 M H_2SO_4 but is about 14 percent dissociated in 0.5 M $NaHSO_4^-$ because the enthalpy change for this partial dissociation is included in our measurements. Compare your estimate for the dissociation of water with the literature value of about +57 kJ/mol.)

☞ **Consider This**

The enthalpies of several common reactions or physical processes can be measured quite accurately with the Styrofoam cup calorimeter. Try measuring:

(a) $Mg(s)$, $Zn(s)$ or $Al(s)$ + $HCl(aq)$

(b) dissolving $NaCl(s)$ in water

(c) dissolving $KNO_3(s)$ in water

(d) melting ice

(e) condensing steam (as in making a latte)

(f) vinegar + baking soda

Compare your results to those found from Hess's law and tabulated enthalpies of formation.

Does stirring the water in your calorimeter change the temperature? Try to duplicate Joule's original experiment to determine the mechanical equivalent of heat.

The Enthalpy of Combustion of a Vegetable Oil

PURPOSE

To explore the idea of food as fuel. To construct a simple calorimeter for measuring the energy produced by burning vegetable oil.

PRE-LAB PREPARATION

From a strictly material viewpoint, human beings are biochemical engines that require both a fuel and an oxidant. The oxidant is oxygen in the air. The fuel is the food we eat, which consists of carbohydrates (starches and sugars), fats, and proteins, plus traces of minerals, vitamins, and other substances we need for good nutrition. The big three (carbohydrates, fats, and proteins) all contain carbon and hydrogen, and most of the carbon and hydrogen we eat in our food is metabolized to carbon dioxide and water. This is equivalent to burning the food and suggests the possibility that we could measure the fuel equivalent of the food we eat by burning it in a calorimeter and measuring the energy produced as heat.

ENERGY NEEDS OF THE BODY

What are the energy requirements of our bodies, and how do our bodies use the energy contained in the food we eat? Careful studies have shown that food energy requirements depend on factors such as age, weight, physical activity, and gender (males require about 10 percent more energy per kilogram of body weight than females).

A surprisingly large fraction of the food we eat is required to sustain basic cell functions necessary for the maintenance of life, such as the ongoing metabolic activities of each cell, the circulation of the blood by the beating of the heart, the rhythmic contraction of the lungs in breathing, and the maintenance of body temperature. These minimum energy needs, called the *basal metabolic rate* (BMR),

must be met before any energy can be used for physical activity or for the digestion of food. As a rough rule of thumb, a college-age male requires about 1.0 kcal per kilogram of body weight per hour and a college-age female requires 0.9 kcal per kilogram per hour. (Problem: Verify that a young 140-lb woman would require 1375 kcal per day for her BMR. Assume that 1 kg = 2.2 lb. *Also note that the nutritional calorie used to express the caloric value of food portions is actually 1000 calories, or 1 kilocalorie.*)

The second component of the total energy requirement is the energy required for physical activity that involves voluntary use of the skeletal muscles. The amount of energy required depends on how many muscles are involved, how much weight is being moved, and how long the activity lasts. Contrary to popular belief, mental effort requires very little extra energy, although it may make you feel tired. The energy requirements for a sedentary (mostly sitting) activity like that of a typist amount to about 20 percent of the BMR. Light activity (such as that of a teacher or student) requires about 30 percent of the BMR. Moderate to heavy activity, such as that of a nurse or athlete, may require energy amounting to 40 to 50 percent of the BMR.

The third component of the energy requirement has to do with digesting our food. When food is eaten, many cells that have been dormant begin to be active, for example, the cells that manufacture and secrete digestive juices and the muscle cells that move food through the intestinal tract. This requirement is about 10 percent of the total kilocalories used for BMR and physical activity.

Now let's go back and add up the daily energy requirements of our hypothetical 140-lb young woman. Her BMR requirement is about 1375 kcal/day. Assuming light activity, we will add 30 percent of the BMR for physical activity, or 412 kcal. Now add 10 percent of the sum (1375 + 412 kcal = 1787 kcal), or 179 kcal, for digestion of food to get a grand total of 1966 kcal/day as her approximate requirement. Of course, these figures are based on several estimates, so her needs might fall within ± 100 kcal/day of the estimated value.

Note that 70 percent (1375 out of 1966 kcal) of her daily food requirement is required just to sustain her BMR. If she increases her physical activity but does not increase her food consumption, she will lose weight. A deficit of 125 kcal/day brings about the loss of body fat at the safe rate of about a pound per month. (A pound of body fat is not pure fat but a mixture of fat, water, and protein, and yields about 3500 kcal when burned. A pound of pure fat would yield about 4040 kcal when burned.)

COMBUSTION CALORIMETRY

The caloric content of food is determined by burning it in a closed metal calorimeter pressurized with oxygen and completely surrounded by water. The calorimeter can be calibrated by burning a substance whose heat of combustion has been accurately determined. The method we will employ to measure the heat (enthalpy) of combustion of a vegetable oil is similar in principle, but the calorimeter we will use is a crude version of the combustion calorimeter used for accurate scientific work.

In this experiment, we will use a lamp that burns oil. The heat from the burning oil will be used to heat water, and the temperature increase of the water will be a measure of the energy produced in the combustion. The enthalpy change for a sample of water that is heated from temperature T_1 to temperature T_2 is given by

$$\Delta H = mC_p \, \Delta T \qquad (1)$$

where ΔH is the enthalpy change in calories (in this example heat absorbed by water), m is the mass of water in grams, C_p is the heat capacity (at constant pressure) in $cal \cdot g^{-1} \cdot K^{-1}$, and ΔT is the temperature change ($\Delta T = T_2 - T_1$) in Kelvin (or Celsius) degrees.

When the oil burns, not all of the energy released finds its way into the water heated in the calorimeter, so it is necessary to determine the *efficiency* of the calorimeter (the fraction of energy that is captured) by burning a substance whose heat of combustion is known. We will use 1-decanol, $CH_3(CH_2)_9OH$, whose enthalpy of combustion has been determined to be 9.9 kcal/g, to calibrate the calorimeter. To do this, we will burn a known mass of 1-decanol under the same conditions used to burn a known mass of vegetable oil. The measured enthalpy change for the vegetable oil will then be divided by the efficiency to get a corrected value that will approximate the true value for the enthalpy of combustion of vegetable oil.

$$\Delta H_{combustion} = \frac{\Delta H_{measured}}{efficiency} \qquad (2)$$

EXPERIMENTAL PROCEDURE

Special Supplies: 12-oz aluminum beverage can (cut the top out with a can opener), three-finger clamp with ring stand, alcohol lamps (filled with a vegetable oil), 0 to 110 °C thermometer, 100-mL graduated cylinder, steel wool, 18-inch-wide heavy duty aluminum foil, size 33 rubber bands, scissors.

Candles can be substituted for the 1-decanol lamps, cutting the number of lamps required in half (see Notes to the Instructor). Votive candles or "tea" candles in metal cups (approximately 4 cm in diameter and 2–5 cm high) are the most suitable.

Chemicals: 1-decanol, vegetable oil (canola, cottonseed, safflower, sunflower, peanut, etc.).

NOTES TO THE INSTRUCTOR: Alcohol lamps with 4 oz (120 mL) capacity and larger-than-ordinary wicks are available from VWR Scientific (1-800-932-5000) as Catalog No. 17805-005. An analytical balance with 300-g capacity is required for weighing the lamps when they are filled.

Put a permanent label on each lamp, specifying the liquid that the lamp contains. If half of the lamps are filled with 1-decanol and the other half filled with vegetable oil, half of the class can do Part 1 (burning 1-decanol) while the other half is doing Part 2 (burning vegetable oil). Then students can exchange lamps to complete the experiment. If kept capped when not in use, lamps can be stored and reused without emptying and refilling.

The number of lamps can be reduced by half by using votive candles or "tea" candles in metal cups to measure calorimeter efficiency in place of lamps containing 1-decanol. If candles are used, take the enthalpy of combustion of the candles to be 10.0 kcal/g.

Increasing the energy capture efficiency. If the room is drafty, the flame may wander, allowing a significant amount of energy to escape. The efficiency of energy capture can be improved by doing two things: (1) Wrap a 5 × 9 inch (13 × 23 cm) sheet of heavy duty aluminum foil around the can to make a skirt that extends about 3 cm below the bottom of the can to capture a greater fraction of energy. The skirt can be held in place with a rubber band. (2) Make a draft shield by folding an 18 × 18 inch (45 × 45 cm) piece of heavy duty aluminum foil in half twice to make a 4-layer thick piece of foil 4.5 × 18 inches. Form the foil into a 4.5 inch high cylinder by wrapping it around a liter bottle. Overlap the ends, holding them in place with a paper clip. At the bottom of the cylinder, cut three slots about 2-cm wide and 2-cm high, folding the 2 cm × 2 cm flaps up to make apertures through which air can enter. Place the draft shield around the lamp, centering the aluminum can over the lamp. When the lamp is lit, the bottom edge of the skirt wrapped around the aluminum can should be positioned just above the tip of the flame. Taking these steps will typically increase the energy capture efficiency to >65 percent and improve the reproducibility of the measurements.

Using a can opener, cut the top out of a 12-oz aluminum beverage can. Mount the can vertically with a three-finger clamp held on a ring stand, shown in Figure 16-1. (To improve energy capture efficiency, make an aluminum foil skirt and draft shield as described in the Notes to the Instructor.)

Parts 1 and 2 can be done in either order. Follow the directions of your instructor. Wipe off any oil on the metal cap surrounding the wick. Check that the wick is flush with the end of the metal tube holding it, and use scissors to trim off any stray threads. (If the wick protrudes from the metal shield, you will get a large, bushy, sooty flame, and combustion will be incomplete.) Light your lamp and

FIGURE 16-1
A simple combustion calorimeter using an aluminum beverage can with the top cut out. An aluminum foil skirt on the can and a draft shield around the lamp (not shown) are recommended to improve the energy capture efficiency. (See Notes to the Instructor.)

see that it burns in a satisfactory manner. If the lamp is difficult to light, or burns with too sooty a flame, try pushing out the wick a bit, trimming it with scissors, then pushing it back in until it is flush with the metal tube.

If you put an aluminum foil skirt on the can, the bottom of the aluminum skirt should be positioned just above the tip of the flame. Blow out the flame, and allow the lamp to cool for a minute or two.

1. CALIBRATION OF THE CALORIMETER. After the lamp has cooled, weigh the 1-decanol lamp (or a candle—see Notes to the Instructor) on the analytical balance, and record the mass to the nearest milligram.

Using your 100-mL graduated cylinder, measure out exactly 100 mL of the water, and pour it into the aluminum can, being very careful not to spill any water on the outside of the can. Stir the water thoroughly with a thermometer; then read the temperature to the nearest 0.2 °C, and record it.

Position the weighed 1-decanol lamp under the can, light the lamp with a match, and carefully center the flame under the bottom of the can. (A draft shield will keep the flame steadier (see Notes to the Instructor). Lift up the draft shield to light the lamp, then quickly lower it into position.)

Stir the water continuously with a thermometer, and when the temperature of the water in the calorimeter has risen about 18–20 °C above room temperature, remove the lamp and blow out the flame. Keep stirring the water in the calorimeter, recording the highest temperature attained to the nearest 0.2 °C.

When the lamp has cooled two or three minutes, reweigh it, and record the mass. You now have the necessary data to calibrate the calorimeter, using the known heat of combustion of 1-decanol (9.9 kcal/g) and the known heat capacity of water.

At the end of the experiment, pour out the water, and look at the bottom of the can. A light coating of soot is normal. If the coat is heavy, the wick should be trimmed, and the measurement repeated. Take care that there is no water on the outside of the can when you start a measurement. Between each measurement, wipe off the excess soot with a paper towel or by scouring it with steel wool. If your instructor so directs, repeat the measurement. With care, the temperature change of the water divided by the mass change of the lamp should agree within 10 percent for the two trials.

2. ENTHALPY OF COMBUSTION OF A VEGETABLE OIL. Record in your report the type of vegetable oil used. Weigh to the nearest milligram, and record the mass of the lamp containing vegetable oil. Using exactly the same procedure as in Part 1, put exactly 100 mL of water in the calorimeter can, and carry out the temperature measurements, heating, and weighing as before. Record the initial and final temperatures of the water; also record the mass of the vegetable oil lamp after it has cooled.

By assuming that the fraction of the heat captured by the calorimeter is the same for the vegetable oil lamp as for the 1-decanol lamp, you now have the necessary data to calculate the heat of combustion of vegetable oil. If your instructor so directs, repeat the measurements to obtain a duplicate set of values. Take care that there is no water on the outside of the can when you start each measurement, and remember to keep the wick trimmed and to wipe off excess soot after each measurement.

When all of the measurements are completed, cap the lamps. Calculate the efficiency of the calorimeter from the data obtained in Part 1 and the enthalpy of combustion of the vegetable oil from the measured efficiency and the data obtained in Part 2.

👉 Consider This

Peanuts, once ignited, burn rather briskly. Burning half a Planter's brand peanut, weighing about 0.5 gram, provides enough energy to heat 100 mL of water about 15 °C. To make a suitable holder for the peanut, wind a conical spiral on the end of a 15-cm length of 20 gauge nichrome wire. (The conical end of a cork borer sharpener makes an ideal form for winding the spiral.)

To determine the heat of combustion of a peanut half, weigh it, place it in the wire holder, ignite it by holding the flame of a match under it until it begins burning, then hold it under the aluminum can containing 100 mL of water. (If you're using a draft shield, you will need to cut a narrow slot in the shield through which you can thrust the wire holder and peanut.) The peanut burns with a large sooty flame and typically leaves behind a blackened bit of ash in the peanut's original shape. (Subtract the mass of the ash to get the net mass of peanut burned.) You can compare the value you get with the nutritional information on the container label.

Suppose that you did not put salad dressing equivalent to one tablespoon of vegetable oil on your salad at dinner. How many fewer kilocalories would you have eaten as fat? Assume that one tablespoon of oil has a volume of 15 mL, a density of 0.92 g/mL, and supplies 8.8 kcal/g of oil when consumed. (Also recall that 1 nutritional calorie is equal to 1 kcal.)

Bicycle riders in the Tour de France (a 2-week, 1500-mile bicycle race) require about 5000 kcal/day to maintain their energy needs, about 2.5 times the energy requirement for a person involved in more normal activities. How many loaves of bread would this be equivalent to, assuming 17 slices per loaf of bread, each slice (40 g) providing about 100 kcal?

The Enthalpy of Combustion of a
Vegetable Oil

NAME

SECTION   LOCKER

INSTRUCTOR   DATE

DATA AND CALCULATIONS

1. Calibration of the Aluminum Can Calorimeter by the Combustion of 1-Decanol

NOTE: If you used a candle in place of a 1-decanol lamp for calibration, assume in step (i) that the enthalpy of combustion of the wax candle is 10.0 kcal/g.

	Trial 1	Trial 2
Data		
(a) Initial mass of lamp (or candle):	_____ g	_____ g
(b) Final mass of lamp (or candle):	_____ g	_____ g
(c) Milliliters of water placed in calorimeter:	_____ mL	_____ mL
(d) Initial temperature of water in calorimeter:	_____ °C	_____ °C
(e) Final temperature of water in calorimeter:	_____ °C	_____ °C
Calculations		
(f) Mass of 1-decanol (or candle) burned:	_____ g	_____ g
(g) Temperature rise of water in calorimeter:	_____ °C	_____ °C
(h) Heat absorbed by water $(\Delta H = mC_p \, \Delta T \times 1 \text{ kcal}/1000 \text{ cal})$ (assume that the density of water is 1.00 g/mL and the C_p for water is 1.00 cal $\cdot$ g^{-1} $\cdot$ K^{-1}):	_____ kcal	_____ kcal
(i) Heat produced by burning $(\Delta H = g_{1\text{-decanol burned}} \times 9.9 \text{ kcal/g})$:	_____ kcal	_____ kcal
(j) Efficiency $= \dfrac{\text{heat absorbed}}{\text{heat produced}}$	_____	_____

(k) Average efficiency = _____

2. Enthalpy of Combustion of a Vegetable Oil

	Trial 1	Trial 2
Data		
Type of vegetable oil used:	_____	_____
(a) Initial mass of vegetable oil lamp:	_____ g	_____ g
(b) Final mass of vegetable oil lamp:	_____ g	_____ g
(c) Milliliters of water placed in calorimeter:	_____ mL	_____ mL
(d) Initial temperature of water in calorimeter:	_____ °C	_____ °C
(e) Final temperature of water in calorimeter:	_____ °C	_____ °C

Calculations   Trial 1   Trial 2

(f) Mass of vegetable oil burned: _____ g   _____ g

(g) Temperature rise of water in calorimeter: _____ °C   _____ °C

(h) Heat absorbed by water

($\Delta H = m C_p \Delta T \times 1$ kcal/1000 cal)
(assume that the density of water is 1.00 g/mL
and the C_p for water is 1.00 cal $\cdot$ g^{-1} $\cdot$ K^{-1}):   _____ kcal   _____ kcal

(i) Calculate the enthalpy of combustion per gram of vegetable oil by using the average efficiency calculated in Part 1.

$$\Delta H_{combustion} = \frac{\text{heat absorbed by water}}{\text{average efficiency}} \times \frac{1}{g_{\text{oil burned}}}$$

                                                   _____ kcal/g   _____ kcal/g

(j) Calculate the percentage relative error, assuming the true value for the enthalpy of combustion of vegetable oil to be 8.84 kcal/g.

$$\text{percentage relative error} = \frac{\Delta H_{comb.} - 8.84 \text{ kcal/g}}{8.84 \text{ kcal/g}} \times 100\%$$

                                                   _____ %   _____ %

☞ Consider This

1. *Suppose that you did not put salad dressing equivalent to one tablespoon of vegetable oil on your salad at dinner. How many fewer nutritional calories would you have eaten as fat? Assume that 1 tablespoon of oil has a volume of 15 mL, a density of 0.92 g/mL, and supplies 8.8 kcal/g of oil when consumed. (Also recall that 1 nutritional calorie is equal to 1 kcal.)*

2. *How many loaves of bread would a bicycle rider in the Tour de France have to eat daily to supply a requirement of 5000 kcal/day? Assume 17 slices of bread/loaf with each slice (40 g) supplying 100 kcal.*

Emission Spectra and the Electronic Structure of Atoms

The Hydrogen Atom Spectrum

PURPOSE

To use a simple spectroscope to determine the wavelengths of the spectral lines emitted by excited hydrogen atoms. To show that these wavelengths fit a pattern of energy states described by a simple formula containing a single constant and integer quantum numbers. To show that atomic spectra can also be used qualitatively to identify elements.

PRE-LAB PREPARATION

The absorption and emission spectra of the elements' gaseous atoms do not look like the continuous rainbow-like spectrum of the heated filament of an incandescent bulb. They show instead many sharp lines separated by regions where there is no emission or absorption. This distinction provides the most convincing evidence that the electrons in atoms do not have a continuous range of energies but only certain discrete values of energy. In reaching this conclusion, studies of the emission spectrum of the simplest atom, hydrogen, have played a crucial role.

The spectroscope, introduced by Bunsen and Kirchhoff about 1859, made it possible to observe the atomic spectra of excited gaseous atoms, and scientists soon discovered that each element had a characteristic spectrum that could be used to identify it (or to detect previously undiscovered elements). These spectral studies produced much information on the spectral lines of the elements, but no real progress was made in explaining the origin of spectra until 1885, when Johann Balmer pointed out that the wavelengths of the lines of the hydrogen atom spectrum were given by a simple empirical formula containing one constant and the squares of small integers. Once this regular pattern was discovered, it became a challenge to human ingenuity to explain these regularities.

In 1913 the Danish physicist Niels Bohr used the notion of the quantum of energy (introduced by Max Planck) and the model of the nuclear atom (published in 1911 by Rutherford) to develop a theory of the behavior of an electron moving around a small, positively charged nucleus. He used this theory to calculate the wavelengths of the spectral lines of the hydrogen atom with remarkable accuracy and showed how the regularities could be explained by assuming that energy was absorbed or emitted only when an electron passed from one energy state to another. Each energy state had an energy given by a constant divided by the square of a small integer, called a quantum number.

Although Bohr's theory was quite successful in interpreting the spectrum of the hydrogen atom, it was not as successful for interpreting the spectra of atoms with more than one electron. This led others, such as Erwin Schrodinger, Werner Heisenberg, and Paul Dirac, to formulate more complete theories, the first of which was published in 1926. Since 1926, the development of these theories has shown that all aspects of atomic and molecular spectra can be explained quantitatively in terms of energy transitions between different allowed quantum states. (A quantum of energy is the smallest amount of energy that can be transferred to or from an atom or molecule. Transferred only in packets of discrete size, called *photons* or *quanta,* the energy is said to be "quantized.")

In this experiment you will use a simple spectroscope to measure the wavelengths of the spectral lines generated by the light energy emitted by several elements when the atoms of those elements have been excited by an electric discharge or a hot flame. A commercial handheld spectroscope (like that shown in Figure 17-6) is constructed with a pre-calibrated scale. Readings from this scale can be converted directly into wavelength readings. If the spectroscope is homemade, the scale of the spectroscope can be calibrated using the known wavelengths of the lines observed in a low-pressure mercury lamp. Once you have a spectroscope with a calibrated scale, you will then measure the wavelengths of hydrogen and some gaseous nonmetal or metallic elements. You will use data for hydrogen to construct quantitatively part of the energy diagram for hydrogen. As you do this experiment, think about the fact that you are following in the footsteps of scientific pioneers whose work ushered in the era of quantum mechanics and earned them Nobel prizes.

WAVES AND DIFFRACTION

Light is electromagnetic radiation. This means that it possesses electric and magnetic properties that vary sinusoidally and in phase with each other, as shown in Figure 17-1.

Because only the electric part of the wave interacts with the electrons in atoms, Figure 17-1 is often simplified to the representation in Figure 17-2. The wavelength of this wave, λ, is defined as the distance between any two repeating portions of the wave. The wavelength of light visible to our eyes is 4×10^{-7} m (blue) to 7.5×10^{-7} m (red). Because these wavelengths are quite small, the units of wavelengths are often given in angstroms (Å), where $1 \text{ Å} = 10^{-10}$ m, or in nanometers (nm), where $1 \text{ nm} = 10^{-9}$ m. Demonstrate for yourself that the limits of the visible spectrum in these units are 4000 Å to 7500 Å, or 400 nm to 750 nm.

Another important property of waves is frequency, ν. The frequency of a wave is the number of wavelengths that pass a given point in a unit of time. The unit of frequency in reciprocal seconds (s^{-1}) is called the Hertz (Hz).

For a given wave, wavelength and frequency are not independent of each other. The higher the frequency, the smaller the wavelength. The reason for this relationship is that all light waves travel at the same velocity. This velocity is the speed of light, c, and is equal to 3.00×10^8 m/s. Some reflection on these three quantities leads to the correct mathematical expression (an analogy with a moving

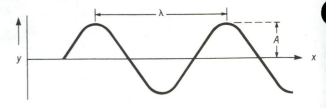

FIGURE 17-2
The amplitude (A) and wavelength (λ) of a light wave.

train in which λ represents the length of one car is helpful):

$$c \text{ (m/s)} = \lambda \text{ (m)} \times \nu \text{ (s}^{-1}) \tag{1}$$

One of the unusual properties of waves is diffraction. All waves display this phenomenon, whether they are ocean waves, light waves, or sound waves. Perhaps you have seen ocean waves striking a small opening or several openings in a breakwater and observed the interesting patterns formed by the waves after they have passed through these openings. One pattern that you might have observed would look like the schematic diagram in Figure 17-3.

These patterns are obtained only when both the openings and the spacing, d, between the openings are of the same magnitude as λ. Thus, in order for diffraction to occur with light waves, the light must pass between slits that are very narrow and very close together. The plastic replica grating used for the homemade spectroscopes was made from a precision-ruled metal grating in which there are 13,800 lines per inch. These plastic gratings are of the transmission type, which means that light does not pass through the grating where the lines have been drawn but rather through the spaces between the lines.

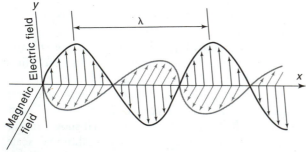

FIGURE 17-1
The electric- and magnetic-field components of a light wave.

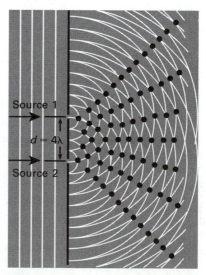

FIGURE 17-3
Diffraction of ocean waves by openings in a barrier.

The quantitative description of the behavior of waves on passing through slits of the appropriate size is the famous Bragg equation:

$$m\lambda = d \sin \theta \qquad (2)$$

The quantity m is a small integer having values $0, 1, 2, 3, \ldots$. The value d is the distance between the lines of the grating. The meaning of the angle θ is displayed in Figure 17-4.

The usefulness of the Bragg equation, as applied to a diffraction grating, is that it shows that the longer the wavelength is, the greater the angle of deviation, θ, will be. In all measurements made in this experiment, you will be observing the first-order, or $m = 1$, diffraction patterns. This means that if light of differing wavelengths strikes the grating, it will be bent to varying angles and in fact separated into its component wavelengths. White light consists of light of all wavelengths. The diffraction pattern of an ordinary tungsten filament lamp therefore is just a lovely rainbow. One of the first observations you should make with your spectroscope is of the light from a fluorescent lamp. In addition to the rainbow pattern observed with the tungsten lamp, you may be able to discern a few discrete lines that are brighter than the general background of the rainbow spectrum. Some of these lines come from the small amount of mercury vapor that is in the fluorescent lamp.

ENERGY LEVELS AND LINE SPECTRA OF ELEMENTS

Our present theory of atomic structure states that electrons in an atom can possess only discrete energy values. As we noted earlier, the word *quantized* is used to describe discrete values. Every element has a characteristic set of energy levels. The energy-level diagram for hydrogen is given in Figure 17-5. The energy levels are characterized by an integer n, which is called the *principal quantum number*. At the temperature of the hydrogen discharge lamp, most of the hydrogen atoms have energy corresponding to the $n = 1$ level, which is called the *ground state*.

If energy is supplied to a hydrogen atom in the $n = 1$ state by an electrical discharge or by heat, a certain fraction of the atoms will absorb this energy and enter the $n = 2$, $n = 3$, or higher levels. These atoms having extra energy are called *excited* atoms. They can lose some of this extra energy in discrete amounts and drop back down to the $n = 3$ or $n = 2$ or ground state, as indicated in Figure 17-5.

The energy that is lost in these transitions appears as a visible quantum of light called a *photon*. It is important to note that all light is quantized in these units, which can be described as both particles and waves, depending on the kind of experiment in which they are observed. Photons possess an energy proportional to the frequency of the light wave. The proportionality constant is Planck's constant, h, which has a value of 6.62×10^{-34} J $\cdot$ s. Thus the electron of a hydrogen atom in an excited energy state can lose energy by emitting a photon that has an energy corresponding to the transition from the n_1 to n_2 energy levels, where n_1 and n_2 represent the quantum numbers of the initial and final energy states and where we have not specified their numerical values.

If a transition takes place from energy level $n_1 \rightarrow n_2$, the absolute magnitude of the energy difference between the n_1 and n_2 levels is related to the wavelength of the photon by the equation

$$|E_{n2} - E_{n1}| = \Delta E = h\nu = \frac{hc}{\lambda} \qquad (3)$$

The last identity is derived from Equation (1), which expresses the relationship between the velocity of light, frequency, and wavelength. The energy, ΔE, can be expressed in joules/photon. This is a very tiny amount of energy. If we multiply this energy in joules/photon by Avogadro's number, we can express the energy in joules (or kJ)/mole of photons. The numbers shown on the energy scale in Figure 17-5 are expressed in kJ/mol.

EXPERIMENTAL PROCEDURE

Special Supplies: Inexpensive spectroscopes like the one shown in Figure 17-6 may be purchased[1] or constructed (Method A), or a bare grating in a darkened room may be used (Method B). The commercial spectroscopes have a calibrated scale that can be directly used to read wavelengths, and

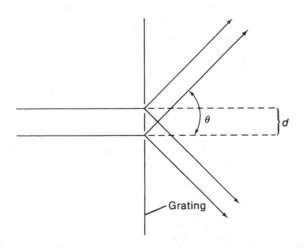

FIGURE 17-4
Diffraction of light by a transmission grating.

---

[1]Science Kit and Boreal Laboratories, Inc., 777 East Park Drive, Tonawanda, NY 14150-6784, 1-800-828-7777. QA Spectroscope, Catalog No. 16525

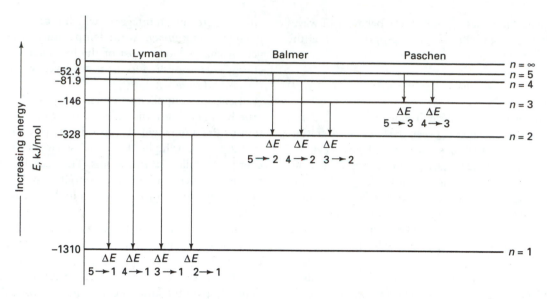

FIGURE 17-5

The energy levels of the hydrogen atom. The numbers on the left are the energy levels expressed in units of kJ/mole. The integers on the right are the principal quantum numbers. The vertical arrows correspond to the energy differences, ΔE, between levels described by the indicated quantum numbers. The energy differences represent observable lines in the hydrogen spectrum. Each observable line corresponds to a spectral transition, that is, the transition from one energy state to another state. Each of the three groups of spectral lines shown here is named for its discoverer.

their higher-quality diffraction gratings give a brighter image. A helium-neon laser (or inexpensive laser pointer) is useful for demonstrating diffraction through a transmission-type plastic replica grating.

Method A: To construct a spectroscope like that shown in Figure 17-7, the following supplies will be needed: a cardboard box with lid (a No. 8 two-piece, set-up mailing box, $6 \times 4.5 \times 2.25$ in., is a convenient size), spray cans of flat black paint (have the laboratory assistant spray the inside of each box with flat black paint a day or two ahead of time; this saves much time and mess), single-edged razor blade, double-edged razor blade, small piece of translucent millimeter graph paper, metric ruler, transmission-type plastic replica grating mounted in a 2×2 in. cardboard slide mount,[2] masking tape.

Method B: The setup shown in Figure 17-8 requires a darkened room in order to see the spectral lines and works best with capillary type gas discharge tubes. A Masonite grating holder and shield, constructed as shown in Figure 17-8, provide a convenient way to observe the spectra. A transmission-type plastic replica grating (see footnote 2) is taped to the grating holder. Black construction paper can be used to make a slit that is taped to the shield (which is painted flat black). A meterstick is used to measure the distance from the slit to the virtual image of the slit on the masonite shield.

Sources: The most effective spectral sources for calibration (mercury spectrum) and examination of the emission spectra of gaseous atoms are inexpensive spectrum tubes excited by a 5000-V, 7-mA current transformer-type power supply. These are available from a number of suppliers.[3] Although more expensive, we recommend a brighter and longer-lived source for viewing the hydrogen atom emission spectrum. It is called a Balmer Gas Discharge Tube,[4] and its ends fit into standard fluorescent tube sockets, making it convenient to mount. An inexpensive power supply for this tube can be made by using a 5000-V, 60-mA neon sign transformer (General Electric Luminous Tube Transformer Model 9T61Y16) with an adjustable autotransformer (Variac) connected to the primary of the transformer. (The Variac is normally operated at 70 V.) Automotive spark-plug wiring is used to make the connections between the secondary winding of the transformer and the spectrum tube. All exposed high-voltage connections should be covered with a heavy coating of silicone sealant.

In order to produce the spectra of different metals, solutions of metal salts can be aspirated into a nebulizer to form a fine mist that is directed into the base of a Bunsen or Fischer burner. Metal salt "pills" or pellets can also be heated on the grid of a Fischer or Meker burner. These can be formed with an inexpensive screw-type press of the kind used to make KBr pellets for taking infrared spectra. (Detailed procedures for constructing and using the flame sources are given in the references cited in the Bibliography.) The chloride salts of lithium,

FIGURE 17-6

A handheld spectroscope for measuring wavelengths (Method A).

[2]Edmund Scientific Co., 101 East Gloucester Pike, Barrington, NJ 08007-1380. Card Mounted Diffraction Viewers

[3]Central Scientific Co., 11222 Melrose Ave., Franklin Park, IL 60131-1364. Catalog No. 87208, Spectrum Tube Power Supply/Holder. A number of different spectrum tubes are available.

[4]Central Scientific Co., 11222 Melrose Ave., Franklin Park, IL 60131-1364. Catalog No. 87206 Balmer Gas Discharge Tube. Catalog No. 87201-000 Gas Discharge Tube Holder

FIGURE 17-7

The construction of a handheld spectroscope for measuring wavelengths by Method A.

FIGURE 17-8

Plan (A) for the construction of a holder for the grating and slit for measuring wavelengths by Method B. The source and power supply are located behind the protective Masonite shield. The slit is a piece of black paper with a 5 mm × 50 mm aperture taped to the Masonite shield. (B) Top view of the setup.

sodium, potassium, calcium, strontium, and copper are recommended. (Some metal vapors, e.g., barium and copper, are toxic, so it is preferable to place the gas burners used to heat the metal salts in a fume hood.)

Tungsten filament bulbs and fluorescent lamps should also be available so that students can compare their spectra (see Part 4 of the Experimental Procedure).

! SAFETY PRECAUTIONS: The power supplies used for the spectrum tubes are sources of high voltage that represent a danger to the unwary user. If the spectrum tube power supply or holder (or both) has exposed electrical connections, a shield like that shown in Figure 17-8 should be placed between the source and the viewer. Cut a rectangular viewing port in the shield so that the source can be viewed through the port.

A low-pressure mercury lamp puts out a large fraction of its energy in the ultraviolet (UV) 254-nm line. This, though not visible, can cause eye damage. This UV light is absorbed strongly by the lens material in eyeglasses and safety goggles. Do not look at the lamp without wearing glasses, and limit the calibration time to 5 to 10 min. For greater safety, a piece of ordinary window glass can be taped over the viewing port on the back of the shield to act as a UV cutoff filter.

In order to prevent the breathing of any metal vapor (barium and copper are toxic), place the gas burners used to heat metal salts in a fume hood.

Your instructor will tell you if you are to use spectroscopes that are already prepared if you are to construct your own (Method A, Figure 17-6 or 17-7), or if you are to use a bare grating in a darkened room (Method B, Figure 17-8).

Demonstrating the diffraction of light using a helium-neon laser. *(Students should be cautioned not to look into the laser beam using their spectroscopes.)* If you have access to an inexpensive helium-neon laser (or laser pointer), the instructor can demonstrate how a monochromatic beam of light is diffracted by a plastic replica transmission grating. Direct the laser beam through the grating (with the laser beam perpendicular to the grating); allow the diffracted beam to fall on a sheet of white paper located about 30 cm away from the grating. You will see the bright undiffracted primary beam (zero order) in the center. Flanking it on either side will be less bright spots of light that are the first-order diffraction; faint spots at twice the angle of the first-order diffraction are the second-order diffracted beams. The instructor may also take this opportunity to explain how the diffracted beam, entering the eye at an angle relative to the direction of the primary beam, appears as if it were coming from a point on the scale of the spectroscope even though no light is falling on the scale. It is an optical illusion. The image viewed is not real and is therefore called a *virtual image*.

1. CONSTRUCTION OF A SPECTROSCOPE.

Method A. If you are to construct a spectroscope like that shown in Figure 17-7, obtain a small cardboard box (6 × 4.5 × 2.25 in.) that has been sprayed on the inside with flat black paint to reduce internal reflections and make the spectra easier to observe. Noting the dimensions given in Figure 17-7, mark the positions of the holes on the outside of the box and cut out the three holes for the *slit*, the *grating*, and the *scale*. Cut the holes through the box and lid simultaneously, using a single-edged razor blade.

Carefully (razor blades are quite sharp) break the double-edged razor blade lengthwise in two. At point A in Figure 17-7, tape the sharp edges facing each other to the outside of the box, with their edges 0.3 to 0.5 mm apart. (No diffraction is obtained from this slit because this opening is many times the wavelength of visible light.) Make sure that the edges of the slit are parallel to each other and perpendicular to the top and bottom of the box.

Observe the grating carefully. You will probably be able to observe some striations on the plastic surface. If you can see these reflections, mount the grating at point B in Figure 17-7, with the lines running perpendicular to the top and bottom of the box. (If you cannot observe these striations, close the box and hold it with the slit pointing toward a fluorescent light. Hold the grating next to the hole at point B and observe the diffraction pattern. The orientation of the grating that produces diffraction patterns on the right and left of the slit is the correct one.) Tape the grating to the inside of the box at point B, with the grating in the correct orientation.

Finally, tape a scale cut from a piece of millimeter graph paper at point C, Figure 17-7. If too much light comes in through the scale for you to see the lines clearly, mask off part of the scale with tape.

Method B. The spectroscope consists of the setup shown in Figure 17-8: a source (capillary spectrum tube), a Masonite shield with a slit taped on the front, a transmission grating taped to a Masonite holder, and a meterstick. The images you see to the left and right of the slit will be virtual, not real, images of the slit (or capillary source). The slit is a 5 × 50 mm rectangular hole cut in the center of a piece of black construction paper that is taped to the front of the Masonite shield. (See Figure 17-8.)

Tape the grating to a Masonite support like that shown in Figure 17-8. (The rectangular holes, 3 × 20 cm, cut in the Masonite grating support and shield are made long in order to allow vertical adjustment of the grating and slit to match the height of the source.) Position the grating about 50 cm from the slit. (The plane of the grating must be parallel to the capillary source or slit and form a light angle with the incident light from the source as shown in Figure 17-9.)

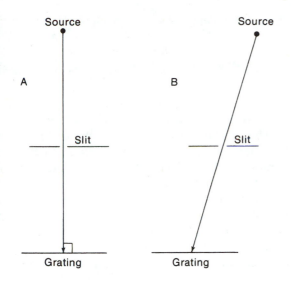

FIGURE 17-9

The angle of the incident light should be 90° with respect to the grating: (A) correct; (B) incorrect.

Make sure you are wearing glasses to protect yourself from UV radiation from the sources. Turn on the mercury source and darken the room. Position your eye directly behind the grating within eyelash distance so that you are looking through the grating toward the source. Look to the right and the left. You should see images of the slit symmetrically displaced on both sides of the slit, corresponding to $m = \pm 1$ in Equation (2). The images on one side may appear slightly brighter because of the construction of the grating. Use the brighter side for all further measurements. If you have a good grating, you may also be able to see the second-order images ($m = \pm 2$) further displaced on both sides of the slit.

For each line you see in the first-order spectrum, record in your report the distance a (distance between the slit and the image of the slit or capillary source on the shield), measured to the nearest millimeter. (See Figure 17-8.) It may be helpful to have a lab partner place a marker such as a pencil at the apparent position of the line. Also measure and record distance b (the distance between the grating and the slit).

2. CALIBRATION OF THE SPECTROSCOPE. Wear glasses at all times while observing lamps through the spectroscope to protect yourself from possible harmful UV radiation. If you are using a commercial spectroscope like that shown in Figure 17-6, you need not construct the calibration graph. Simply verify that the reading for the lines observed in the mercury lamp corresponds to the known wavelengths given in Table 17-1. Each major division on the scale of the spectroscope marked by a numeral corresponds to an increment of

TABLE 17-1

Wavelengths of the Mercury Spectrum Lines Used for Spectroscope Calibration

Color	Wavelength (nm)
Violet	404.7
Blue	435.8
Green	546.1
Yellow	579.0

100 nm, e.g., a reading of 4.2 on the scale corresponds to a wavelength of 420 nm (4200 Å).

If you are making your own spectroscope, you will use the emission lines of mercury as a "standard" to calibrate the spectroscope. Look at the spectrum of a mercury lamp. You should be able to see three or four lines. The colors and wavelengths of these lines are shown in Table 17-1.

Measure the positions of the four most prominent lines on the scale as accurately as possible (nearest millimeter). These four lines appear as two pairs of lines separated by an appreciable gap. The lines of each pair are separated by about 30 nm; the gap between the pairs of lines is about 110 nm. (Some may have difficulty seeing the less bright violet line of the violet and blue pair at 405 and 436 nm.) Record the average of several observations. Do you see any additional lines besides these four? If so, comment on them in the report form, and measure their positions on the scale.

The following precautions and directions will enable you to measure reproducibly the positions of lines with handheld spectroscopes (Method A). (a) The lamp should be far enough away from the spectroscope (at least 25 to 30 cm) so that the direction of propagation of the light is well defined or collimated (Figure 17-10). If you do not observe this precaution, you will find that whenever you move your eye the spectral lines will

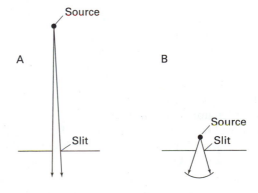

FIGURE 17-10

For the light to be well collimated when a handheld spectroscope is used (Method A), the slit should be at least 25 to 30 cm from the source. (A) well-collimated light; (B) poorly collimated light.

appear to wander rather than remaining stationary on the scale. (b) On the other hand, the lamp should not be too far away (more than 40 cm), or the intensity of light falling on the slit will be too low for you to observe the spectrum clearly. (c) The light must be incident on the grating at 90° if the angle of diffraction is to satisfy the equation given previously (see Figure 17-9).

On the graph paper provided in your report form, draw a calibration graph of wavelength versus scale reading. Do this during your laboratory period. In the event that a good straight line is not obtained, measure the positions of the lines again.

You will use your linear graph to determine the wavelength of unknown lines. First, measure the position of the unknown line on the scale of your spectroscope; then, read from your calibration graph the wavelength that corresponds to the measured scale position.

3. OBSERVATIONS AND MEASUREMENTS OF THE HYDROGEN SPECTRUM.
Measure the scale position of the observable lines of the hydrogen emission spectrum. Normally at least three lines can be seen: violet, green, and red. From your measurements and your calibration graph, determine the wavelengths of these lines.

Look up the accepted literature values of the wavelengths of these lines in the *CRC Handbook of Chemistry and Physics* (see the Bibliography). Calculate your relative error as a percentage.

$$\text{percentage relative error} =$$
$$\frac{\text{your value} - \text{accepted value}}{\text{accepted value}} \times 100\% \qquad (4)$$

Convert your values of wavelengths for the hydrogen atom spectrum to units of kJ/mol. First calculate the energy of one photon in units of J from the relation $\Delta E = hc/\lambda$, where h is Planck's constant (6.626×10^{-34} J $\cdot$ s), c is the speed of light (3.00×10^{-8} m/s), and λ is the wavelength (m). Then multiply the result by Avogadro's number ($N_A = 6.023 \times 10^{23}$) to express the energy in J/mol of photons. Finally, divide by 1000 to express the result in units of kJ/mol. For each wavelength corresponding to a line in the hydrogen atom spectrum, the calculated energy difference in kJ/mol should correspond to a *difference* between two energy states shown in Figure 17-5. For example, the red line of hydrogen corresponds to the difference between -328 kJ/mol and -146 kJ/mol, which is 182 kJ/mol.

Draw that portion of the hydrogen atom energy level diagram that you have measured, using energy units of kJ/mol. Make your figure to scale on the graph paper provided.

Using a form of the Rydberg equation,

$$\Delta E = h\nu = \frac{hc}{\lambda} = R_H \left(\frac{1}{n_2^2} - \frac{1}{n_1^2} \right) \qquad (5)$$

calculate for each of the three lines you measured a value for the Rydberg constant, R_H, in units of kJ/mol of photons. This constant relates the energy change (kJ/mol) of a given transition to the principal quantum numbers for the transition. The values of the quantum numbers (n_1 and n_2) for each of the lines you observed can be obtained from Figure 17-5. Keep in mind that each line you observed corresponds to a transition between two energy states. Each state has an energy and a quantum number shown in Figure 17-5 (n_1 is the quantum number of the initial state, while n_2 is the quantum number of the final state, so $n_1 > n_2$ for the $n_1 \to n_2$ transition).

THE IONIZATION ENERGY OF THE HYDROGEN ATOM
The transition from $n = 1$ to $n = \infty$ describes the ionization of the hydrogen atom, the energy required to completely remove the electron from a hydrogen atom. The energy difference corresponding to this difference is equal to the Rydberg constant. Thus the value of the Rydberg constant in units of kJ/mol can be read directly from Figure 17-5 as 1310 kJ/mol, the difference in energy between the $n = 1$ and $n = \infty$ levels.

☞ Consider This
Obtain other spectrum tubes containing various gaseous elements such as helium, neon, argon, or oxygen. Observe the emission spectra of the gases, and note and record the scale positions of the lines you see. (In the case of some gases such as neon, there may be too many lines to record. In this case describe the wavelength region[s] where most of the lines fall, and give the corresponding predominant color of the lines.) Compare this information to that recorded in the CRC Handbook.

Obtain a variety of metal salts, e.g., sodium, potassium, calcium, copper, lithium, strontium chlorides. Ask your instructor for techniques for obtaining bright flames on a Bunsen or Fischer burner. (See the Instructor's Manual.*) Determine whether you can detect the presence of one or more of these salts in a "unknown" solution.*

Observe the spectrum of the light from a tungsten filament lamp and from a fluorescent lamp by using your spectroscope. How are the spectra similar? Different? How does the fluorescent lamp spectrum resemble that of the mercury lamp in Part 2 of the experiment?

How are the ideas that are explored in this experiment used to identify elements in the sun and stars? How about planets that only reflect light? How does the Doppler Effect (or red-shift) come into play in this setting? How are these observations linked to the Big Bang Theory?

BIBLIOGRAPHY

Bernard, M. "Spectroscopic Cation Analysis Using Metal Salt Pills," *J. Chem. Educ.*, **1980,** *57:* 153.

CRC Handbook of Chemistry and Physics. Published annually by the Chemical Rubber Company, Cleveland.

Edwards, R. K., Brandt, W. W., and Companion, A. A. "A Simple and Inexpensive Student Spectroscope," *J. Chem. Educ.*, **1962,** *39:* 147.

Harris, S. P. "Letter to the Editor," *J. Chem. Educ.*, **1962,** *39:* 319.

Logan, K. R. "Some Experiments in Atomic Structures," *J. Chem. Educ.*, **1974,** *51:* 411.

Smith, G. D., Sanford, C. L., and Jones, B. T. "Continuous Liquid-Sample Introduction for Bunsen Burner Atomic Emission Spectrometry," *J. Chem. Educ.*, **1995,** *72:* 438–440.

(e) Look up the literature value of R_H _____. Compute your percentage error for each determination.

Your wavelength (nm)	Experimental value of R_H	Percentage error

QUESTIONS

1. Using Equation (5), the Rydberg equation, show that the energy of the transition from the lowest (ground) state ($n_1 = 1$) to the highest energy state ($n_2 = \infty$) is given by $\Delta E(kJ/mol) = -R_H$, the Rydberg constant. Therefore, the magnitude of this constant is equal to the ionization energy of one mole of hydrogen atoms, in units of kJ/mol.

2. The images of the lines that you see appear to be light reflected from the scale. Are the images real, that is, does light really fall on the scale at the point where you see the image? (Hint: You may want to consult an elementary physics book to answer this question. Look up the term *virtual image.*)

3. When you observe the spectra of the metal salts, how do you know that the emission comes from the metal atoms and not from the atoms of the anions (chlorine, for example)? Suggest one or more experiments that might provide an answer to this question.

Consider This

Obtain other spectrum tubes containing various gaseous elements such as helium, neon, argon, or oxygen. Observe the emission spectra of the gases, and note and record the scale positions of the lines you see. (In the case of some gases such as neon, there may be too many lines to record. In this case describe the wavelength region(s) where most of the lines fall, and give the corresponding predominant color of the lines.) Compare this information to that recorded in the CRC Handbook.

Obtain a variety of metal salts, e.g., sodium, potassium, calcium, copper, lithium, strontium chlorides. Ask your instructor for techniques for obtaining bright flames on a Bunsen or Fischer burner. (See the Instructor's Manual.*) Determine whether you can detect the presence of one or more of these salts in a "unknown" solution.*

Observe the spectrum of the light from a tungsten filament lamp and from a fluorescent lamp by using your spectroscope. How are the spectra similar? Different? How does the fluorescent lamp spectrum resemble that of the mercury lamp in Part 2 of the experiment?

How are the ideas that are explored in this experiment used to identify elements in the sun and stars? How about planets that only reflect light? How does the Doppler Effect (or red-shift) come into play in this setting? How are these observations linked to the Big Bang Theory?

Ionic and Covalent Bonding

Conductivity of Solutions of Ionic and Covalent Compounds

PURPOSE

To use electrical conductivity as a way of discriminating between substances with ionic and covalent bonding. To compare the electrical conductivity of substances in the pure solid (or liquid) state and in solution. To measure the effects of solvent properties on the dissociation of substances with polar covalent bonding, such as HCl. To measure changes in conductivity that result from the reaction of ionic substances to form products that are weakly ionized or insoluble.

PRE-LAB PREPARATION

IONIC, POLAR COVALENT, AND COVALENT BONDING

When an active metal reacts with an active nonmetallic element, the nonmetal atom becomes negatively charged when one or more electrons transfer from the metal, which in turn becomes positively charged. In general, when atoms act cooperatively in this fashion, each element winds up with a closed shell electron configuration, like the electronic configuration of the noble gases of Group VIII. For the reaction of a sodium atom and a chlorine atom, we can represent the changes in the valence shells of the atoms by electron-dot formulas, for example,

$$Na\cdot \ + \ \cdot \ddot{\underset{\cdot\cdot}{Cl}}: \ \longrightarrow \ Na^+ \ :\ddot{\underset{\cdot\cdot}{Cl}}:^-$$

Such electrically charged atoms are called *ions,* and in the product Na^+Cl^-, the Na^+ ion has the same electron configuration as a neon atom, while the Cl^- ion has the electron configuration of argon. The bonding force between ions, due primarily to the attraction of unlike electric charges, is called an *ionic bond.*

However, when two elements of similar electronegative character react, they form electron pair bonds, mutually shared by both of the atomic nuclei. For example,

$$H\cdot \ + \ \cdot H \ \longrightarrow \ H\!:\!H \quad \text{or} \quad H_2$$

$$2H\cdot \ + \ \cdot \ddot{O}: \ \longrightarrow \ :\ddot{\underset{H}{O}}\!:\!H \quad \text{or} \quad H_2O$$

Such a bond is called a *covalent bond.* In the second example, the water molecule, the oxygen atom is more electronegative than hydrogen, so the O—H bond is slightly polar, or partially ionic, in character. We could describe the water molecule as having *polar covalent* bonding. Note that the two types of bonds found in Na^+Cl^- and molecular H_2 are really two extremes of a continuum— the purely ionic bond at one extreme and the purely covalent bond at the other. In between are the polar covalent or partially ionic bonds.

STRUCTURE AND BOND TYPE

Substances with covalently bonded atoms generally have discrete or separate molecules packed together in the crystal structure of the solid.[1] If soluble in water or other suitable solvent, these substances dissolve to form a solution of electrically neutral molecules, and the resulting solutions do not conduct an electric current. Examples of such *nonelectrolytes* are common table sugar (sucrose, $C_{12}H_{22}O_{11}$) and acetone (CH_3COCH_3).

Salts, as well as *acids* and *bases,* have either ionic bonds or bonds that are quite polar. In solid salts, the ions are

---

[1]Omitted from discussion here are substances called *network solids,* which form covalently bonded giant molecular networks such as silica (SiO_2) and carbon in the form of diamond, as well as metals and intermetallic compounds.

packed together so that ions of positive charge are surrounded by ions of negative charge and vice-versa. As a result, ions form the structural units of the crystal. When melted by heat, vacancies or holes are introduced into the rigid crystal structure, allowing the ions to move about so that the melted salt becomes fluid and a good electrical conductor. When these substances dissolve in water, the ions likewise separate as independently moving particles, and the solutions are electrical conductors. So, in equations throughout this manual, we write the formulas of such substances in solution, or in the molten state, as separate ions. Examples of such *electrolytes* are sodium hydroxide (NaOH), which dissociates into Na^+ and OH^- ions, and potassium sulfate (K_2SO_4), which dissociates into K^+ and SO_4^{2-} ions in a 2:1 ratio.

Strong acids and bases and most salts ionize completely in dilute aqueous solution. (In fact, we may regard this as a definition of a strong acid or base—it is a substance that is completely dissociated into ions in solution.) In solutions of weak or less active acids and bases, a substantial fraction of the dissolved substance is present in molecular form. So, although the total concentration may be high, the concentration of ions is low. Partial ionization (or dissociation) accounts for their lower conductivity. Salts that are only slightly soluble, even though they may be completely dissociated into ions in solution, have low electrical conductivity because of their limited solubility.

THE ROLE OF THE SOLVENT IN DISSOCIATION REACTIONS

Most of the solution reactions we have studied have been in aqueous solution. However, it is important to remember that nonaqueous solvents are often used and that the dissociation of acids, bases, and salts depends very much on the properties of the solvent. For example, the dissociation of an acid involves the transfer of a proton from the acid to the solvent. Therefore, the extent of dissociation of an acid will depend on the intrinsic basicity of the solvent. Although acetic acid is only partially dissociated in water, it is completely dissociated in a more basic solvent, such as liquid ammonia.

$$HC_2H_3O_2 + H_2O \rightleftharpoons H_3O^+ + C_2H_3O_2^-$$
partial reaction
$$HC_2H_3O_2 + NH_3 \longrightarrow NH_4^+ + C_2H_3O_2^-$$
complete reaction

The dielectric constant of the solvent and the solvation energy of ions in a solvent are also important. If two ions of opposite charge are placed in a vacuum, the force between them is given by Coulomb's law,

$$force = \frac{k\,q_1 q_2}{r^2}$$

where k is a constant, q_1 and q_2 are the charges on the ions, and r is the distance between the centers of the ions. When the two ions are immersed in a material medium, the force between them is reduced in inverse proportion to the *dielectric constant,* ϵ, of the medium, as expressed in the equation

$$force = \frac{k\,q_1 q_2}{\epsilon\,r^2}$$

Therefore, the total energy expended in separating the ions will decrease as the dielectric constant increases. That means that a solvent with a large dielectric constant will tend to promote the dissociation of an ionic solute more than will a solvent with a small dielectric constant. The hydrocarbons (with $\epsilon < 5$) do not promote the dissociation of ionic solutes. Water (with $\epsilon \cong 80$) is a good solvent for ionic solutes.

Solvation energy is also an important factor. When an ionic solute dissociates in a solvent, the total energy of the reaction may be thought of as composed primarily of two terms: (1) the energy expended in the separation of the positive and negative ions of the crystal lattice (called the *lattice energy*), and (2) the *solvation energy* liberated by the association of the ions with solvent molecules; solvation energy is the result of the interaction between the electric charges of the ions and the electric dipoles of the solvent molecules. When there is enough solvation energy to offset the energy expended in separating the positive and negative ions, the ionic solute will dissociate in the solvent.[2] A polar solvent, therefore, will tend to promote the dissociation of an ionic solute.

IONIC EQUATIONS

In describing a reaction in which ions are either the reactants or products, we often focus our attention on the essential changes, ignoring "spectator" ions that are not participating in the chemical reaction. This abbreviated description we will call the *net ionic equation.*

For example, the reaction of aqueous NaOH with aqueous HNO_3 could be written in full as

$$Na^+ + OH^- + H_3O^+ + NO_3^- \rightarrow Na^+ + NO_3^- + 2H_2O$$

which we will call the *total ionic equation.* Note that in this reaction neither the Na^+ nor the NO_3^- has reacted; both appear as separate particles on each side of the equation. When these spectator ions are omitted, the

---

[2]A more careful examination of the factors that determine the solubility of a solute must include the effect in changes in the solvent structure that occur when ions are introduced, causing polar solvent molecules to cluster around the ions. The entropy change of the system is a measure of this effect, which can be strong enough that salts such as NaCl will dissolve even though the lattice energy is greater than the solvation energy.

equation that expresses the essential change is the *net ionic equation*:

$$OH^-(aq) + H_3O^+(aq) \rightarrow 2H_2O$$

(Note that the net ionic equation is still a balanced equation with respect to charge and the number of each kind of atom. The spectator ions that are not shown are also still contributing to the electrical conductivity of the solution.)

Typical ionic reactions are those in which ions unite to form a weakly ionized or insoluble substance. For example, if a weakly acidic substance such as ammonium chloride, NH_4Cl, reacts with a base such as sodium hydroxide, $NaOH$, the net change is expressed by

$$NH_4^+(aq) + OH^-(aq) \rightarrow NH_3(aq) + H_2O$$

A typical example of ions uniting to form an insoluble substance is expressed by the equation

$$Ca^{2+}(aq) + CO_3^{2-}(aq) \rightarrow CaCO_3(s)$$

It makes no difference whether the Ca^{2+} is derived from calcium chloride, calcium nitrate, or any other soluble calcium salt. Carbonate ion, CO_3^{2-}, could be obtained equally well from sodium carbonate, ammonium carbonate, or any soluble substantially-ionized carbonate salt.

Often the net result of an ionic reaction is determined by a competition between two largely undissociated substances. For example, carbonic acid, H_2CO_3, is formed when a strong acid acts on a carbonate salt:

$$CaCO_3(s) + 2H_3O^+ \rightarrow Ca^{2+} + H_2CO_3(aq) + 2H_2O(l)$$

Carbonic acid readily dehydrates to CO_2 and water, so the net ionic reaction is

$$CaCO_3(s) + 2H_3O^+ \rightarrow Ca^{2+} + CO_2(g) + 2H_2O$$

Thus this reaction is driven to the right by two factors: (1) the production of a very weakly dissociated substance, H_2O, and (2) the escape of volatile CO_2 gas.

THE EXPERIMENTAL METHOD: ELECTRICAL CONDUCTIVITY

In this experiment, we will measure the electrical conductivity of pure liquid substances and dissolved substances to determine if there is any solvent-assisted ionic dissociation. We will also try to determine the net result of any reaction that may take place when solutions of ionic substances are mixed.

Solutions of ionic-type substances conduct electric current by the movement of their ions under the influence of an electric field. The faster the ion moves, the greater its ion mobility. At the solution/electrode interfaces, the circuit is completed by electron transfer reactions, because in metal wires the only charge carriers are electrons.[3] For a given applied voltage, the amount of current depends on the concentration of ions and, to a lesser extent, on differences in individual ion mobilities. (Hydrogen ion has an unusually high mobility—five to eight times that of many other ions. To better understand why, see the discussion in Consider This at the end of the experiment.) Figure 18-1 shows a simple apparatus for the comparison of electrical conductivities.

We will compare the electrical conductivities of a number of solutions and of the products formed after some chemical reactions, and use the data we obtain to interpret the character of the solutions and the course of any reactions. We will interpret these results in terms of the net ionic equations for the reactions.

EXPERIMENTAL PROCEDURE

Special Supplies: Conductivity apparatus. (Use a commercial conductivity meter or an apparatus constructed as described in the notes that follow and shown in Figure 18-1.)

Chemicals: Pure 17 M acetic acid, 6 M acetic acid, 0.1 M acetic acid, methanol (CH_3OH), 6 M HCl, 0.1 M HCl, toluene, anhydrous HCl in toluene produced by bubbling HCl gas through dry toluene (*if gaseous HCl is not available, you may substitute a solution prepared by dissolving pure acetic acid in toluene to make a 1 M solution*), 0.1 M $HgCl_2$, 0.1 M NaOH, 0.1 M NH_3, 0.1 M H_2SO_4, 0.1 M $Ba(OH)_2$, 0.1% thymol blue indicator solution, sucrose (s), NaCl(s), $CaCO_3(s)$ (marble chips), zinc metal (mossy).

....

Notes on the Construction and Use of the Conductivity Apparatus Shown in Figure 18-1

1. Build the circuit in a small aluminum chassis box, with the power cord, fuse, and pilot light on the rear, the banana jacks on the side, and the push-button switch on top. Mount the chassis box on a piece of ¾-inch plywood through which holes have been drilled to allow two pairs of 14-gauge copper wire electrodes (connected in parallel) to protrude about 15 cm. Space the wires about 1 cm apart. All but the bottom 5 cm of the copper wire should be insulated. Fasten a length of ½-inch OD aluminum rod on the underside of the plywood so that the apparatus can be supported by a clamp holder on a ring stand. Connect the aluminum chassis to the green ground lead of the three-wire power supply cord.

---

[3]With alternating current (AC), the net chemical reaction at the electrode surface is diminished, and at high frequency the chemical reaction is practically eliminated.

FIGURE 18-1

An apparatus for comparing the electrical conductivities of various solutions.

2. The voltage across the electrodes is about 6 V AC rms (18 V peak-to-peak) when the push-button switch is depressed, so be careful not to touch the electrodes while the switch is depressed. The 5-ohm, 5-watt resistor limits the current to a safe value if the electrodes are inadvertently shorted.

3. An inexpensive digital multimeter with a direct-current (DC) multirange ammeter provides a suitable readout. Two to four current ranges covering 50 microamperes (μA) to 200 or 300 milliamperes (mA) will work well.

4. To make a measurement, adjust the ammeter to th highest current range (200–300 mA). Put enough of the

test substance in a 30-mL beaker to give a depth of about 1 cm. Lower the electrodes into the beaker so that they are immersed to a depth of 1 cm.

Then depress the push-button switch to get a current reading. Switch the ammeter to the most sensitive current range that can be used without driving the meter off scale. The solution conductivity is directly proportional to the current. Record the current reading, and return the ammeter to the highest current range.

5. Between each measurement, rinse the electrodes with distilled water, and dry them with tissue. As an extra precaution, unplug the apparatus while cleaning the electrodes. Periodically clean the electrodes with steel wool, and rinse with water.

••

WASTE COLLECTION: Mercuric chloride, $HgCl_2$, is toxic. After you have made measurements of the conductivity of 0.1 M $HgCl_2$, pour the solution into a waste bottle. *Do not pour it down the drain.* Likewise, toluene and solutions of HCl (or acetic acid) in toluene should be poured into waste containers, not down the drain.

1. ELECTROLYTES AND NONELECTROLYTES.

If you will be using the conductivity apparatus shown in Figure 18-1, read over the preceding notes describing how to use the apparatus. If you will use a commercial conductivity meter, follow your instructor's directions. When you are confident that you understand how to use the conductivity meter, measure and record the conductivity of the following *substances and/or their solutions in water.* Determine if they are largely, moderately, or poorly ionized, or if they are essentially nonelectrolytes. (For an apparatus like that shown in Figure 18-1, a nonelectrolyte will give a very small current, less than a few microamperes. A completely dissociated strong electrolyte, such as 0.1 M HCl, will typically give a current of more than 100 mA.) Test the pure substance first; then add some water to it and retest. In each case, the magnitude of the current will be a measure of the conductivity of the solid or solution, indicating whether or not the substance contains ions that are free to move.

(a) To begin, test distilled (or deionized) water and tap water. What causes the difference in conductivity? Then test the following substances and their aqueous solutions: methanol (CH_3OH), glacial acetic acid (CH_3COOH), sucrose (table sugar, $C_{12}H_{22}O_{11}$), sodium chloride (NaCl), and *dilute solutions only* (not the pure substances) of 0.1 M HCl, 0.1 M NaOH, and 0.1 M $HgCl_2$. Be sure to dispose of the mercuric chloride, $HgCl_2$, in the waste container reserved for this purpose. *Do not pour it down the drain—it is toxic.*

(b) Observe the *effect of the solvent* on the ionization of HCl in toluene and in water. First test a little pure toluene (C_7H_8) in a thoroughly dry beaker, and then a solution of HCl in toluene. (*Use the solution already on your reagent shelf,* prepared by bubbling anhydrous HCl through dry toluene—do *not* add aqueous concentrated HCl to toluene.) Then add 5 mL of deionized water to this solution, mix it well, and repeat the test with the electrodes immersed farther into the lower aqueous layer. (*If HCl in toluene is not available, you may substitute a 1 M acetic solution in toluene; the effect is less dramatic because acetic acid is not a strong acid in water like HCl.*) Be sure to dispose of the toluene solutions in the waste container reserved for this purpose.

(c) Compare the chemical behavior of 6 M HCl and 6 M acetic acid with your conductivity data. Put three or four marble chips into each of two clean beakers. Add 6 M HCl to one beaker and 6 M acetic acid to the other. Compare the rate of evolution of CO_2 gas. Put one or two pieces of mossy zinc metal into another pair of beakers, add 6 M HCl and 6 M acetic acid, and compare the rates of evolution of hydrogen (H_2) gas. Record your observations, and write balanced chemical reactions for each of the four reactions you observed.

2. TYPICAL IONIC REACTIONS.

You will experiment with some ionic reactions to determine the nature of their products. Through conductivity tests of the separate reactants and of the mixture after reaction, you will discover whether the acids, bases, and salts concerned are largely ionized (strong electrolytes) or only moderately ionized (weak electrolytes). For each reaction, write the total ionic equation and the net ionic equation that best describes the reaction taking place. This will require careful thinking on your part. Use the following procedures.

(a) *0.01 M HCl with 0.01 M NaOH.*[4] These solutions can be prepared by diluting 5 mL of each of the 0.1 M solutions to 50 mL. It is an advantage to use both pairs of electrodes connected in parallel (see Figure 18-1) by placing one pair of electrodes in 0.01 M HCl and simultaneously placing the other pair in an equal volume of 0.01 M NaOH contained in a separate beaker. Then mix the solutions, divide them into two equal portions, and again place the two solutions simultaneously in contact with the two electrode pairs. (This compensates for the dilution effect of mixing two solutions.) Why is the con-

_____

[4]Quite dilute 0.01 M solutions are suggested so that you can interpret moderate conductivity changes more easily.

ductivity of the product solution lower than that of the reactant solutions?

(b) *0.1 M acetic acid, HC₂H₃O₂, with 0.1 M NH₃.[5] As in Part 2a, test equal volumes of the separate solutions, mix them, divide the solution into two equal parts, then re-measure the conductivity. Record your measurements.*

(c) *0.1 M H₂SO₄ with 0.1 M Ba(OH)₂.* Measure and record the conductivity of each solution by itself. Then add two or three drops of 0.1% thymol blue indicator solution to the sulfuric acid solution, and add 0.1 M Ba(OH)₂ solution one drop at a time to the sulfuric acid solution, stirring continuously. The sulfuric acid will be exactly neutralized when the indicator changes from its pink form through its yellow (acid) form to its blue (basic) form. If you add too much Ba(OH)₂, overshooting the end point, add 0.1 M H₂SO₄ drop by drop, stirring continuously, until the *exact* end point—shown by a sharp color change of the indicator—is reached. Now, measure and record the conductivity of this mixture. Write balanced ionic equations to describe the reactions that take place.

👉 Consider This

Solutions of strong bases like NaOH and strong acids like HCl have much higher conductivity than a solution of the same concentration of NaCl. By examining a number of examples of this, scientists have discovered that it is the hydronium ion (H_3O^+) and hydroxide ion (OH^-) that are responsible for the higher conductivity.

Conductivity is related to the mobility of an ion (how fast it moves in solution under the influence of an electric field). In order to get from point A to point B in solution, most ions have to physically move from point A to point B,

dragging along the cluster of water molecules that surrounds each ion. However, for hydronium ions, the whole hydronium ion does not need to move in order to transfer a positive charge from point A to point B. The way this works is shown in the following diagram:

Imagine that there is a hydronium ion at point A on the left end of a chain of water molecules. The water molecules are loosely associated by hydrogen bonding, as shown by the dotted lines representing the hydrogen bonds. The positive charge can be transferred from the hydronium ion at point A to the water molecule at point B by having three protons hop from one molecule to the next as indicated by the three arrows. After transfer of the three protons, the water molecule at point B would be transformed into a hydronium ion (H_3O^+), so the charge associated with a hydronium ion would have moved from point A to point B without physical movement of the whole hydronium ion originally at point A. This rapid proton hopping (or proton transfer) process allows charge to move along much faster than physical movement of the whole ion, such as a Na^+ or Cl^- ion.

Now, you draw a similar diagram, starting with a hydroxide ion on the right at point B, that shows how the −1 charge associated with a hydroxide ion could move from point B to point A, by shifting protons left to right from one water molecule to the next. (Hydronium ions and hydroxide ions under the influence of the same electric field would move in opposite directions.)

---

[5]Solutions in water are sometimes formulated as NH₄OH, called ammonium hydroxide. Because the evidence for the existence of such a molecule in solution is ambiguous, we will simply refer to a solution of ammonia in water as "aqueous ammonia solution."

Ionic and Covalent Bonding

Conductivity of Solutions of Ionic and Covalent Compounds

NAME		
SECTION		LOCKER
INSTRUCTOR		DATE

OBSERVATIONS AND DATA

1. Electrolytes and Nonelectrolytes

(a) Record the conductivity measurement of each *substance and solution* tested. Then, after noting the range of measurements made (about five orders of magnitude), classify the substances and solutions as containing essentially no ions (lowest values), a few ions (middle values), or many ions (highest values).

	Conductivity	Category		
		No ions	Few ions	Many ions
Distilled $H_2O(l)$				
Tap H_2O				
Methanol $CH_3OH(l)$				
$CH_3OH(aq)$				
Glacial acetic acid, $CH_3COOH(l)$				
Aqueous acetic acid, $CH_3COOH(aq)$				
Sucrose, $C_{12}H_{22}O_{11}(s)$				
Aqueous sucrose, $C_{12}H_{22}O_{11}(aq)$				
$NaCl(s)$				
$NaCl(aq)$				
0.1 M $HgCl_2(aq)$				
0.1 M HgCl(aq)				
0.1 M $NaOH(aq)$				

(i) Write the equation for the reaction that forms the few (but important) ions in pure water.

(ii) Why is tap water more conductive than distilled water?

(iii) Is there any evidence that methanol forms ions either in the pure state or when dissolved in water? Would you classify methanol as a strong electrolyte, weak electrolyte, or nonelectrolyte?

(iv) Why does acetic acid form ions when it is dissolved in water but not when it is in the pure (glacial) form? Write an equation to describe the formation of ions in aqueous acetic acid.

(v) Compare your measurements for $CH_3OH(aq)$, $CH_3COOH(aq)$, and $HCl(aq)$. Would you classify aqueous acetic acid as a strong electrolyte, weak electrolyte, or nonelectrolyte? What species are present in aqueous acetic acid? In glacial acetic acid?

	Species present
$CH_3COOH(aq)$	
$CH_3COOH(l)$	

(vi) Does sucrose behave like a strong electrolyte, weak electrolyte, or nonelectrolyte when it is dissolved in water? What species are present in $C_{12}H_{22}O_{11}(aq)$?

(vii) The models of solid NaCl describe it as consisting of Na^+ cations and Cl^- anions. If this is the case, why is NaCl(s) not an electrolyte?

(viii) Explain the distinctly different behavior in the conductivity of NaCl(s) and NaCl(aq). What species are present in each of these substances?

	Species present
NaCl(aq)	
NaCl(s)	

(ix) Is $HgCl_2(aq)$ a strong electrolyte, weak electrolyte, or nonelectrolyte? What species are present in this solution? Is this consistent with the statement in the Pre-Lab Preparation that "salts ionize completely in dilute aqueous solution"?

(x) What species are present in 0.1 M HCl and NaOH solutions? Write equations to describe the reactions that yield these species.

	Species present	Equation
HCl(*aq*)		
NaOH(*aq*)		

(b) The effect of the solvent

	Conductivity	Category		
		No ions	Few ions	Many ions
Toluene, $C_7H_8(l)$				
HCl (*in toluene*) (or CH_3COOH *in toluene*)				
Aqueous layer after mixing with HCl (*in toluene*) (or CH_3COOH *in toluene*)				

Why is the conductivity of HCl (or CH_3COOH) different in the two solvents? What causes this difference?

(c) Correlating chemical and conductivity behavior

(i) Compare the rates of reaction of $CaCO_3(s)$ with 6 M acetic acid and 6 M hydrochloric acid. The reaction of $CaCO_3$ with an acid, HA, is

$$2HA + CaCO_3(s) \rightarrow Ca^{2+} + 2A^- + CO_2(g) + H_2O$$

(ii) Compare the rates of reaction of Zn(s) with 6 M CH_3COOH and 6 M HCl. The reaction of zinc with an acid, HA, is

$$2HA + Zn(s) \rightarrow Zn^{2+} + 2A^- + H_2(g)$$

(iii) Is there a correlation between the rates of the chemical reactions of 6 M CH_3COOH and 6 M HCl and the conductivities observed for $CH_3COOH(aq)$ and $HCl(aq)$ in part 1a? Suggest an explanation.

2. Typical Ionic Reactions

(a) *0.01 M HCl with 0.01 M NaOH.* The relative conductivities of the solutions tested are as follows.

0.01 M HCl/0.01 M NaOH measured in parallel: _____

Mixture measured in parallel: _____

Do the reactants exist as molecules or ions? _____

Does the conductivity change indicate the creation of more ions or a decrease in the number of ions?

_____

The total ionic equation for the reaction is _____

The net ionic equation for the reaction is _____

Interpret any changes in the conductivity of the solutions, before and after mixing, in accordance with the preceding equations:

(b) *0.1 M CH$_3$COOH with 0.1 M NH$_3$.* The relative conductivities of the solutions tested are as follows.

0.1 M CH$_3$COOH/0.1 M NH$_3$ measured in parallel: _____

Mixture measured in parallel: _____

Do the reactants exist as molecules or ions? _____

Does the conductivity change indicate the creation of more ions or a decrease in the number of ions?

_____

The total ionic equation for the reaction is _____

The net ionic equation for the reaction is _____

Interpret any changes in the conductivity of the solutions, before and after mixing, in accordance with the preceding equations:

(c) *0.1 M H$_2$SO$_4$ with 0.1 M Ba(OH)$_2$.* The relative conductivities of the solutions tested are as follows.

0.1 M H$_2$SO$_4$: _____ 0.1 M Ba(OH)$_2$: _____ Mixture: _____

The total ionic equation for the reaction is _____

The net ionic equation for the reaction is _____

Interpret any changes in the conductivity of the solutions, before and after mixing, in accordance with the preceding equations:

☞ **Consider This**

In the CONSIDER THIS section at the end of the experiment we discussed a way to explain the large mobility of hydronium ion (H_3O^+) and suggested that a similar proton transfer (or proton hopping) mechanism might explain the large mobility of hydroxide ion (OH^-).

Let's begin with a diagram showing a hydroxide ion at point B and a water molecule at point A:

$$
\begin{array}{c}
\qquad\qquad H \\
\qquad\qquad \backslash \\
O-H\cdots O-H\cdots O-H\cdots O^- \\
/\| \qquad\qquad / \qquad\qquad \|\backslash \\
H\ A \qquad\qquad H \qquad\qquad B\ H
\end{array}
$$

Add curved arrows to the diagram above to show how proton transfer from one water molecule to the next could lead to the transfer of ($-$) charge from point B to point A, effectively transforming the hydroxide ion at point B into a water molecule, and the water molecule at point A into a hydroxide ion.

Ions move, thereby transferring charge, under the influence of an electric field. Add lines at each end of the diagram to represent the electrodes in a conductivity apparatus.

Would the charge on the electrode at point A have to be positive or negative to attract a negatively charged hydroxide ion (OH^-)? _____

Would the charge on the electrode at point B have to be positive or negative to attract a positively charged hydronium ion (H_3O^+)? _____

What should the polarity of the electrodes be in order to induce proton transfer from one water molecule to the next, moving from left to right (point A to point B), considering that protons are positively charged particles?

   Charge on the electrode at point A (positive or negative?): _____

   Charge on the electrode at point B (positive or negative?): _____

An electric current is a flow or movement of charged particles. In a metal wire, electrons are the charge carriers, hopping from atom to atom. In solution, it is the movement of ions in solution that constitutes the electric current. Note that both positive and negative ions contribute to the electric current in solution.

Explain in simple language why ions move and in which direction they move in the presence of an electric field created between two electrodes having opposite charges.

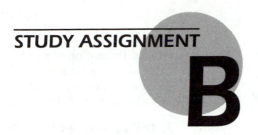

STUDY ASSIGNMENT

B

Writing Lewis Structures

PURPOSE

To learn a procedure for drawing the Lewis structures of covalently bonded molecules (and molecular ions). To use these structures to describe the bonding and charge distribution in the molecules.

INTRODUCTION

When atoms of elements combine with one another to form molecules like water, H_2O, or molecular ions like nitrate, NO_3^-, or ammonium, NH_4^+, the bonding can be described as one in which electrons are simultaneously attracted to two or more atomic nuclei. This picture can be simplified by imagining that the filled inner electron shells of an atom neutralize much of the nuclear charge of an atom, producing an atom core with a residual effective nuclear charge. Then we can approximately describe the chemical bond between two atoms as the result of mutual attraction between the outer valence electrons of the atoms and the effective nuclear charge of the atom cores. In this simplified picture we need consider only the outer shell valence electrons.

In order for there to be a strong mutual attraction between two atoms, there must be vacancies in the valence orbitals of at least one atom. Strong covalent (shared-electron) bonds are not formed between atoms with filled valence orbitals. When the outer orbitals of two atoms are completely filled, the electrons of one atom cannot penetrate close enough to the nucleus of the other atom to be strongly attracted. Consequently, the interactions between the two atoms will be very weak. This is typical of the interatomic forces between atoms of a noble gas, such as between two argon atoms.

In the early 1900s, before the invention of quantum mechanics or the notion of what are now called *orbitals,*

G. N. Lewis recognized the tendency of atoms with partially filled valence shells to form chemical bonds. He first published his ideas in 1916; a fuller account appeared in his book *Valence and the Structure of Atoms and Molecules,* published in 1923. Lewis's most important idea was that bonding can take place between two atoms when they share electrons. He visualized the outer electron shell of an atom as an octet of electrons arranged at the eight corners of a cube. For example, two fluorine atoms, each with a vacancy in the valence shell, could fill their vacancies by sharing their electrons in such a way that both atoms would have a filled octet, as shown in Figure B-1.

In this study assignment we will present a set of simple rules for drawing Lewis structures for molecules and ions and make use of them in deducing some properties of the chemical bonds. The Lewis structures give an approximate picture of how the electrons are arranged in the molecule.

The Lewis structure is useful for more than just a description of covalent bonding. With the addition of a few more simple rules, the shape of a molecule can be predicted. From the shape and qualitative distribution of electrons, we can make predictions about the polarity of the molecule and whether or not it will have a permanent dipole. From the shape of the molecule, we may also be able to deduce how the orbitals of the atoms have combined to form the chemical bonds. Although these features will not be discussed in this study assignment, you will need to be able to draw Lewis structures before you can go on to apply the valence shell electron-pair repulsion (VSEPR) rules to predict molecular shapes.

While learning and using the rules for drawing Lewis structures, remember the basic idea of covalent bonding: Each bond is formed by sharing a pair of electrons between two atoms. Generally, enough electron-pair bonds are formed to fill the outer orbital vacancies. For

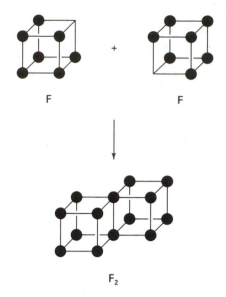

FIGURE B-1

G. N. Lewis imagined an octet of valence electrons occupying the eight corners of a cube. A fluorine atom with seven valence electrons has one electron position vacant. A bond between two fluorine atoms is formed by sharing two electrons so that the octets on both atoms become filled.

the nonmetals of the first two periods, these orbitals are the *s* and *p* orbitals, which can hold two and six electrons, respectively. Thus, the filled orbitals will accommodate eight electrons, giving rise to the *octet rule,* which states that each atom will have a share of eight electrons, with an electron configuration like that of a Group VIII noble gas. The octet rule results not so much from any special stability of the noble gas electron configuration but because bond formation proceeds until all the vacant orbitals are filled to their maximum capacity.

The exceptions to the octet rule follow from the nature of atomic orbitals as well. The hydrogen atom can share a maximum of two electrons because the filled outer 1*s* orbital contains a maximum of two electrons, and quantum number restrictions dictate that 1*p* orbitals cannot exist. Elements in the third period can sometimes share more than eight electrons by making use of vacant 3*d* orbitals, whose energy lies a little above the energy of the 3*s* and 3*p* orbitals.

Noble gases normally form no covalent bonds because their atoms lack outer orbital vacancies that can be occupied by electrons from another atom. Also, the noble gases are not good electron donors because their nuclei strongly attract their valence electrons. Thus they have little tendency to share their electrons. The heavier noble gases will form chemical bonds only with atoms, such as fluorine and oxygen, that are very electronegative. They can form bonds with fluorine by making use of vacant *d* orbitals that lie at higher energies than the filled *s* and *p*

orbitals. Promotion of *s* and *p* electrons to the *d* orbitals creates outer orbital vacancies, allowing the sharing of electrons and the formation of stable bonds.

A METHOD FOR WRITING LEWIS STRUCTURES

In this study assignment we will first formulate some rules for drawing Lewis structures. We will then restate each rule and discuss details of its application and possible exceptions. We will work out several examples illustrating each rule, then work out sample structures by using the rules step by step. Finally, we will discuss some of the finer points in order to ensure that the structures we draw are accurate representations of the bonding in the molecules. You should then be able to apply the rules to draw Lewis structures for molecules and molecular ions listed at the end of this study assignment.

RULES FOR WRITING SIMPLE LEWIS STRUCTURES

1. Write the chemical formula for the molecule (or ion) and determine the total number of valence electrons in the molecule.

2. Draw the skeletal arrangement of the molecule showing single bonds connecting the atoms.

3. Assume that each bond in the skeleton requires two valence electrons (an electron-pair bond). After subtracting two electrons for each bond from the total number of valence electrons, assign the remaining electrons to give each atom an octet, or share of eight electrons.

4. If after each atom has been given a share of eight electrons, additional electrons remain, assign the extra electrons to the central atom of the molecule.

5. If there are not enough electrons to give each atom a share of eight electrons, then form multiple bonds between atoms by moving electron pairs to form double (or triple) bonds.

RULE 1   DETERMINE THE TOTAL NUMBER OF VALENCE ELECTRONS

The number of valence electrons in a molecule is simply the sum of the valence electrons of the atoms that make up the molecule. The number of valence electrons for each atom can be determined either by noting the main group to which the atom belongs or by counting across the periodic table from the left (ignoring the transition metals). Thus hydrogen in Group IA has one electron, silicon in Group IVA has four valence electrons, all of the halogens (F, Cl, Br, and I) in Group VIIA have seven, and the noble gases have eight valence electrons.

For each of the following molecules, the total number of valence electrons is shown.

H_2 $\quad$ $(2 \times 1e^-) = 2e^-$

PCl_5 $\quad$ $(5e^-) + (5 \times 7e^-) = 40e^-$

CH_4 $\quad$ $(4e^-) + (4 \times 1e^-) = 8e^-$

H_2O_2 $\quad$ $(2 \times 1e^-) + (2 \times 6e^-) = 14e^-$

CO_2 $\quad$ $(4e^-) + (2 \times 6e^-) = 16e^-$

N_2O_4 $\quad$ $(2 \times 5e^-) + (4 \times 6e^-) = 34e^-$

If the molecule is a molecular ion, then the charge on the ion modifies the electron count. Because electrons have a negative charge, one electron is added to the total for each negative charge on the ion and one is subtracted for each positive charge. For each of the following molecular ions, the total number of valence electrons is shown.

OH^- $\quad$ $(6e^-) + (1e^-) + (1e^-) = 8e^-$

(One electron has been added for the -1 charge.)

NH_4^+ $\quad$ $(5e^-) + (4 \times 1e^-) - (1e^-) = 8e^-$

(One electron has been subtracted for the $+1$ charge.)

NO_2^- $\quad$ $(5e^-) + (2 \times 6e^-) + (1e^-) = 18e^-$

SO_4^{2-} $\quad$ $(6e^-) + (4 \times 6e^-) + (2e^-) = 32e^-$

NO^+ $\quad$ $(5e^-) + (6e^-) - (1e^-) = 10e^-$

PO_3^{3-} $\quad$ $(5e^-) + (3 \times 6e^-) + (3e^-) = 26e^-$

$C_2H_3O_2^-$ $\quad$ $(2 \times 4e^-) + (3 \times 1e^-) + (2 \times 6e^-) +$ $(1e^-) = 24e^-$

RULE 2 DRAW THE BOND SKELETON OF THE MOLECULE

Rule 2 is easy to apply if the structure consists of a central atom to which the surrounding atoms are bonded, because the central atom is customarily written first. Thus the following molecules have the bond skeletons shown:

There are some exceptions to this rule, but they are easily recognized, and their structures are determined by noting a consequence of the rules for drawing Lewis structures: An atom will tend to form one bond for each electron it needs to fill its valence orbitals. Thus hydrogen forms only one bond and is always found on the periphery of a molecule. Group VIA atoms, such as oxygen, tend to form two bonds (two electrons are required to complete the octet). Group VA atoms tend to form three bonds, and so on. The skeleton should be drawn so that each atom forms $8 - n$ bonds. Eight is the total capacity of the s and p orbitals, while n is the number of valence electrons that each atom has. The difference, $8 - n$, gives the number of vacancies in the valence orbitals of an atom. As we will see in the final Lewis structure, most of the time each atom will form $8 - n$ bonds.

In the following formulas the central atom is not written first, but consideration of the number of bonds that an atom is able to form will often lead to the correct bond skeleton.

Propene, C_3H_6

or

or

or

For molecules like HCN, H_2CO, N_2O, O_3, or C_3H_6, where more than one bond skeleton is possible, we will need more information about the structure of the molecule in order to draw the correct structure. (The bond skeleton of HCN is implied by the way the formula is written, and for the other molecules one of the structures may later emerge as the best one.)

Bond skeletons that have three-membered rings, such as

and

for ozone and propene, respectively, can often be discarded because they involve a great deal of strain or distortion of normal bond angles. In general, do not draw a structure that contains a three-atom ring unless you know that the molecule contains such a ring.

RULE 3   ASSIGN THE REMAINING ELECTRONS TO GIVE EACH ATOM AN OCTET

Each bond that is drawn in a skeleton represents a pair of electrons. The number of valence electrons left to assign is the total number, determined by Rule 1, minus the number of electrons required to form the bonds in the skeleton, which is twice the number of bonds shown in the skeleton structure drawn according to Rule 2.

$$\text{remaining electrons} = \text{total valence electrons} - (2 \times \text{number of bonds})$$

TABLE B-1

Where possible, assign electrons so that each atom has an octet (a share of eight electrons).

Molecule	Number of valence electrons	Skeleton	Remaining electrons	Complete octets
HF	8	H—F	6	H—$\ddot{\text{F}}$:
H_2O	8	H—O—H	4	H—$\ddot{\text{O}}$—H
CCl_4	32	(skeleton)	24	(structure)
PF_3	26	(skeleton)	20	(structure)
SF_4	34	(skeleton)	26	(structure) (2 electrons unassigned)
O_2	12	O—O	10	:$\ddot{\text{O}}$—$\ddot{\text{O}}$: (insufficient electrons)

It generally works best to assign the remaining electrons in pairs to each of the atoms in the skeleton structure, starting with the outer atoms and ending with the central atom. (Remember that a hydrogen atom can share no more than two electrons, not eight.) (See Table B-1.)

RULE 4   IF THERE ARE ADDITIONAL ELECTRONS, PLACE THEM ON THE CENTRAL ATOM

If all of the electrons are assigned by Rule 3 and each atom (except hydrogen) has a share of eight electrons, then the Lewis structure is complete. In some instances, as in the case of SF_4 in Table B-1, there are still electrons to be assigned after all octets are complete. In these cases the extra electrons are assigned to the central atoms as lone pairs. This gives the central atom an *expanded octet*, as shown by the completed Lewis structure shown for SF_4 in Table B-2.

TABLE B-2

Put "extra" electrons on the central atom giving it an expanded octet (more than eight electrons).

Molecule	Number of valence electrons	Skeleton	Remaining electrons	Lewis structure
SF_4	34		26	
XeF_2	22	F—Xe—F	18	:F̈—Xe—F̈:
ICl_4^-	36		28	

TABLE B-3

If there are not enough electrons to complete all of the octets, then move lone pairs to form multiple bonds between the atoms until each atom has a share of eight electrons.

Molecule	Number of valence electrons	Skeleton	Rule 3 structure	Lewis structure[a]
O_2	12	O—O	:Ö—Ö	Ö=Ö
H_2CO	12			
HCN	10	H—C—N	H—C—N̈:	H—C≡N:

[a]Note that the application of the rules leads to a structure that implies that O_2 would have no unpaired electrons, whereas O_2 is known to be paramagnetic with two unpaired electrons.

The explanation for this violation of the octet rule is that in these instances the central atoms have a *d* subshell with empty orbitals that has the same principal quantum number as the *s* and *p* valence orbitals and is thus close in energy to the *s* and *p* valence shell orbitals. The atomic orbitals from this *d* subshell can hold some of the electrons and thus participate in the bonding in the molecule. For molecules like PCl_5 and SF_6 the octet of the central atom must be expanded to contain all of the valence electrons of the atoms. (See also the examples for XeF_2 and ICl_4^- in Table B-2.)

RULE 5   IF THERE ARE NOT ENOUGH ELECTRONS TO COMPLETE ALL OF THE OCTETS, THEN MOVE ELECTRON PAIRS TO FORM MULTIPLE BONDS BETWEEN ATOMS

In the example of the O_2 molecule shown under Rule 3, there are insufficient electrons to complete each atom's octet. In such a situation, two atoms can share more than one pair of electrons between themselves, thus forming multiple bonds. If two pairs are shared between the same two atoms, a double bond is formed; three shared pairs between two atoms constitute a triple bond. Measurements of bond energies and bond lengths show that these multiple bonds are real. The bonds in a double or triple bond are stronger and shorter than in a single bond.

In practice, to complete the octets for the structures of H_2CO and HCN shown in the discussion of Rule 2 and for O_2 shown in Rule 3, we must move some lone pairs on atoms into position between that atom and an adjacent atom with an incomplete octet. Thus for O_2, H_2CO, and HCN we complete the Lewis structures by forming double or triple bonds. (See Table B-3.)

EXCEPTIONS TO THE OCTET RULE

Although the octet rule (based on filling each orbital vacancy so that each atom has a noble gas electron configuration) is the central rule in drawing Lewis structures, there are some instances where the best description of a molecule contains atoms that do not have an octet of electrons. A brief summary and explanation of these situations may help you to recognize circumstances when the octet rule is likely to be broken.

1. *Hydrogen.* For hydrogen atoms the maximum valence orbital capacity is two electrons. This is because no $1p$ orbital exists, owing to quantum number restrictions.

2. *Beryllium and Boron.* In compounds of beryllium and boron, experiments show that when these atoms form two and three bonds, respectively, the bonds are single bonds. Thus BeF_2 and BCl_3 have these Lewis structures:

even though strict application of the rules would predict that they would have the following structures:

These possible structures are improbable (high-energy) electron distributions because they put positive charge on the electronegative atoms F and Cl. (In the following subsection we show how the formal charges indicated on the atoms are calculated.)

3. *Expanded Octets.* The central atom of some molecules may have more than an octet of electrons. This occurs in cases where the central atom has more than four atoms bonded to it, as in PCl_5 and SF_6, or where there are more than enough electrons to satisfy the octets of all of the atoms, as in XeF_2. These extra electrons are assigned to the central atom as lone pairs (Rule 4). An expanded octet can occur only in atoms that have empty *d* orbitals in their valence shell. (Note that 2*d* valence orbitals do not exist, so a nitrogen atom cannot have an expanded octet, whereas a phosphorus atom may.)

4. *Odd-Electron Molecules.* There are a few molecules that contain an odd number of electrons; NO, NO_2 and CH_3 are examples. Molecules like these, which contain an unpaired electron, are known as *free radicals* and tend to be quite reactive. It is impossible to form octets from an odd number of electrons, so the Lewis structures of radicals contain an atom that has only seven electrons instead of an octet.

FORMAL CHARGE

Some of the Lewis structures that you draw will contain atoms with a formal charge; that is, if we assume that the electrons in the covalent bonds are shared equally between the atoms, there will be a net charge on some of the atoms in the structure. In the case of molecular ions like NH_4^+ and SO_4^{2-} there will inevitably be at least one atom with a formal charge, but formal charges may also exist in neutral molecules.

To determine the formal charge of each atom in a structure, we divide each electron-pair bond in half and assign an electron from each bond to each of the two atoms bonded together. Counting one electron for each electron-pair bond and two electrons for each lone pair, we determine the total number of electrons assigned to each atom. We then compare this number to the number of valence electrons that the neutral atom would have. If it has one more than the number of valence electrons, we say that it has a formal charge of -1. If it is short by one electron, its formal charge is $+1$. Thus for NH_4^+ we would assign one electron from each bond to the hydrogen atoms and the others to the nitrogen atom. The nitrogen atom therefore would be assigned four electrons, one short of its normal valence of five, and the nitrogen atom would have a formal charge of $+1$. The hydrogen atoms each have one electron, their normal number of valence electrons, and thus have a zero formal charge. Note that adding all the for-

mal charges gives the net charge of the molecular ion (or zero in the case of a neutral molecule).

$$\left(\begin{array}{c}\text{formal}\\\text{charge}\end{array}\right) = \left(\begin{array}{c}\text{group}\\\text{number}\end{array}\right) - \left(\begin{array}{c}\text{number}\\\text{of bonds}\end{array}\right) - \left(\begin{array}{c}\text{number of}\\\text{unshared electrons}\end{array}\right)$$

Thus for the chlorate ion, ClO_3^-, the Lewis structure can be written as

and the formal charges would be

chlorine formal charge $= 7 - 3 - 2 = +2$

oxygen formal charge $= 6 - 1 - 6 = -1$

Again, note that addition of the formal charges on the atoms gives the net charge on the molecular ion.

We stress that the formal charge is just that: it is calculated by formal rules and may not always accurately reflect the true distribution of electron density. However, the formal charges usually correctly show the tendency of atoms to have either an excess or a deficiency of electron charge when compared to the neutral atoms.

Knowledge of the formal charges in a structure can help to eliminate an unlikely Lewis structure. In general, a good Lewis structure does not contain adjacent atoms with formal charges of the same sign. Three possible Lewis structures for the nitric acid molecule, HNO_3, are

The first structure has a $+1$ formal charge on the adjacent O and N atoms and would not be a reasonable structure for HNO_3. The other structures are better (lower-energy) representations of the bonding in nitric acid.

The $8 - n$ bond rule that was used to write the molecular skeleton is also the result of a formal counting of electrons. If an atom forms $8 - n$ bonds, it will have a formal charge of zero.

RESONANCE

Applying our rules for drawing Lewis structures to the substance sulfur dioxide, SO_2, gives two possible structures:

$$^{\ominus}:\ddot{O}-\overset{..}{\underset{\oplus}{S}}=\ddot{O} \quad \text{or} \quad \ddot{O}=\overset{..}{\underset{\oplus}{S}}-\ddot{O}:^{\ominus}$$

implying the existence of two different types of S—O bonds in this molecule. (A similar situation arises in the nitric acid example we discussed earlier.) Experiments show that the two S—O bonds are actually identical. The best representation of the SO_2 molecule is an average or superposition of these two structures, represented by

$$^{\ominus}:\ddot{O}-\overset{..}{\underset{\oplus}{S}}=\ddot{O} \longleftrightarrow \ddot{O}=\overset{..}{\underset{\oplus}{S}}-\ddot{O}:^{\ominus}$$

These structures are called *resonance structures*. They occur whenever there are two or more Lewis structures that have identical positions for the nuclei but have different electron arrangements. Resonance structures often exist in a situation where identical atoms are bonded to the same atom and there is a double bond in the structure, but you should look for the possibility of resonance before settling on any one Lewis structure.

We emphasize that a real molecule has only one stable electron configuration. The molecule does not flip back and forth between two electron configurations, but has a configuration that is equivalent to what we would obtain by superimposing all of the resonance structures.

SUMMARY

A systematic method for drawing Lewis structures is summarized by the following procedures.

1. Determine the total number of valence electrons in the molecule.

(a) Sum the valence electrons of each atom in the molecule.

(b) Add an electron for each negative charge, or subtract one for each positive charge on a molecular ion.

2. Draw the bond skeleton of the molecule.

(a) Often the central atom of the molecule is written first in the chemical formula, then all other atoms are bonded directly to it.

(b) If the central atom is not written first, make use of the fact that atoms tend to form $8 - n$ bonds where n is the number of valence electrons of the neutral atom.

3. Assign the remaining electrons to give each atom an octet of electrons. H, Be, and B do not form an octet.

4. If there are more than enough electrons to give each atom an octet, place the additional electrons on the central atom to form an expanded octet.

5. If there are not enough electrons to complete all of the octets, then move lone pairs to form multiple bonds between atoms until each atom has a share of eight electrons.

6. Determine the formal charge of each atom in the resulting structures), and eliminate any structures with unlikely charge distributions (adjacent atoms with the same charge).

7. Look to see if there are possible equivalent resonance structures.

Writing Lewis Structures

NAME

SECTION   LOCKER

INSTRUCTOR   DATE

EXERCISES

Draw the Lewis structure(s) for each of the molecules and molecular ions listed below. Follow the procedures given in the summary at the end of the study assignment. Include the formal charges in the Lewis structures. Look for equivalent resonance structures and write the number of equivalent structures.

Molecule	Number of valence electrons	Bond skeleton	Remaining electrons	Lewis structures with formal charges	Resonance structures? How many?
CF_4					
NH_3					
OCl_2					
PO_4^{3-}					
NH_4^+					
C_2H_6					

Molecule	Number of valence electrons	Bond skeleton	Remaining electrons	Lewis structures with formal charges	Resonance structures? How many?
C_2H_4					
C_2H_2					
CO					
XeF_4					
ClO_4^-					
I_3^-					
NO_2^-					

Molecule	Number of valence electrons	Bond skeleton	Remaining electrons	Lewis structures with formal charges	Resonance structures? How many?
NO_3^-					
CO_3^{2-}					
H_2O_2					
N_2H_4					
CH_3NH_2					
BrO_3^-					

QUESTIONS

1. The bond energies (the energy required to separate molecules into atoms) and the bond lengths are shown for the molecules N_2, O_2, and Cl_2.

Molecule	Bond energy (kJ/mol)	Bond length (nm)
N_2	945	0.110
O_2	498	0.121
Cl_2	243	0.199

Draw the Lewis structures of these molecules. What is the bond order (the number of bonds) in each molecule?

(a) Bond orders: N_2 _____ O_2 _____ Cl_2 _____

State the correlation between the

(b) bond energies and the bond orders

(c) bond lengths and the bond orders of the molecules

2. The molecules PF_5 and AsF_5 exist, but the analogous molecule NF_5 does not. Why not?

3. For the molecule N_2O there are five possible Lewis structures that satisfy the octet rule. Three have the N—N—O bond skeleton and two would have the N—O—N skeleton.

(a) Draw the five Lewis structures.

(b) By considering the formal charges, can you suggest which structures might be eliminated?

(c) Which structures have the smallest formal charges?

(d) Because oxygen is more electronegative than nitrogen, it is more likely to have a negative formal charge. Which structure would then be the best description of N_2O?

(e) Are the structures that have the N—O—N bond skeleton resonance forms of the structures with the N—N—O skeleton? Explain.

4. Hydrogen isocyanide, HNC, will convert to hydrogen cyanide, HCN, over time. Use their Lewis structures to explain this.

Models of Molecular Shapes
VSEPR Theory and Orbital Hybridization

PURPOSE

To build models of molecules of the type AX_mE_n, following the rules of the valence shell electron-pair repulsion (VSEPR) method. (A stands for the central atom, X represents an atom bonded to the central atom, and E represents the lone pair[s] of electrons on the central atom.) To use the models to help visualize the geometry of hybrid orbitals of the central atom and the geometric arrangement of the atoms. To use models to help predict if the molecule will have a dipole moment. To explore the connections among Lewis structures, VSEPR theory, and the localized bond orbital description of a covalent chemical bond.

INTRODUCTION

Chemists use symbols and models to help them think about the chemistry of atoms and molecules. For example, the symbols in a chemical equation, such as $2H_2 + O_2 \rightarrow 2H_2O$, provide a visual image of the different kinds of atoms that participate in a chemical reaction, making it easier to see at a glance the chemical composition of the reactants and products.

Chemists also use mathematical equations to model the functional relationships among several physical variables represented by symbols. For example, the ideal gas equation, $PV = nRT$, concisely models the relation between the pressure, volume, number of moles, and temperature of an ideal gas. Relationships between the variables P, V, T, and n may also be represented by pictures, called *graphs*, where values of one variable are plotted versus another.

We often draw pictures of orbitals to represent the distribution of electrons about the nucleus of a hydrogen atom. Physical models of atoms and molecules are also useful for visualizing how atoms are connected in three-dimensional space—what we call the *molecular geometry* of a molecule.

In fact, using computers and computer graphics software, we can now perform accurate calculations of molecular geometry that can be translated into visual images on a video terminal. Such images can be rotated so that we can view them from any angle and are useful for understanding the three-dimensional shapes of molecules, particularly those that are large and complex, such as enzymes. Figure 19-1 presents two images of the molecule ammonia (NH_3) produced by molecular modeling programs. The ball-and-stick model clearly shows how the atoms are connected by chemical bonds and the molecule's overall geometry. The space-filling model provides a more accurate impression of the molecule's shape and how the molecule might look if we could magnify its size and see the electron clouds that define its size and shape. The surface plotted is called an *isodensity surface* because it represents a surface having constant an electronic charge density that includes about 98 percent of the molecule's electronic charge.

Let's review some of the models we use in talking about the electronic structure of atoms and the chemical bonds between atoms.

ELECTRONIC STRUCTURE OF ATOMS

Most general chemistry texts show two ways of representing the electronic structure of an atom, using the $1s^2 2s^2 2p^6$ . . . notation or using boxes or circles to represent hydrogen atom–like orbitals. For example, the electronic structure of a carbon atom in its lowest energy state, called the ground state, can be represented by the notation $1s^2 2s^2 2p^2$ or by the orbital diagram

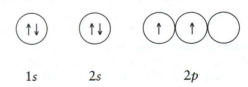

$1s \qquad 2s \qquad\qquad 2p$

FIGURE 19-1

Two different models of the molecule ammonia, NH_3. In A the ball-and-stick model shows the geometric arrangement of the atoms and lone pair of electrons clearly and corresponds most closely to the way we draw the Lewis structure for the NH_3 molecule. The space-filling model shown in B gives a more accurate picture of the overall shape and distribution of electronic charge density in the molecule.

COVALENT CHEMICAL BONDS

Then we were introduced to the intuitive insight of G. N. Lewis, who postulated the idea that a covalent bond between two atoms can be described as the result of the mutual attraction of a pair of negatively charged electrons and the positively charged cores of the atoms. (In this context, an atom core is a positively charged nucleus surrounded by filled inner electron shells.) Later, quantum chemistry provided an interpretation explanation of why electrons tend to group in pairs. This behavior is closely related to a property of electrons that we call *spin*. Recall that the electrons belonging to a particular orbital of a hydrogen-like atom have the same quantum numbers except for the spin quantum number. Electrons belonging to the same orbital have opposite spins. We say they are *spin paired*. Roughly speaking, we may think of an orbital in an atom's valence shell that is filled with two spin-paired electrons as an impenetrable charge cloud that occupies a certain volume of space and excludes other electrons from this space. Some authors use the

term *electron pair domains* or *electron charge clouds* to describe these regions of space occupied by two electrons with paired spins.

The number of *valence electrons* (outer shell electrons) largely determines how many bonds an atom can form and the shapes of the resulting molecules. For example, in the second period of the periodic table (Li to Ne) the outer valence shell has a capacity for eight electrons ($2s$ and $2p$ orbitals). When the valence shell is half-filled or less (Li to C), the number of bonds formed by the atom is equal to the number of valence electrons. When the valence shell is half-filled or greater (C to Ne) the number of bonds formed is equal to the number of *orbital vacancies*. (The number of orbital vacancies is just equal to the number of electrons that would have to be added to the atom to fill its valence shell.) Thus, Li, Be, and B typically form 1, 2, and 3 covalent bonds, respectively, while N, O, F, and Ne typically form 3, 2, 1, and 0 covalent bonds, respectively. The carbon atom, having a half-filled valence shell, is able to form the maximum number of bonds (normally four).

LEWIS STRUCTURES

To represent a chemical bond, we learn how to draw Lewis structures, which show the electrons grouped in pairs. Lewis structures are useful for predicting things like the number of lone pairs and bonding pairs and the *bond order* (single, double, or triple bond); such information is in turn correlated with bond energy and bond length. However, we cannot deduce the molecular shape (or geometry) from the Lewis structure alone.

VSEPR THEORY

Building on the idea of a Lewis structure, we observe that the shapes of molecules can be explained by assuming (a) that valence electrons group together in *bonding pairs* (single or multiple bonds) and *lone pairs* and (b) that the bonding pairs or lone pairs of electrons tend to repel one another. A simple set of rules summarizing this behavior is called the valence shell electron-pair repulsion (*VSEPR*) theory, which can be used to make predictions about the molecular geometry of molecules.

VSEPR rules are very simple. A chemical bond is viewed as the result of the mutual attraction of two positively charged atom cores for one (or more) pairs of electrons that occupy a volume of space largely located between the two atoms. The bonding and lone pairs of electrons on a central atom arrange themselves so as to balance these attractive bonding forces with the repulsive forces that exist between the electron pairs in filled valence orbitals. The magnitude of the repulsive interactions, in order of decreasing repulsion, is assumed to be: Lone Pair—Lone Pair (LP—LP) > Lone Pair—

A   Tetrahedral structure of CCl₄

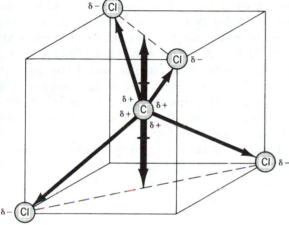

B   Canceling bond dipoles in CCl₄

FIGURE 19-2

The symmetry of the tetrahedral structure of CCl₄ shown in A causes the bond dipole moments to cancel one another as shown in B. Thus, CCl₄ has a zero resultant dipole movement.

the overlap can be improved by combining the *s* and *p* orbitals on an atom to form a new set of orbitals called *hybrid orbitals*. Thus the combination of an *s* orbital with the three *p* orbitals on a carbon atom forms a new set of four sp^3 hybrid orbitals that show the tetrahedral geometry characteristic of carbon. In other words, this particular way of combining the orbitals on the central carbon atom provides a way to explain or rationalize the known tetrahedral structure of carbon compounds, in which carbon is bonded to four other atoms. The process of combining the orbitals of an atom is called *orbital hybridization*. The number of hybrid orbitals formed is always equal to the number of atomic orbitals that are combined. The hybrid orbital description helps us make better predictions (or explanations) of molecular geometry. It should be emphasized here that we will use the notion of orbital hybridiza-

tion only in a qualitative way to explain molecular shapes. However, in making quantitative computations of the properties of molecules, the mathematical functions describing the atomic orbitals are always combined in a way that minimizes the total energy of the molecule.

LIMITATIONS OF MODELS AND THE ATOMIC ORBITAL CONCEPT

It is important to remember that models are not the real thing. In particular, when you make models of molecules you should keep in mind the following:

1. The notion of an orbital in a many-electron atom is a somewhat artificial but useful concept based on the notion that electrons in an atom with many electrons behave somewhat like the single electron in a hydrogen atom. Therefore atomic orbitals in atoms with many electrons are sometimes said to be *hydrogen-like* orbitals. Atomic orbitals in atoms with many electrons are constructed so that when we add up the electron density of the orbitals we get the correct electron density for an atom or molecule as determined by comparison with the experiment.

2. The concept of orbital hybridization provides a way to explain the geometry (arrangement in three-dimensional space) of the chemical bonds. Thus, the pictures we draw of covalent bonds in molecules are often based on hybrid orbitals. Why is this? In simple language, it is because our brains don't add or integrate the visual images of the electronic charge density of the pure atomic *s*, *p*, and *d* atomic orbitals very accurately. As a result, we often use hybrid orbitals (mixtures of *s*, *p* and *d* atomic orbitals) rather than native (unhybridized) *s*, *p*, and *d* atomic orbitals, because hybrid orbitals help us draw pictures of the molecular geometry that are in better agreement with the facts. It's also true that the labels used to designate hybrid orbitals are firmly entrenched in the chemistry curriculum. So you will encounter the hybrid orbital labels in later courses, particularly in organic chemistry, which emphasizes the chemistry of carbon compounds.

3. Aside from the conceptual limitations of the orbital concept, the physical models of molecules you will build also have limitations; that is, there are many ways in which they fall short of adequately representing real molecules, particularly their dynamic character. As you build the models, think about how they differ from real molecules. You will be asked to note some of the limitations of the models in your conclusion to the experiment.

EXPERIMENTAL PROCEDURE

Special Supplies: Molecular model kits. (Prentice-Hall Molecular Model Set for General Chemistry, available from Aldrich Chemical Co., P.O. Box 355, 1001 West Saint Paul Ave., Milwaukee, WI 53233, tel 800-558-9160, *are a convenient size and are durable.) Each student (or pair of students) should have a minimum of one 6-hole octahedral atom center, one 5-hole trigonal bipyramid atom center and one 4-hole tetrahedral atom center to represent the central A atoms, six 1-hole atom centers to represent the bonded (X) atoms and six medium-length gray plastic connectors that will represent the bonding and lone-pair electrons.*

NOTES TO INSTRUCTOR: *If you use Prentice-Hall molecular models it should be noted that the 6-hole octahedral atom centers and 5-hole trigonal bipyramid atom centers are contained only in the General Chemistry model sets and are not contained in the Organic Chemistry model sets. However, it is useful to also have one Organic Chemistry model set to provide spare (white) one-hole atom centers, as they are easily lost. One general chemistry set can serve two students working individually or four students working in pairs.*

Other activities combine well with model building. For example, we sometimes have students build VSEPR models only through the AX₄ family, view a film or video on infrared spectroscopy, take an infrared spectrum of a gas such as carbon dioxide or methane, and view animations of molecular vibrations using a molecular modeling program, such as Spartan (from Wavefunction, Inc.) These activities show students the dynamic character of molecules. The absorption of infrared light by molecules can be related to the greenhouse effect and to the notion that each molecule has a unique infrared spectrum that serves as an identifying "fingerprint." Or, if you have enough computers and molecular modeling programs available, students can build visual images of the molecules they are studying rather than working with physical models. Space-filling (isodensity-surface) renderings also give students a more realistic idea of molecular shapes.

The models are constructed by joining the atom centers with gray connectors. A gray connector without a (white) X atom joined to it will be considered to be a lone-pair of electrons on the central A atom. Thus, the gray connectors that link atoms will represent either bonding electrons or lone-pair electrons on the central (A) atom, depending on whether or not a (white) X atom is connected to the central atom.

Following the directions given in the Report form, draw Lewis structures and build models of each assigned molecule. As you proceed, fill in the table on the first page of the Report form. The models can be grouped into five families, each family having a characteristic maximum number of electron pairs (bonding pairs + lone pairs). For each model, begin with the appropriate central atom.

Total number of bonds and lone pairs. The total number of bonds and lone pairs on the central atom determines the VSEPR family. This number also represents the maximum number of electron pairs on the central atom and the maximum number of atoms that can be bonded to the central atom. For each VSEPR family, the

sum of the number of bonds and the number of lone pairs remains constant. Within each VSEPR family there are one or more VSEPR classes, in which atoms are replaced by lone pairs of electrons.

ORBITAL HYBRIDIZATION. Associated with each VSEPR family is an idealized geometry and orbital hybridization. The relationships among the five VSEPR families, total number of bonds and lone pairs, orbital hybridization and idealized geometries are summarized in Table 19-1.

VSEPR CLASS, MOLECULAR GEOMETRY, AND DIPOLE MOMENT. For each VSEPR family there will be one or more VSEPR classes given by the general formula AX_mE_n, in which atoms have been replaced by lone pairs of electrons. Each of the thirteen VSEPR classes shown in Table 19-1 has a molecular geometry with a particular name that describes the arrangement of the atoms. Note that the number of bonds and lone pairs determine the idealized geometry, but if lone pairs are present on the central atom, these are ignored when naming the molecular geometry. In other words, only the geometric arrangement of the atoms is considered when naming the molecular shape.

The overall symmetry of the molecule will determine whether or not the molecule has a permanent dipole moment.

BIBLIOGRAPHY

Gillespie, R. J., and Hargittai, I. *The VSEPR Model of Molecular Geometry,* Allyn & Bacon, 1991.

Pauling, L. *The Nature of the Chemical Bond,* 3rd edition, Cornell University Press, 1960, Chapter 4.

☞ Consider This

Almost every model that chemists use has features designed to highlight certain aspects of the system being modeled. As a result, any model distorts other aspects of the system. Critique the models used in this experiment by answering these questions.

1. *In Figure 19-1 A (p. 236), how are bond pairs of electrons represented? What features of bond pairs are well represented by this "stick" model? Which features of electrons are greatly distorted here?*

2. *What features of lone-pair electrons are well represented in Figure 19-1 B? What is meant by the dots in the figure?*

3. *Compare Figure 19-1 A and 19-1 B. In what ways is B an improvement?*

In your text you can probably find the ammonia molecule represented by a structure such as

$$H - N - H$$
$$|$$
$$H$$

What features of molecular structure does this represent? Which features are ignored by this structure?

Can you invent other ways to represent the structures of molecules that are more realistic or less confusing than the ones you see in your text?

There are a few molecules with coordination numbers beyond the six-coordinate examples used here, such as IF_7. Apply the principles that you learned about VSEPR and orbital hybridization to describe the geometry of this molecule and the bonding within it.

2. The AX₃ Family (sp^2 orbital hybridization)

The VSEPR AX₃ class (the first member of the AX₃ family)

Example 2:   BF₃   Experimentally it is determined that BF₃ is a trigonal planar molecule with all B—F bonds identical, and all F—B—F angles equal to 120°.

(a) Draw the Lewis structure for the BF₃ molecule:

3

$\dfrac{21}{24}$

$|\overline{F} - B - \overline{F}|$
$\quad\quad |\underline{F}|$

(b) Show, by filling in the orbital diagrams for the *valence* electrons of the B atom in its ground and hybrid orbital states, how s and p orbitals can be combined to form a set of three half-filled sp^2 hybrid orbitals.

Electronic configuration of B (Z = 5   ) $1s^2 2s^2 2p^1$

ground state ⇒ ___sp^2___ hybrid orbitals

(c) What atomic orbitals are combined to give sp^2 hybrid orbitals? $1s + 2p$

(d) How many sp^2 hybrid orbitals are formed by hybridization? 3

(e) What is the geometric arrangement of the sp^2 hybrid orbitals? trigonal planar

Construct a model of the sp^2 hybrid orbitals, arranging them as shown below.

Make a model of BF₃ by inserting three gray connectors into the set of three holes on a (brown) atom center that form a trigonal planar set 120° apart. Connect three white balls, representing fluorine atoms, to the three gray connectors inserted in the (brown) atom center. As before, each connector represents a bonding pair of electrons, with one of the electrons in each bond coming from the boron atom and the other coming from a fluorine atom.

(f)   Now fill in the report table for BF_3.

The AX_2E VSEPR class (a member of the AX_3 family)

Example 3: $SnCl_2$ Experimentally it is determined that in the gas phase this is a bent (nonlinear) molecule with a Cl—Sn—Cl angle of about 99°. (The smaller-than-anticipated bond angle might be attributed to the tendency of lone-pair electrons to repel more strongly the pairs of bonding electrons and to the large size of the tin atom, which lowers the strength of the Cl—Cl repulsions because of the greater distance between the chlorine atoms.) Both bonds are equal in length and strength.

(a)   Draw the Lewis structure for the $SnCl_2$ molecule:

$$Sn \quad 4$$
$$2Cl \quad \underline{14}$$
$$18$$

$$|\overline{Cl} = Sn - \overline{C}l|$$

(b)   By filling in the orbital diagrams for the *valence* electrons of the Sn atom in its ground and hybrid orbital states, show how *s* and *p* orbitals can combine to give a set of three sp^2 orbitals, one orbital containing a lone pair of electrons and two half-filled bonding orbitals.

Electronic configuration of Sn (Z = _____) $\underline{5s^2 5p^2}$

ground state ⇒ $\underline{sp^2}$ hybrid orbitals

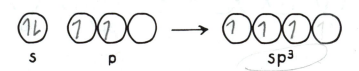

(Note that another orbital diagram might be drawn in which the *s* and *p* orbitals of tin are mixed to form four equivalent sp^3 hybrid orbitals. This would allow the formation of four bonds, and indeed tetrahedral $SnCl_4$ molecules are known to exist.)

Remove a white ball from one of the gray connectors of the previous BF_3 model. This gray connector now represents a lone pair, with the brown atom center representing a tin atom and the two remaining white balls representing chlorine atoms. Assume that the lone-pair electrons tend to repel the two remaining bonding pairs more than the bonding pairs repel each other. What might this be expected to do to the Cl—Sn—Cl bond angle?

(c)   Now fill in the report table for $SnCl_2$.

3. The AX₄ VSEPR family (sp^3 orbital hybridization)

The VSEPR AX₄ class (the first member of the AX₄ family)

Example 4: CH₄ Experimentally it is determined that all C—H bonds are identical, and all H—C—H bond angles are 109°.

(a) Draw the Lewis structure for the CH_4 molecule:

$$H-\underset{|}{\overset{H}{C}}-H$$

(b) By filling in the orbital diagrams for the *valence* electrons of the C atom in its ground state and its hybrid orbital state, show how *s* and *p* orbitals can be combined to form a set of four half-filled sp^3 hybrid orbitals.

Electronic configuration of C (Z = 6) $2s^2 2p^2$

ground state ⇒ _____sp^3_____ hybrid orbitals

(c) What atomic orbitals are combined to give sp^3 hybrid orbitals?

$2s + 2p$

(d) How many sp^3 hybrid orbitals are formed by hybridization? 4

(e) What is the geometric arrangement of the sp^3 hybrid orbitals? tetrahedral

Construct a model of the sp^3 hybrid orbitals by inserting four gray connectors into the four holes of a (black) tetrahedral atom center, as shown below:

Make a model of CH_4 by fastening white atom centers, representing H atoms, to each of the four gray connectors on the tetrahedral atom center.

(f) Now fill in the report table for CH_4.

More about the VSEPR method for predicting molecular shapes. *The relative magnitude of the repulsive forces between bonding pairs of electrons and lone-pair electrons influences molecular shapes.* Suppose one or more of the sp^3 orbitals is filled with a lone pair of electrons. When only one lone pair is present, the repulsions between a lone pair and a bond pair (LP—BP) are greater than those between two bond pairs (BP—BP). When more than one lone pair is present, the repulsions between two lone pairs (LP—LP) are even greater than the BP—LP repulsions. The magnitude of the repulsions therefore decreases in the order LP—LP $\gg$ LP—BP $>$ BP—BP, and these repulsions determine the geometry and bond angles of the molecule. In particular, structures with 90° LP—LP interactions involve such large repulsive forces that they become much less stable than structures that allow larger angles between the lone pairs.

The VSEPR AX₃E class (a member of the AX₄ family)

Example 5: NH_3 Experimentally it is determined that all N—H bonds are equivalent and all H—N—H bond angles are 107.3°.

(g)   Draw the Lewis structure for the NH_3 molecule:

(h)   By filling in the orbital diagrams for the *valence* electrons of the N atom in its ground and hybrid orbital states, show how it can hybridize to give hybrid orbitals whose geometry is represented by the model (one orbital filled with a lone pair of electrons plus three half-filled bonding orbitals).

Electronic configuration for N (Z = 7   ) $\underline{2s^2 2p^3}$

ground state ⇒ $\underline{\quad Sp^3 \quad}$ hybrid orbitals

Make a model of the NH_3 molecule by removing a white atom from one of the gray connectors of the CH_4 model. This connector now represents a lone pair of electrons, with the central atom representing a nitrogen atom and the three white atom centers representing hydrogen atoms. Let the molecule rest on the lab bench on the triangle of three hydrogen atoms. What name would you use to describe the geometry of the molecule considering only the positions of the atoms?

(i)   If you assume that LP—BP repulsion is greater than BP—BP repulsion, what would you expect to happen to the bond angles between the remaining bonding orbitals?

(j)   Now fill in the report table for NH_3, referring to your model of the ammonia molecule.

The VSEPR class AX$_2$E$_2$ (a member of the AX$_4$ family)

Example 6:   *H$_2$O*   Experimentally it is determined that both O—H bonds are equivalent and the H—O—H bond angle is 104.5°.

(k) Draw the Lewis structure for the H$_2$O molecule:

$$H - \bar{\underline{O}} - H$$

(l) By filling in the orbital diagrams for the *valence* electrons of the O atom in its ground and hybrid orbital states, show how *s* and *p* orbitals can be combined to give a set of four *sp^3* orbitals (two orbitals filled with lone pairs of electrons and two half-filled bonding orbitals).

Electronic configuration of O (Z = 6) $2s^2 2p^2$

ground state ⇒ sp^3 hybrid orbitals

Make a model of water, H$_2$O, by removing a white atom from the NH$_3$ model. Now there will be two gray connectors without atoms, representing the two lone pairs of electrons on the water molecule. Assume that the central atom now represents an oxygen atom and that the two white balls remaining represent hydrogen atoms. What name would you use to describe the geometric arrangement of the atoms?

(m) Why might you expect the H—O—H bond angle (104.5°) to be smaller than the H—N—H bond angle (107.3°)?

(n) Now fill in the report table for H$_2$O.

4. The VSEPR AX$_5$ family (*sp^{3}d* orbital hybridization)

Atoms of the main-group elements of the third and following periods can accommodate more than eight electrons around the nucleus. In other words, more than four orbitals are available—because of the availability of vacant *d* orbitals whose energy is not much different from the energy of the *s* and *p* orbitals in the same period.

The VSEPR AX$_5$ class (the first member of the AX$_5$ family)

Example 7:   *PCl$_5$*   Experimentally it is found that three P—Cl bonds lie in a plane making angles of 120° to one another. The other two P—Cl bonds are slightly longer and are perpendicular to the three bonds in the trigonal plane.

(a) Draw the Lewis structure for the PCl$_5$ molecule (note that the P atom will have a share of ten electrons):

P   5
5 Cl 35
     ——
     40

$$|\overline{Cl}|$$
$$|\overline{Cl}| - P - \overline{Cl}||$$
$$|\overline{Cl}|\quad|\overline{Cl}||$$

(b) By filling in the orbital diagrams for the *valence* electrons of the P atom in the ground and hybrid orbital states, show how s, p, and d orbitals can be combined to form a set of five sp^{3}d hybrid orbitals.

Electronic configuration of P (Z = 15) $3s^2 3p^3$

ground state ⇒ sp^3d hybrid orbitals

⑪ ①①① ○○○○○ → ①①①①① ○○○○
S   p   d   sp³d   d

(c) What atomic orbitals are combined to give sp^{3}d hybrid orbitals?

3s, 3p, 3d

(d) How many sp^{3}d hybrid orbitals are formed by hybridization?

5

(e) What is the geometric arrangement of the orbitals?

trigonal bipyramid

Construct a model of the sp^{3}d hybrid orbitals, by adding five gray connectors to the trigonal bipyramid atom center, as shown below:

Make a model of PCl$_5$ by connecting five white atoms, representing Cl atoms, to the trigonal bipyramid atom center you constructed in the previous step.

(f) Now fill in the report table for PCl$_5$. Think carefully about the symmetry of the atoms in the trigonal plane and the two atoms perpendicular to the trigonal plane. Would you expect the bond moments to cancel, giving a net dipole moment of zero? Or would you expect the molecule to have a nonzero dipole moment?

The VSEPR AX₄E class (a member of the AX₅ family)

Example 8:   SF₄   Experiments show that this molecule has a seesaw shape. The F—S—F bond angle of the fluorine atoms in the equatorial positions is about 101° (less than the 120° angle of the idealized geometry), while the F—S—F bond angle of the two axial fluorine atoms is 169° (less than the 180° angle of the idealized geometry). Both of these distortions from the ideal trigonal bipyramid geometry might be explained as resulting from repulsion of the bonding orbitals by the lone-pair electrons in the equatorial position.

(g)   Draw the Lewis structure for the SF₄ molecule:

S 6
F 28
———
34

(h)   By filling in the orbital diagrams for the *valence* electrons of the S atom in its ground and hybrid orbital states, show how s, p, and d orbitals can be combined to give a set of five sp^3d hybrid orbitals (one lone pair plus four half-filled bonding orbitals).

Electronic configuration of S (Z = 14) $3s^2 3p^4$

ground state ⇒ Sp^3d hybrid orbitals

Make a model of the molecule SF₄ by removing a white atom from the PCl₅ model. The empty gray connector now represents a lone pair of electrons, and the remaining 4 white atoms representing the 4 F atoms, all connected to the central S atom. Notice that the geometry that results depends on which bonding orbital is replaced by a lone pair of electrons. The lone pair can be located in the plane containing three orbitals (called the equatorial positions) or in one of the two perpendicular orbitals (called the axial positions). How many different models are possible, considering that the equatorial and axial positions are not equivalent? _____

(i)   The following diagrams show the two possible geometric arrangements, in which lines represent the bonding orbitals and the lone pair is represented by a lobe (or rabbit ear). Next to each diagram, list the number of LP—LP, LP—BP, and BP—BP repulsions. Count only repulsions at 90° angles, ignoring repulsions at angles greater than 90°.

①   2 LP - BP
                           4 BP - BP

②   3 LP - BP
                           3 BP - BP

(j) Which is the preferred geometric arrangement, and what is its molecular geometry? Explain why it is the preferred arrangement.

(k) Now fill in the report table for SF_4.

The VSEPR AX$_3$E$_2$ class (belonging to the AX$_5$ family)

Example 9: BrF_3 Experimentally determined geometry shows T-shaped molecules. The bond angles between the axial and equatorial F atoms is about 85°, less than the 90° that would be expected in the idealized structure.

(l) Draw the Lewis structure for the BrF_3 molecule:

Br 7
3F 21
    ——
    28

$|\overline{F}-\overset{\displaystyle .}{\underset{\displaystyle |\overline{F}|}{\overline{Br}}}-\overline{F}|$

(m) By filling in the orbital diagrams for the *valence* electrons of the Br atom in its ground and hybrid orbital states, show how *s*, *p*, and *d* orbitals can be combined to form a set of five *sp*^{3}*d* hybrid orbitals (two orbitals filled with lone pairs plus three half-filled bonding orbitals).

Electronic configuration of Br (Z =   ) _____

ground state ⇒ _____ hybrid orbitals

s   p   d   sp³d   d

Construct a model of the BrF_3 molecule by removing a white atom from the SF_4 model, creating a model with 3 bonds and 2 lone pairs. The remaining 3 white atoms will represent the 3 F atoms. Assuming that the 2 lone pairs can be in equatorial or axial positions, how many different geometric arrangements are possible? (Hint: Consider the possible arrangements to be equatorial-equatorial, axial-equatorial, and axial-axial.) _____

(n) The following diagrams show the possible geometric arrangements, using lines to represent the 3 bonding orbitals and lobes (or rabbit ears) to represent the two lone pairs. Next to each diagram list the number of LP—LP, LP—BP, and BP—BP repulsions. Again, count only repulsions at 90° angles, ignoring repulsions at angles greater than 90°.

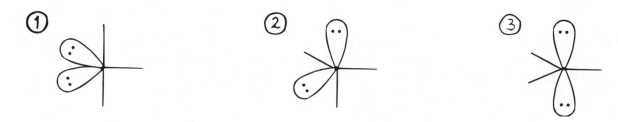

(o) Which is the preferred form, and what is its geometric arrangement? Explain why it is the preferred arrangement.

(p) Now fill in the report table for BrF_3.

The VSEPR class AX$_2$E$_3$ (a member of the AX$_5$ family)

Example 10: *XeF$_2$* Experiments show that this molecule has a linear geometry.

(q) Draw the Lewis structure for the XeF$_2$ molecule (the Xe atom will have a share of 12 electrons):

$$\overline{|F} - \overline{Xe} - \overline{F|}$$

(r) By filling in the orbital diagrams for the *valence* electrons of the Xe atom in its ground and hybrid orbital states, show how s, p, and d orbitals can be combined to form a set of five sp^3d hybrid orbitals (three lone pairs plus two half-filled bonding orbitals).

(Note in this hybrid orbital model how the promotion of an electron from the filled p orbitals to the empty d orbitals creates orbital vacancies [half-filled bonding orbitals] that can overlap with similar half-filled bonding orbitals of the F atoms to form a Xe—F bond.)

Electronic configuration of Xe (Z =   ) _____

ground state ⇒ _____ hybrid orbitals

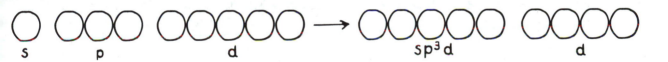

Construct a model of XeF$_2$ by removing a white atom from the model for BrF$_3$. The three empty gray connectors represent three lone pairs while the two remaining white atoms represent fluorine atoms on the central Xe atom.

(s) How many different geometric arrangements are possible? (Hint: The 2 fluorine atoms can be located in equatorial-equatorial, axial-equatorial, or axial-axial positions.) _____

(t) The following diagrams show the possible geometric arrangements in trigonal bipyramid geometry. Next to each diagram list the number of LP—LP, LP—BP, and BP—BP repulsions. Again, count only repulsions at 90° angles, ignoring repulsions at angles greater than 90°.

①

②

③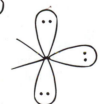

(u) Which is the preferred geometric arrangement? Explain why it is the preferred arrangement.

(v) Now fill in the report table for XeF$_2$.

5. The VSEPR AX$_6$ family (*sp*^{3}*d*^{2} orbital hybridization)

The VSEPR AX$_6$ class (the first member of the AX$_6$ family)

Example 11:   SF$_6$ This molecule is experimentally determined to be octahedral. All bonds are equivalent and all bond angles are 90°.

(a) Draw the Lewis structure for the SF$_6$ molecule. (The S atom will have a share of 12 electrons.)

(b) By filling in the orbital diagrams for the *valence* electrons of the S atom in its ground and hybrid orbital states, show how *s*, *p*, and *d* orbitals can be combined to form a set of six *sp*^{3}*d*^{2} hybrid orbitals (six half-filled bonding orbitals).

Electronic configuration of S (Z = _____) _____

ground state ⇒ _____ hybrid orbitals

(c) What atomic orbitals are combined to give *sp*^{3}*d*^{2} hybrid orbitals?

_____

(d) How many *sp*^{3}*d*^{2} hybrid orbitals are formed? _____

(e) What is the geometric arrangement of the *sp*^{3}*d*^{2} hybrid orbitals? _____

Construct a model of the *sp*^{3}*d*^{2} hybrid orbitals as shown below.

Make a model of SF$_6$ by fastening a white atom center to each of the six gray connectors representing the *sp*^{3}*d*^{2} hybrid orbitals.

(f) Now fill in the report table for SF$_6$.

The VSEPR class AX$_5$E (a member of the AX$_6$ family)

Example 12:   IF$_5$ Experiments show that the molecule has a square pyramidal molecular geometry with bond angles of 90°.

(g) Draw the Lewis structure for the IF$_5$ molecule (the I atom will have a share of 12 electrons):

(h) By filling in the orbital diagrams for the *valence* electrons of the I atom in the ground and hybrid orbital states, show how s, p, and d orbitals can be combined to form a set of six sp^3d^2 hybrid orbitals (one lone pair plus five half-filled bonding orbitals).

Electronic configuration of I (Z =   ) _____

ground state ⇒ _____ hybrid orbitals

Make a model of IF$_5$ by removing a white atom from the SF$_6$ model. The empty gray connector represents the lone pair of electrons on the I atom, and the five remaining white atom centers represent the F atoms. Does it make any difference which orbital is replaced by a lone pair?

(i) Considering only the positions of the atoms, what name is given to the IF$_5$ molecular geometry? _____

(j) Now fill in the report table for IF$_5$.

The VSEPR class AX$_4$E$_2$ (a member of the AX$_6$ family)

Example 13:   XeF$_4$ Experiments show that this molecule has a square planar geometry with bond angles of 90°.

(k) Draw the Lewis structure for the XeF$_4$ molecule (the Xe atom will have a share of 12 electrons):

255

(l) By filling in the orbital diagrams for the *valence* electrons of the Xe atom in the ground and hybrid orbital states, show how *s*, *p*, and *d* orbitals can be combined to form a set of six sp^3d^2 hybrid orbitals (two lone pairs plus 4 half-filled bonding orbitals).

Electronic configuration of Xe (Z =   ) _____

ground state ⇒ _____ hybrid orbitals

Xe 8
4F $\frac{28}{36}$

S   p   d   → sp^3d^2 d

Make a model of the XeF$_4$ molecule by removing a white atom center from the model for IF$_5$. There will be two empty gray connectors representing the two lone pairs of electrons while the remaining four white atom centers represent the F atoms in XeF$_4$.

(m) How many different models are possible? _____

Next to the diagram for each model, list the total number of LP—LP, LP—BP, and BP—BP repulsions. Ignore repulsions at angles greater than 90°.

(n) Which is the preferred model and what is its geometric arrangement? Explain why it is the preferred arrangement?

(o) Complete the report table for XeF$_4$.

END OF LAB ACTIVITY

(a) Check that you have completed all of the entries for the table of molecular geometry on the first page of the report, summarizing your results.

(b) Disassemble the models, and return them to their storage boxes.

(c) Answer the following questions.

QUESTIONS

1. In what way are the models used in this experiment inaccurate representations of molecules?

2. In what way are the models used in this experiment accurate representations of molecules?

3. For all of the molecular structures belonging to the AX_5 family (trigonal bipyramid sp^3d orbital hybridization) in which the central atom has one or more lone-pair electrons, what is the preferred location of the lone pairs, the *equatorial* or the *axial* position?

4. Note that the different VSEPR families do not have the same number of members. (The members of each family are the VSEPR classes that belong to that family.) For example, the next VSEPR class of the AX_6 family after AX_4E_2 would be the hypothetical AX_3E_3. This would require a central atom with nine valence electrons to combine with three electrons contributed by the three X atoms to give a total of 12 electrons on the central A atom. Why would you expect that the VSEPR class AX_3E_3 would not exist?

Consider This

Almost every model that chemists use has features designed to highlight certain aspects of the system being modeled. As a result, any model distorts other aspects of the system. Critique the models used in this experiment, by answering these questions.

1. *In Figure 19-1 A (p. 236), how are bond pairs of electrons represented? What features of bond pairs are well represented by this "stick" model? Which features of electrons are greatly distorted here?*

2. *What features of lone-pair electrons are well represented in Figure 19-1 B? What is meant by the dots in the figure?*

3. *Compare Figure 19-1 A and 19-1 B. In what ways is B an improvement?*

In your text you can probably find the ammonia molecule represented by a structure such as

$$H-N-H$$
$$|$$
$$H$$

What features of molecular structure does this represent? Which features are ignored by this structure?

Can you invent other ways to represent the structures of molecules that are more realistic or less confusing than the ones you see in your text?

There are a few molecules with coordination numbers beyond the six-coordinate examples used here, such as IF$_7$. Apply the principles that you learned about VSEPR and orbital hybridization to describe the geometry of this molecule and the bonding within it.

Intermolecular Forces
Slime Gel: Making and Killing Slime

PURPOSE

To make "slime," a gel of poly(vinyl alcohol) and borax. To see how the physical properties of linear polymers can be altered by forming intermolecular bonds between the chains.

PRE-LAB PREPARATION

If in some cataclysm, all of scientific knowledge were to be destroyed, and only one sentence passed on to the next generation of creatures, what statement would contain the most information in the fewest words? I believe it is the atomic hypothesis *(or the* atomic fact, or whatever you wish to call it) that all *things are made of atoms—little particles that move around in perpetual motion, attracting each other when they are a little distance apart, but repelling upon being squeezed into one another. In that one sentence, you will see, there is an enormous amount of information about the world, if just a little imagination and thinking are applied.*[1]

Richard P. Feynman

Feynman's description of two particles attracting each other when they are a little distance apart but repelling each other upon being squeezed into one another is graphically shown in Figure 20-1. One curve represents the *energy of interaction*, the other represents the *force of interaction* between the two particles. Net repulsive energy and force are taken as positive; net attractive force and energy are taken as negative. The zero value corresponds to separation of the two particles to a large (infinite) distance apart.

If we represent the potential energy of interaction by the symbol U, the interaction force is just the (negative)

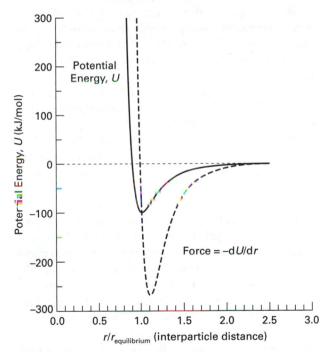

Potential Energy, U and Force vs. Interparticle Distance

FIGURE 20-1

The potential energy, U, and interaction force between two particles as a function of the normalized distance between the particles, $r/r_{equilibrium}$. The potential energy is a minimum and the force is zero at the equilibrium separation, where $r/r_{equilibrium} = 1.00$ and $r = r_{equilibrium}$. As the particles are squeezed together both the potential energy and the interaction force approach large positive values, reflecting the repulsion between the particles. As the distance between the particles becomes large, both the potential energy and interaction force approach zero.

[1]Richard P. Feynman, Robert B. Leighton, Matthew Sands *The Feynman Lectures on Physics,* Vol I, Addison-Wesley Publishing Co., Inc., 1963, pp. 1–2.

derivative of the potential energy curve with respect to the interparticle distance, r—that is, force = $-dU/dr$. The interparticle distance is plotted as a normalized or dimensionless parameter, in which the interparticle distance, r, is divided by the equilibrium interparticle distance, $r_{equilibrium}$. At the point where $r/r_{equilibrium} = 1$, the energy is at its minimum value, corresponding to the trough of the energy curve, and the force is zero. The depth of the trough below the $U = 0$ line is a measure of the energy binding the two particles together (100 kJ/mol in this example). As r becomes smaller than $r_{equilibrium}$ ($r/r_{equilibrium} < 1$), the potential energy and repulsive force increase rapidly as the two particles are squeezed together (U and the force become more positive). As r becomes larger than $r_{equilibrium}$ ($r/r_{equilibrium} > 1$), the potential energy and the force both rapidly approach zero, illustrating that the intermolecular forces between neutral particles typically operate over a very short range.

What is an atom? A picture of an atom has emerged from early twentieth century investigations as having a small, unimaginably dense nucleus comprised of neutral particles (*neutrons*) and positively charged particles (*protons*). The positively charged nucleus is surrounded by an approximately spherical cloud of negatively charged particles with much smaller mass, called *electrons*. This electron cloud defines the volume of space that the atoms occupy when they are squeezed together and in effect determines the size of the atom.

When atoms get close enough together, they can form stable aggregates that we call *molecules*. If the forces holding the atoms together are sufficiently large, we say that the atoms have formed a chemical bond, in which the atoms are held together in a stable configuration that resists change.

When molecules are close to one another, they also tend to stick together, attracted by weak short-range forces. If there were no *intermolecular forces* between molecules, all molecules would be gases. Molecules cannot exist in a liquid or solid form unless there are intermolecular forces holding them together. Without these forces there would be no liquid water or ice, no solid wood, nothing to hold the strands of protein together in the tissues and muscles of our bodies. In short, life as we know it could not exist.

At the root of atomic and molecular structure is the mutual attraction of particles with opposite charge, what are called *electrostatic* or *electrodynamic forces,* depending on whether the particles are stationary or moving. The fundamental kind of electric force binding two hydrogen atoms to an oxygen atom to form a water molecule is the same as the force that holds water molecules together to form liquid water. In this sense there is no fundamental distinction between a chemical process, such as the formation of water by reaction of molecular hydrogen and oxygen, and a physical process, such as the evaporation of

liquid water to form gaseous water. All can be viewed as resulting from the mutual interaction of positively charged nuclei and negatively charged electrons. So the distinction between a physical and chemical process is one of degree and not of kind. Physical processes involve weaker intermolecular (electric) forces and chemical processes involve stronger interatomic (electric) forces.

Weak intermolecular electric forces are responsible for many of the familiar bulk properties of liquids and solids, including viscosity (the resistance to flow), the volatility or vapor pressure of liquids and solids, the melting point and energy of fusion (melting), the boiling point and energy of vaporization, the solubility of one kind of molecule in another, the adsorption and catalytic activity of surfaces, adhesion (related to the properties of adhesives and glues), surface tension, friction and lubrication, the formation of monomolecular films and bilayer membranes, detergent action, the formation of emulsions and foams, gelation, hydrogen bonding, and specific interactions in synthetic or biological systems (often called *molecular recognition*), which are dependent on the shapes and charge distribution in interacting molecules. For example, enzymes in the body exert their catalytic effects on chemical reactions by binding a molecule in a way that facilitates or speeds up the reaction rate of the molecule. Perhaps you can use your own thinking and imagination to add to the list.

In this experiment we will study how altering the intermolecular forces between polymer chains has a dramatic effect on the physical properties of the polymer solution. The term *polymer* comes from the Greek (*poly,* many, + *meros,* part) and refers to a large molecule made up of many parts. *Macromolecule* is a term synonymous with polymer. Polymers are made by linking together simple molecules called *monomers* ("single parts"). If the monomers add together so that the polymer contains all of the atoms of the monomer molecules, the reaction is called *addition polymerization*. If small by-product molecules such as water are formed during the polymerization process, the reaction is called *condensation polymerization*.

In this experiment we will study a synthetic polymer, made by addition polymerization of the monomer vinyl acetate, followed by hydrolysis of the acetate group to produce an alcohol (—OH) group. The resulting polymer is called poly(vinyl alcohol) or PVA. It is a water soluble linear polymer with a carbon atom backbone having a predominant pattern of —OH groups on alternate carbon atoms, as shown in Figure 20-2.

THE GELATION OF POLY(VINYL ALCOHOL) WITH BORAX

Adding a molecule that produces cross-linking interactions between the chains of a linear polymer produces a structure with less flexibility. In the limit where the cross-

FIGURE 20-2
Poly(vinyl alcohol), or PVA, is a linear polymer in which the pre-dominant structure is a carbon atom backbone with hydroxy ($-OH$) groups on alternate carbon atoms. The PVA polymer strands have an average molecular weight of about 78,000 corre-sponding to an average of about 1800 $-[CH_2\text{-}CH\text{-}OH]-$ monomer units (MW = 44) in each strand.

FIGURE 20-3
A tetrahydroxyborate ion, $B(OH)_4^-$, forms a cross-link between two strands of poly(vinyl alcohol). Each tetrahydroxyborate ion can react with four $-OH$ groups. The B—O bonds are labile, rapidly and continuously breaking and reforming.

linking bonds, once formed, are very strong and do not let go, a rigid three-dimensional structure will be formed. Examples of this would include epoxy resins and vulcan-ized rubber.

If the cross-link bonds are weaker or are not perma-nently formed, something can form that is between a liq-uid and solid, having properties of a viscous gel. It can flow, but only slowly. Intermolecular links of this kind are formed when a solution of borax (sodium tetraborate decahydrate, $Na_2B_4O_7 \cdot 10H_2O$, FW 381.4) is added to a solution of PVA. When dissolved in water, sodium tetrab-orate forms an equimolar mixture of boric acid and sodium tetrahydroxyborate ion, $B(OH)_4^-$

$$Na_2B_4O_7(s) + 7H_2O \rightleftharpoons$$
$$2B(OH)_3(aq) + 2NA^+(aq) + 2B(OH)_4^-(aq) \quad (1)$$

The concentration of $B(OH)_4^-$ in solution can be deter-mined by titration of $B(OH)_4^-$ with a strong acid such as sulfuric acid:

$$2Na^+ + 2B(OH)_4^- + H_2SO_4 \rightarrow$$
$$2B(OH)_3(aq) + 2Na^+ + SO_4^{2-} \quad (2)$$

From the overall stoichiometry of Equations (1) and (2), you can see that one mole of sulfuric acid titrates the two $B(OH)_4^-$ ions produced by one mole of borax.

When a solution containing tetrahydroxyborate ion, $B(OH)_4^-$, is added to an aqueous solution of poly(vinyl alcohol), the borate ion forms crosslinks between the alcohol groups on adjacent polymer chains, as shown in Figure 20-3. These look like covalent bonds, and in fact they are polar covalent bonds, but they have an additional important property—they are *labile*. This means that they are continually and rapidly breaking and reforming. This is very much like the bonding in water between hydrogen and oxygen atoms. The polar covalent bonds in water are fairly strong, but there is a tendency for water molecules to dissociate to a slight extent, forming H_3O^+ ions and OH^- ions. This process of breaking and reform-

ing bonds is so rapid that if you put a small amount of D_2O (where deuterium is the form of hydrogen with mass 2, having an extra neutron in the nucleus) into a large amount of ordinary water, the deuterium form of hydro-gen is rapidly distributed throughout all of the water molecules so that most of the deuterium labeled water will be in the form of HDO, with very few water mole-cules existing as D_2O. Similarly, when $B(OH)_3$ labeled with ^{18}O (a form of oxygen with two extra neutrons in the nucleus) is added to a solution containing $B(OH)_4^-$ ion, the ^{18}O label is rapidly and uniformly distributed among all of the boron-containing species.

When the crosslinks are labile (rapidly breaking and reforming), the linear polymer chains can still move past one another but not as easily as when there is no cross-linking, so a viscous gel forms that still retains some liq-uid-like properties. However, if the gel is stirred very rapidly, the gel becomes more viscous—the harder you stir, the more viscous it becomes. In other words, it does not behave like a normal liquid. In fact, the gel has three distinct responses to shearing and stressing. With slow shearing (stirring) it flows like a highly viscous fluid, such as thick molasses. At somewhat faster shearing it behaves like a piece of rubber. Fast shearing causes it to fracture when it is pulled apart suddenly.

NATURAL POLYMERS
Natural linear polysaccharide (polysugar) polymers can form similar gels with borax. A plaything called Slime®, marketed by the Mattel Corporation, is a gel formed by mixing a solution of the natural polysaccharide *guar* with borax. Guar is extracted from a leguminous plant native to India.

Other natural linear polysaccharides include agar (extracted from algae), carrageenan (extracted from sea-weed), and starch (extracted from corn or potatoes). These natural products are widely used to thicken and give a smooth texture to foods and cosmetics and also find uses as adhesives and as sizing for silk and paper.

Natural polymers are of crucial importance to living organisms. Without natural structural polymers such as the *cellulose* in plants or the *proteins* that constitute the muscle and sinew of animals, no living plant or animal could stand upright against the pull of gravity. We would all be creatures of the sea, where our delicate membranes could be supported by floating in water. Plant and animal *enzymes,* which catalyze all vital biochemical reactions, are polymers of amino acids. Finally, deoxyribonucleic acid (DNA), the very stuff of the genetic inheritance of all plants and animals, is a polymer of nucleic acids.

EXPERIMENTAL PROCEDURE

Special Supplies: 1-mL microburet (see the Introduction to the lab manual), polyethylene transfer pipets with 0.5 mL and 1.0 mL calibration marks, wooden stirrers (tongue depressors or popsicle or "craft" sticks), plastic sandwich bags, Handi-Wrap film.

Chemicals: 4% (40 g/L) poly(vinyl alcohol) solution (PVA powder, J.T. Baker, 99.0–99.8% fully hydrolyzed, MW 77,000–79,000); (guar gum is a suitable and less expensive substitute for PVA); 4% (40 g/L) borax, sodium tetraborate decahydrate ($Na_2B_4O_7 \cdot 10H_2O$) solution; 3.0 M H_2SO_4; 1.0 M NaOH; 0.1% methyl red indicator.

NOTE TO INSTRUCTOR: *PVA powder dissolves slowly. It must be sprinkled in the water in small increments with vigorous stirring, sufficient to produce a small vortex, otherwise lumps will form that are difficult to dissolve. Warming the water to 75–80 °C helps dissolve the powder faster. A large stirrer/hot plate, set to low heat with a large magnetic stirring bar, provides both vigorous stirring and warming. If the PVA solution is heated on a hot plate set on high with inadequate stirring, the solution is likely to scorch, producing a very disagreeable odor. Allow 24–48 hours to prepare the PVA solution, and get it fully hydrated and free of lumps.*

Use the same precautions if guar gum is substituted for poly(vinyl alcohol).

CLEAN UP AND WASTE DISPOSAL

At the end of the laboratory, wash your hands well to remove any residue of PVA or borax. Water can be used to clean up any residue that may accidentally get on your clothing or the benchtop. The gel may be discarded in the trash.

If your instructor permits you to take the gel with you from the laboratory, put it in a plastic sandwich bag. Be careful not to create messes by getting it on clothing or furniture. Do not leave it where small children may play with it or try to eat it. Finally, note that handling the gel will eventually contaminate it, leading sometimes to the growth of molds. The gel may be disposed of by dissolving it in water or putting it in the trash.

1. TITRATION OF BORAX WITH SULFURIC ACID.

Borax, sodium borate decahydrate, is an efflorescent salt,

meaning that it loses water when exposed to air of 50 percent or less relative humidity. Therefore, it is necessary to titrate the borax solution in order to determine its concentration accurately. Put 10 mL of the 4% borax solution into a 25-mL volumetric flask. (The 4% borax solution can be measured in a 10-mL graduated cylinder, or you can weigh out 10.0 g into the 25-mL flask, because 10.0 mL $\cong$ 10.0 g).

Add 10 drops of 0.1% methyl red indicator, and titrate the resulting yellow solution with 3.0 M H_2SO_4 in a 1-mL microburet to the first appearance of a permanent red color. (See the Introduction to the lab manual for a description of the 1-mL microburet, made from a 1-mL tuberculin syringe.) Repeat the titration using a second 10-mL sample of 4% borax solution. From the volume and concentration of sulfuric acid, calculate the molar concentration of borax. (One mole of H_2SO_4 titrates one mole of borax as shown by Equations [1] and [2]). Compare with the expected theoretical molarity expected for 4.0% borax, 0.105 M.

2. MAKING "SLIME," A GEL OF POLY(VINYL ALCOHOL) AND BORAX.
Place 50 mL of a 4% aqueous solution of poly(vinyl alcohol) in a 250-mL beaker. Add 10 drops of 0.1% methyl red indicator to the PVA solution. Stirring continuously, add 1.0 mL of 4% borax solution. It will be sufficiently accurate to measure the volume of borax solution using a polyethylene transfer pipet having a 1-mL calibration mark. Something stiff and wide, like a wooden popsicle or craft stick or a half-inch wide stainless steel scoop, makes a better stirrer than a glass rod (recall mixing cans of paint). Continue vigorous stirring until the mixture forms a homogeneous gel. If the mixture separates into two phases (a gel and liquid), try scraping the gel off the stirrer and using a back and forth stirring motion straight across the beaker to chop up the gel, rather than a circular stirring motion. With time and sufficient stirring, the solution will form a homogeneous gel. The resulting gel has a pH of about 9, causing the methyl red indicator to appear yellow.

Note and record the characteristics of the gel, such as its viscosity (resistance to flow). Try stirring it slowly, then as fast as you can. Is there a difference in the way the gel behaves? Can you pour the gel from one container to another? What happens to the gel when you place a lump of it in the center of a clean watch glass or clean piece of plastic film (such as Handi-Wrap)?

Now add, with stirring, a second 1.0 mL portion of 4% borax solution to the gel. Stir until the gel is homogeneous. Note and record any change in the properties of the gel. Is it more viscous? Or less? Is it still as easily stirred?

Finally, add with stirring a third 1.0 mL portion of 4% borax solution to the gel. As before, note and record any changes in the properties of the gel. After preparing the gel, observe its unusual visco-elastic properties. When completely gelled, the material can be handled with the bare hands. (This is the fun part.) Try rolling the gel into a ball, then seeing how long it takes for the ball to flatten out when placed on a clean watch glass or piece of Handi-Wrap film. Try pulling on the gel at different speeds, slowly then very rapidly. What happens when you try to pull the gel apart suddenly? Can you knead pieces of the gel together by working them with your fingers? Note and record how the gel behaves under these different conditions.

3. KILLING AND RESURRECTING SLIME GEL.

From the borax titration you did in Part 1, you know the volume of 3.0 M H_2SO_4 that it took to neutralize 10 mL of borax solution. Therefore, it should take 0.30 times this volume to neutralize the total of 3.0 mL of borax added to the gel. Using your 1-mL microburet, add precisely this volume of 3.0 M H_2SO_4 to the gel. (It should be a small volume, in the range 0.10–0.15 mL.) In principle, this is precisely the amount of H_2SO_4 required to react with all of the $B(OH)_4^-$ ion that was present in the borax solution. If it is tetrahydroxyborate ion that forms the cross links in the gel, what would you expect to happen if the ion is converted to $B(OH)_3$? Stir the gel until it is again homogeneous. If you have added enough H_2SO_4 to neutralize the $B(OH)_4^-$, the methyl red indicator should be red, like the endpoint in the titration. During the intermediate stage of mixing, before the mixture is homogeneous, you may see lumps of yellow gel in a red solution. If so, keep stirring and mixing until the solution is homogeneous.

Note and record any changes in the properties of the gel resulting from the addition of the H_2SO_4. Is the gelation process reversible? To test this, add 2.0 mL of 1.0 M NaOH, measured with a polyethylene transfer pipet having a 1-mL calibration mark. This is more than enough to neutralize the added H_2SO_4. The methyl red indicator should again be yellow. Stir the gel again. Does the gel regain its high viscosity? Note and record your observations.

☞ Consider This

Let's try to imagine what a solution of poly(vinyl alcohol) or PVA looks like on a microscopic scale. First, show that the average strand of PVA (MW 78,000) has about 1800 monomer units. If each unit contains two carbon atoms, and if each carbon atom has a diameter of about 154 picometers, how long would the PVA strand be if stretched out in a straight line, like a strand of uncooked spaghetti?

Next, if we had one liter of PVA solution spread out in a sheet whose thickness was equal to the length of a PVA strand, what would be the area of the sheet? If we divide

this area by the number of PVA strands in one liter of 4% PVA solution (about 3×10^{20}), we get the average area per strand. If we think of this average area/strand as a little square containing, on average, the cross-section area of one strand of PVA, the square root of the area of this little square would give the edge length of the square. This edge length would be equal to the average distance between strands of PVA. (Imagine each little square containing a cross-section area of a PVA strand located at the center of the square.) Show that this edge length is about 2.4 nanometers, or about ten times the diameter of a PVA strand assumed to be about 0.2–0.3 nanometers. So the strands of PVA must be fairly close together in solution but not as tightly packed as individual strands in a bowl of cooked spaghetti. It would be more like spaghetti strands being cooked in a pot of water, if the strands were stirred so that they were uniformly distributed throughout the pot.

Now imagine that tetrahydroxyborate ion forms labile bonds between the strands of PVA. We talked before about the lability of —OH bonds in water being based on the fast reaction that results in the ionization of water:

$$H_2O + H_2O \rightleftharpoons H_3O^+ + OH^- \quad (K_w = 10^{-14}) \quad (3)$$

An analogous equilibrium involving $B(OH)_3$ and $B(OH)_4^-$ ion would be the fast reaction that results in the ionization of $B(OH)_3$:

$$B(OH)_3 + H_2O \rightleftharpoons$$
$$H_3O^+ + B(OH)_4^- \quad (K_a \cong 10^{-9}) \quad (4)$$

In the equilibrium of Equation (4), $B(OH)_3$ acts as a weak Lewis acid (an electron pair acceptor). The continual rapid breaking and reforming of B—OH bonds could account for the lability of the bonds between tetrahydroxyborate and PVA strands shown in the chemical equilibrium in Figure 20-3. These labile bonds tend to stick the PVA strands together, while still allowing them to slowly slide past one another.

Calculate the molar concentration of —OH groups in 4% PVA solution, assuming 40 g/L PVA and a monomer molecular weight of 44 g/mol. Show that one mole of borax provides enough $B(OH)_4^-$ to form bonds with eight moles of —OH groups. Considering this information, how many mL of 4% (0.1 M) borax solution would be required to form bonds with all of the —OH groups in 50 mL of 4% PVA solution?

BIBLIOGRAPHY

Cassassa, E. Z., Sarquis, A. M., and Van Dyke, C. H. "The Gelation of Polyvinyl Alcohol with Borax: A Novel Class Participation Experiment Involving the Preparation and Properties of a 'Slime'," *J. Chem. Educ.,* **1986,** *63:* 57–60.

McLaughlin, K. W., Wyffels, N. K., Jentz, A. B., and Keenan, M. V. "The Gelation of Poly(Vinyl Alcohol) with $Na_2B_4O_7 \cdot 10H_2O$: Killing Slime," *J. Chem. Educ.,* **1997,** *74:* 97–99.

Sarquis, A. M. "Dramatization of Polymeric Bonding Using Slime," *J. Chem. Educ.,* **1986,** *63:* 60–61.

Sarquis, A. M. (ed.) *Chain Gang: The Chemistry of Polymers,* Terrific Science Press, Center for Chemical Education, Miami University Middletown, 4200 E. University Blvd., Middletown, Ohio 45042-3497, http://www.muohio.edu.

Walker, J. "Serious Fun with Polyox, Silly Putty, Slime, and Other Non-Newtonian Fluids," *Scientific American,* **1978,** *239* (5): 186.

Woodward, L. *Polymers All Around You,* Terrific Science Press, *ibid.*

Consider This

Show that the average strand of PVA (MW 78,000) contains about 1800 monomer units.

Calculate the approximate length of a PVA strand assuming that each monomer unit contributes two carbon atom diameters to the length (1 carbon atom diameter = 154 picometers).

Show that 1.00 L of 4% (40 g/L) PVA contains about 3×10^{20} PVA strands.

Assuming that 1.00 L of 4% PVA solution is spread out to form a thin sheet of thickness equal to the average length of a PVA strand, calculate the total area of the liquid sheet. Then calculate the average area of the sheet of solution per strand of PVA. Assuming this area is a little square, the square root of this area gives the edge length of the square which is approximately the average distance between PVA strands. Show that this distance is about ten times the estimated diameter of a PVA strand (0.2–0.3 nanometers).

Calculate the molar concentration of —OH groups in 4% PVA solution, assuming 40 g/L PVA and a monomer molecular weight of 44 g/mol. Show that one mole of borax provides enough $B(OH)_4^-$ to form bonds with eight moles of —OH groups. Considering this information, how many mL of 4% (0.1 M) borax solution would be required to form bonds with all of the —OH groups in 50 mL of 4% PVA solution?

Liquids and Solids

The Vapor Pressure and Enthalpy of Vaporization of Water

The Enthalpy of Fusion of Water

PURPOSE

To determine the vapor pressure of water as a function of temperature. To use the vapor pressure data to make a Clausius-Clapeyron plot (Ln[P] versus $1/T$). To determine the enthalpy of vaporization of water from the slope of the Ln(P) versus $1/T$ plot. To measure the enthalpy of fusion of ice, using a Styrofoam cup calorimeter.

INTRODUCTION

HEAT AND MOLECULAR MOTION

Our accumulated knowledge about the effect of heat provides a rather detailed picture of what happens when you heat (or transfer energy to) a substance. On an atomic or molecular scale, heating increases the motion of the atoms. If the atoms are linked by chemical bonds to form a molecule, we can focus our attention not on individual atoms but on the molecule as a whole.

If we heat a molecule, it moves about faster. This motion of the center of mass of the molecule is called *translational* motion (meaning motion from place to place). Gaseous molecules have a substantial fraction of their total energy stored as kinetic energy of translation. They move about rapidly, bouncing off one another and the walls of their container.

Molecules can also *rotate* about their centers of mass, and as a gas is heated, the molecules rotate faster. Energy is stored in the rotating molecules, just as energy is stored in a rotating flywheel.

Individual atoms in molecules are linked by chemical bonds that are "springy," constraining the atoms that are bonded together to move like masses connected by springs. As the molecules are heated, the energies and amplitudes of their *vibrations* increase.

To summarize, the energy we put into a molecule is stored as *translational, rotational,* and *vibrational* motions of the molecules.

When we cool a substance, we remove energy from it, and molecular motions slow down. We reach the lowest temperature possible, called absolute zero, when we have removed as much energy as possible from the substance. At this temperature molecular motion nearly ceases.

As we remove energy from gaseous molecules, at some temperature they will condense into a liquid and eventually into a solid, because there are attractive forces between molecules caused by dipole-dipole (or induced-dipole) attraction. To separate the molecules in a solid to the distances characteristic of those in a gas, we must do work against the opposing attractive forces. (Recall that *work = force × distance.*) This work is the energy required to melt the solid and vaporize the liquid.

Now imagine the following experiment. Suppose we begin adding energy to ice at a uniform rate while measuring the temperature. For water, a plot of temperature versus energy (or time) would look like Figure 21-1. As we warm the solid, the molecules begin to move. On average, the molecules in the solid are not going anywhere. They simply begin to bounce around in the cage formed by their surrounding neighbors. In effect, they are like gas molecules confined to a very small container not much larger than the molecule itself. At the same time, the amplitude of vibration of the atoms bonded together increases, and the molecules may begin to rotate in their small cages. A few molecules at the surface of the ice may acquire enough kinetic energy to break free from the solid to become gas molecules. The number of gas molecules may be large enough to produce a measurable pressure. This pressure is called the *vapor pressure* of the solid.

At the temperature we call the *melting point,* the amplitude of the motions increases to the point where the

FIGURE 21-1

The *temperature* versus *energy* heating curve for one mol of water.

forces between neighboring molecules are no longer suffi-cient to hold them in fixed positions. The regular structure of the solid breaks down, forming "holes" (or vacancies) in the solid. The molecules can move more freely, jump-ing from hole to hole. This produces a phase change—the solid becomes a *liquid.* The liquid has a dramatically lower viscosity than the solid, so the material adopts the shape of the container.

On average, the molecules of the liquid have more energy than those of the solid. The energy required to change the solid into a liquid is called the *heat of fusion* or the *enthalpy of fusion.*

Because liquid molecules have a greater average energy than molecules in the solid, a greater number of them can escape into the gas phase, so the vapor pressure of the liquid is greater than that of the solid. As we continue to heat the liquid, a greater number of molecules acquire sufficient energy to escape from the liquid, so the vapor pressure increases as the temperature of the liquid increases. At the temperature we call the *normal boiling point,* where the vapor pressure is equal to 1 atm, all of the energy added goes into a second phase change, called *vaporization.* If the

container is not closed, the vapor pressure cannot exceed 1 atm, so the liquid begins to boil, forming bubbles of vapor that escape from the liquid. If we keep adding energy, we eventually convert all of the liquid to the gas.

The data showing the gas, liquid, and solid equilibria can be displayed in a diagram called a *phase diagram.* The vari-ables plotted are the pressure and temperature. The phase diagram for water between −20 and +20 °C is shown in Figure 21-2. The lines on the phase diagram represent the temperatures and pressures at which two phases can coexist in equilibrium. The unique point where all three phases (gas, liquid, solid) can coexist is called the *triple point.*

Now recall the experiment where we imagined that we began to heat a solid initially at a low temperature. Figure 21-1 shows how the temperature changes as we add energy to the substance, while Figure 21-2 shows how the vapor pressure of ice and liquid water change with temperature.

LATENT HEAT

When you heat a substance, will a thermometer immersed in the substance always indicate a temperature increase? Not necessarily. Figure 21-1 shows that while

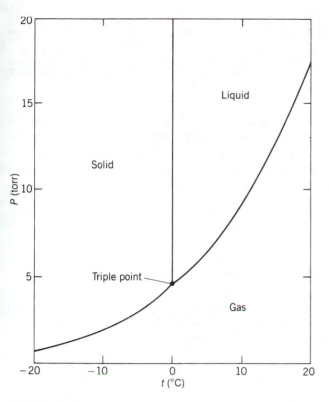

FIGURE 21-2

The *pressure* versus *temperature* phase diagram for water in the range −20 °C to +20 °C. The lines show where two phases can coexist in equilibrium. At the triple point, all three phases coexist in equilibrium.

the solid is melting or the liquid is vaporizing, the temperature remains constant. Heat that produces a phase change such as the melting of ice (or the boiling of water) without producing a temperature change is called *latent* heat. A thermometer, therefore, can be used to measure temperature changes but cannot always be used to measure energy changes.

VAPOR PRESSURE AND THE CLAUSIUS-CLAPEYRON EQUATION

Look again at Figure 21-2. Note that the vapor pressure of the ice and liquid water does not increase in a linear fashion with temperature. It increases in a more rapid, or *exponential*, fashion. An equation that fits quite well the vapor pressure data shown in Figure 21-2 for the solid-gas and liquid-gas equilibria is the *Clausius-Clapeyron* equation:

$$P = A \times e^{-\Delta H/RT} \qquad (1)$$

where ΔH is the enthalpy of *sublimation* (J/mol) for the *solid-gas* equilibrium or the enthalpy of *vaporization* for the *liquid-gas* equilibrium; R is the gas constant

(8.314 J · mol⁻¹ · K⁻¹); T is the Kelvin temperature (K); and A is a constant with the units of pressure.

By taking the natural logarithm of both sides of Equation (1), we can convert the equation to a linear form:

$$\text{Ln}(P) = \text{Ln}(A) - \frac{\Delta H}{R}\left(\frac{1}{T}\right) \qquad (2)$$

A plot of $\text{Ln}(P)$ versus $1/T$ gives a straight line whose slope is equal to $-\Delta H/R$. (See Figure 21-3.) Note that the slope for the ice vapor pressure line is steeper than that for the water vapor pressure line. This is because the enthalpy of sublimation of the solid is greater than the enthalpy of vaporization of the liquid by an amount of energy equal to the enthalpy of fusion:

$$\Delta H_{\text{sublimation}} = \Delta H_{\text{fusion}} + \Delta H_{\text{vaporization}} \qquad (3)$$

If we have vapor pressure data for the liquid, we can determine the enthalpy of vaporization of the liquid. If we have vapor pressure data for the solid, we can deter-

Ln(P) vs. 1/T(K)

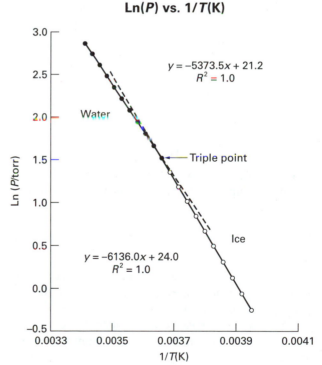

FIGURE 21-3

A Ln(P) versus 1/T (Clausius-Clapeyron) plot for water in the range −20 °C to +20 °C. The slope for the water line segment is −5373 K. The slope for the ice line segment is −6136 K. The value of $R^2 = 1.0$ indicates a good fit to a straight line. The enthalpy of vaporization can be calculated from the slope of the water line segment and the enthalpy of sublimation from the slope of the ice line segment: $\Delta H = -R \times$ slope.

mine the enthalpy of sublimation of the solid. The difference between them gives the enthalpy of fusion of the solid, as shown by Equation (3).

THE ENTHALPY OF FUSION OF WATER

There is another way to determine the enthalpy of fusion of solid water (ice): making use of the temperature change of a known amount of liquid water when ice is added to it. This method requires a knowledge of the heat capacity of water. The experiment can be carried out by using a simple calorimeter made of Styrofoam coffee cups and a thermometer to measure the temperature changes.

It takes energy to melt ice—initially at 0 °C—and to warm the water from the melted ice to the final temperature of the water in the calorimeter. That energy comes from the warm water initially placed in the calorimeter. Using the conservation of energy principle, we can write

$$\Delta H_{\text{calorimeter}} + \Delta H_{\text{warm water}} + \Delta H_{\text{ice water}} + \Delta H_{\text{fusion}} = 0 \quad (4)$$

We want to calculate the enthalpy of fusion of the ice. Assume that the enthalpy change of the calorimeter (Styrofoam cups plus thermometer) is negligible (zero), and that the enthalpy change of the warm water in the calorimeter is given by

$$\Delta H_{\text{warm water}} = m\, C_P\, \Delta T \quad (5)$$

where m is the mass of the warm water put in the calorimeter; C_P is the heat capacity of water at constant pressure ($4.18\ \text{J} \cdot \text{g}^{-1} \cdot \text{K}^{-1}$); and ΔT is the temperature change of the warm water, $T_{\text{final}} - T_{\text{initial}}$.

The enthalpy change of the melted ice water is given by a similar expression.

$$\Delta H_{\text{ice water}} = m\, C_P\, \Delta T \quad (6)$$

where m is the mass of the added ice; C_P is the heat capacity of the melted ice water (as before); and ΔT is the temperature change of the melted ice water, $T_{\text{final}} - 0\ °C$.

Thus, we can calculate each of the first three terms in Equation (4), which enables us to determine the remaining unknown term, ΔH_{fusion}.

EXPERIMENTAL PROCEDURE

Special Supplies: If available, an electric stirrer/hot plate and magnetic stirring bar (egg shaped, 40 mm length); 7 mm × 300 mm stirring rods, if hand stirring is to be used; 1 mL × 0.01 mL disposable borosilicate glass serological pipets (Corning 7079-1N or Kimble 72120-11100); ⅛ inch ID × ¹⁄₃₂ inch wall thickness silicone rubber tubing; rubber septa (size 4 natural rubber septa available from the Aldrich Chemical Co. as catalog num-

ber Z10,073-0); 5-mL disposable plastic Luer-tip syringes to use as filling devices; 1-L tall-form Berzelius beakers (Corning 1060-1L or Kimble 14030-1000); ¾ inch width water-resistant labeling tape (TimeMed tape or equivalent, or electrician's tape); crushed ice; 6-oz Styrofoam cups; 0–110 °C thermometer (or, if available, a digital thermometer reading to 0.1 °C, such as the Hanna CheckTemp 1); string or 22-gauge copper wire.

! SAFETY PRECAUTIONS: Determining the enthalpy of vaporization of water involves heating one liter of water in a beaker. If the beaker is supported on an iron ring over a burner, make sure that a second ring or beaker clamp is used to secure the beaker so that it cannot tip over.

NOTES TO INSTRUCTOR: Only students working very efficiently can complete all three parts of the experiment and the report in a 3-hour lab period. We suggest three possible options. Allow two lab periods to complete the whole experiment. Assign students Parts 1 and 2 only. Or, have students omit Part 1 (the gathering of vapor pressure data for water) and do parts 2 and 3, using vapor pressure data for water in the range 50 to 75 °C from Table 3 of the Appendix.

A stirrer/hot plate and digital thermometer are conveniences. If a stirrer/hot plate is not available, the solution can be heated on a simple hot plate or over a bunsen burner while stirring by hand with a stirring rod. Likewise, a digital thermometer allows more accurate and convenient temperature measurements, but an ordinary thermometer can be used, estimating the temperature to the nearest 0.5 °C.

If the serological pipets are precut and lightly fire polished beforehand, students can move ahead more quickly to the vapor pressure measurements. (Any cotton plugs in the pipets should be removed. This is most easily done by connecting them to a water tap and forcing them out by water pressure.) In our experience, polystyrene serological pipets are troublesome when used with water, because water does not wet the polystyrene, making it very difficult to remove trapped air bubbles.

1. THE VAPOR PRESSURE OF WATER. Assemble the microscale vapor pressure measuring device as shown in Figure 21-4. You will need two 1-mL borosilicate glass serological pipets (with 0.01 mL graduations), a size 4 rubber septum (or stopple), a 6-cm length of silicone rubber tubing (⅛ inch ID × ½ inch wall thickness), and two 10-cm lengths of tape (TimeMed or electrician's tape).

Take one of the 1-mL serological pipets, and, using a glass knife or glass tubing cutter or triangular file, score it precisely at the 0 mark and the 0.85 mL mark. In turn, moisten the score, and grasp the tubing with both thumbs opposite the score. Apply pressure with your forefingers pushing against your thumbs. The glass pipet will normally break cleanly with only light pressure. Then lightly fire polish each end where the pipet has been cut, just enough so that the sharp edge is just slightly rounded so that it will not cut the rubber tubing. (200-grit sandpaper can also be used to take off the sharp edge of the glass.) Take the second pipet, and make a single score at the 0.85 mL mark. Snap off the bottom end of the pipet, and lightly fire polish the end of the pipet.

Next cut a 6-cm length of silicone rubber tubing, and push each end on to the lower ends of the pipets to make a connecting loop. The ends of the rubber tubing should come up to about the 0.82 mL mark. When you face the assembly, make sure that you can easily read the graduation numbers on the pipets.

Now cut a 10-cm length of TimeMed or electrician's tape, and lay it on the bench top with the adhesive side up. Lay the connecting rubber loop in the middle of the tape, with the two pipets parallel and spaced about 3 cm apart and with the top edge of the tape just even with the ends of the rubber tubing. Then, fold over the ends of the tape, so that the adhesive sides of the tape are sealed to each other to make a semirigid connection at the bottom of the pipets.

Next turn the assembly upside down, and insert the untaped end of the shorter pipet into a beaker containing deionized water. Apply suction to draw the water up into the two pipets. (This can be done with a 5-mL Luer-tip syringe and short length of silicone rubber tubing connected to the end of the longer pipet.) When the assembly is turned upright, the water level should be somewhere between the 0.35 and 0.40 mL marks in both pipets. Add or remove water as necessary. Take care to eliminate all bubbles in the pipets and connecting rubber tubing loop before proceeding to the next step.

Next get a rubber septum, and force the smaller diameter end (3.4 mm ID) of the septum onto the unsealed end of the shorter pipet. This makes a tight seal, trapping a small volume of air and water vapor. Make sure that the septum is forced on as far as it will go, so that it seals off the end of the pipet. (The small end of the septum will come down to about the 0.04 mL mark.) Then take another 10-cm piece of tape, adhesive side up as before, and lay the pipets down on the tape with the two pipets parallel and the top edge of the tape just below the rubber septum, as shown in Figure 21-4A. Fold over the ends of the tape to make a second connection. Now you can use the longer pipet end as a handle, and the whole assembly will be semirigid.

Fill a tall-form 1-L beaker nearly to the brim with water. If available, a stirrer/hot plate is convenient for making measurements, along with a Teflon-covered magnetic stirring bar. (We find that an egg-shaped bar gives good stirring and is less prone to flopping around.) If you are going to heat the beaker over a gas burner and stir by hand, place an iron ring or beaker clamp around the upper portion of the beaker to reduce the risk of it tipping over. (**Caution:** Hot water can produce very painful burns.) Use a sturdy stirring rod, about 300 mm in

FIGURE 21-4

(A) Enlarged view of the construction of the microscale apparatus for measuring vapor pressure. (B) Overall view of the apparatus.

length, to do the stirring, and be careful not to strike the fragile bulb of the thermometer.

Put the microscale vapor pressure assembly in the beaker up against the side of the beaker so that it will be out of the way of the stirrer if you are using a magnetic stirrer/hot plate. (A piece of tape can be used to hold it upright in place.) Thorough stirring is necessary to ensure constant temperature throughout the beaker before each volume and temperature reading is made. Using string or copper wire, suspend the thermometer so that the bulb is located approximately at the interface between the water and trapped air in the pipet, and so you can easily see the temperature graduations. (A digital thermometer is even more convenient.)

Stir the solution until the temperature reading is constant; then record the initial volume of air and water vapor to the nearest 0.005 mL. (Read the volume at the bottom of the curved meniscus, as you do when reading the volume in a graduated cylinder or buret.) Rapidly heat the water until the volume of trapped gas is about 0.8 mL (this typically requires a temperature of about 80 °C if the initial volume at room temperature is about 0.4 mL). Then turn down the heat (or turn it off), and stir the solution nearly continuously (if you are stirring by hand), pausing only to take nearly simultaneous readings of the volume of trapped gas and temperature. Don't allow the water bath to cool too rapidly—about a degree per minute is satisfactory. Take readings about every 5 °C as the water bath cools to about 50 °C, stirring when you are not taking a reading. Cold water or small amounts of ice may be added, followed by thorough stirring, to obtain temperatures below 60 °C more quickly. You can use a small beaker to dip out water so that the 1-L beaker does not overflow.

When the temperature reaches 50 °C, record the last readings of temperature and volume. Then wrap a towel around the beaker to protect your hands, and carry it to a sink. Without disturbing the vapor pressure apparatus, cautiously add cold tap water, allowing the water in the beaker to overflow into the sink. After a minute or so, carefully lift out the vapor pressure apparatus, taking care to keep it upright, and fill the 1-L beaker with a slurry of crushed ice and water. Place the vapor pressure apparatus in the ice bath, and stir until the temperature is below 3 °C (preferably less than 1 °C). When the temperature is nearly constant, take readings of the temperature and the volume of trapped gas. Be careful to take this reading accurately because it corresponds to the volume of air in the apparatus, a crucial quantity in subsequent calculations.

Read and record the barometric pressure and temperature of the barometer. **Calculations:** If we assume that the vapor pressure of water at 3 °C is negligible in comparison to atmospheric pressure, we can calculate the number of moles of air trapped in the graduate, n_{air}, by assuming that air behaves as an ideal gas:

$$n_{air} = \frac{PV_{3°C}}{RT} \tag{7}$$

where P is the atmospheric pressure (in torr); $V_{3°C}$ is the volume (in milliliters) measured at the lowest temperature (<3 °C); R is the gas constant (6.237×10^4 mL $\cdot$ torr $\cdot$ mol^{-1} $\cdot$ K^{-1}), and T is the Kelvin temperature of the ice water.

For each of the temperature-volume measurements made between 75 °C and 50 °C, calculate the partial pressure (in torr) of air in the graduate from the equation

$$P_{air} = n_{air} \times \frac{RT}{V} \tag{8}$$

For each temperature, we obtain the vapor pressure of water from Dalton's law of partial pressures:

$$P_{water} = P_{total} - P_{air} \tag{9}$$

where P_{total} is equal to the barometric pressure, ignoring the small pressure exerted on the column of air by the weight of water caused by the difference in the water heights in the two arms of the apparatus as the gas expands. (If so desired, a correction for this effect can be easily made, since it is a linear function of the volume reading of the gas.)

2. THE ENTHALPY OF VAPORIZATION OF WATER.
Using the vapor pressure data obtained in Part 1, calculate the natural logarithm of the pressure, Ln(P) (torr), and $1/T$ (K) for each data point, retaining four significant figures. Plot Ln(P) versus $1/T$. Draw the best straight line through the points. Calculate the slope of the plot, using two points on the line that are far apart. (A spreadsheet program, such as Microsoft Excel or other graphing program, also may be used to make the plot and to determine the best fit straight line and slope, if you have access to the program and know how to use it. Figure 21-3 is an example of a plot produced using Excel.)

$$\text{slope} = \frac{\text{Ln}(P)_2 - \text{Ln}(P)_1}{(1/T)_2 - (1/T)_1} \tag{10}$$

(Note that the slope is a negative number, with units of Kelvins.) Calculate the enthalpy of vaporization from the relation

$$\Delta H_{vap} \text{ (J/mol)} = -R \times \text{slope} (K) \tag{11}$$

taking $R = 8.314$ J $\cdot$ mol^{-1} $\cdot$ K^{-1}.

3. THE HEAT OF FUSION OF ICE.
Weigh to the nearest 0.1 g and record the mass of a pair of nested 6-oz Styrofoam cups to be used as a calorimeter. Place 100 mL of water in a 250-mL beaker, and warm it to about 10 °C above room temperature. Carefully pour all of the warm water into the Styrofoam cup calorimeter. Reweigh, and record the mass to the nearest 0.1 g. Put a digital thermometer in the calorimeter, or suspend a 0 to 100 °C thermometer in the calorimeter by means of a string or copper wire from an iron ring on a ring stand to reduce the risk of breaking the thermometer. (See Figure 21-5.) Stir the warm water with the thermometer until the reading is constant. Record the temperature to the nearest 0.2 °C.

Blot about 25 g of crushed ice in a triple layer c absorbent paper towels to remove as much of the adhering water as possible. Using a stirring rod, brush the ice

FIGURE 21-5

A Styrofoam cup calorimeter. The beaker provides extra stability so that the nested cups won't tip over.

into the calorimeter. Stir until all of the ice has melted, noting the approximate temperature. If needed, add more (blotted) ice until the temperature is about 10 °C below room temperature. Carefully note the lowest temperature attained, and record it to the nearest 0.2 °C. To get the

mass of ice added, reweigh the calorimeter, and record the mass to the nearest 0.1 g.

Calculations: Calculate the enthalpy of fusion of ice (in joules per gram) from your data, using Equations (4) to (6). Also calculate the enthalpy of fusion in units of joules per mole, and compare your results with the accepted experimental value.

👉 Consider This

You can use the microscale vapor pressure apparatus to determine the vapor pressure of other pure volatile liquids, such as alcohols (methanol, ethanol, 2-propanol, etc.) or of solutions, where the effect on the vapor pressure of adding a nonvolatile solute to a volatile solvent can be observed.

Water has a considerably higher boiling point and lower vapor pressure than other molecules of similar or larger molar mass. For example, dinitrogen (N_2), methane (CH_4), methanol (CH_3OH), acetone ($O{=}C[CH_3]_2$), and dimethyl ether (CH_3OCH_3) all have lower boiling points and higher vapor pressures than water at the same temperature. How might this be explained?

If the intermolecular forces between molecules of a substance were very small, what effect would you expect this to have on the boiling point and vapor pressure of the substance? Give one or two specific examples of molecules where you might expect the intermolecular forces to be very small. What would you look for? Should the atoms of the molecules be large or small? Why might this matter? Should the electronegativity difference of the atoms bonded to one another be large or small? Why might this matter?

Use the information presented in Figure 21-3 to calculate a value for the enthalpy of fusion of water. Compare it to the literature value of 6009 Joules/mol.

BIBLIOGRAPHY

De Muro, J. C.; Margarian, H.; Mikhikian, A.; No, K. H.; Peterson, A. R., "An Inexpensive Microscale Method for Measuring Vapor Pressure, Associated Thermodynamic Variables, and Molecular Weights," *J. Chem. Educ.* **1999,** *76:* 1113–1116.

Levinson, G. S., "A Simple Experiment for Determining Vapor Pressure and Enthalpy of Vaporization of Water," *J. Chem. Educ.* **1982,** *59:* 337.

Liquids and Solids

The Vapor Pressure and Enthalpy of Vaporization of Water

The Enthalpy of Fusion of Water

NAME

SECTION   LOCKER

INSTRUCTOR   DATE

DATA AND CALCULATIONS

1. The Vapor Pressure of Water

Data

V (mL)	t (°C)	Barometric pressure (torr)
	(t < 3 °C)	

Calculations

Calculate the moles of air in the graduated cylinder, using the temperature-volume data at the lowest temperature (t < 3 °C).

$$n_{air} = \frac{PV}{RT}$$

where P is the barometric pressure (in torr), V is the volume (**in milliliters**) at ice-water temperature, $R = 6.237 \times 10^1$ mL · torr · mol^{-1} · K^{-1}, and T is the Kelvin temperature of the ice water.

n_{air}: _____ mol

Calculate each entry in the following table, using your volume-temperature data and these equations:

$$T(K) = t(°C) + 273.2$$

$$P_{air} = \frac{n_{air}RT}{V}$$

(Calculate P_{air} in units of torr, using the value of n_{air} from the previous step.) Then calculate P_{water} using Dalton's law:

$$P_{water} = P_{total} - P_{air}$$

where P_{total} is the barometric pressure. (Calculate P_{water} in units of torr.)

V (mL)	T (K)	P_{air} (torr)	P_{water} (torr)

2. The Enthalpy of Vaporization of Water

Data

Use your data from Part 1. (If you did not do Part 1, your instructor will direct you to a source of vapor pressure–temperature data.)

Calculations

Write the values of the vapor pressure of water and the temperature from the table in Part 1 (or another source, if your instructor so directs) in the following table, and calculate for each pair of pressure-temperature values the log of the pressure and the reciprocal of the temperature, retaining four significant figures.

P (torr)	T (K)	$Ln(P)$	$\dfrac{1}{T}$

Plot $Ln(P/torr)$ versus $1/T(K)$, and draw the best straight line through the points; that is, draw the line so that as many points as possible lie near the line and so that there are about as many positive as negative deviations for individual points. Or better, if you have access to a computer with a spreadsheet program such as Excel, use the graphing capabilities of the spreadsheet to make a plot of your data points and add a best fit linear trendline.

Ln(*P/torr*) for water versus $1/T$ (in reciprocal kelvins, K^{-1}) in the range 50–75 °C

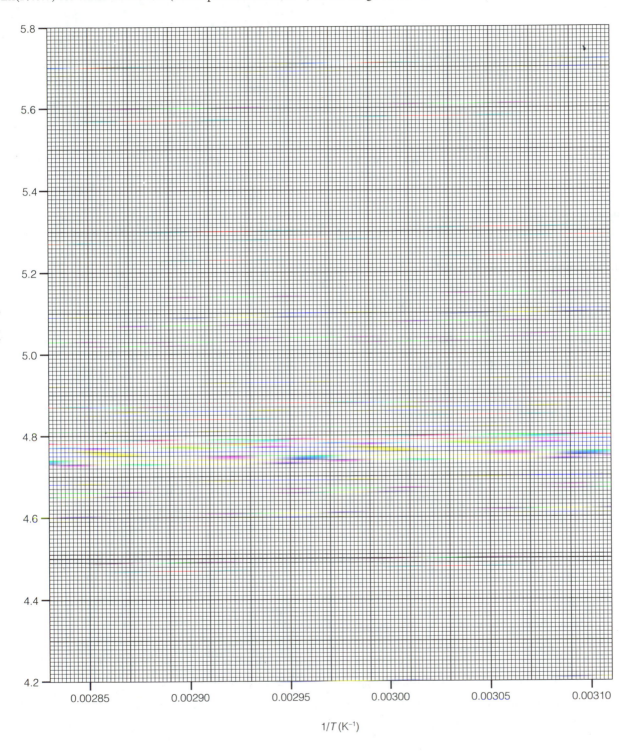

Calculate the enthalpy of vaporization from the slope of the Clausius-Clapeyron plot. Choose two points that lie on the straight line you have drawn through your data points on the Ln(P) versus $1/T$(K) plot, preferably points that are near each end of the line. Calculate the slope of the line from the equation

$$\text{slope} = \frac{\text{Ln}(P)_2 - \text{Ln}(P)_1}{(1/T)_2 - (1/T)_1}$$

Calculate the enthalpy of vaporization from the equation

$$\Delta H_{vap} = -R \times \text{slope}$$

_____ J/mol

The accepted value for ΔH_{vap} is 42.6 kJ/mol. (The value given in most textbooks of 40.7 kJ/mol is the enthalpy of vaporization at 100 °C. The value given here is the average value for the temperature range 50–75 °C.)

$$\text{Percentage difference} = \frac{\text{your value} - \text{accepted value}}{\text{accepted value}} \times 100\% =$$

_____ %

3. The Enthalpy of Fusion of Water

Calculations

Mass of empty Styrofoam cups: _____ g

Mass of Styrofoam cups + 100 mL warm water: _____ g

Initial temperature of warm water: _____ °C

Mass of cups + warm water + ice: _____ g

Final temperature of water + melted ice: _____ °C

Data

From your data, calculate the enthalpy of fusion of ice using Equations (4) to (6).

$$\Delta H_{\text{calorimeter}} + \Delta H_{\text{warm water}} + \Delta H_{\text{ice water}} + \Delta H_{\text{fusion}} = 0$$

$$0 + m_{ww}\, C_P\, \Delta T_{ww} + m_{iw}\, C_P\, \Delta T_{iw} + \Delta H_{\text{fusion}} = 0$$

where m_{ww} is the mass of the warm water, ΔT_{ww} the temperature change of the warm water (see Equation [5]), m_{iw} is the mass of the ice water (mass of ice added), and ΔT_{iw} the temperature change of the ice water (see Equation [6]). Show your calculation.

Calculated heat of fusion of ice in joules per *gram* of water, ΔH_{fusion}: _____ J/g

Calculate the heat of fusion of ice in joules per *mole* of water (i.e., for 18.0 g of water).

Calculated molar heat of fusion of ice, ΔH_{fusion} _____ J/mol

The accepted value of $\Delta H_{\text{fusion}} = 6.01$ kJ/mol.

Calculate the percentage difference between your value and the accepted value.

$$\text{Percentage difference} = \frac{\text{your value} - \text{accepted value}}{\text{accepted value}} \times 100\% =$$

_____ %

👉 **Consider This**

1. *Water has a considerably higher boiling point and lower vapor pressure than other molecules of similar or larger molar mass. For example, dinitrogen (N_2), methane (CH_4), methanol (CH_3OH), acetone ($O=C[CH_3]_2$), and dimethyl ether (CH_3OCH_3) all have lower boiling points and higher vapor pressures than water at the same temperature. How might this be explained?*

2. *If the intermolecular forces between molecules of a substance were very small, what effect would you expect this to have on the boiling point and vapor pressure of the substance? What properties would you look for in a molecule in order to have a low boiling point and high vapor pressure? Should the atoms of the molecules be large or small? Why might this matter? Should the electronegativity difference of the atoms bonded to one another be large or small? Why might this matter? Give two or three examples of molecules where you might expect the intermolecular forces to be very small.*

3. *Use the information presented in Figure 21-3 to calculate a value for the enthalpy of fusion of water. Compare it to the literature value of 6009 Joules/mol.*

4. *Why would contact with steam at 100 °C produce a more severe burn than contact with liquid water at the same temperature?*

5. *Orange growers often spray water on their trees to protect the fruit in freezing weather. Explain how the energy of the water → ice phase transition could provide protection from freezing weather.*

Colligative Properties
The Molar Mass of a Soluble Substance by Freezing Point Depression

PURPOSE

To explore one of the colligative properties of a solution, the freezing point depression. To use this property to determine the molar mass of an unknown substance.

PRE-LAB PREPARATION

When a nonvolatile solute is dissolved in a solvent, the solution properties change from those of the pure solvent. Relative to the pure solvent, the solution has a lower vapor pressure, a lower freezing point, and a higher boiling point. These effects are called *colligative* properties of the solution because they are linked together by a common feature: They all depend primarily on the ratio of the number of solute particles to the number of solvent particles. The size of the particles or whether they are molecules or ions makes little difference; it is the *relative number of particles* that is important.[1]

Vapor pressure lowering, freezing point depression, and boiling point elevation—all of these characteristics are observed in both aqueous and nonaqueous solutions. For example, a solution of a nonvolatile solute like sucrose or ethylene glycol in water shows the same kind of effects as a solution of aspirin dissolved in cyclohexanol or sulfur in naphthalene.

We make practical use of the colligative properties of an aqueous solution of ethylene glycol by putting it in the cooling system of an automobile. Because antifreeze—an aqueous solution of ethylene glycol—has both a lower freezing point and a higher boiling point than pure water, the solution protects the car's cooling system against both freezing and boiling over. Boil-over protection is increased by using a spring-loaded pressure cap, which allows the system to operate above atmospheric pressure. Increases in pressure further raise the boiling point of the solution.

The origin of the freezing point depression and boiling point elevation of an aqueous solution can be understood with the setup shown in Figure 22-1. The phase diagram shown in Figure 22-2 describes what takes place in the apparatus. This diagram shows that solvent vapor pressure lowers when a nonvolatile solute is dissolved in the solvent. Note that the vapor pressure curve for the solution (shown by the dotted curve) is lower than the vapor pressure curve for pure water (shown by the solid curve). Recall that the boiling point is the temperature at which the equilibrium vapor pressure equals the atmospheric pressure. The equilibrium vapor pressure of the solvent over the solution is *less* than that of the pure solvent at the same temperature. Therefore, the solution must be heated to a higher temperature than the pure solvent in order to reach atmospheric pressure. That means that the boiling point of the solution is higher than that of the pure solvent. Increasing the pressure above atmospheric pressure, as is done when you use a pressure cap on your auto radiator, raises the boiling point even further. (If you drew a horizontal line at $P = 2$ atm, it would intersect the vapor pressure curve at a higher temperature.)

When the temperature of most solutions is lowered to the freezing point, the solid that separates out is mainly pure solvent. Lowering the vapor pressure of a solvent by adding a nonvolatile solute causes the equilibrium vapor

---

[1]These statements are quantitatively true for ideal solutions, in which the forces between molecules are small and for molecules having moderate molecular weights (< 500 g/mole). For real solutions, except at great dilution, there is always some deviation. For polar or ionic substances or for solutions of high molecular weight polymers, this deviation may be considerable. In particular, ions cannot move independently because of the strong electric forces between them. When measuring the colligative properties, we therefore try to use as dilute a solution as will allow satisfactory experimental precision.

Pure water

Vapor pressure of pure water: 23.8 torr at 25°C

1 molal ethylene glycol in water

Vapor pressure of water above the solution: 23.4 torr at 25°C

FIGURE 22-1

An illustration of the phenomenon of vapor pressure. If a beaker of a volatile liquid such as water is placed in an evacuated chamber, some of the liquid evaporates to fill the chamber with vapor. The pressure of this vapor is determined solely by the temperature of the liquid. When a nonvolatile solvent, such as ethylene glycol, is added to the pure solvent, the resulting solution has a vapor pressure less than the pure solvent at each temperature.

pressure curve to intersect the sublimation pressure curve at a lower temperature than the curve for the pure solvent (see Figure 22-2). The freezing point curves for the solution and the pure solvent begin at the intersections of the sublimation pressure and vapor pressure curves, that is, at the *triple points* (shown in Figure 22-2), where all three phases (solid, liquid, and gas) simultaneously coexist. (The liquid phase is either the pure solvent or the solution containing added solute.) Therefore, the freezing point curve for the solution lies at a *lower* temperature than the freezing point curve of the pure solvent.

The freezing point depression effect has several important scientific uses. It allows us to compare the freezing point depressions of a known and unknown substance in the same solvent in order to determine the molar mass of the unknown. This type of comparison is called *cryoscopy*, the stem *cryo-* coming from the Greek word meaning "icy" or "cold." In this experiment you will determine the molecular weight of an unknown dissolved in cyclohexane by the cryoscopic method.

Cryoscopy is also used in the clinical laboratory to measure the total concentration of solutes in urine, making use of the fact that the freezing point depression is proportional to the sum of the concentrations of

all dissolved particles. The urine that initially is formed within the kidney has a much lower concentration of solutes than the urine that leaves the kidney. This is because the kidney conserves body water by recovering much of the water in urine as it passes through the kidney tubules. The kidney's ability to concentrate urine is one of this organ's most important functions, and when the renal tubules are damaged, this ability is one of the first functions to be lost. Thus a cryoscopic measurement of the solute content of urine is an indication of kidney function.

RAOULT'S LAW

The equation that quantitatively describes vapor pressure lowering is called *Raoult's law*. It is expressed

$$P_1 = X_1 P_1^0 \qquad (1)$$

where P_1 is the vapor pressure of the solvent above the solution, P_1^0 is the vapor pressure of the pure solvent, and X_1 is the mole fraction of the solvent in the solution. The mole fraction is a dimensionless concentration unit that expresses the fraction of the total number of molecules (or moles) that are solvent molecules:

$$X_1 = \frac{n_1}{n_1 + n_2} \qquad (2)$$

where n_1 is the number of moles of solvent and n_2 is the number of moles of solute. This justifies the earlier statement that the colligative properties depend only on the relative numbers of solvent and solute molecules.

It is convenient in working with the colligative properties of solutions to introduce the *molality concentration scale* for solutes. We define the molality, m, of a solute as the number of moles of the solute per 1000 g of solvent:

$$\text{molality} = \frac{\text{moles of solute}}{1000 \text{ g of solvent}}$$

$$= \frac{\text{moles of solute}}{\text{kilogram of solvent}} \qquad (3)$$

For example, a solution prepared by dissolving 1.00 mol KCl (74.55 g) in 1.00 kg of water is 1.00 molal in KCl. The "particle molality" would be 2.00 molal, because KCl dissociates into two ions, K^+ and Cl^-.

For dilute solutions the molality is proportional to the mole fraction. Both molality and mole fraction are temperature-independent concentration scales, in contrast to the molar concentration, which changes with temperature because of expansion or contraction of the solution volume.

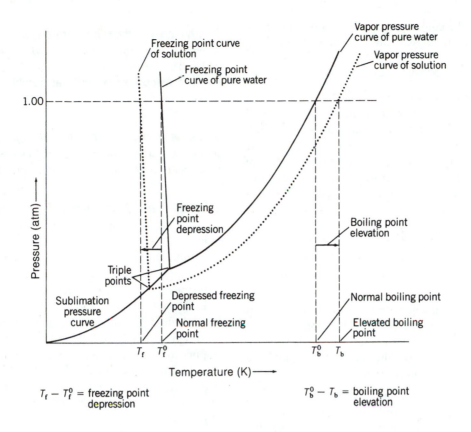

FIGURE 22-2

Phase diagrams for pure water (solid line), and for water containing a nonvolatile solute (dotted line). The presence of the solute lowers the vapor pressure of the solvent. The reduced vapor pressure of the solvent gives rise to an *increase* in the boiling point of the solution relative to pure water, $T_b - T_b^0 > 0$, and to a *decrease* in the freezing point of the solution relative to pure water, $T_f - T_f^0 > 0$. (When the temperature of an aqueous solution is lowered to the freezing point, the solid that separates out is almost always pure ice, so the sublimation pressure curve is assumed to remain unchanged.) The extent of vapor pressure lowering is exaggerated in the diagram to show the effects more clearly.

THE FREEZING POINT DEPRESSION

The equation describing the freezing point depression is

$$\Delta T_f = K_f m$$

The symbol ΔT_f represents the freezing point depression: $\Delta T_f = T_f^0 - T_f$ where T_f^0 is the freezing point of the pure solvent and T_f is the freezing point of the solution. This quantity is directly proportional to the concentration of particles expressed as the molality, m, of the particles in the solution. The proportionality constant, K_f, is called the *molal freezing point depression constant*. By using a solute of known molar mass, we can determine K_f for a particular solvent. The molality (and molar mass) of an unknown compound can then be determined by measuring the freezing point depression of a solution of the unknown in the same solvent.

Note the phrase *particle molality* used in discussing the molality of an ionic solute, KCl. If the solute does not dissociate into two or more particles, the particle molality equals the molality of the solute. This will be true for all

the solutions we will study in this experiment, solutions of *p*-dichlorobenzene and organic compounds dissolved in cyclohexane. The distinction between particle molality and solute molality is important to keep in mind when dealing with aqueous solutions. A 0.1 m KCl solution would lower the freezing point of water approximately twice as much as a 0.1 m in sucrose solution. (Sucrose, ordinary table sugar, does not dissociate in water.) Does this help to explain why $CaCl_2$ is often used in northern climates to melt ice on roads and sidewalks?

The freezing point depression for a 1 m solution in water (the molal freezing point constant) is 1.86 °C, and the corresponding boiling point rise (the *molal boiling point constant*) is 0.52 °C. The same principles apply for solutions in other solvents. For cyclohexane, which we will use in this experiment, the freezing point is 6.5 °C, with an ideal molal freezing point constant of 20.2 °C/m.

These constants apply to measurements in quite dilute solution. For the more concentrated solutions (0.5 m) that we must use, the freezing point constant may deviate slightly from the ideal value. This is the reason we will first determine the constant under the conditions of the experiment, and then use this value for determining molar mass.

CALCULATIONS OF MOLAR MASS

Either boiling point or freezing point may be used to determine the molar mass of a soluble substance. For the latter, it is necessary only to determine the freezing point of a solution containing a known mass of the solute in a known mass of the solvent and to compare this with the freezing point of the pure solvent, as follows.

1. Use the measured freezing point depression and the molal freezing point constant in Equation (1) to determine the molality of the unknown cyclohexane solution.

2. This molality is equal to the moles of unknown divided by the kilograms of the solvent, cyclohexane; so if we multiply the molality by the mass of cyclohexane (in kilograms), we will know the number of moles of unknown that the solution contains.

3. The molar mass is the number of grams per mole, so dividing the mass of the unknown by the number of moles (which we just determined in Step 2) gives the molar mass of the unknown.

EXPERIMENTAL PROCEDURE

Special Supplies: Thermometer (1 °C divisions are suitable; electronic thermometers with 0.1 °C resolution are more convenient to read and give better precision),[2] 27 cm of $\frac{1}{16}$-in. diameter brass rod (brazing rod available from welding supply companies is ideal), 20 mm × 150 mm test tube, No. 2 solid stopper, No. 2 one-hole stopper.

Chemicals: Cyclohexane, p-dichlorobenzene, solid organic compounds for use as unknowns for molar mass determination, salt (NaCl; see following note), ice.

NOTES TO INSTRUCTOR: If an electronic thermometer with 0.1 °C resolution is used, a smaller 15 × 125 mm test tube may be used, omitting the rubber stopper, and the amount of cyclohexane may be decreased to 5 mL and the solute mass to 0.25–0.30 g.

Some grades of salt (NaCl) contain anticaking additives that cloud the ice-water solution, making it more difficult to see the crystals of cyclohexane as they form. It is best to use a grade of salt that does not contain these additives.

---

[2]Moderately priced electronic digital thermometers (cost, approximately $30) with a 4-in. stainless steel temperature sensing probe and 1 meter cable, a range of −50 to +150 °C, and temperature resolution of 0.1 °C, are available from Hanna Instruments Inc., Highland Industrial Park, 584 Park East Drive, P.O. Box 849, Woonsocket, RI 02895, tel (401) 765-7500 (Model Checktemp 1C).

1. THE FREEZING POINT OF CYCLOHEXANE.

Figure 22-3 illustrates the simple apparatus required. Using a sharp razor blade, a 45° pie-shaped sector is cut longitudinally from a No. 2 one-hole rubber stopper. The split stopper easily holds a thermometer in position without danger of breakage. A circular stirrer, fashioned from $\frac{1}{16}$-in. diameter brazing rod (or 14-gauge copper wire), is fitted through the notch at the side of the stopper.

Weigh the empty 20 × 150 mm test tube with the solid stopper to ±0.01 g on a balance. (You can stand the test tube up in a beaker.) Add about 10 mL of cyclohexane (C_6H_{12}), and weigh the test tube apparatus again. Accurately weigh (to the nearest milligram) about 0.5 to 0.6 g of p-dichlorobenzene, $C_6H_4Cl_2$ (molar mass 147.0), on a creased sheet of glazed weighing paper.

Clamp the 20 × 150 mm test tube with the cyclohexane in a water bath (550 mL of ice water in a 600-mL

20 × 150 mm test tube

600-mL beaker

Water and ice
Wire stirrer

FIGURE 22-3
Freezing point apparatus for determination of molar mass.

beaker). The bottom of the test tube should be about 2 cm from the bottom of the beaker. Place the thermometer and stirrer in position, with the thermometer bulb about 2 to 3 mm above the bottom of the test tube, threaded through the ring of the stirrer. This arrangement ensures good stirring around the thermometer bulb. Stir the cyclohexane briskly and continuously. When crystals of solid cyclohexane begin to form and the temperature is constant, read the thermometer to the nearest 0.1 C°, and record this value. You may need to wipe condensation from the outside of the beaker in order to read the thermometer clearly. Warm the cyclohexane to room temperature; then repeat the cooling operation until you are certain of the freezing point (as accurately as your thermometer can be read). When reading the thermometer, be sure to have your eye on a level with the mercury meniscus, to avoid parallax in the reading.

2. THE MOLAL FREEZING POINT CONSTANT OF CYCLOHEXANE.

Remove the stopper and attached thermometer and stirrer (avoiding loss of cyclohexane), just far enough to add the sample of p-dichlorobenzene. Weigh the glazed weighing paper to obtain, by difference, the weight of the p-dichlorobenzene added. Warm the mixture, and stir it to dissolve the solute completely. Pour most of the water out of the 600-mL beaker, refill the beaker with ice, add 50 mL (approximately 65 g) of solid NaCl, and refill the beaker with water, leaving enough room for the test tube. Stir the mixture of salt and ice water to dissolve as much salt as possible, then lower the test tube into the ice–salt water mixture and clamp it in place. Now stir briskly as before, and observe as accurately as possible the steady temperature at which a *small* amount of solid solvent is in equilibrium with the solution. There may be a small amount of supercooling before crystallization. Study Figure 22-4, noting especially that when a larger amount of solvent is frozen out, the remaining solution is more concentrated and there-

fore gives a declining freezing point. This is why readings must be taken as soon as the first crystals appear in the solution, when only a very small amount of the solvent has frozen solid.[3] Repeat this freezing point measurement several times by warming the solution and allowing it to cool until you are certain of this equilibrium temperature, to within 0.1 °C if possible. When you have finished the measurements, pour the solution into the waste cyclohexane bottle provided. *Do not pour it into the sink.* Remove the last trace of solution by rinsing the tube with cyclohexane in the fume hood and drying the tube with a gentle stream of air in the hood.

From these data, calculate the molal freezing point constant for cyclohexane—that is, the change in freezing point caused by 1 mol of p-dichlorobenzene per kilogram of cyclohexane. Compare your value with the literature value of 20.2 °C/m. (The literature value is an idealized value, extrapolated to zero concentration of solute.) If sufficient time is available, repeat the determination with a fresh cyclohexane sample and another weighed sample of p-dichlorobenzene.

3. THE MOLAR MASS OF AN UNKNOWN SOLID.

Obtain from your instructor a sample of an unknown organic material. The freezing point measurements for pure cyclohexane and for the unknown dissolved in the cyclohexane are made exactly as in Part 1, using a 0.5-g sample of unknown. From these data, calculate the molar mass of the unknown, using the molal freezing point constant for cyclohexane determined in Part 2. Be sure to dispose of the cyclohexane solution in the proper container when you are finished. If your instructor so directs, repeat the measurement with a second sample of unknown.

👉 Consider This

Is the Freezing Point Depression a linear function of molality? Make a series of different concentrations of p-dichlorobenzene in cyclohexane, and graph the freezing point temperature versus concentration.

Why is the molality concentration scale used in the context of colligative properties and not the more common molarity scale? Use the data in the CRC Handbook of Chemistry and Physics *in a section entitled "Concentrative Properties of Aqueous Solutions: Conversion Tables") to plot ΔT_f versus molality and versus molarity in order to explore these relationships.*

FIGURE 22-4
A typical freezing point crystallization curve.

---

[3] For precise determination, readings are taken every 30 to 60 s during the cooling process. A curve similar to Figure 22-4 is plotted, and the exact freezing point determined by extrapolation, as illustrated.

Most everyone has participated in the fun of making homemade ice cream. (If you haven't, do it soon!) A key feature of the process takes advantage of a colligative property. Describe why it is necessary to add salt to the ice that surrounds the ice-cream container in order to freeze the ice cream inside. This is an interesting system to explore. Place a quantity of ice in a Styrofoam cup calorimeter (See Experiment 14 or 15.) Monitor the temperature as you add salt to the system. How can the temperature drop if no heat can enter the system? How low can you get the temperature to go? Convert this temperature to °F. (Note: This was one of Fahrenheit's calibration points as he developed the thermometer. Why is the normal freezing point of pure water a better calibration point?)

Cool a clear plastic bottle of a soft drink in the freezer until it just begins to freeze. Take the cap off the bottle. (Over the sink is a good place to do this.) What happens? Explain in terms of the colligative properties that you explored in this experiment. Can you design an experiment to determine the carbon dioxide concentration in a soft drink using the Freezing Point Depression property?

What has a lower freezing point, a diet soda or a regular version of the same soda? Why? Can you use the calorie content listed on the can to estimate the sugar concentration in the soda, then calculate an expected freezing point for the soda? What complications come into play?

Colligative Properties
The Molar Mass of a Soluble Substance by Freezing Point Depression

NAME

SECTION   LOCKER

INSTRUCTOR   DATE

1. Experimental Data

Sample or unknown no. _____

	p-dichlorobenzene		Unknown	
	Sample 1	Sample 2	Sample 1	Sample 2
Mass of cyclohexane + test tube	g	g	g	g
Mass of test tube	g	g	g	g
Mass of cyclohexane	g	g	g	g
Mass of sample + weighing paper	g	g	g	g
Mass of weighing paper	g	g	g	g
Mass of unknown	g	g	g	g
Freezing point of cyclohexane	°C	°C	°C	°C
Freezing point of solution	°C	°C	°C	°C

Use the graph below to plot the data from one of your cooling curves if so directed by your instructor.

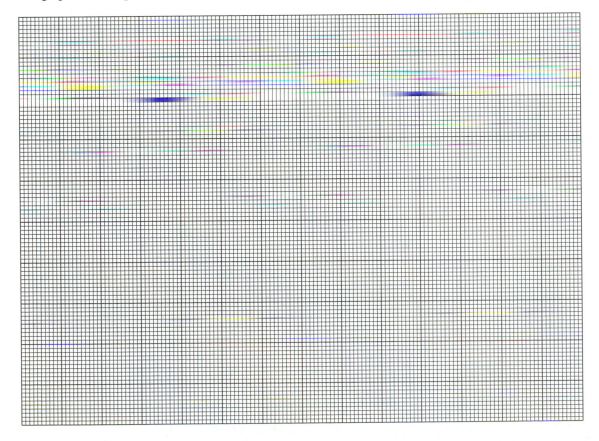

2. Freezing Point Constant (Show a sample calculation.)

	Calculations	p-dichlorobenzene	
		Sample 1	Sample 2
Freezing point depression $\Delta T_f = T_f^0 - T_f$		°C	°C
Moles of solute, p-dichlorobenzene		mol	mol
Molality of solution $= \dfrac{\text{moles of solute}}{\text{kilograms of solvent}}$		mol/kg	mol/kg
Molal freezing point constant, from $\Delta T_f = K_f m$		°C · kg/mol	°C · kg/mol

3. The Molar Mass of an Unknown Solid (Show a sample calculation.)

	Calculations	Unknown	
		Sample 1	Sample 2
Freezing point depression $\Delta T_f = T_f^0 - T_f$		°C	°C
Molality of unknown solution, from $\Delta T_f = K_f m$		mol/kg	mol/kg
Molality of unknown solution, from molality of solution $= \dfrac{\text{moles of solute}}{\text{kilograms of solvent}}$		mol	mol
Molar mass of unknown $= \dfrac{\text{grams of unknown}}{\text{moles of unknown}}$		g/mol	g/mol
Percentage error $= \dfrac{\text{your value} - \text{actual value}}{\text{actual value}} \times 100\%$		%	%

Approval of Instructor: Your unknown is _____

QUESTIONS ON EXPERIMENTAL ERROR

1. (a) Assume that you have a solution of an unknown solute in cyclohexane. If the solution has a freezing point depression of 9.50 °C, what is the molality of this solution? (The molal freezing point constant of cyclohexane is 20.2 °C/m.)

_____ m

  (b) Since the thermometer used has markings for each Celsius degree, assume you can be certain of the freezing point depression only to within ±0.2 °C. If the freezing point depression is measured as 9.7 °C, then what molality would be reported for this solution?

_____ m

  (c) What is the relative error in the molality that is introduced by this limitation in the thermometer? (Hint: The relative error in the molality will be the same as the relative error in the freezing point depression.)

$$\left(\text{relative error} = \frac{\text{measured value} - \text{actual value}}{\text{actual value}} \times 100\% \right)$$

_____ %

2. The relative error can be reduced if a more precise thermometer is used. Assume that with the more precise thermometer you can determine the temperature to within ±0.02 °C.

  (a) If the measured freezing point depression with such a thermometer is 9.52 °C, what molality is determined for this solution?

_____ m

  (b) What is the relative error in the molality as determined with this more precise thermometer?

_____ %

PROBLEMS

1. Cyclohexanol, $C_6H_{11}OH$, is sometimes used as the solvent in molecular weight determinations by freezing point depression. If 0.253 g of benzoic acid, C_6H_5COOH, dissolved in 12.45 g of cyclohexanol, lowered the freezing point of pure cyclohexanol by 6.55 °C, what is the molal freezing point constant of this solvent?

2. Since the freezing point of a solution depends on the relative number of *particles*, what would you calculate to be the freezing point of 0.1 *m* solutions in water of (a) NaCl, (b) BaCl$_2$? Assume that these salts are 100% ionized in solution. (Compare your answers with the actual respective freezing points: -0.348 °C and -0.470 °C. The difference is due to the decreased activity of the ions. Because of the attractive forces between the positive and negatively charged ions, they do not move completely independently of one another.)

(a) _____

(b) _____

3. How many grams of each of the following per kilogram of water in your car radiator are needed to give equal protection against freezing down to -10 °C? (a) Methyl alcohol, CH$_3$OH, b.p. 64.6 °C, (b) ethylene glycol, C$_2$H$_4$(OH)$_2$, b.p. 197.2 °C. In spite of higher cost, what advantage does ethylene glycol possess over methyl alcohol as a winter antifreeze and/or summer coolant?

(a) _____

(b) _____

☞ **Consider This**

Is the Freezing Point Depression a linear function of molality? Make a series of different concentrations of p-dichlorobenzene in cyclohexane, and graph the freezing point temperature versus concentration.

Why is the molality concentration scale used in the context of colligative properties and not the more common molarity scale? Use the data in the CRC Handbook of Chemistry and Physics *in a section entitled "Concentrative Properties of Aqueous Solutions: Conversion Tables") to plot* ΔT_f *versus molality and versus molarity in order to explore these relationships.*

Most everyone has participated in the fun of making homemade ice cream. (If you haven't, do it soon!) A key feature of the process takes advantage of a colligative property. Describe why it is necessary to add salt to the ice that surrounds the ice-cream container in order to freeze the ice cream inside. This is an interesting system to explore. Place a quantity of ice in a Styrofoam cup calorimeter (See Experiment 14 or 15.) Monitor the temperature as you add salt to the system. How can the temperature drop if no heat can enter the system? How low can you get the temperature to go? Convert this temperature to °F. (Note: This was one of Fahrenheit's calibration points as he developed the thermometer. Why is the normal freezing point of pure water a better calibration point?)

Cool a clear plastic bottle of a soft drink in the freezer until it just begins to freeze. Take the cap off the bottle. (Over the sink is a good place to do this.) What happens? Explain in terms of the colligative properties that you explored in this experiment. Can you design an experiment to determine the carbon dioxide concentration in a soft drink using the Freezing Point Depression property?

What has a lower freezing point, a diet soda or a regular version of the same soda? Why? Can you use the calorie content listed on the can to estimate the sugar concentration in the soda, then calculate an expected freezing point for the soda? What complications come into play?

23

Some Examples of Chemical Equilibria
Le Châtelier's Principle

PURPOSE

To observe a number of interesting and colorful chemical reactions that are examples of chemical systems at equilibrium. To see how these systems respond to changes in the concentrations of reactants or products or to changes in temperature. To see that the direction of the shift in the equilibrium tends to at least partially offset the change in conditions, a principle first clearly stated by Le Châtelier.

PRE-LAB PREPARATION

Any chemical equation that describes an actual chemical reaction sums up the result of an experiment (more often, several experiments). Someone had to measure out the reactants, carry out the reaction, identify the products, and measure quantitatively the number of moles of products produced per mole of reactant consumed. The last step is the writing of a balanced chemical equation that concisely summarizes the experimental observations.

Many chemical reactions go essentially to completion, but some do not, stopping at some point of equilibrium that lies between no reaction and an essentially complete reaction. A state of equilibrium is the point of balance where the rate of the forward reaction equals the rate of the backward reaction. For these reactions, the same point of equilibrium can be reached from either end of the reaction, by mixing together either the reactants or the products, clearly indicating that chemical reactions can go either forward or backward.

In principle, every chemical reaction is a two-way reaction, but if the point of equilibrium greatly favors the reactants, we say that there is *no reaction*. If the point of equilibrium greatly favors the products, we say the reaction is *complete*. For either of these two extremes it is dif-

ficult to measure experimentally the concentrations of all of the reactants and products at equilibrium. In the first case (reactants greatly favored), the concentration of products is practically zero. In the second case (products greatly favored), no significant quantity of the reactants will be produced when the products are mixed together. Thus, when the equilibrium greatly favors the products, the reaction appears to go in only one direction: *reactants → products.*

Nearly every chemical reaction consumes or releases energy. (We call these, respectively, *endothermic* or *exothermic* reactions.) For these reactions, we can regard energy as if it were a reactant or product, and by adding energy to a system (by heating it) or removing energy from a system (by cooling it), we can produce a shift in the point of chemical equilibrium. The concentrations of reactants and products change to reflect the new equilibrium.

There is an added complication that must be considered. Not all chemical reactions are rapid. In fact, some are so slow that you might be fooled into thinking that no reaction occurs, whereas the equilibrium point, when it is finally reached, might greatly favor the products. The rate at which a chemical reaction reaches equilibrium will be taken up when you study the kinetics of chemical reactions.

For reactions that rapidly reach equilibrium, the point of equilibrium may be approached from either the side of the reactants or the side of the products, which emphasizes the dynamic nature of chemical reactions. The reactions you will study in this experiment are all rapid reactions, so you will be able to observe almost immediately the effects of changing the concentration of either reactants or products. In particular you will observe that when you change the concentration of a reactant or product, the point of equilibrium shifts in a direction that tends to offset the change. This behavior can be summarized in a general principle that was first fully stated in 1884 by the French chemist Henri Louis Le Châtelier: *A chemical reaction that is displaced from equilibrium by a change in conditions (concentrations, temperature, pressure, volume) proceeds toward a new equilibrium state in the direction that at least partially offsets the change in conditions.*

This introductory experiment is designed to show qualitatively several important features of chemical equilibria. In subsequent experiments you will learn about the quantitative aspects of chemical equilibria, measuring the equilibrium constants for a variety of reactions, and making calculations of equilibrium concentrations from previously measured equilibrium constants.

EXPERIMENTAL PROCEDURE

Special Supplies: Electric hot plate, crushed ice, polyethylene transfer pipets (25 drops/mL).

Chemicals: 1 M K_2CrO_4, 3 M H_2SO_4, 6 M NaOH, 0.1% methyl orange indicator, 0.1% phenolphthalein indicator, 6 M HCl, 0.1 M CH_3COOH (acetic acid), 1 M $NaCH_3COO$ (sodium acetate), 0.1 M NH_3, 1 M NH_4Cl, 0.1 M $Fe(NO_3)_3$, 0.1 M KSCN (potassium thiocyanate), 0.15 M anhydrous $CoCl_2$, in methanol (20 g anhydrous $CoCl_2$ per liter of methanol; if only the hydrated salt $CoCl_2 \cdot 6H_2O$ is available, it must be dried before use), concentrated (12 M) HCl, saturated (5.4 M) NaCl, 0.1 M $BaCl_2$, 0.1 M $CaCl_2$, 0.5 M $H_2C_2O_4$ (oxalic acid), 0.5 M $K_2C_2O_4$ (potassium oxalate), 6 M NH_3.

! SAFETY PRECAUTIONS: Be careful when handling concentrated (12 M) HCl and 6 M NaOH. They cause skin and eye burns. Wear eye protection.

Ba^{2+} is toxic, and CrO_4^{2-} and $Cr_2O_7^{2-}$ are known to be carcinogenic. Rinse them off immediately if they contact your skin. Dispose of them in waste containers. Do not pour them down the drain.

Take care when heating the methanol solution of $CoCl_2$. Methanol vapors are toxic and flammable, so the heating must be done by placing the test tube in a beaker of hot water at 65 to 70 °C on a hot plate in the fume hood.

1. THE SHIFTING OF EQUILIBRIA IN ACID-BASE REACTIONS; THE COMMON ION EFFECT.

(a) *The Chromate Ion–Dichromate Ion Equilibrium.* Yellow chromate ion reacts with hydrogen ion first to form hydrogen chromate ion and then, by condensation and loss of H_2O, orange dichromate ion:

$$2CrO_4^{2-} + 2H_3O^+ \rightleftharpoons 2HCrO_4^- + 2H_2O \rightleftharpoons Cr_2O_7^{2-} + 3H_2O \quad (1)$$

At present, we need consider only the overall reaction,

$$2CrO_4^{2-} + 2H_3O^+ \rightleftharpoons Cr_2O_7^{2-} + 3H_2O \quad (2)$$
$$\text{yellow} \qquad\qquad\qquad \text{orange}$$

To 3 mL of 1 M K_2CrO_4 in a test tube add several drops of 3 M H_2SO_4. Mix this, and observe any change. Now add several drops of 6 M NaOH, stirring continuously, until a change occurs. Again add H_2SO_4. Interpret the observed changes. How did the equilibrium shift in response to the added reagents—toward the formation of yellow chromate ion or toward the formation of orange dichromate ion? Explain how the addition of hydroxide ion exerts an effect, even though it doesn't appear in the overall equation.

(b) *Weak Acid–Base Indicator Equilibria.* In Experiment 18 you compared the relative concentrations of molecules and ions in weak acids and weak bases. Let us now see how the chemical equilibria shift when the concentrations of these ions or molecules are changed. First we will observe the effects of strong acid and strong base on indicators, which are themselves weak acids and bases. The chemical equation for the dissociation of the indicator methyl orange can be written as

$$HIn + H_2O \rightleftharpoons H_3O^+ + In^- \quad (3)$$
$$\text{red} \qquad\qquad\qquad \text{yellow-orange}$$

protonated HIn form

deprotonated In$^-$ form

where HIn represents the protonated (acid) form of the indicator (red in color) and In$^-$ represents the deprotonated (or base) form of the indicator (yellow-orange in color). Methyl orange indicator changes color around pH 4, which corresponds to a hydronium ion concentration of 10^{-4} M. (For a fuller discussion of indicators and the pH scale, see your text.)

Other acid-base indicators change color at different hydronium ion concentrations. Phenolphthalein, whose acid form is colorless, changes to the pink base form around pH 9, which corresponds to a hydronium ion concentration of 10^{-9} M.

First, in order to observe the effects of acid and base on indicators, add a drop of methyl orange to 3 mL of water. Then add 2 drops of 6 M HCl followed by 4 drops of 6 M NaOH. Repeat the experiment, using phenolphthalein indicator in place of the methyl orange. Record your observations.

(c) *Weak Acid–Weak Base Equilibria.* Now we will use the indicators to observe changes involving weak acids and bases that are themselves colorless. To each of two 3-mL samples of 0.1 M CH_3COOH (acetic acid), add a drop of methyl orange. To one of the samples add 1 M $NaCH_3COO$ (sodium acetate), a few drops at a time, with mixing. Compare, and record the colors in the two test tubes. The added salt, sodium acetate, has an ion in common with acetic acid, a weak acid that dissociates in water to give acetate ion.

$$CH_3COOH + H_2O \rightleftharpoons H_3O^+ + CH_3COO^- \quad (4)$$

Note that adding acetate ion produces a change in the color of the indicator. The color change indicates that the indicator has been changed from its acid to its base form, which in turn must mean that the hydrogen ion concentration became smaller when the sodium acetate was added. Explain your observations in terms of the equilibria shown in Equations (3) and (4). The effect on the dissociation of acetic acid that is produced by adding sodium acetate is called the *common ion effect*.

To each of two 3-mL samples of 0.1 M NH_3 add a drop of phenolphthalein indicator. Note and record the color. To one sample add 1 M NH_4Cl, a few drops at a time, with mixing. To the other add 6 M HCl, a drop at a time, with mixing. In each case, note any changes in the solution's color and odor.

Write the equation for the reaction of NH_3 with water to form NH_4^+ and OH^-. Interpret the results in terms of the changes of H_3O^+ concentration (shown by the color change of the indicator) and the equilibrium for the reaction of NH_3 with water. Explain clearly how the equilibria shift when NH_4^+ ions (from NH_4Cl) and H_3O^+ (from HCl) are added. Write the net ionic equation for the reaction of NH_3 with HCl.

2. COMPLEX ION EQUILIBRIA.
It is common for cations (especially those with $+2$ or $+3$ charge) to attract negatively charged ions (or neutral molecules with lone pairs of electrons) to form aggregates called *complexes*. If the resulting aggregate has a net charge, it is called a *complex ion*. The composition of these complexes may vary with the proportion and concentration of reactants.

(a) *The Thiocyanatoiron(III) Complex Ion.* This ion, sometimes called the *ferric thiocyanate* complex ion, is formed as a blood-red substance described by the following equilibrium equation.[1]

$$Fe^{3+} + SCN^- \rightleftharpoons Fe(SCN)^{2+} \quad (5)$$

In a 100-mL beaker add 3 mL of 0.1 M $Fe(NO_3)_3$ to 3 mL of 0.1 M KSCN. Dilute by adding 50 to 60 mL of water until the deep-red color is reduced in intensity, making further changes (either increases or decreases in color) easy to see. Put 5 mL of this solution in a test tube; add 1 mL (about 25 drops) of 0.1 M $Fe(NO_3)_3$. To a 5-mL portion in a second test tube, add 1 mL of 0.1 M KSCN. To a third 5-mL portion, add 5 to 6 drops of 6 M NaOH. ($Fe[OH]_3$ is very insoluble.) Put a fourth 5-mL portion in a tube to serve as a control. Compare the relative intensity of the red color of the thiocyanato complex in each of the first three test tubes to that of the original solution in the fourth tube. Interpret your observations, using Le Châtelier's principle and considering the equilibrium shown in Equation (5).

(b) *The Temperature-Dependent Equilibrium of Co(II) Complex Ions.* The chloro complex that consists of cobalt(II), $CoCl_4^{2-}$ is tetrahedral and has a blue color. The aquo complex of cobalt(II), $Co(H_2O)_6^{2+}$, is octahedral and has a pink color. (Figure 23-1 shows the geometry of the tetrahedral and octahedral complexes.) There is an equilibrium between the two forms in aqueous solution, and because the conversion of one form to another involves a considerable energy change, the equilibrium is temperature dependent:

$$CoCl_4^{2-} + 6H_2O \rightleftharpoons$$
$$4Cl^- + Co(H_2O)_6^{2+} + energy \quad (6)$$

Le Châtelier's principle applied to Equation (6) predicts that if energy is removed (by cooling the system), the equilibrium tends to shift toward the aquo complex, because a shift in this direction produces some energy, thus partly offsetting the change.

Put 3 mL of 0.15 M $CoCl_2$ (in methanol) into a 13×100 mm test tube. (Methanol is used as a solvent so that you can observe the effects of adding water.) Using a dropper, add *just enough* water to the blue methanol solution to change the color to that of the pink aquo complex. Divide the pink solution into two equal portions in two 13×100 mm test tubes. Add concentrated (12 M) HCl one drop at a time to one test tube until you observe a color change. Record your observations. Heat the test tube containing the other portion of the pink solution in a beaker of hot water (65 to 70 °C) on a hot plate in a fume hood. (**Caution:** Methanol vapors are toxic and flammable. If a fume hood is not available, use an inverted funnel connected to an aspirator [as shown in Figure 5-1] to remove the methanol vapor.) You should

---

[1] For simplicity, we use the unhydrated formulas. This makes no difference in our equilibrium consideration, however, since water is always present at high constant concentration. With the addition of SCN^- to hydrated iron(III) ion, $Fe(H_2O)_6^{3+}$, a substitution of SCN^- for H_2O occurs, with possible formulas such as $Fe(H_2O)_5(SCN)_2^+$, $Fe(H_2O)_4(SCN)_2^+$, $Fe(H_2O)_3(SCN)_3$, ..., $Fe(SCN)_6^{3-}$.

FIGURE 23-1

The cobalt(II) complexes with chloride ion and water have different molecular geometries and different colors.

note a color change. (If you do not, you probably added too much water to the original methanol solution. Try again.) The color change is reversible. Cooling the solution in an ice bath will restore the original pink color. Repeat the cycle of heating and cooling to verify this. Record your observations, and interpret them by applying Le Châtelier's principle to the equilibrium shown in Equation (6).

3. THE EQUILIBRIA OF SATURATED SOLUTIONS.

(a) *Saturated Sodium Chloride.* To 4 mL of saturated (5.4 M) sodium chloride in a 13 × 100 mm test tube, add 2 mL of concentrated (12 M) HCl. Mix, observe, and record your observations. The Cl^- concentration in the original solution is 5.4 M (the same as the NaCl concentration). Calculate the total Cl^- concentration that the solution would have after adding 2 mL of 12 M HCl, and explain what you observed in terms of the saturated solution equilibrium.

(b) *Saturated Barium Chromate.* To 3 mL of 0.1 M $BaCl_2$, add 5 to 6 drops of 1 M K_2CrO_4, stir, and then add 10 to 12 drops of 6 M HCl. Record your observations. Write an equation for the reaction of Ba^{2+} with CrO_4^{2-}. Using the equation you have written, explain how the addition of HCl shifts the equilibrium of the reaction. (Recall your observations in Part 1 as summarized by Equation [2].)

4. APPLICATION OF THE LAW OF CHEMICAL EQUILIBRIUM TO SOLUBILITY EQUILIBRIA.

In qualitative analysis, Ca^{2+} is usually precipitated as insoluble calcium oxalate:

$$Ca^{2+} + C_2O_4^{2-} \rightleftharpoons CaC_2O_4(s) \tag{7}$$

What conditions will make the precipitation as complete as possible? Since Ca^{2+} is the unknown ion, it is desirable to drive the reaction as far as possible to the right by obtaining the maximum $C_2O_4^{2-}$ concentration possible. Would it be better to use a substance that completely dissociates, such as the soluble salt $K_2C_2O_4$, or a substance that forms only a slight amount of oxalate ion, such as the weak acid $H_2C_2O_4$? The dissociation equilibria are

$$K_2C_2O_4 \rightleftharpoons 2K^+ + C_2O_4^{2-} \tag{8}$$

and

$$H_2C_2O_4 + H_2O \rightleftharpoons H_3O^+ + HC_2O_4^-$$
$$+$$
$$H_2O$$
$$\Updownarrow$$
$$H_3O^+ + C_2O_4^{2-} \tag{9}$$

Would it be better to make the solution acidic or basic to achieve the maximum $C_2O_4^{2-}$ concentration? Test your reasoning by the following experiments.

Mix 4 mL of 0.1 M $CaCl_2$ with 4 mL of deionized water, and place equal portions in three 13 × 100 mm test tubes. To tube 1 add 0.3 mL (6 to 7 drops) of 0.5 M $H_2C_2O_4$, and to tube 2 add 0.3 mL (6 to 7 drops) of 0.5 M $K_2C_2O_4$. Compare the results. In your records refer to these experiments as solution 1 and solution 2, respectively.

To tube 1 add 0.5 mL (10 drops) of 6 M HCl, and mix. (For purposes of your records, refer to this new mixture as solution 3.) Explain the results. Now, again to the solution in tube 1 add a slight excess of 6 M NH_3, and mix. (Refer to this mixture as solution 4.) Is the precipitate that forms $Ca_2C_2O_4$ or $Ca(OH)_2$? ($Ca[OH]_2$ could be a possible product because the reaction of NH_3 with water generates some OH^- ion.) To find out if the precipitate is $Ca(OH)_2$, add a few drops of 6 M NH_3 to the third tube that you prepared (the one that contains diluted 0.1 M $CaCl_2$ solution). (For reference, call this solution 5.) Does a precipitate form? From this result what conclusion can you draw about the identity of the precipitate in solution 4? Explain all of your results in terms of the equilibria shown in Equations (7), (8), and (9) and the reaction with acid of NH_3:

$$NH_3 + H_3O^+ \rightleftharpoons NH_4^+ + H_2O \tag{10}$$

Would the acidity of the solution be more important in

the precipitation of the salt of a strong acid or in the precipitation of the salt of a weak acid? To put this question another way, is H_3O^+ more effective in competing with the metal ion for the *anion of a strong acid* or the *anion of a weak acid*?

☞ Consider This

When two chemical equilibria have a species in common, we say that they are "linked." Changes affecting one equilibrium will also affect the other, if the equilibria are linked. Give an example of a system in which two linked equilibria coexist. Next, give an example of a chemical system in which two independent equilibria, not linked, simultaneously coexist.

BIBLIOGRAPHY

Berger, T. G.; Mellon, E. K.; and Dreisbach, J., "A Multistep Equilibria-Redox-Complexation Demonstration to Illustrate Le Châtelier's Principle," *J. Chem. Educ.* **1996,** *73,* 783.

Bodner, G. M., "On the Misuse of Le Châtelier's Principle for the Prediction of the Temperature Dependence of the Solubility of Salts," *J. Chem. Educ.* **1980,** *57,* 117–119.

Jensen, W. B., "Le Châtelier's Principle, Solubility, and Déjà Vu," *J. Chem. Educ.* **1987,** *64,* 287–288 (Letter to the editor).

OBSERVATIONS AND DATA

1. The Shifting of Equilibria in Acid-Base Reactions. The Common Ion Effect

(a) *The Chromate Ion–Dichromate Ion Equilibrium*

Rewrite the equation for this equilibrium.

Describe the color changes on the addition of

H_2SO_4

NaOH

Interpret these observations in terms of the equilibrium described by Equation (2) and Le Châtelier's principle.

How does OH^- exert an effect?

(b) *Weak Acid–Base Indicator Equilibria*

Describe the effects of adding HCl and NaOH to

methyl orange

phenolphthalein

Interpret these effects in terms of Equation (3) and Le Châtelier's principle.

(c) *Weak Acid–Weak Base Equilibria*

Complete the equation for the dissociation equilibrium of acetic acid in water.

$$CH_3COOH + H_2O \rightleftharpoons$$

Explain the observed changes when 1 M sodium acetate is added to 0.1 M acetic acid (+ methyl orange).

Complete the equation for the reaction of ammonia with water.

$$NH_3 + H_2O \rightleftharpoons$$

Describe any observed odor or color changes when you add the following to 0.1 M NH_3 (+ phenolphthalein):

$$NH_4Cl$$

$$HCl$$

In which direction, left or right, does each reagent above shift the equilibrium for the reaction of NH_3 with water? Explain fully.

Write the net ionic equation for the reaction of NH_3 with HCl.

2. Complex Ion Equilibria

(a) *The Thiocyanatoiron(III) Complex Ion*

Rewrite the equation for this equilibrium.

Compare the intensity of the red color on addition of each of the following with the intensity of the color in tube 4. Interpret your observations in terms of the equation describing the equilibrium and Le Châtelier's principle.

$Fe(NO_3)_3$

KSCN

NaOH (Assume that the brown precipitate is formed by the reaction: $Fe^{3+} + 3OH^- \rightarrow Fe[OH]_3[s]$.)

(b) *The Temperature-Dependent Equilibrium of Co(II) Complex Ions*

Rewrite the equation for this equilibrium.

Describe what happens when you add 12 M HCl to the pink (aquo) complex.

Describe what happens when you heat the pink (aquo) complex and when you cool it.

Interpret these results in terms of the equation describing the equilibrium and Le Châtelier's principle.

3. The Equilibria of Saturated Solutions

(a) *Saturated Sodium Chloride*

Describe what happens when 12 M HCl is added to saturated (5.4 M) NaCl.

What is the Cl$^-$ concentration

(i) In the original solution?

(ii) After addition of the 12 M HCl (before any precipitate forms)?

Explain this behavior in terms of the Cl$^-$ concentrations in the two solutions, and in terms of the equilibrium equation $NaCl(s) \rightleftharpoons Na^+(aq) + Cl^-(aq)$.

(b) *Saturated Barium Chromate*

Write the equation for the equilibrium established when K_2CrO_4 and $BaCl_2$ solutions are mixed, and explain the changes observed when this solution is treated with 6 M HCl.

4. Application of the Law of Chemical Equilibrium to Solubility Equilibria

Solutions mixed	Relative amounts of precipitate, if any
(1) $CaCl_2(aq)$ + $H_2C_2O_4(aq)$	
(2) $CaCl_2(aq)$ + $K_2C_2O_4(aq)$	
(3) Results of (1) + 6 M HCl	
(4) Results of (3) + 6 M NH_3	
(5) $CaCl_2(aq)$ + 6 M NH_3	

Answer questions (a) and (d) relating these data on the relative amounts of $CaC_2O_4(s)$ formed, or dissolved, under various conditions to the following linked equilibrium equations. (The equilibria are said to be linked because each equilibrium shares at least one species in common with another equilibrium.)

$$Ca^{2+} + C_2O_4^{2-} \overset{\boxed{1}}{\rightleftharpoons} CaC_2O_4(s)$$
$$+$$
$$H_3O^+ + NH_3 \overset{\boxed{3}}{\rightleftharpoons} NH_4^+ + H_2O$$
$$\boxed{2} \Updownarrow$$
$$HC_2O_4^- + H_3O^+ \overset{\boxed{4}}{\rightleftharpoons} H_2C_2O_4 + H_2O$$
$$+$$
$$H_2O$$

(a) Account for the difference in behavior of $H_2C_2O_4$ and $K_2C_2O_4$ in tubes 1 and 2. Specifically, why does one solution form so much more $CaC_2O_4(s)$?

(b) From this array of linked equilibria, explain the effect of adding HCl on these equilibria. (Note that H_3O^+ participates in equilibria 2, 3, and 4.) Why does the addition of HCl change the amount of $CaC_2O_4(s)$?

(c) Consider solutions 4 and 5, where NH_3 was added. Does $Ca(OH)_2(s)$ form when NH_3 is added to a Ca^{2+} solution (solution 5)? How does ammonia affect the amount of $CaC_2O_4(s)$ present in solution 4?

(d) Why does the H_3O^+ concentration have more influence on the precipitation of the salt of a weak acid, such as $CaC_2O_4(s)$, than on the precipitation of the salt of a strong acid, such as $AgCl(s)$ or $CaSO_4(s)$?

☞ Consider This

When two chemical equilibria have a species in common, we say that they are "linked." Changes affecting one equilibrium will also affect the other, if they are linked. Give an example of a system in which two linked equilibria coexist. Next, give an example of a chemical system in which two independent equilibria, not linked, simultaneously coexist.

Determination of an Equilibrium Constant by Spectrophotometry
The Iron(III)-Thiocyanate Complex Ion

PURPOSE

To confirm, by using the method of continuous variation, that the iron(III) ion forms a 1:1 complex with thiocyanate ion. To use a spectrophotometer to determine the amount of complex at equilibrium, enabling you to calculate the equilibrium constant for the formation of the complex.

PRE-LAB PREPARATION

When two or more substances react to form a particular set of products, we generally want to know two things about the chemical reaction: (1) the chemical composition of the product(s) and (2) the extent of reaction.

The extent of reaction is measured by the equilibrium constant for the reaction. If the equilibrium constant is very small (or zero), we say that essentially there is no reaction. If the equilibrium constant is very large, we say that the reaction goes to completion.

For many chemical reactions the point of equilibrium lies between the two extremes of no reaction and complete reaction. In these circumstances it is an interesting challenge to determine the composition of the equilibrium mixture of reactants and products and to calculate the equilibrium constant for the reaction.

The reaction we will consider is the formation of a complex ion by reaction of Fe(III) ion (ferric ion, Fe^{3+}) with SCN^- (thiocyanate ion) to form a red-colored complex ion:

$$Fe^{3+} + SCN^- \rightleftharpoons Fe(SCN)^{2+}$$

metal + ligand   metal-ligand
     ion   complex ion   (1)

We will use the method of continuous variation to confirm that one ferric ion reacts with one thiocyanate ion to form a single product, called a *complex ion*. In this context the term *complex ion* does not imply that the product is especially complicated. Rather, it means that the reaction product is an aggregate (or complex) of an Fe^{3+} ion and a SCN^- ion.

There is a universal tendency for positively charged metal ions to form complexes with negatively charged ions (e.g., SCN^-) and with neutral molecules (e.g., water and ammonia) that have molecular dipoles (those in which one side of the molecule being negative and the other side positive). This tendency can be understood as the result of an attractive force between the positively charged metal ions and negatively charged ions (or the negative end of a molecular dipole). Therefore, when you dissolve a metal salt in water, the metal ions will form complexes either with the water, with the anions of the salt, or with any other anions or neutral molecules having lone-pair electrons that may be added. The ions or neutral molecules bound to the metal ion are called *ligands* (from the Latin word *ligare*, to bind).

When it has a net charge, the complex often remains in solution as a complex ion. The $Fe(SCN)^{2+}$ complex ion we are studying behaves this way. The equilibrium is called a *homogeneous* equilibrium because all of the reactants and products are present together in *one phase*, the solution.

Some neutral complexes are very insoluble and precipitate out of solution. An example of this kind of behavior is the precipitation of insoluble solid $BaSO_4$ when a solution of Na_2SO_4 is added to a solution of $BaCl_2$. The equilibrium between a solid insoluble salt and its constituent ions is called a *heterogeneous* equilibrium because the system involves two phases: a solid phase, which consists of the pure insoluble salt, and a liquid phase, which contains the constituent ions of the salt in solution.

We make the distinction between the two kinds of equilibria, *complex formation* and *precipitation reactions*, on the basis of whether or not the equilibrium is homogeneous or heterogeneous. The driving force for both these kinds of reactions is the same, namely, the attraction between a positively charged metal ion and a negatively charged anion (or neutral molecule with lone-pair electrons).

THE METHOD OF CONTINUOUS VARIATION

Determining the stoichiometry of a chemical reaction by the method of continuous variation is based on a simple idea. Keep the total number of moles of reactants constant and vary their ratio, while measuring the number of moles of product produced by the reaction. The maximum amount will be formed when the mole ratio of the reactants is precisely the same as the ratio of the coefficients in the equation that describes the net chemical reaction.

We can use a simple analogy to explain how this works. It's based on the notion of a *limiting reactant,* a concept that you have already studied. Suppose that you are preparing hot dogs by putting frankfurters in buns. Each hot dog requires one frankfurter and one bun, so we can write the "reaction" for the formation of a hot dog in the following way:

$$\text{frankfurter} + \text{bun} \rightarrow \text{hot dog} \qquad (2)$$

Now suppose that the sum of the number of frankfurters and buns is constant, for example, 10. How many hot dogs could we make if we had 2 frankfurters and 8 buns? Just 2 hot dogs with 6 buns left over; the frankfurters represent the limiting reactant. When would we be able to produce the greatest number of hot dogs? If we had 5 frankfurters and 5 buns, we could make 5 hot dogs with no frankfurters or buns left over.

There is a special way of graphing or plotting the data to determine the ratio that represents the "stoichiometry" of our reaction, a 1:1 ratio of frankfurters and buns. This graph is constructed by plotting the number of hot dogs that we can make against the fraction of frankfurters, where we define the fraction of frankfurters, F, as

$$F = \frac{\text{number of frankfurters}}{\text{number of frankfurters} + \text{number of buns}} \qquad (3)$$

Defined this way, the fraction of frankfurters ranges from 0 to 1. In Figure 24-1 we have plotted the number of hot dogs that we can make versus the fraction of frankfurters for the case where the number of frankfurters plus the number of buns equals 10. We can make the greatest number of hot dogs when the fraction of frankfurters equals 0.5. This ratio corresponds to an equal number of frankfurters and buns, 5 in this example. So from the plot

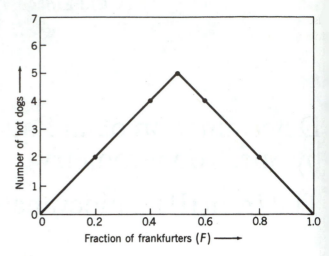

FIGURE 24-1

A graph of the number of hot dogs that can be made versus the fraction of frankfurters for the case where the number of frankfurters plus the number of buns equals 10. The maximum occurs when the fraction of frankfurters equals 0.5. At this point the number of frankfurters and buns are both equal to 5 and the frankfurter-to-bun ratio is 1.0.

we can determine the "stoichiometry" of our "reaction." To make a hot dog, we require one frankfurter for each bun.

We will use the method of continuous variation just described to confirm the 1:1 stoichiometry of the $Fe(SCN)^{2+}$ complex. While holding the sum of the moles of Fe(III) and moles of SCN^- constant, we will vary the mole fraction of Fe(III) as we measure the amount of $Fe(SCN)^{2+}$ complex formed.

There are two further points to consider. First, what if the reaction does not go to completion when we mix the reactants? It turns out that we still find that the maximum yield of products comes when the reactants are mixed in the ratio corresponding to the stoichiometry of the complex, provided that only one chemical reaction exists at equilibrium in the range of concentrations being studied. (In the system we are studying, it has been found that only one complex is formed under the conditions we will use. However, if we increase the ratio of SCN^- to Fe[III] much above 4:1, a second complex, having two SCN^- ions to each Fe[III], begins to form.) If the reaction does not go to completion, we will obtain a plot that is rounded at the top rather than sharply peaked like a triangle. This rounding occurs because at equilibrium we have a significant amount of both reactants and products present together, so less product is formed than if the reaction were 100 percent complete.

The second point that needs to be made is that we must have some way of measuring the amount of product formed. We will do this by spectrophotometry, making se of the fact that the Fe(III)-thiocyanate complex formed is red and absorbs visible light, whereas the reactants, Fe^{3+} and SCN^- ions, are essentially colorless and therefore do not absorb light in the region of the spectrum where the complex does.

WHY ARE TRANSITION METAL ION COMPLEXES OFTEN COLORED?

The absorption of light is something like a chemical reaction in the sense that it involves an encounter (or collision) between two things: the absorbing molecule (or ion) and a packet of electromagnetic energy that we call a *photon*. Associated with the energy of the photon is a characteristic wavelength, λ, and frequency, v. The product of the frequency times the wavelength of the photon is equal to the speed at which the photon travels, the speed of light, c:

$$\lambda v = c \qquad (4)$$

The *energy* of the photon is related to the frequency and wavelength of the photon by the equation

$$E = hv = \frac{hc}{\lambda} \qquad (5)$$

where E is the energy of the photon and h is Planck's constant.

In order for there to be a high probability that a photon will be absorbed, the photon's energy must match the difference between two energy states of the molecule. For visible light the energy states correspond to those of the electrons in the molecule—in particular, to the energy difference between the energy states of the outermost valence electrons (d-orbital energy states) belonging to the Fe(III) ion in the complex ion.

We may be used to thinking that the five $3d$ orbitals of an Fe atom (or ion) all have the same energy, but that is true only for a free atom (or ion) in the gaseous state. The presence of any ion or molecule bound to an Fe^{3+} ion alters the energy of the d orbitals, producing energy differences between them. Relative to the free ion, electrons in d orbitals whose lobes of electron density point toward the ligands along the x, y, or z axes are generally raised in energy, while electrons in d orbitals pointing between the ligands are generally lowered in energy. This creates an energy difference between the d orbitals, as shown in Figure 24-2. So transition metal ions such as Fe(III), which have partially filled d orbitals, can absorb photon energy if the photon's energy approximately matches the difference in energy of the d orbitals. In the process, an electron absorbs the photon energy, going from a filled or half-filled d-orbital energy state to a higher, energy-

FIGURE 24-2

(A) Six negative charges arranged octahedrally along a set of x, y, z Cartesian reference axes create an electrostatic field that splits the M^{+n} metal ion d-orbital energy levels. (B) An energy level diagram showing how the five d orbitals are split into two groups by an electrostatic field with octahedral symmetry. When a photon is absorbed, an electron is promoted from the lower to the upper d-orbital energy level.

empty or half-filled d-orbital energy state. Such an electronic transition is not possible if all the d orbitals are filled with pairs of electrons having opposite spins, as in the Zn^{2+} ion. Thus simple Zn^{2+} salts are usually colorless, whereas many transition metal ions having partially filled d orbitals absorb visible light and therefore display a variety of colors.

Groups I, II, and III metal ions, such as Na^+, Ca^{2+}, and Al^{3+}, do not absorb visible light because the spacing between their electronic energy levels is much larger than the spacing between the d-orbital energy states in transition metal ions. As a result, the absorption occurs in the ultraviolet (which our eyes cannot see).

MEASUREMENT OF THE COMPLEX CONCENTRATION BY SPECTROPHOTOMETRY

An instrument designed to measure the amount of light absorbed at different wavelengths is called a *spectrophotometer*. Figure 24-3 is a diagram of a widely used spectrophotometer that consists of a light source, an entrance

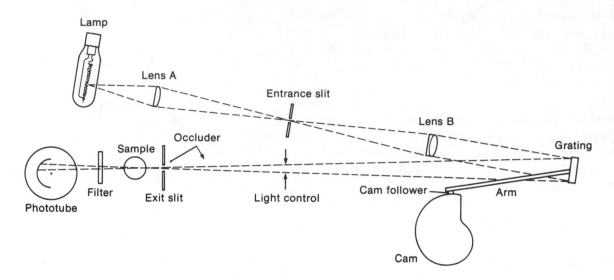

FIGURE 24-3
Schematic optical diagram of the Bausch & Lomb Spectronic 20.

slit, collimating lenses that form a parallel beam of light, a reflection grating to disperse (spread out) the light into its component wavelengths, an exit slit to isolate a narrow band of wavelengths of the dispersed light, a sample cell holder, and a detector to measure the amount of light that passes through the sample.

The light detectors used in most spectrophotometers provide an electrical signal that is proportional to the radiant power, P. We may think of the radiant power as a measure of light intensity proportional to the number of photons arriving per second at the detector.

In the Spectronic 20, which has only a single beam, the measurement of the light absorption by the sample is made by positioning the grating to give the desired wavelength of light. The phototube current is read on a meter that is first adjusted to read 0 when no light is falling on the detector. Then a tube containing only the reference solution (the solvent medium or water) is inserted in the sample cell and the light control knob is adjusted so that the meter reads its maximum value (100 percent transmittance of the light). Then a tube containing the sample solution is placed in the cell holder. If the sample solution absorbs some of the light, less light will arrive at the detector and the meter will read some number less than 100 percent transmittance. The transmittance, T (which ranges between 0 and 1), is defined by the equation

$$T = \frac{P}{P_0} \qquad (6)$$

where P_0 is proportional to the number of photons per second passing through the reference tube, and P is pro-

portional to the number of photons per second passing through the sample tube, as shown schematically in Figure 24-3. The percentage transmittance (which ranges between 0 and 100) is defined by the equation

$$\%T = 100T \qquad (7)$$

Therefore the transmittance, T, represents the fraction of photons that pass through the sample molecules without being absorbed, as shown in Figure 24-4.

The probability that a photon will be absorbed at a particular point in the cell is proportional to both the concentration of absorbing molecules and the number of photons arriving at the point. The absorption of photons causes the radiant power, P, to decline exponentially as it passes through the absorbing sample:

$$T = \frac{P}{P_0} = 10^{-\epsilon bc} \qquad (8)$$

We have previously defined P_0 and P (see Figure 24-4). In Equation (8) the constant ϵ is called the *molar absorptivity* (it has units of liters per mole per centimeter), b is the cell *path length* in centimeters, and c is the *concentration* of the absorbing sample in moles per liter. Taking the logarithm (to the base 10) of both sides, we obtain a quantity, called the *absorbance, A,* that is proportional to the concentration of the sample:

$$A = \text{Log}_{10}\left(\frac{P_0}{P}\right) = \epsilon bc \qquad (9)$$

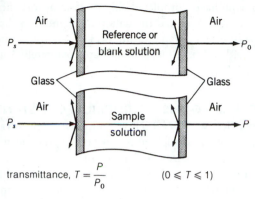

transmittance, $T = \dfrac{P}{P_0}$ $\quad (0 \leqslant T \leqslant 1)$

percentage transmittance, $\%T = \dfrac{P}{P_0} \times 100$ $\quad (0 \leqslant \%T \leqslant 100)$

FIGURE 24-4

The measurement of transmittance (or percentage transmittance). The radiant power from the source is P_s. When the reference or blank solution is in the cell, the radiant power that emerges from the cell is P_0. When the sample solution is in the cell, the radiant power that emerges is P. The use of a reference or blank solution compensates for reflection losses at the air-glass and glass-solution interfaces as well as absorption by the glass and solvent, to ensure that you measure only absorption due to the sample.

Spectrophotometers like the Spectronic 20 have both a linear $\%T$ scale and a nonlinear (logarithmic) absorbance scale. The relationship between these scales is shown in Figure 24-5.

We will determine the concentration of the absorbing $Fe(SCN)^{2+}$ complex by measuring the absorbance of the complex, either reading the $\%T$ or reading the absorbance directly. If you measure the $\%T$, it is a simple matter to compute the absorbance from the measured $\%T$ readings by using a calculator:

$$A = \text{Log}_{10}\left(\frac{P_0}{P}\right) = \text{Log}_{10}\left(\frac{100}{\%T}\right) \qquad (10)$$

Knowing the value of the molar absorptivity and the cell path length b (or their product, ϵb), we can calculate the

FIGURE 24-5

The relationship between absorbance (A), transmittance (T), and percentage transmittance ($\%T$). The absorbance scale is a nonlinear (logarithmic) scale. The transmittance scales are linear.

concentration (in moles per liter) of the absorbing complex from the simple relation

$$c = \frac{A}{\epsilon b} \qquad (11)$$

EXPERIMENTAL PROCEDURE

Special Supplies: Spectrophotometer, spectrophotometer cells (15 × 125 mm test tubes for the Spectronic 20), two burets (or 10-mL graduated cylinders), plastic wrap.

Chemicals: 0.00200 M KSCN in 0.50 M HNO_3 solution + 0.5 g/L sulfamic acid, 0.00200 M $Fe(NO_3)_3$ in 0.5 M HNO_3 solution + 0.5 g/L sulfamic acid[1], $NaHCO_3(s)$.

! SAFETY PRECAUTIONS: The stock solutions of KSCN and $Fe(NO_3)_3$ contain 0.50 M HNO_3, which is corrosive. Wipe up any spills immediately, and if you get any solution on your skin, immediately rinse it off with water.

WASTE COLLECTION: When you have finished making absorbance measurements, empty all of the solutions into an 800-mL beaker, neutralize with sodium bicarbonate until the solution is just basic to litmus or pH paper, and discard the neutralized solution down the drain.

1. ABSORBANCE MEASUREMENT. Obtain seven 15 × 125 mm test tubes. Label the tubes from 1 to 7. In each test tube put the quantities of 0.00200 M KSCN in 0.5 M HNO_3 solution and 0.00200 M $Fe(NO_3)_3$ in 0.5 M HNO_3 solution shown in Table 24-1. *The solution volumes must be measured as accurately as possible.* It is convenient to dispense the two stock solutions from burets. If burets are not available, use two 10-mL graduated cylinders, adjusting the volumes precisely with a dropper. Note that in each tube the sum of the concentrations of the thiocyanate ion and the Fe(III) ion is constant and is equal to 0.00200 M. This fulfills the condition required for using the method of continuous variation discussed earlier.

Thoroughly mix the solution in each tube by covering the mouth of the tube with a small piece of plastic wrap to protect your thumb and inverting the tube several times. If you are using a Spectronic 20, adjust

---

[1]When the KSCN and $Fe(NO_3)_3$ solutions are made up, add 0.5 g/L of sulfamic acid ($^+H_3NSO_3^-$) to each solution to eliminate nitrogen oxides from the nitric acid, which might oxidize thiocyanate ion. (The sulfamic acid does not interfere with the Fe[III]/SCN$^-$ reaction.) In each stock solution the 0.50 M HNO_3 is present to suppress hydrolysis of the Fe^{3+} ion and to maintain a solution of constant ionic strength so that the equilibrium constant will not change. At low ionic strengths, the value of the equilibrium constant is strongly dependent on the ionic strength of the solution.

TABLE 24-1

In the method of continuous variation, the total solution volume is kept constant.

Tube number	0.00200 M solution (mL)	
	KSCN	Fe(NO$_3$)$_3$
1	10	0
2	8	2
3	6	4
4	5	5
5	4	6
6	2	8
7	0	10

the spectrophotometer wavelength to 450 nm, and set the 100% T value with a tube containing pure water. Measure the absorbance (or %T) of each solution at 450 nm, following the procedures given by your instructor for the particular instrument used in your laboratory. (If your instrument measures only %T, you can calculate the absorbance of each solution by using Equation [10].) Save the seven solutions you have prepared until you have calculated and plotted the absorbance values as described in Part 2.

2. CONFIRMING THE MOLE RATIO OF THE COMPLEX.

Determine the ratio of Fe^{3+} to SCN^- in the complex by plotting the absorbance of the seven solutions you measured versus the mole fraction of Fe(III), where the mole fraction of Fe(III) in each solution is calculated from the relation[2]

$$\text{mole fraction Fe(III)} = \frac{\text{mL Fe(III) sol'n}}{\text{mL Fe(III) sol'n} + \text{mL KSCN sol'n}} \quad (12)$$

Draw a smooth curve through the points you have plotted. Show the plot to your instructor. If it is satisfactory, determine the mole fraction of Fe(III) where the maximum value of absorbance occurs. What mole ratio of iron(III) to thiocyanate ion does this mole fraction of Fe(III) correspond to?

---

[2]The mole fraction of Fe(III) is fundamentally defined as mol Fe(III)/ (mol Fe[III] + mol KSCN), but because the number of moles of each substance is equal to the volume times the molarity of the solution, and because the molarities of the solutions are equal, the mole fraction is equal to the volume relationship of Equation (12).

When you have completed all of your absorbance measurements and plotted the data points, you may discard the solutions into an 800-mL beaker. Neutralize the solutions with sodium bicarbonate (approximately 5 g) and discard them down the drain.

3. CALCULATING THE EQUILIBRIUM CONSTANT FOR THE REACTION.

Three unknown concentrations appear in the equilibrium constant expression for Equation (1), so three independent equations are required to determine these unknown concentrations. The three equations we will use are the equilibrium constant expression

$$K = \frac{[Fe(SCN)^{2+}]}{[Fe^{3+}][SCN^-]} \quad (13)$$

and two material balance equations, which account for all the Fe(III) and thiocyanate-containing species in the solution:

$$c_{Fe} = [Fe^{3+}] + [Fe(SCN)^{2+}] \quad (14)$$

$$c_{SCN} = [SCN^-] + [Fe(SCN)^{2+}] \quad (15)$$

The material balance equations state that all of the Fe(III)-containing species and all of the thiocyanate-containing species must add up to the amounts of Fe(III) and thiocyanate ion originally put into the solutions. For each solution, c_{Fe} and c_{SCN} are given by the expressions

$$c_{Fe} = 0.00200 \text{ M} \times \frac{\text{mL Fe(NO}_3)_3 \text{ sol'n}}{10.0 \text{ mL}} \quad (16)$$

and

$$c_{SCN} = 0.00200 \text{ M} \times \frac{\text{mL KSCN sol'n}}{10.0 \text{ mL}} \quad (17)$$

We will calculate values of the equilibrium constant for just three solutions, those in tubes 2, 4, and 6. We start by using Equation (11) to calculate the concentration of the $Fe(SCN)^{2+}$ complex in each of the three solutions from the measured absorbances of the solutions.

....

NOTE: If you use conventional cells of square cross section with 1-cm path length, assume that $\epsilon = 4700$ L $\times$ mol^{-1} $\times$ cm^{-1} and $b = 1.00$ cm. If you use a Spectronic 20 with 15 $\times$ 125 mm test tubes as sample cells, take the product ϵb in Equation (11) to be equal to 6120 L $\times$ mol^{-1}.

....

Next, for each solution, calculate the total concentration of Fe(III), c_{Fe}, from Equation (16) and the total concentration of thiocyanate ion, c_{SCN}, from Equation (17). Now, for each solution you know the concentration of the $Fe(SCN)^{2+}$ complex and the total concentrations, c_{Fe} and c_{SCN}.

Next, for each solution, calculate the equilibrium concentration of iron(III) ion, $[Fe^{3+}]$, using Equation (14) and the equilibrium concentration of thiocyanate ion, $[SCN^-]$, using Equation (15). This gives for each solution the three equilibrium concentrations you need to calculate a value of the equilibrium constant K from Equation (13).

Finally, calculate the average value of the equilibrium constants you calculated for the three solutions in tubes 2, 4, and 6.

👉 Consider This

Suppose that you took the contents of tube number 4, containing equal amounts of 0.0020 M Fe(III) and 0.0020 M KSCN solution, and diluted it by a factor of two by adding an equal volume of 0.5 M HNO$_3$. (It's important to add 0.5 M HNO$_3$ rather than water, in order to keep the ionic strength constant. Otherwise, the value of the equilibrium constant would change because its value depends on the ionic strength, as described in the articles by Frank and Oswalt [1947] and Cobb and Love [1998] listed in the Bibliography.)

When the solution has been diluted by a factor of two, what would you expect to happen to the concentration of $Fe(SCN)^{2+}$ as measured by the absorbance at 450 nm? Would it decrease by a factor of two? Try it and see. Can you explain why it decreases approximately a factor of four, rather than a factor of two?

Thiocyanate is found in human saliva. You could determine the thiocyanate of your own saliva following the detailed procedure described by Lahti et al. (1999), which involves using a spectrophotometric measurement of the red color of the iron(III)-thiocyanate complex formed when Fe(III) is mixed with saliva in acid solution.

BIBLIOGRAPHY

Cobb, C. L., and Love, G. A. "Iron(III) Thiocyanate Revisited: A Physical Chemistry Equilibrium Lab Incorporating Ionic Strength Effects," *J. Chem. Educ.,* **1998,** *75:* 90–92.

Frank, H. S., and Oswalt, R. L. "The Stability and Light Absorption of the Complex Ion FeSCN^{++}," *J. Am. Chem. Soc.,* **1947,** *69:* 1321–1325.

Lahti, M.; Vilpo, J.; Hovinen, J. "Spectrophotomorphic Determination of Thiocyanate in Human Saliva," *J. Chem. Educ.,* **1999,** *76:* 1281–1282.

Ramette, R. W. "Formation of Monothiocyanatoiron(III): A Photometric Equilibrium Study," *J. Chem. Educ.,* **1963,** *40:* 71–72.

REPORT	Determination of an Equilibrium Constant by Spectrophotometry	NAME	
24	The Iron(III)-Thiocyanate Complex Ion	SECTION	LOCKER
		INSTRUCTOR	DATE

DATA AND CALCULATIONS

1. Absorbance Measurements

Record the absorbance, measured at a wavelength of 450 nanometers (nm), for each solution. If the spectrophotometer you use does not read absorbance directly, record the %T reading and calculate the absorbance by using Equation (10). Also calculate for each solution the mole fraction of Fe(III) by using Equation (12).

Test tube number	%T	Absorbance	Mole fraction Fe(III)
1			
2			
3			
4			
5			
6			
7			

2. Confirming the Mole Ratio of the Complex

For each of the seven solutions whose absorbance was measured in Part 1, plot the absorbance versus the mole fraction of Fe(III). Draw a smooth curve through the seven points, and determine the value of the mole fraction of Fe(III) where the absorbance is a maximum. From this value determine the nearest-integer value of the Fe(III)-to-SCN⁻ ratio.

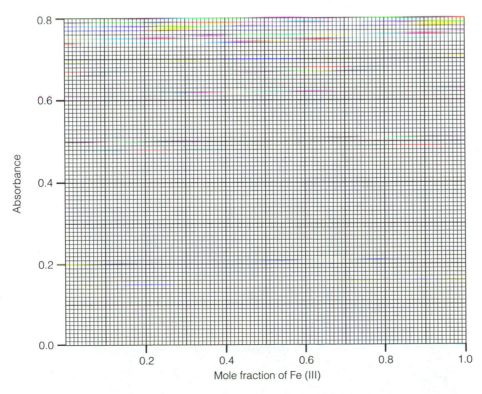

What is the value of the mole fraction of Fe(III) where the absorbance is a maximum?

What is the nearest-integer value of the Fe(III)-to-SCN$^-$ ratio?

3. Calculating the Equilibrium Constant for the Reaction

For the solutions in test tubes 2, 4, and 6, calculate values of the equilibrium constant following the procedure outlined in Part 3 of the Experimental Procedure.

	Test tube number		
	2	4	6
Absorbance			
[Fe(SCN)$^{2+}$] from Eq. (11)	mol/L	mol/L	mol/L
c_{Fe} from Eq. (16)	mol/L	mol/L	mol/L
c_{SCN} from Eq. (17)	mol/L	mol/L	mol/L
(Fe^{3+}) from Eq. (14)	mol/L	mol/L	mol/L
(SCN$^-$) from Eq. (15)	mol/L	mol/L	mol/L
K from Eq. (13)	L/mol	L/mol	L/mol
	Average value of K for the three solutions: —————— L/mol		

Show all of the calculations for the solution in test tube 4.

☞ Consider This

Suppose that you took the contents of tube number 4, containing equal amounts of 0.00200 M Fe(III) and 0.00200 M KSCN solutions, and diluted it by a factor of two by adding an equal volume of 0.5 M HNO$_3$. (It's important to add 0.5 M HNO$_3$ rather than water, in order to keep the ionic strength constant. Otherwise, the value of the equilibrium constant would change because its value depends on the ionic strength, as described in the articles by Frank and Oswalt [1947] and Cobb and Love [1998] listed in the Bibliography.)

When the solution has been diluted by a factor of two, what would you expect to happen to the concentration of Fe(SCN)$^{2+}$ as measured by the absorbance at 450 nm? Would it decrease by a factor of two? Try it and see. Can you explain why it decreases approximately a factor of four, rather than a factor of two?

25

Factors Affecting the Rates of Chemical Reactions

PURPOSE

To study a number of factors that influence chemical reaction rates, including (1) concentration, (2) the nature of the chemical reactants, (3) the surface area in a heterogeneous reaction involving reacting solids and liquids, (4) the temperature of the reacting system, (5) the presence of catalysts, and (6) the effects of diffusion in a poorly mixed solution of chemical reactants.

PRE-LAB PREPARATION

At present it is safe to say that we do not yet understand in detail every factor that influences the rate of a chemical reaction. Moreover, we cannot explore in two or three hours of experimentation everything that is known about the subject, so you should regard this experiment as an introduction to several of the more important factors that influence the rates of chemical reactions.

We have omitted from this experiment the study of some factors that are known to influence reaction rates. For example, we will not discuss the effect of light on chemical reactions. (The interaction of light photons with molecules is an important subdiscipline of chemistry called *photochemistry.*) We might also discuss the effects of molecular geometry and *stereochemistry,* for it is well known that the shapes of molecules influence reaction rates. Reaction rates in solution also are affected by the nature of the solvent. These are all subjects you will encounter in more advanced courses in chemistry.

MEASURING THE RATE OF A CHEMICAL REACTION

Let's begin by defining what we mean by the rate of a chemical reaction. The simplest and most fundamental definition of reaction rate is either the number of moles

of reactants that disappear per unit time or the number of moles of products that appear per unit time. According to this definition the units of reaction rate would be moles per second (mol/s).

If the volume of the reacting system is constant (or nearly constant), it is convenient to define the reaction rate in moles per unit volume per unit time, for example, $mol \cdot L^{-1} \times s^{-1}$. This approach allows us to measure reaction rates as changes in concentration per unit time, which is convenient when we study reactions in solution (where volume changes are usually small) or in the gas phase (where the reactants and products are confined to a reaction flask of fixed volume).

To measure the rate of a chemical reaction, we must make use of chemical or physical properties of the system, properties that allow us to determine changes in the number of moles (or concentration) per unit time. As an example, the rate of a reaction that evolves a gas can be measured by measuring the volume of gas produced per second. If a reactant or product has a characteristic color, we can measure the intensity of the color (proportional to concentration) at a particular wavelength in an instrument called a *spectrophotometer.*

Another method, one that we will use in this experiment, requires us to measure the time required for a fixed amount of a reactant to disappear. This is called a *clock reaction,* and we will use this method to study the reaction of hydrogen peroxide with iodide ion.

$$H_2O_2 + 2I^- + 2H^+ \rightarrow I_2 + 2H_2O$$

The method works like this: A fixed amount of sodium thiosulfate ($Na_2S_2O_3$) is added to the reaction mixture, along with H_2O_2, KI, starch, and an acetic acid–sodium acetate buffer. (The buffer keeps the hydrogen ion concentration nearly constant.) As the hydrogen peroxide reacts with iodide ion, iodine (I_2) is produced. As the I_2 is formed, it reacts very rapidly with thiosulfate ion to reform iodide ion (thus keeping the iodide ion concentration constant).

$$2\,S_2O_3^{2-} + I_2 \rightarrow 2\,I^- + S_4O_6^{2-}$$

When the last bit of thiosulfate ion is used up, the I_2 concentration increases. In the presence of iodide ion molecular I_2 forms the triiodide complex ion,

$$I_2 + I^- \leftrightharpoons I_3^-$$

which in turn forms a dark blue-black complex with the added starch. The appearance of the blue color signals that all of the thiosulfate (and a corresponding amount of H_2O_2) has been used up.

The time required for a fixed amount of thiosulfate (and H_2O_2) to react is measured. The faster the reaction, the shorter the reaction time, so the reaction rate and the reaction time are inversely related. The amount of thiosulfate is deliberately made small so that not more than 10 percent of the H_2O_2 reacts. Thus, during the course of the reaction, the concentrations of H_2O_2, I^-, and H^+ remain nearly constant, approximately equal to their initial values. This allows us to study the effects of concentration by varying the initial concentration of one or more reactants.

EFFECTS OF CONCENTRATION

In order to react, molecules or ions must encounter one another. These encounters (or collisions) are often violent enough to produce a rearrangement of the atoms and the formation of new products. The more molecules there are per unit volume, the more likely they are to collide with one another; so increasing the concentrations of the reactants generally increases the rate of reaction. However, the dependence of reaction rate on the concentrations of the reactants is intimately linked to the reaction pathway, and in some instances an increase in the concentration of a reactant has no effect (or even a slowing effect) on the rate of reaction. Experiment 26 involves a quantitative study of the dependence of reaction rate on the concentrations of the reactants.

INFLUENCE OF THE NATURE OF THE REACTANTS

The alkali metals all react with water. The net reaction is the same for each metal:

$$M + H_2O \rightarrow M^+(aq) + OH^-(aq) + \frac{1}{2}H_2(g)$$

where M represents Li, Na, K, Rb, or Cs. The rate of reaction increases noticeably in the sequence lithium → cesium. Lithium reacts briskly, but cesium reacts violently, producing enough heat to ignite the hydrogen produced. The metals that react most rapidly have the lowest melting points and the smallest ionization energies.

The reactions of the halogens with hydrogen show the opposite progression. The lightest element, fluorine, reacts explosively with hydrogen at room temperature, whereas iodine reacts slowly.

EFFECT OF SURFACE AREA IN A HETEROGENEOUS REACTION

In a reaction between a solid and a gas (or liquid), the reactants can get together only at the surface of the solid. The larger the surface area of the solid, the greater the number of atoms available to react. Unmilled wheat doesn't burn very well, but when ground to fine flour and dispersed in air, it forms an explosive mixture. When this mixture explodes, it produces so much force that it can destroy huge concrete grain elevators.

EFFECT OF TEMPERATURE

Chemical reactions involve the rearrangement of atoms into new configurations. Chemical bonds are broken, and new bonds are formed. This process necessarily involves the stretching of bonds and the distortion of bond angles. This is exactly what happens when you put more energy into a molecular system. There is an increase in the amplitude of the vibrations of atom-atom stretching and bending, putting the atoms into configurations where they can more easily react. In addition, the kinetic energy of the molecules increases. They move about faster, so they collide with greater force, producing the bond-length and bond-angle distortions characteristic of reacting molecules. So a molecule at a higher temperature has higher energy and a higher probability of reacting.

Dramatic temperature effects are often observed with reactions that are exothermic. The reaction of paper with the oxygen in air is extremely slow at room temperature, but as the paper is heated it suddenly bursts into flame. The temperature at which this happens is called the *ignition temperature*. At this temperature the energy produced by the chemical reaction cannot be dissipated rapidly enough to keep the reactant from heating up. This self-heating process results in a dramatic rise in the temperature and the rate of reaction. This is characteristic of combustions and chemical explosions.

EFFECT OF CATALYSTS

A catalyst increases the rate of a chemical reaction but is not permanently altered in the net chemical reaction.

Catalysts generally function by interacting with one or more reactants to form an intermediate that is more reactive. In effect, catalysts change the reaction pathway so that less energy is required for the reaction to take place. Many enzyme catalyst molecules are very specific, catalyzing the reaction of a single kind of substrate molecule. Other catalysts are more general. For example, finely divided palladium metal catalyzes the rather nonspecific addition of H_2 to $C=C$ bonds found in different kinds of molecules.

EFFECTS OF DIFFUSION AND MIXING

In the gas phase, molecules are moving quite rapidly. Nitrogen molecules at room temperature have an average speed of about 300 m/s (about 650 mph). But nitrogen molecules released in a corner of a room wouldn't get to the other side in a fraction of a second. This is because they collide so often with other molecules that they must jostle their way to the other side, like a person fighting through a crowd at a football game.

If it's difficult to travel rapidly in the gas phase, imagine how hard it is in a liquid, where each molecule is surrounded by a cage of molecules. Just to move a distance roughly equal to the size of a molecule requires that the molecule push and shove its neighbor out of the way. So if two different reactant solutions are not thoroughly mixed, the reaction time can be extended by minutes until the random motions of the slowly diffusing molecules produce a more uniform distribution of reacting molecules. This is an important consideration to keep in mind when studying chemical reactions in solution (or when making up a reagent solution). Thorough mixing is essential.

EXPERIMENTAL PROCEDURE

Special Supplies: Mortar and pestle, apparatus for collecting gases shown in Figure 25-1, 2.5×2.5 cm squares of cotton cloth (muslin).

Chemicals: 3% (0.9 M) H_2O_2, solution A (containing in the same solution 0.2 g/L starch,[1] 0.5 M CH_3COOH, 0.05 M $NaCH_3COO$, 0.30 M KI, and 0.030 M $Na_2S_2O_3$), Mg metal turnings, Ca metal turnings, 0.1% phenolphthalein indicator, marble chips, 1 M HCl, steel wool, iron nails (approx. 4 cm long), methanol, 50:50 (v:v) 2-propanol-water, 1 M $CuCl_2$, 3 M $CuCl_2$, 3 M $FeCl_3$, $KMnO_4(s)$, 0.005 M $KMnO_4$, 0.5 M $H_2C_2O_4$ (oxalic acid), 6 M H_2SO_4.

! SAFETY PRECAUTIONS: Part 4 of this experiment involves igniting small quantities of flammable solvents. If you follow the directions, this is no more hazardous than lighting a Bunsen burner. It is recommended that the flammable solvents be kept in the fume hood and that the

requisite amounts be dispensed to you there. *There must be no open flames in the hood where the solvents are stored.* After obtaining your small amounts of solvent, you should take them to the lab bench where the experiment is to be carried out. *The containers of flammable solvents must never be taken to the lab bench.* If a fume hood is not available, the instructor should demonstrate the experiments in parts 4b and 4c.

1. EFFECTS OF CONCENTRATION ON REACTION RATES.
Place 5 mL of solution A in a 50-mL beaker. Add 5 mL of 3% H_2O_2, mix, and note the number of seconds required for the solution to turn blue. Record the reaction time.

Repeat the procedure using 5 mL of solution A, 4 mL of water, and 1 mL of 3% H_2O_2. Record the reaction time. Is it longer? Which reactant concentrations have been changed? What is the relationship between the reaction time and the concentration of H_2O_2?

2. THE INFLUENCE OF THE NATURE OF REACTANTS.
Obtain three or four small pieces of magnesium metal turnings and about the same amount of calcium metal turnings. Add enough water to each of two 15×150 mm test tubes to fill them about one-third full. Place the magnesium turnings in one test tube and the calcium turnings in the other. Both of these metals are in the same chemical family in the periodic table. Which metal is more reactive with water?

To determine what gas is produced, invert an empty test tube for about 15–30 s over the mouth of the tube in which the metal is reacting more vigorously. Quickly move the mouth of the inverted tube near the flame of a Bunsen burner. Is the gas flammable? What is the chemical composition of the gas?

Look closely at the slower-reacting metal as you agitate the test tube. Is any gas being produced? Add a drop of phenolphthalein indicator solution to each test tube containing the metals. What other product is formed besides a gas? Write chemical equations describing the reactions of the metals with water.

Is there a correlation of the reaction rates with the ionization potentials of these metals (is it easier to remove electrons from magnesium or calcium)?

3. THE INFLUENCE OF SURFACE AREA IN A HETEROGENEOUS REACTION.
(a) Set up the apparatus shown in Figure 25-1, using a 125-mL Erlenmeyer flask, a 13×100 mm test tube to collect the gas, and a large beaker or plastic basin. Fill the test tube with tap water, and, holding your thumb or finger over the mouth, invert it in the beaker of water over the end of the bent gas delivery tube.

---

[1]Put 0.2 g starch in 200 mL of boiling water, and heat for 5 min. Dilute the solution to about 800 mL, add the other reagents in the order given, allow to cool, and dilute to 1 L.

FIGURE 25-1
A simple apparatus for measuring the rate of evolution of a gas. Invert a test tube filled with water over the gas delivery tube in the beaker; then measure the time required to fill the test tube with gas.

Place 3 g of coarse marble chips ($CaCO_3$) in the 125-mL Erlenmeyer flask. Add 25 mL of water; then add 25 mL of 1 M HCl. Immediately insert the one-hole stopper in the flask, and record the time required to fill the test tube with gas. What gas is produced?

Rinse the Erlenmeyer flask with tap water, and place the used marble chips in the waste receptacle provided. *Do not discard the marble chips in the sink.* Obtain another 3 g of marble chips, and using a mortar and pestle, pulverize them to the size of grains of sugar. Using a rolled-up sheet of paper as a funnel, transfer the granular material to the 125-mL Erlenmeyer flask. Fill a test tube with water, and invert it over the delivery tube as before. Add 25 mL of water to the flask, followed by 25 mL of 1 M HCl, and stopper immediately. Record the time required to fill the test tube with gas. Save this apparatus for use in Part 4. Is there a correlation between the surface area of the marble and the reaction rate? Write an equation describing the chemical reaction between $CaCO_3$ and HCl.

(b) Place a small iron nail (cleaned so that it is free of rust) in a 20-mm diameter test tube. In a second 20-mm test tube place a 0.6-g ball of steel wool, using a glass rod to push it to the bottom of the test tube. Add 10 mL of 1 M $CuCl_2$ to each test tube, and watch them closely. Observe and record any color changes, and note in which tube they occur more rapidly. Is a new solid substance forming? What do you think it is? Agitate each test tube, and feel the lower portion of the tubes. Is one tube warmer than the other? What happened to the steel wool? Would the same thing eventually happen to the nail? Write an equation describing the chemical reaction that takes place.

4. THE INFLUENCE OF TEMPERATURE. (a) Measure 5 mL of solution A (used in Part 1) into a test tube. Measure 4 mL of water and 1 mL of 3% H_2O_2 into a second test tube. Place both test tubes in a 250-mL beaker of warm water (50 °C). After about 3 min, pour solution A and then the H_2O_2 solution into a 50-mL beaker, stir, and note the time. When the blue color appears, note and record the elapsed time. Compare the reaction time with that observed for the second reaction mixture you studied in Part 1.

(b) *Ignition Temperature* Review the Safety Precautions. In a fume hood, place 10 to 20 drops of methanol in a 30-mL beaker. Remove the beaker to your lab bench. Heat the end of a glass stirring rod at the tip of the inner blue cone of a Bunsen burner until the glass starts to soften. Quickly touch the heated end to the methanol. A chemical reaction should begin; if it does not, reheat the rod, and try again. *Stay at arm's length from the beaker. Do not touch the beaker until it has cooled.* Describe the reaction that takes place. Write a balanced chemical equation for the reaction. Why doesn't methanol burst into flames at room temperature when exposed to air?

(c) Obtain a small square of cotton cloth (muslin) and a book of matches. (The cotton cloth should be tightly woven, about 2 to 3 cm on a side, and cleanly cut, so there are no ragged edges.) Working in a hood, place about 10 mL of 50:50 (v:v) 2-propanol-water in a 50- or 100-mL beaker. Using crucible tongs, grasp the square by one corner, and dip it into the 2-propanol-water solution. Lift the cloth, and allow the excess solution to drain until no more solution drips off the cotton. Place a small watch glass over the solution in the beaker. Carry the cotton, still held in the tongs, to your lab bench. Have a lab partner light a match and bring it just near enough to the square of cotton (held in the tongs) to ignite the 2-propanol-soaked cotton. Make certain that no other person is within arm's length of the flaming cloth. When the flame dies out, carefully note the condition of the cloth. How can the 2-propanol burn without igniting the cotton?

5. EFFECT OF CATALYSTS ON THE RATE OF DECOMPOSITION OF H_2O_2.

Using the same apparatus as in Part 3a, place 20 mL of 3% H_2O_2 in the flask. Fill a 13 × 100 mm test tube with water, and invert it over the gas delivery tube. Then add 20 drops of 3 M $CuCl_2$ to the flask, replace the stopper, continuously swirl the contents, and note and record how long it takes to fill the test tube with the evolved gas. The gas coming over first will contain some air that was present in the flask. After two or three minutes, test the gas in the Erlenmeyer flask by inserting a glowing splint into the flask. Does the gas support combustion? What is the identity of the gas?

Rinse the flask and repeat the experiment, adding only the H_2O_2 solution to the flask. Continuously swirl the contents of the flask. Does the gas evolve more slowly? After a few minutes' wait, you should be able to tell whether the $CuCl_2$ catalyzes the reaction.

Now fill another test tube with water, and place it over the gas delivery tube. Remove the stopper just long enough to add 2 drops of 3 M $FeCl_3$. Insert the stopper, and continuously swirl the solution. Note and record the time required to fill the test tube with gas.

Rinse the flask. Add to it 20 mL of 3% H_2O_2, 1 drop of 3 M $CuCl_2$, and 1 drop of 3 M $FeCl_3$, in that order. Quickly replace the stopper, continuously swirl the solution in the flask, and record the time required to fill a 13 × 100 mm test tube with gas.

Repeat the procedure using 20 mL of 3% H_2O_2, 5 drops of 3 M $CuCl_2$, and 5 drops of 3 M $FeCl_3$. Quickly stopper the flask, and continuously swirl the contents of the flask. Note and record the time required to fill a 13 × 100 mm test tube with gas.

6. THE IMPORTANCE OF MIXING REACTANTS.

(a) Half-fill two 13 × 100 mm test tubes with water. Drop a small crystal of solid $KMnO_4$ into each tube. Put one tube aside in a test tube rack or beaker, and let it remain undisturbed. Swirl the second tube to dissolve the crystal of $KMnO_4$. Observe the crystal in the undisturbed tube every few minutes, and note how long it takes for the crystal to dissolve, then to spread (diffuse). Record your observations.

(b) Add 10 mL of 0.50 M $H_2C_2O_4$ (oxalic acid) to 10 mL of 6 M H_2SO_4, and mix thoroughly. Place 8 mL of this solution into each of two 15 × 125 mm test tubes. Measure out two separate 8-mL portions of 0.005 M $KMnO_4$. Add one portion of $KMnO_4$ solution to one of the test tubes containing oxalic acid, holding the test tube at a 45° angle and carefully pouring the $KMnO_4$ down the side of the test tube to avoid mixing the two solutions. Note the time. Pour another 8 mL of $KMnO_4$ into the second test tube, stopper the tube, mix by inverting the tube several times, and note the time. As the reactions proceed, MnO_4^- will oxidize the oxalic acid to CO_2 and will itself be reduced to colorless Mn^{2+} ion. Record the times required for the solutions in the two test tubes to become completely colorless.

☞ Consider This

Sit and observe a campfire. How are the factors discussed in this experiment evident in the burning of a wood fire? (You should be able to find examples of each factor, with the possible exception of catalysis.)

The process of cooking food involves chemical reactions (charring and denaturing of proteins are typical examples), physical processes (water vaporization), and the energy transfer issues of radiation, conduction, and convection. In a typical kitchen, find examples that illustrate the factors listed in this experiment.

The growth of a plant is a series of chemical reactions. Are these reactions temperature dependent? Design an experiment to test this factor. Why does the growth rate decrease above certain temperatures?

Is rate of plant growth dependent on concentration? Design experiments to test the fertilizer-concentration dependence. Why does an excess of fertilizer cause a decrease in the growth rate? (You may need to consult Experiment 22 for hints.)

Can you measure the CO_2 dependence of the growth rate for plants? This factor is a significant component of the discussion about global warming. Why?

Design an experiment to measure the rate of the chemical reaction known as rusting,

$$Fe(s) + O_2(g) \rightarrow Fe_2O_3 \cdot n\,H_2O(s)$$

Can you determine the temperature dependence? The oxygen concentration dependence?

OBSERVATIONS AND DATA

1. Effects of Concentration on Reaction Rates

Record the number of seconds required for the solution to turn blue after the addition of

5 mL 3% H_2O_2 to 5 mL of solution A   _____ s

4 mL H_2O + 1 mL 3% H_2O_2 to 5 mL of solution A   _____ s

How are the reaction time and the initial concentration of H_2O_2 related?

2. The Influence of the Nature of the Reactants

Record what you observe when Mg and Ca are added to water.

Mg _____

Ca _____

Which metal is more reactive? _____

What do you observe when the collected gas is brought near the flame of a Bunsen burner?

_____

What are the colors of the solutions when phenolphthalein indicator is added? _____

Are the solutions acidic or basic? _____

Write balanced chemical equations for the reaction(s) of Mg and Ca with water.

Mg + $H_2O \rightarrow$ _____

Ca + $H_2O \rightarrow$ _____

Consult your text or the *CRC Handbook of Chemistry and Physics* to determine which metal (Ca or Mg) has the smaller sum of ionization energies for removal of two electrons.

	Mg	Ca
First ionization energy		
Second ionization energy		
Sum		

From your observations, how are reaction rates and ionization energies correlated?

3. The Influence of Surface Area in a Heterogeneous Reaction

(a) $CaCO_3 + HCl$

What gas is produced in the reaction? _____

Record the times required to fill a test tube with gas for

coarse $CaCO_3$ _____ s

pulverized $CaCO_3$ _____ s

Write a balanced chemical equation for the reaction of $CaCO_3$ with HCl.

$CaCO_3(s) + HCl(aq) \rightarrow$ _____

Is the reaction rate faster for the coarse or pulverized $CaCO_3$? Explain.

(b) $Fe + CuCl_2$

Describe the reactions of $CuCl_2$ solution with iron in the form of

a nail _____

steel wool _____

What new substances are formed? _____

Does the reaction evolve heat? _____

Which reaction is faster? _____

What evidence is there that the iron nail is reacting? _____

Write a balanced chemical equation for the reaction of Fe with Cu^{2+}.

$Cu^{2+} + Fe \rightarrow$ _____

4. The Influence of Temperature

(a) *H₂O₂/I–clock reaction*

Record the reaction time for the addition to solution A of

4 mL H_2O + 1 mL 3% H_2O_2 _____ s

Compare the reaction time with that observed at room temperature for the second reaction mixture you studied in Part 1. (This second reaction solution in Part 1 had the same initial concentrations of reactants.)

What was the effect of increasing the temperature on

the reaction time? _____

the rate of reaction? _____

(b) *Ignition temperature*

Describe what happened when you brought the heated glass rod into contact with the methanol.

Write a balanced chemical equation for the reaction of methanol and oxygen.

$CH_3OH + O_2 \rightarrow$ _____

Why doesn't methanol burst into flames at room temperature when exposed to air?

(c) *Cotton–2-propanol ignition*

Describe what happened when you ignited the cotton soaked in 2-propanol. How can the 2-propanol burn without igniting the cotton?

5. Effect of Catalysts on the Rate of Decomposition of H_2O_2

Record the times required to fill a test tube with gas for each of the H_2O_2–catalyst reaction mixtures.

20 drops of 3 M $CuCl_2$	_____ s
No added catalyst	_____ s
2 drops of 3 M $FeCl_3$	_____ s
1 drop 3 M $CuCl_2$ + 1 drop 3 M $FeCl_3$	_____ s
5 drops 3 M $CuCl_2$ + 5 drops 3 M $FeCl_3$	_____ s

Does the gas evolved support combustion?

Give the identity of the gas. _____

Write a balanced chemical equation for the decomposition of hydrogen peroxide into water and oxygen.

$H_2O_2 \rightarrow$ _____

Is there any evidence that the combination of Cu^{2+} and Fe^{3+} is a more effective catalyst than either alone?

6. The Importance of Mixing Reactants

(a) *Solution and diffusion of $KMnO_4$*

Describe what you observe for the tubes containing
an undisturbed crystal of $KMnO_4$

_____

a crystal of $KMnO_4$ swirled until it dissolves

_____

(b) *$KMnO_4$–$H_2C_2O_4$ reaction*

Record the times required for complete reaction of the $KMnO_4$ with $H_2C_2O_4$ for the

unmixed solution   _____ s

mixed solution   _____ s

Summarize your conclusions about the effects of mixing and the rate of diffusion in solution.

QUESTIONS

1. Imagine that you have a cube of $CaCO_3$ whose edge length is 1 cm. What is the surface area of the cube (in square centimeters)? Now imagine that you pulverize it to form small cubes of edge length 0.05 cm. How many of the little cubes form? What is the surface area of each little cube? What is the new surface area (in square centimeters) of the pulverized $CaCO_3$? If the rate of reaction of HCl with $CaCO_3(s)$ is proportional to the surface area of the $CaCO_3$, how much faster would the HCl react with the pulverized $CaCO_3$ than it did with the original large cube?

2. A stoichiometric mixture of two volumes of H_2 and one volume of O_2 in a flask will not react if undisturbed, but any spark or small flame will cause a violent explosion, shattering the flask. How do you interpret this rate of behavior?

☞ **Consider This**

Sit and observe a campfire. How are the factors discussed in this experiment evident in the burning of a wood fire? (You should be able to find examples of each factor, with the possible exception of catalysis.)

The process of cooking food involves chemical reactions (charring and denaturing of proteins are typical examples), physical processes (water vaporization), and the energy transfer issues of radiation, conduction, and convection. In a typical kitchen, find examples that illustrate the factors listed in this experiment.

The growth of a plant is a series of chemical reactions. Are these reactions temperature dependent? Design an experiment to test this factor. Why does the growth rate decrease above certain temperatures?

Is rate of plant growth dependent on concentration? Design experiments to test the fertilizer-concentration dependence. Why does an excess of fertilizer cause a decrease in the growth rate? (You may need to consult Experiment 22 for hints.)

327

Can you measure the CO_2 dependence of the growth rate for plants? This factor is a significant component of the discussion about global warming. Why?

Design an experiment to measure the rate of the chemical reaction known as rusting,

$$Fe(s) + O_2(g) \rightarrow Fe_2O_3 \times n\, H_2O(s)$$

Can you determine the temperature dependence? The oxygen concentration dependence?

26

The Rate of a Chemical Reaction
Chemical Kinetics

PURPOSE

To determine the rate law for a chemical reaction, specifically

$$H_2O_2 + 2I^- + 2H^+ \rightarrow 2H_2O$$

by observing how changing the concentrations of H_2O_2, I^-, and H^+ affects the rate of reaction. To see the effect of temperature on the reaction rate. To observe the effect on the reaction rate of adding a catalyst, molybdate ion.

PRE-LAB PREPARATION

Two important questions may be asked about a chemical reaction: (1) *How far or how completely do the reactants* interact to yield products, and (2) *how fast* is the reaction? "How far?" is a question of chemical equilibrium. The study of chemical systems at equilibrium is the realm of chemical energetics, more commonly called *chemical thermodynamics*. The second question—"How fast?"—is the realm of *chemical kinetics*, the subject of this experiment. For elementary reversible reactions, there is a direct relation between the equilibrium constant and the rate constants for the forward and reverse reactions. Furthermore, a study of the factors affecting the rate often reveals important information about the reaction pathway or the mechanism of the chemical reaction.

In this experiment we will see how changing the concentrations of each of the reactants affects the rate and yields the rate law for the reaction. We will measure the effect of a catalyst and of temperature on the rate of the reaction. This should give you a better understanding of the particular reaction being studied and increase your understanding of the more complete discussion of these problems in your text. Let us first discuss some of the questions to be considered.

1. *How does the concentration of each of the reacting substances affect the rate?* For a homogeneous[1] chemical reaction, it is usually possible to express the rate as an algebraic function of the concentrations of the reactants. For many chemical reactions, this algebraic function is simply the product of the concentrations of the reactants, each concentration being raised to a power called the *order* of the reaction, with respect to that reactant. As an example, suppose that we have a reaction whose overall stoichiometry is

$$2A + B \rightarrow P \tag{1}$$

and whose *rate law* can be expressed by an algebraic function of the type

$$\text{rate} = k[A]^x[B]^y \tag{2}$$

The proportionality constant k, called the *specific rate constant*, is the rate when the concentrations of A and B are both 1.0 mol/L. Brackets around the symbols A and B mean "concentrations of" A and B.

Right at the beginning, we emphasize that the values of the exponents x and y that appear in the rate expression cannot be predicted from the overall stoichiometry and must be determined by experiment. In other words, *rate expressions based on the stoichiometric equation are seldom correct*. This is because the reaction pathway (or reaction mechanism) seldom involves the simultaneous collision of all reactants to yield products in a one-step

---

[1]A *homogeneous* reaction is one occurring uniformly throughout *one phase* (e.g., a solution), while a *heterogeneous* reaction is one occurring at the interface of *two phases* (e.g., a solid and a liquid, or a gas and a solid). The latter reaction rate depends on the extent of subdivision and surface area of the phases and on adsorption and other phenomena.

TABLE 26-1

Rate Expressions for Different Reaction Paths for the Overall Reaction 2A + B → Products

Reaction pathway[a]	Rate law or rate expression	Reaction order with respect to reactant A	Reaction order with respect to reactant B	Total reaction order ($x + y$)
Pathway 1: 2A + B → P	rate = $k[A]^2[B]$	2	1	3
Pathway 2: 2A → C (*slow*) B + C → P (*fast*)	rate = $k[A]^2$	2	0	2
Pathway 3: A + B → D + E (*slow*) A + D → E (*fast*) 2E → P (*fast*)	rate = $k[A][B]$	1	1	2

[a]The symbols A and B represent reactants, and P represents the products of the stoichiometric reaction. The symbols C, D, and E represent chemical *intermediates* that are formed in the elementary reactions. Intermediates have only a fleeting existence and do not appear as final products of the reaction.

reaction, which we will call an *elementary reaction*.[2] The stoichiometric equation more often represents the sum of several elementary reactions, at least one of which may be distinctly slower than the others and is therefore the rate-controlling reaction. The rate expression therefore corresponds to the rate for the slow step in the total reaction. Also keep in mind that in writing a rate expression like that of Equation (2), we have assumed that the rate of the backward reaction is negligible.

To illustrate how the rate law depends on the reaction pathway, Table 26-1 presents three different hypothetical reaction mechanisms, each having the overall stoichiometry 2A + B → P, where A and B represent the reactant molecules, and P represents the products. The rate expression for each reaction pathway is shown, together with the reaction orders with respect to A and B. The terminology here uses the designation x and y for the order of the reaction with respect to the reactant concentrations, [A] and [B]. The *total reaction order* is the sum of the individual reaction orders, $x + y$. Note that for each reaction pathway, the sum of all the elementary reactions gives the stoichiometric equation, but that the reaction orders *do not* necessarily correspond to the stoichiometric coefficients. For the first reaction pathway, where the elementary reaction is identical to the stoichiometric reaction, doubling the concentrations of A and B would increase the rate by a factor of 8 ($2^2 \times 2^1 = 8$); for the second and third pathways, the rate would be increased by a factor of 4.

Determining the rate law is often one of the first steps in a kinetic investigation. This task can often be

simplified by using a small concentration of one reactant in the presence of 20- to 100-fold larger concentrations of all the other reactants. Under these conditions, the reactants that constitute the larger concentration remain essentially constant, so that the total order of the reaction is reduced to the order of the reactant that constitutes the small concentration. For example, the first reaction mechanism shown in Table 26-1 has the rate law given by rate = $k[A]^2[B]$. If the concentration of A is made 50 times larger than the concentration of B, the rate expression is given by rate = $k'[B]$, where $k' = k[A]^2$, the concentration of A being treated as a constant because it changes only four percent during the course of the reaction. The total third-order reaction can now be treated as a *pseudo–first-order* reaction with respect to the reactants. This is sometimes called the "swamping" technique. A related technique, called the *initial rate* method, is to measure the rate of a reaction over a period of time short enough so that less than five to ten percent of the reactants is consumed. The reactant concentrations can be treated approximately as constants, and the reaction order with respect to each reactant can be determined by varying the concentration of only one constituent at a time and measuring the corresponding rate change.

2. *How does the temperature at which a reaction occurs affect the rate?* Figure 26-1 shows a plot of reaction rate versus temperature for three different classes of reactions. Reactions in Class I (most ordinary chemical reactions in the gas phase and in solution) show a regular exponential increase over a temperature range of 50 to 100 °C. Class II behavior is typical of a combustion or explosion reaction. Class III behavior is characteristic of many enzyme-catalyzed reactions: At lower temperatures the reaction rate increases with temperature, but at higher temperatures it decreases as the configuration of the enzyme catalyst is altered and enzyme

[2]The *reaction pathway* (or *reaction mechanism*) of a chemical reaction is composed of a set of molecular reactions, which we call elementary reactions, or steps, and which are written just like overall chemical reactions. You must be careful to distinguish between an elementary reaction and the stoichiometric chemical equation. Only for a single-step reaction, in which the elementary reaction is identical to the overall stoichiometric reaction, will the exponents appearing in the rate expression be the same as the coefficients appearing in the stoichiometric equation.

denaturation occurs. The reaction we will study in this experiment belongs to Class I, often called Arrhenius-type behavior after the chemist who proposed an explanation to account for it.

When we compare measured rates of chemical reactions in the gas phase with the collision rates calculated from the kinetic gas theory, it becomes clear that only a small proportion of the many molecular collisions result in reaction.[3] In most collisions, the molecules simply rebound like billiard balls, without change. Only molecules with the proper geometric orientation and sufficient relative momentum collide with sufficient energy to "break and make" chemical bonds, momentarily forming an "activated complex" and eventually forming stable products. The proportion of molecules that have a given energy can be calculated from the Maxwell-Boltzmann distribution law; this proportion increases at higher temperatures as shown in Figure 26-2. We will call the minimum net energy that the colliding molecules must possess in order to react the *activation energy*, E. The situation is somewhat analogous to an attempt to roll a ball up from a mountain valley, over a pass, and down to another valley. The energy necessary to get the ball up to the top of the pass is the activation energy.

3. *How does a catalyst affect a reaction?* A catalyst is a substance that produces a new reaction pathway with a lower activation energy than the original uncatalyzed reaction pathway. A lower activation energy for the catalyzed reaction means that at a given temperature, a larger fraction of the molecules will possess enough energy to react; thus the catalyst increases the rates of *both* the forward and backward reactions without affecting the equi-

librium constant for the reaction. Although the catalyst is not consumed in the overall reaction, it may participate in the formation of intermediates and then be reformed in subsequent steps to produce the final products.

THE EXPERIMENTAL METHOD

In this experiment we will study the rate of oxidation of iodide ion by hydrogen peroxide, which proceeds according to the net stoichiometric reaction

$$H_2O_2 + 2I^- + 2H^+ \rightarrow I_2 + 2H_2O \tag{3}$$

By varying the concentrations of each of the three reactants (H_2O_2, I^-, and H^+), we will determine the order of the reaction with respect to each reactant and the rate law of the reaction, which is of the form

$$rate = \frac{-\Delta[H_2O_2]}{dt} = k[H_2O_2]^x[I^-]^y[H^+]^z \tag{4}$$

In this formula x, y, and z must be experimentally determined and are most often integers (0, 1, 2, etc.) or half-integers. We will use an acetic acid–acetate ion buffer to control the H^+ concentration at a small constant value, around 10^{-5} M. By using a solution containing known concentrations of acetic acid and acetate ion, we can readily calculate the H^+ concentration from the equilibrium constant expression for the dissociation of acetic acid, given by

$$K_a = \frac{[H^+][C_2H_3O_2^-]}{[HC_2H_3O_2]} \tag{5}$$

We will also examine the effect of temperature on the reaction rate and the effect of a catalyst (3.3×10^{-4} M Mo [VI]).

---

[3]We assume in this experiment that the arguments based on kinetic gas theory also apply qualitatively to chemical reactions in solution.

I   II   III

Rate

Temperature

FIGURE 26-1

Effect of temperature on reaction rate. Class I: exponential (Arrhenius) behavior typical of most simple inorganic and organic reactions. Class II: typical of explosion reactions. Class III: typical of enzyme-catalyzed reactions. [After A. A. Frost and R. G. Pearson, *Kinetics and Mechanism*, 2nd ed., Wiley, New York, Chapter 2, Figure 7, Copyright © 1961.]

To measure the reaction rate we will employ a clever variation of the initial rate method. A small amount of sodium thiosulfate and starch indicator is added to the reaction mixture. Thiosulfate ion does not react at an appreciable rate with any of the reactants, but it does react rapidly with iodine, according to the equation

$$2 S_2O_3^{2-} + I_2 \rightarrow 2 I^- + S_4O_6^{2-} \qquad (6)$$

As a result, iodine is reduced back to iodide ion as fast as it is formed until all of the thiosulfate is used up. At this point the solution suddenly turns blue because the iodine concentration rapidly increases to the point where the I_2 forms an intense blue complex with the starch indicator. The time required to consume a fixed amount of thiosulfate is very reproducible, and since we measure this time, this type of reaction is often called a *clock reaction.*

While thiosulfate ion is present in solution, both reactions (3) and (6) are simultaneously taking place, and, by adding them together, we see that the net result in solution is given by

$$H_2O_2 + 2 S_2O_3^{2-} + 2 H^+ \rightarrow 2 H_2O + S_4O_6^{2-} \qquad (7)$$

You should be careful to note that Equation (7) expresses only the net result of the two reactions. Hydrogen peroxide is not reacting directly with $S_2O_3^{2-}$, but the result is the same as if it did, and two moles of thiosulfate are consumed per mole of hydrogen peroxide reacted. Knowing the amount of thiosulfate we put in the reaction mixture, we can calculate the reaction rate from equations that can be approximated by

$$\text{rate} \cong \frac{-\Delta[H_2O_2]}{\Delta t} = -\frac{1}{2}\frac{\Delta[S_2O_3^{2-}]}{\Delta t} \qquad (8)$$

and

$$\text{rate} \cong \frac{\frac{1}{2}[VM_{\text{thiosulfate}}/V_{\text{tot}}]}{\Delta t} \qquad (9)$$

where $VM_{\text{thiosulfate}}$ is the number of millimoles of thiosulfate added ($V_{\text{mL}} \times M_{\text{mol/L}}$), V_{tot} is the total volume (in milliliters) of the reaction mixture, and Δt is the time required for all of the thiosulfate to react.

To make the expression for the rate of reaction a positive quantity, we insert the minus sign, because $\Delta[H_2O_2]/\Delta t$ is a negative quantity due to the fact that the hydrogen peroxide concentration is decreasing with time. The factor of ½ is inserted because hydrogen peroxide disappears half as fast as does thiosulfate, as shown by the stoichiometry of Equation (7).

FIGURE 26-2
The Maxwell-Boltzmann distribution law. At a higher temperature (curve T_2), the total number of molecules (shaded area under this curve) with energies greater than E_a, the critical energy necessary for reaction, is much larger at temperature T_2 than at the lower temperature (curve T_1).

Equation (8) makes the assumption that only a small amount of hydrogen peroxide is consumed (say, less than 3% of the total). Our rate measurement will be a very close approximation to the exact expression and will accurately measure the initial rate of reaction of the hydrogen peroxide.

Now we are in a position to understand how the thiosulfate plays an important role in the rate measurement. First, it allows us to make a simple measurement of the reaction rate by measuring the time required for a given amount of thiosulfate to be consumed under conditions where only a small fraction of the hydrogen peroxide is consumed. It also reacts rapidly with the reaction product I_2 as it forms, regenerating I^-. This keeps the I^- concentration constant and suppresses any backward reaction (the reverse of Equation [3]), so that we can be confident that we are measuring only the rate of the forward reaction.

EXPERIMENTAL PROCEDURE

Special Supplies: Buret, 5-mL graduated pipets, 10-mL graduated pipets, thermometer, timer (a digital watch or a clock with a sweep second hand is suitable). The optional standardization procedure requires the foregoing and a 1-mL pipet.

• •

NOTE: All solutions for this experiment should be made up in high-quality distilled or deionized water. Traces of metal ions, particularly copper or iron salts, catalyze the decomposition of H_2O_2.

• •

Chemicals: 0.8 to 0.9 M H_2O_2 containing 0.001 M H_2SO_4 (the small amount of acid stabilizes the H_2O_2 solution and has a negligible effect on the rate measurements; a fresh 3% H_2O_2 solution sold in drugstores is about 0.9 M and can be used for the H_2O_2 solution, which must be standardized, preferably on the day the solution is to be used), 0.050 M sodium thiosulfate ($Na_2S_2O_3$), 0.050 M KI, 0.050 M acetic acid–sodium acetate buffer (0.05 M in $HC_2H_3O_2$ and 0.05 M in $NaC_2H_3O_2$), 0.30 M acetic acid ($HC_2H_3O_2$), 0.1% starch solution (freshly made), 0.010 M

Mo(VI) catalyst (containing 1.76 g/L of ammonium heptamolybdate [NH$_4$]$_6$Mo$_7$O$_{24}$ · 4H$_2$O). The H$_2$O$_2$ standardization procedure requires, in addition, KI(s), 2 M H$_2$SO$_4$, and 3% ammonium heptamolybdate catalyst solution in a dropping bottle. If a 1.00-mL microburet is used for the standardization titration, a 2.50 M Na$_2$S$_2$O$_3$ solution will also be required.

NOTES TO INSTRUCTOR: If students are expected to complete the experiment and write-up in a 3-hour laboratory period, they should be provided with standardized hydrogen peroxide solution and omit Part A of the experiment. Alternatively, you may wish to have students do Part A and replicate each reaction mixture measurement, allotting a second laboratory period for the write-up and a discussion of the results.

If students make only a single reaction time measurement for each reaction mixture, a "wild" result will spoil the calculations. To avoid this, we suggest that you gather the reaction time data (or have the students write it on the board) so that a wild result can be spotted and the measurement repeated if necessary.

Two students may work together advantageously in this experiment, one student mixing the solutions and the other noting the time to the exact second. *However, each student should record the data completely* before leaving the laboratory.

A. STANDARDIZATION OF THE H$_2$O$_2$ SOLUTION

The instructor will tell you if you are to perform the standardization. For best results, the H$_2$O$_2$ solution should be standardized the same day it is to be used. The concentration should be in the range 0.8 to 0.9 M in order to yield conveniently measurable reaction times.

To perform the standardization, first fill a buret with 0.050 M Na$_2$S$_2$O$_3$. Then pipet exactly 1.00 mL of the approximately 0.8 M H$_2$O$_2$ solution into a 125-mL Erlenmeyer flask containing 25 mL of water.

Add 10 mL of 2 M H$_2$SO$_4$, 1 g of solid KI, and 3 drops of 3% ammonium molybdate catalyst.

Swirl until the KI dissolves, and immediately titrate the brown iodine solution that forms with 0.050 M sodium thiosulfate until the brown color begins to fade to yellow.

At this point, add 2 mL of 0.1% starch indicator, and titrate to the disappearance of the blue color of the starch-iodine complex. The endpoint is very sharp and should require 32 to 36 mL of 0.050 M Na$_2$S$_2$O$_3$ for the conditions specified.

If a 1.00-mL microburet is used, the concentration of sodium thiosulfate titrant should be increased to 2.50 M, with the titration requiring 0.64–0.76 mL for the conditions specified.

Calculate the exact concentration of the hydrogen peroxide stock solution, which should be in the range 0.8 to 0.9 M.

B. REACTION RATE MEASUREMENTS

Six reaction mixtures will provide the information necessary to determine the effects of the concentrations of H$_2$O$_2$, I$^-$, and H$^+$ on the rate of the reaction, as well as the effect of temperature and a catalyst. Table 26-2 specifies the temperature and reagent volumes to be used for each reaction mixture.

Make up each reaction mixture in a 250-mL beaker, adding the reactants for each reaction mixture in the order in which they appear in Table 26-2 (reading from left to right). Read and record the temperature to the nearest 0.2°C. Graduated cylinders (100, 50, and 10 mL) can be used to measure the volumes of water, buffer, KI, starch, acetic acid, sodium acetate, and catalyst. Special care should be taken to measure the volumes as accurately as possible. If you performed the standardization, the 0.050 M Na$_2$S$_2$O$_3$ solution can be conveniently dispensed from a buret; otherwise use a 5-mL graduated pipet. The H$_2$O$_2$ is conveniently dispensed from a 10-mL graduated pipet (or 10-mL graduated cylinder).

TABLE 26-2
Composition of the Reaction Mixtures

Reaction mixture	Temperature (°C)	Water (mL)	0.05 M HC$_2$H$_3$O$_2$ NaC$_2$H$_3$O$_2$ buffer (mL)	0.3 M HC$_2$H$_3$O$_2$ (mL)	0.05 M KI (mL)	0.1% starch (mL)	0.05 M Na$_2$S$_2$O$_3$ (mL)	0.01 M Mo(VI) catalyst (mL)	0.8 M H$_2$O$_2$ (mL)
1	R.T.[a]	75	30	0	25	5	5	0	10
2	R.T.	80	30	0	25	5	5	0	5
3	R.T.	50	30	0	50	5	5	0	10
4	R.T.	30	30	45	25	5	5	0	10
5	R.T. + 10	75	30	0	25	5	5	0	10
6	R.T.	70	30	0	25	5	5	5	10

[a]R.T. stands for room temperature.

To avoid contamination of the reagents, it is best to dispense each solution from a separate pipet or graduated cylinder. Put labels on them so you don't confuse them. You must be especially careful not to get any contaminant in your hydrogen peroxide stock solution. Iodide ion catalyzes the decomposition of H_2O_2 into oxygen and water, and saliva contains substances that catalyze decomposition of H_2O_2, so do not pipet any solutions by mouth. If you use pipets, draw up the solutions with a rubber bulb. *Do not return any reagent solutions to the stock bottles.*

The first step in each reaction measurement is to place in a clean 250-mL beaker the specified volumes of each reagent listed to the left of the vertical line in Table 26-2. (The beakers *must* be clean. If they are not, wash them with detergent, and rinse them thoroughly with tap water; then give them a final rinse with deionized water, and shake off the excess. Do not attempt to dry them with a paper towel; it may leave contamination.) The beaker now contains all the reactants except H_2O_2. The timer or stopwatch is started when the hydrogen peroxide is added (with vigorous stirring) and stopped when the solution turns blue. The final volume of each reaction mixture is 150 mL. Knowing this fact and the amount of each reactant added, you can easily compute the initial concentration of each reactant.

You will run a duplicate measurement on the first reaction mixture in order to check your technique. If time permits (and your instructor so directs), make duplicate measurements for each reaction mixture.

Inaccuracies most often arise when mistakes are made in measuring out the reactants (or forgetting to add a reactant!), or when the solution is inadequately stirred while the hydrogen peroxide is being added. You must also be careful not to get any contaminant in your hydrogen peroxide stock solution, since many substances catalyze its decomposition into water and oxygen.

MIXTURE 1. Place all reactants except the H_2O_2 in a 150-mL beaker. Note and record the temperature of the solution. Arrange your watch or have your timer ready.

Then, using a pipet, add exactly 10 mL of H_2O_2 solution, and stir continuously. Start the timer when half of the H_2O_2 solution has drained from the pipet, and continue stirring until you are certain the solution is thoroughly mixed.

Watch the solution carefully for the sudden appearance of the blue color, and stop the timer when it appears. (The total elapsed time should be of the order of 2 to 4 min.)

Discard the solution and rinse the beaker.

Repeat the measurement using a fresh solution, if your instructor so directs. The times should agree within a few seconds.

MIXTURE 2. Prepare solution 2, and note and record the temperature, which should be within 0.5 °C of the temperature recorded for mixture 1. (The temperature may be adjusted if necessary by cooling or heating in a water bath contained in a larger beaker.)

When the timer is ready, add 5 mL of H_2O_2 solution. Since the concentration of H_2O_2 is less than in mixture 1, you may expect that the reaction time will be at least as great as that for mixture 1.

MIXTURE 3. Prepare solution 3. Note and record the temperature, which should be the same as for mixtures 1 and 2 (adjust if necessary). Note that the concentration of KI has been doubled over that contained in mixture 1, so the reaction time may be shorter.

Carry out the measurement as described for mixture 1.

MIXTURE 4. Carry out the measurement as described for mixture 1. This solution contains ten times as much acetic acid, therefore the H^+ concentration is ten times larger. If the reaction time is the same as for mixture 1, what does this suggest about the reaction order with respect to H^+?

MIXTURE 5. Prepare solution 5, which has the same composition as solution 1, and place it in a beaker of hot water.

Stir until the temperature is about 10 to 12 °C above that of mixture 1; then add 10 mL of H_2O_2 solution, start the timer, and note and record the temperature.

When the solution turns blue, stop the timer, and again note and record the temperature. (The average of the two readings is taken as the temperature of the reaction mixture.)

MIXTURE 6. This solution has the same composition as mixture 1, except that it contains 5.0 mL of 0.010 M Mo(VI) catalyst. Mixture 6 is adjusted to the same temperature as mixture 1, 10 mL of H_2O_2 is added, and the reaction time is recorded.

C. INTERPRETATION OF THE DATA
1. CALCULATING THE REACTION ORDERS.
When we want to measure the order of a reaction with respect to a reactant, we vary the initial concentration of that reactant and keep constant the temperature and the initial concentrations of all other reactants.

For example, note that in reaction mixtures 1 and 2 only the concentration of H_2O_2 has been changed. (See Table 26-2.) To get the reaction order with respect to H_2O_2, we first note that the reaction rate is inversely proportional to the reaction time, Δt, as shown by Equation (9). Because we put the same amount of thiosulfate ion in each reaction mixture, the concentration terms cancel out when we take the ratio. Therefore the ratio of the reaction rates for any two reaction mixtures will be equal to the inverse of the ratio of the reaction times.

$$\frac{\text{rate}_1}{\text{rate}_2} = \frac{\Delta t_2}{\Delta t_1} \tag{10}$$

Starting with Equation (4), we can write expressions for the rates of reaction in reaction mixture 1:

$$\text{Rate}_1 = k[H_2O_2]_1^x [I^-]^y [H^+]^z \tag{11}$$

and reaction mixture 2

$$\text{Rate}_2 = k[H_2O_2]_2^x [I^-]^y [H^+]^z \tag{12}$$

Dividing Equation (11) by Equation (12), we obtain an expression for the ratio rate$_1$/rate$_2$.

$$\frac{\text{rate}_1}{\text{rate}_2} = \frac{\Delta t_2}{\Delta t_1} = \left(\frac{[H_2O_2]_1}{[H_2O_2]_2}\right)^x \tag{13}$$

Note that when we write the equation for the ratio of the rates, all of the concentrations that have been held constant in the two reaction mixtures cancel out, leaving only the ratio of the concentrations of H_2O_2 (raised to the power of the reaction order, x). In addition, if the temperature is constant, the specific rate constant, k, that appears in Equations (11) and (12) will also cancel out.

Because we know the reaction times and the concentrations of H_2O_2 in reaction mixtures 1 and 2, we can calculate the unknown reaction order, x, with respect to H_2O_2. The nearest-integer value of the reaction order can usually be determined by inspection, but if the logarithms of both sides of Equation (10) are taken, we can calculate x from the resulting equation.

$$x = \frac{\log(\Delta t_2/\Delta t_1)}{\log([H_2O_2]_1/[H_2O_2]_2)} \tag{14}$$

The same method is used to obtain the reaction order with respect to I^- (using the data for reaction mixtures 3 and 1) and the reaction order with respect to H^+ (using the data for reaction mixtures 4 and 1).

2. EFFECT OF TEMPERATURE. In reaction mixtures 5 and 1, the concentrations of all the reactants are the same, but the temperature of mixture 5 is greater than that of mixture 1. As before, the ratio of the rates is given by the inverse ratio of the reaction times. Because all the concentration terms are the same, they cancel out; but the specific rate constants do not cancel because the specific rate constant changes with temperature. Therefore we have

$$\frac{\text{rate}_5}{\text{rate}_1} = \frac{\Delta t_1}{\Delta t_5} = \frac{k_{T_5}}{k_{T_1}} \tag{15}$$

where k_{T_5} and k_{T_1} are the specific rate constants at the temperatures T_5 and T_1 of reaction mixtures 5 and 1, respectively.

3. EFFECT OF A CATALYST. Using the measured reaction times, we can calculate the ratio of the reaction rates from the data for reaction mixtures 6 and 1.

$$\frac{\text{rate}_6(\text{catalyzed})}{\text{rate}_1(\text{uncatalyzed})} = \frac{\Delta t_1}{\Delta t_6} \tag{16}$$

How many times faster is the catalyzed reaction rate? What is the molar concentration of the Mo(VI) catalyst in the reaction mixture?

Consider This

The growth of a plant is the result of a series of chemical reactions. Is rate of plant growth dependent on concentration? Design experiments to test fertilizer-concentration dependence. Why does an excess of fertilizer cause a decrease in the growth rate? (You may need to consult Experiment 22 for hints.)

Are the reactions involved in plant growth temperature dependent? Design an experiment to test this factor. Why does growth rate decrease above certain temperatures?

Can you measure CO_2 dependence of the growth rate for plants? This factor is a significant component of the discussion about global warming. Why?

Design an experiment to measure the rate of the chemical reaction known as rusting,

$$Fe(s) + O_2(g) \rightarrow Fe_2O_3 \cdot n\,H_2O(s)$$

Can you determine temperature dependence? Oxygen concentration dependence?

BIBLIOGRAPHY

Bamford, C. H., and Tipper, C. F. H. (eds.) *Comprehensive Chemical Kinetics*, Vol. 6, Elsevier Publishing Co., New York, 1972, p. 406. A summary discussion with references of the reaction of hydrogen peroxide with halide ions.

King, E. L. *How Chemical Reactions Occur*, W. A. Benjamin, Inc., Menlo Park, CA, 1963, pp. 36–38, 80–82.

The Rate of a Chemical Reaction

Chemical Kinetics

NAME

SECTION   LOCKER

INSTRUCTOR   DATE

A. STANDARDIZATION OF THE H_2O_2 SOLUTION

If your instructor so directs, carry out the standardization procedure in duplicate.

Data	Trial 1	Trial 2
Buret final reading		
Initial reading		
Net volume of $Na_2S_2O_3$		
(0.050 M for a 50-mL buret)		
(2.50 M for a 1.00-mL buret)		

Calculate the H_2O_2 concentration, which should be in the range 0.8 to 0.9 M. If it is outside this range, consult your instructor.

B. REACTION RATE MEASUREMENTS

Make a single measurement on each solution, unless your instructor directs otherwise. If a measurement is spoiled for any reason, the procedure should be repeated.

Reaction Mixture	Trial 1		Trial 2	
	Time (s)	Temperature (°C)	Time (s)	Temperature (°C)
1				
2				
3				
4				
5				
6				

C. INTERPRETATION OF THE DATA

1. Calculating the Reaction Orders

(a) Order of the reaction with respect to hydrogen peroxide, H_2O_2

Reaction mixture	Time (s)	H_2O_2 concentration in the reaction mixture (mol/L)
1		
2		

Using Equation (14), calculate the reaction order and round off to the nearest integer.

(b) Order of the reaction with respect to iodide ion, I^-

Reaction mixture	Time (s)	I^- concentration in the reaction mixture (mol/L)
1		
3		

Using Equation (14), calculate the reaction order and round off to the nearest integer.

(c) Order of the reaction with respect to H^+

Taking K_a for acetic acid as 1.8×10^{-5} M, calculate the H^+ concentration from Equation (5), using the known concentrations of acetic acid and sodium acetate contained in reaction mixtures 1 and 4. (In mixture 4 be sure that you account for the acetic acid contained in both the buffer and the added 0.3 M acetic acid.)

Reaction mixture	Time (s)	Acetic acid concentration (mol/L)	Sodium acetate concentration (mol/L)	Calculated H^+ concentration (mol/L)
1				
4				

Show your calculations of the H^+ concentration.

Using Equation (14), calculate the reaction order and round off to the nearest integer.

2. Effect of Temperature

Reaction mixture	Time (s)	Temperature (°C)
1		
5		

Using Equation (15), calculate the reaction rate ratio for reaction mixtures 5 and 1. When the temperature is increased, does the reaction rate decrease or increase?

$$\frac{\text{rate}_5}{\text{rate}_1} =$$

3. Effect of a Catalyst

Reaction mixture	Time (s)	Mo(VI) concentration in the reaction mixture (mol/L)
1		
6		

Using Equation (16), calculate the ratio of the rates of the catalyzed and uncatalyzed reactions.

$$\frac{\text{rate (catalyzed)}}{\text{rate (uncatalyzed)}} =$$

QUESTIONS

1. Calculating the Reaction Orders

(a) From your measurements, what are the reaction orders with respect to each of the three reactants H_2O_2, I^-, and H^+?

H_2O_2 _____ I^- _____ H^+ _____

(b) What is the rate expression (rate law) for the uncatalyzed reaction?

(c) What is the total reaction order?

(d) Is the rate law consistent with the idea that the mechanism of the reaction is $H_2O_2 + 2I^- + 2H^+ \rightarrow I_2 + 2H_2O$ (all in one step)? Explain why or why not.

(e) The following reaction pathway has been proposed for the pH range 3.5 to 7:

$$H_2O_2 + I^- \rightarrow OH^- + HOI$$

$$H^+ + OH^- \rightleftharpoons H_2O$$

$$HOI + H^+ + I^- \rightarrow I_2 + H_2O$$

Add the equations above. Do they give the overall reaction?

According to your results, which of the steps would be the slow, rate-determining step?

If the proposed mechanism agrees with the experimental rate law, does this prove that the mechanism is correct and that no other pathway is possible? Defend your answer.

2. Effect of Temperature

Do your measurements for Part C.2. confirm the oft-quoted rule of thumb that "the rate of a chemical reaction approximately doubles for every 10 °C temperature increase?"

PROBLEMS

1. Considering the volumes and the concentrations used in mixture 1, what percentage of the moles of H_2O_2 present have been consumed during the timed reaction? Calculate the number of moles of H_2O_2 present initially, and determine how many have reacted with iodide when the thiosulfate runs out. How does the I^- concentration change during this same time? Why?

2. Using Equations (9) and (11), calculate the specific rate constant, k, from your data for reaction mixture 1. (Use the integer reaction orders you determined for x, y, and z in the rate expression of Equation [11].) Compare your value with the reported value of $0.0115 \, L \cdot mol^{-1} \cdot s^{-1}$ at 25 °C for the uncatalyzed reaction.

$$k = \underline{\hspace{4cm}}$$

3. It has been found that in acid solution (greater than 1 M H^+) the rate expression for the reaction of H_2O_2 with I^- is of the form: rate $= k_2[H_2O_2][I^-][H^+]$ where $k_2 = 0.25 \, L^2 \cdot mol^{-2} \cdot s^{-1}$ at 25 °C. Therefore the rate law over the whole range of acidity from very acid solutions to a neutral solution is given by a *two-term* rate expression:

$$\text{rate} = k_1[H_2O_2][I^-] + k_2 [H_2O_2][I^-][H^+]$$

where k_1 has the value of $0.0115 \, L \cdot mol^{-1} \cdot s^{-1}$ at 25 °C. Compare the relative magnitudes of these two terms under the conditions employed in your rate measurements for mixture 1. (You should find that the second term makes a negligible contribution to the reaction rate at H^+ concentrations smaller than 3×10^{-4}.)

Consider This

The growth of a plant is the result of a series of chemical reactions. Is rate of plant growth dependent on concentration? Design experiments to test fertilizer-concentration dependence. Why does an excess of fertilizer cause a decrease in the growth rate? (You may need to consult Experiment 22 for hints.)

Are the reactions involved in plant growth temperature dependent? Design an experiment to test this factor. Why does growth rate decrease above certain temperatures?

Can you measure CO_2 dependence of the growth rate for plants? This factor is a significant component of the discussion about global warming. Why?

Design an experiment to measure the rate of the chemical reaction known as rusting,

$$Fe(s) + O_2(g) \rightarrow Fe_2O_3 \cdot n\,H_2O(s)$$

Can you determine temperature dependence? Oxygen concentration dependence?

EXPERIMENT

27

The pH Scale and Acid-Base Titrations
Standardization of NaOH Solution
The Titration of Vinegar and Stomach Antacids

PURPOSE

To understand the meaning of pH. To learn the difference between the H_3O^+ concentration and the total acidity of an acid. To learn the technique of acid-base titration by using a buret to measure volume and an acid-base indicator to determine the end point of the titration. To quantitatively determine the neutralization capacity of various commercial antacid tablets.

PRE-LAB PREPARATION

A quick look in the cabinets of the kitchen, laundry, or bathroom of the average home would turn up a number of substances that are acids or bases. Acidic substances we might find include citrus juices, vinegar, carbonated soft drinks, and toilet bowl cleaners. We might also find substances that are bases: baking soda (sodium bicarbonate), detergents, household ammonia, chlorine-based bleaches, drain cleaners, and stomach antacid tablets.

Studies of our own body fluids would also reveal that some are acidic and others are nearly neutral or slightly basic. The contents of the stomach can be very acidic—the cells lining the walls of the stomach secrete hydrochloric acid. Human urine can be slightly acidic or basic, depending on what you may have last eaten or the state of your health. Human blood, tears, and saliva are usually just slightly basic.

DEFINITIONS OF ACIDS AND BASES

In about 1887 S. Arrhenius proposed that substances that produce hydrogen ion in water be called acids and substances that produce hydroxide ion in water be called bases. Since that time, the concept of an acid or base has been extended and redefined to explain more fully the terms *acidity* and *basicity* and the role of the solvent in acid-base equilibria.

In 1923 J. N. Brønsted and T. M. Lowry independently proposed the definition of an acid as a proton donor and of a base as a proton acceptor. This definition has proved useful because it explicitly points out the role played by the solvent (often water) in acid-base chemistry. The dissociation of water is described by the equilibrium

$$H_2O + H_2O \rightleftharpoons H_3O^+ + OH^- \tag{1}$$

$$K_W = [H_3O^+][OH^-] = 1.0 \times 10^{-14} \text{ at } 25\,°C \tag{2}$$

In this equilibrium, water plays the role of both an acid and a base. (Such substances are often called *amphiprotic*.) Water can act as both an acid and a base because it can form two Brønsted acid-base pairs: H_3O^+/H_2O and H_2O/OH^-.

In pure water the concentrations of H_3O^+ and OH^- are equal:

$$[H_3O^+] = [OH^-] = \sqrt{K_W}$$

$$= 1.0 \times 10^{-7} \text{ M at } 25\,°C \tag{3}$$

When $[H_3O^+] = [OH^-]$, we say the solution is neutral (neither acidic nor basic).

What happens when we add a little vinegar (whose active ingredient is acetic acid) to pure water? The acetic acid donates a proton to a water molecule, thus creating additional H_3O^+ ions:

$$\underset{(HA)}{CH_3\overset{\displaystyle O}{\overset{\|}{C}}OH} + H_2O \rightleftharpoons H_3O^+ + \underset{(A^-)}{CH_3\overset{\displaystyle O}{\overset{\|}{C}}O^-} \tag{4}$$

343

The equilibrium constant expression for this reaction is

$$K_a = \frac{[H_3O^+][A^-]}{[HA]} = 1.76 \times 10^{-5} \tag{5}$$

where K_a represents the acid dissociation constant for acetic acid, a weak monoprotic acid represented by the symbol HA. The anion of the acid, acetate ion, is represented by the symbol A^-. So the acetic acid acts as a proton donor, and the water acts as a proton acceptor. Since the H_3O^+ concentration increases, the OH^- concentration must decrease, because their concentrations are linked by the equilibrium constant expression shown in Equation (2).

Suppose that we add a little household ammonia to pure water. The ammonia reacts with water as a Brønsted base, accepting a proton from the water and leaving OH^- ions in solution:

$$NH_3(aq) + H_2O \rightleftharpoons NH_4^+(aq) + OH^-(aq) \tag{6}$$

The equilibrium constant expression for this reaction is

$$K_b = \frac{[NH_4^+][OH^-]}{[NH_3]} = 1.8 \times 10^{-5} \tag{7}$$

Note that water acts as a base (proton acceptor) with respect to acids that are stronger acids than water, but it acts as an acid (proton donor) with bases that are stronger bases than water. The equilibria shown in Equations (1), (4), and (6) are examples of the amphiprotic behavior of water.

THE pH SCALE

Typical H_3O^+ and OH^- concentrations in solution range from large to very small values (from approximately 10 M to 10^{-14} M). It would be impossible to represent such a range of concentrations on a linear graph because the concentrations would either have to be compressed into an impossibly tiny length on the low end—or because we would have to have an astronomically long piece of graph paper. In a situation like this, we resort to using a logarithmic scale, which assigns an equal length on the scale to each power of 10 (or decade) of the H_3O^+ concentration.

There is another important reason why a logarithmic scale is used. A hydrogen ion-sensing device called a pH meter (which employs an electrochemical cell to measure the $[H_3O^+]$) gives a voltage output that depends on the logarithm of the hydrogen ion concentration. So the digital readout (or deflection of the meter needle of the pH meter) is proportional to $Log[H_3O^+]$.

Therefore, for both practical and theoretical reasons, it is useful to define a variable called the pH, where

$$pH = -Log[H_3O^+] \tag{8}$$

and the inverse relation

$$[H_3O^+] = 10^{-pH} \tag{9}$$

We take the symbol "p" to mean *take the negative logarithm of the quantity that immediately follows*. So pK_a means $-Log\ K_a$, and so on. Including the minus sign in the definition is a convenience. Most of the concentrations of interest are less than 1 M, so if the choice had been made to define pH as $+Log[H_3O^+]$, we would be required always to carry along a minus sign with the numerical values. However, including the minus sign in the definition of pH has one adverse consequence that is sometimes confusing. We often call the pH scale an acidity scale, but as the acidity of a solution increases, the pH decreases.

The neutral point on the aqueous pH scale (pH = 7.00 at 25 °C) is a consequence of the magnitude of the equilibrium constant for the dissociation of water. (At temperatures other than 25 °C in water or in other solvents, the neutral point would not be at pH 7.00.)

THE DISTINCTION BETWEEN pH AND TOTAL ACIDITY

If an acid completely dissociates in solution, the hydrogen ion concentration (and the pH) will be a measure of the total amount of acid in the solution. For a strong acid

$$[H_3O^+] = C \tag{10}$$

$$pH = -Log\ C = pC \tag{11}$$

where C is the molar concentration of strong acid in solution. This relation will be true for common strong acids such as $HClO_4$, HCl, and HNO_3.

But if the acid is a weak monoprotic acid, it will be incompletely dissociated, and an approximation of the hydrogen ion concentration (or pH) of the solution is given by the expressions

$$[H^+] = \sqrt{K_a C} \tag{12}$$

$$pH = -\log \sqrt{K_a C} = \frac{1}{2}pK_a + \frac{1}{2}pC \tag{13}$$

where K_a is the acid dissociation constant of the weak acid, and C is the molar concentration of the acid. (Take pencil and paper and derive Equations [12] and [13], consulting your text as necessary.) Equations (12) and (13) show that in a weak monoprotic acid, the hydrogen ion concentration is not equal to C and the pH is not equal to pC. For a weak acid, K_a is generally $\ll 1$, so $[H_3O^+]$ $\ll C$. Contrast these results with the simple expressions shown in Equations (10) and (11) for a strong acid.

ACID-BASE TITRATIONS

We can't directly determine the concentration, C, of a weak acid just by measuring the pH. Equation (13) shows that we also need to know the acid's dissociation constant, K_a. But we can determine C (which we call the *total acidity* or the *stoichiometric concentration*) by measuring the amount of a strong base (such as NaOH) required to react completely with a sample of the acid. This is called an *acid-base titration*. The NaOH is dispensed from a buret, so if we know the concentration and the volume of NaOH required to titrate the acid, we can accurately calculate the total amount of weak acid in the solution.

At the end point in the titration of a weak acid with a base, such as acetic acid (CH_3COOH, which we will abbreviate as HA) with NaOH, the following relationship holds:

$$\text{number of moles of acid} = \text{number of moles of base} \quad (14)$$

This is because the net ionic equation describing the reaction is

$$HA + OH^- \rightarrow A^- + H_2O \quad (15)$$

and the equilibrium constant of this neutralization reaction is extremely large ($K = K_a/K_W \cong 10^9$), indicating that it proceeds essentially to completion. Because one mole of NaOH reacts with one mole of HA, we can calculate the moles of HA from the relation

$$\text{moles HA} = \text{moles NaOH}$$
$$= M_{NaOH} \times V_{NaOH} \quad (16)$$

So, we can determine the concentration of acetic acid in a sample of vinegar by titration with a NaOH solution of known concentration.

A plot of the pH of the solution *versus* the milliliters of NaOH added is called a *titration curve*. An example of a titration curve for a sample of vinegar (a dilute solution of acetic acid) titrated with NaOH is shown in Figure 27-1. Note that when the last bit of acid is titrated, the pH *increases* dramatically, indicating a rapid *decrease* in the hydrogen ion concentration. This rapid change signals the end point of the titration.

Either a pH meter or an acid-base indicator can be used to detect the end point of the titration. An acid-base

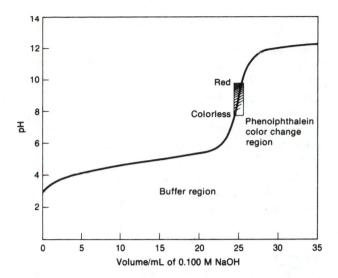

FIGURE 27-1

Titration curve for approximately 3 mL of vinegar diluted to 25 mL in water and titrated with 0.1 M NaOH. Phenolphthalein (or thymol blue) is a suitable indicator for detecting the end point of the titration.

indicator is an organic dye molecule that is itself a weak acid. The acid and base forms of the indicator have different colors, so when it goes from its acid form to its base form, the indicator shows a color change. The pH range where this color change takes place is governed by the dissociation constant (K_a) of the acid indicator molecule. If we represent the protonated indicator molecule by the symbol HIn, we can write the equation for the dissociation of the indicator as follows:

$$HIn + H_2O \rightleftharpoons H_3O^+ + In^- \quad (17)$$

$$K_a = \frac{[H_3O^+][In^-]}{[HIn]} \quad (18)$$

If we take the logarithm of both sides of Equation (18), we obtain

$$\log K_a = \log [H_3O^+] + \log \frac{[In^-]}{[HIn]} \quad (19)$$

which we can rearrange to the following form:

$$pH = pK_a + \text{Log} \frac{[In^-]}{[HIn]} \quad (20)$$

Now, take pencil and paper and prove to yourself that Equation (20) can be obtained from Equations (18) and (19). Then prove to yourself that when pH = $pK_a - 1$, $[In^-]/[HIn] = 0.1$, and that when pH = $pK_a + 1$, $[In^-]/[HIn] = 10$.

The pencil and paper exercise shows us that the indicator changes from being mostly in its acid form (HIn) to being mostly in its base form (In⁻) in an interval of about 2 pH units, centered on the pK_a of the indicator. This is called *the pH interval for the color change of the indicator.* (Table 5 of the Appendix gives the colors and pH intervals of a number of indicators.)

In order to have the color change of the indicator accurately signal the end point, we must choose an indicator whose pK_a is about the same as the pH at the end point of the titration, as shown in Figure 27-1. Also note in Figure 27-1 that at the end point for the titration, the solution has a composition and pH that are identical to those of a sodium acetate solution having the same concentration.

Acid-base indicators have such an intense color that only a small amount of the indicator is necessary to produce a readily visible color. So the contribution of the indicator acid, HIn, to the total acidity of the solution is negligible.

STANDARDIZATION OF THE NaOH TITRANT

The acid-base titration procedure we have just described requires that we know the concentration of the NaOH titrant in order to be able to determine the concentration of the acid that we are titrating. The accurate determination of the concentration of the NaOH titrant is called a *standardization* procedure. It works like this. We weigh out a sample of an acid available in very pure form, called a *primary standard* grade of acid, dissolve it in water, and titrate the acid with our NaOH titrant. Knowing the mass of the acid used, its formula weight, and the volume of NaOH required to neutralize it, we can determine the concentration of the NaOH titrant. We will use a pure grade of potassium hydrogen phthalate, KHP, whose formula and structure are

Potassium hydrogenphthalate (KHP)

$KC_8H_5O_4$, FW = 204.23 g/mole

KHP is a weak carboxylic acid, like acetic acid, so we can use the same indicator to titrate KHP as we use for the titration of acetic acid in vinegar.

TITRATION OF STOMACH ANTACIDS

In this experiment we will determine the neutralization capacity of commercial stomach antacid tablets. These substances are bases, so direct titration would require an acid titrant. However, many of them do not readily dissolve in water and react rather slowly with acids. To circumvent this difficulty we will use a back-titration procedure. An excess of a strong acid (HCl) is added to the antacid tablet, and the mixture is heated to ensure complete reaction. Some of the HCl is consumed by reaction with the antacid tablet. The antacid and HCl solution is then titrated with NaOH to determine how much HCl remains unreacted. Subtracting the amount of HCl remaining from that originally added gives the amount of HCl that the antacid tablet consumed. This gives an accurate measurement of the tablet's neutralization capacity. It also resembles the way the tablet reacts with HCl in the stomach. Typical titration curves for the back titration of antacid and HCl solutions with NaOH are shown in Figure 27-2. The indicator thymol blue (which is diprotic and therefore has two color transition regions) is a good choice for most antacids.

Antacid tablets are probably one of the most widely used self-prescribed medicines. They are taken to relieve the medically undefined conditions of heartburn or acid indigestion and sour stomach. Although such gastric distress is often attributed to excess production of HCl, sometimes the latter is not responsible for the symptoms. For instance, gastric HCl production may be less than normal in diseases such as superficial chronic gastritis and carcinoma of the stomach. The common condition popularly called "heartburn" is thought to be caused by regurgitation of the gastric contents into the esophagus, and acid indigestion may be merely due to overeating or to an irritating food.

Television commercials stress the competition among various brands of antacids to see which has the greatest neutralization capacity, but in actuality if the pH in the stomach rises too high, the entire digestive process may be hindered. The digestion of proteins is catalyzed by the enzyme pepsin. Its proteolytic activity is inhibited if the pH of the stomach contents rises any higher than the natural buffering of the proteins in the food itself would allow it to go (around pH 4). Therefore, an antacid should only neutralize an amount of HCl in excess of what the body of a healthy individual normally secretes following a meal. Studies show that HCl produced by gastric stimulation varies widely from individual to individual, ranging from about 1 to 22 mmol of HCl per hour, but the average figure for "excess" production is only about 10 mmol of HCl per hour greater than that of normal HCl production. Therefore, one dose of an antacid product should react with no more than about 10 mmol of HCl per hour. For many people the upper limit should be less. This raises the question whether people who regularly consume the doses recommended on the labels of various antacid products are in fact interfering with normal digestion. The doses neutralize much more than 10 millimoles of HCl, so the body's natural response raises the pH of the stomach contents above pH 4.

FIGURE 27-2

Back-titration of unreacted HCl with standardized 0.3 M NaOH after addition of 5.0 mL of 3.0 M HCl to antacid tablets of various compositions. The indicator thymol blue has two color transitions, at the pHs shown approximately by the Ts.

EXPERIMENTAL PROCEDURE

Special Supplies: *Parts 1, 2, and 3: Microscale:* 1 mL buret; *Macroscale:* 25 or 50-mL buret.

Part 2: Distilled white vinegar (5% by weight acetic acid). Dark-colored vinegars are difficult to titrate because their color obscures the indicator end point.

Part 3: Stomach antacids (Tums Antacid Chewable Tablets, Regular; Phillips Milk of Magnesia tablets; etc.), 5-mL pipet, polyethylene transfer pipet.

Chemicals: *Parts 1, 2, and 3:* 0.1% thymol blue indicator.

Part 1 (optional standardization): Pure, dry potassium hydrogen phthalate (KHP).

Parts 1 and 2 (standardization and vinegar titration): *Microscale:* 1.00 M NaOH; *Macroscale:* 0.1 M NaOH.

Part 3 (antacid titration): 3.00 M HCl; *Microscale:* 6.00 M NaOH; *Macroscale:* 0.30 M NaOH.

NOTES TO INSTRUCTOR: If you prefer to use 25- or 50-mL burets instead of 1-mL burets, in Part 2 use a 2.50 mL (or 2.50 g) sample of vinegar titrated with 0.100 M NaOH. Also in part 2, back-titrate the sample with 0.30 M NaOH in a 25- or 50-mL buret in place of 6.0 M NaOH in a 1-mL buret.

In Part 3 it is preferable to use antacid tablets that will neutralize 10 to 11 mmol of HCl. Tablets with a larger amount of active ingredient will require a proportionate increase in the volume of added 3.00 M HCl. Use a volume of 3.00 M HCl that will leave about 4 to 5 mmol excess HCl after reaction with the tablet.

! **SAFETY PRECAUTIONS:** *Parts 1 and 2:* Sodium hydroxide is corrosive. Wear eye protection. Do not touch or handle the chemicals with your bare hands. Clean up any spills immediately.

Part 3: If available, use a hot plate to heat the flask of HCl containing an antacid tablet. If you use a Bunsen burner, be sure that the flask is clamped so that it cannot be overturned. Boiling hot solution that splashes on you from the bench top can inflict serious burns.

WASTE DISPOSAL: Collect all wastes in a single beaker, neutralize with sodium bicarbonate, and flush down the drain.

1. STANDARDIZATION OF NaOH WITH POTASSIUM HYDROGEN PHTHALATE (OPTIONAL).

Your instructor will tell you whether you will be using microscale or macroscale burets. Read the section of the Introduction describing how to use burets.

Microscale: Obtain three clean 50-mL Erlenmeyer flasks. Tare each flask in turn, and weigh into each flask about 0.160 grams of pure, dry potassium hydrogen phthalate (KHP), recording the mass of each sample to the nearest milligram. Add 20 mL of deionized water to each flask, and swirl to dissolve the KHP. Add 5 drops of 0.1% thymol blue indicator to the dissolved KHP.

Fill a 1-mL microburet (see the Introduction at the front of the lab manual) with approximately 1.0 M NaOH, taking care to expel all bubbles from the microburet before beginning the titration. Record the initial volume reading of the microburet, and titrate the first solution to the appearance of the first permanent blue color. Then record the final volume reading of the buret. Repeat the titration with the remaining two samples of KHP.

Macroscale: Obtain three clean 250-mL Erlenmeyer flasks. Tare each flask in turn, and weigh into each flask about 0.400 g of pure, dry potassium hydrogen phthalate (KHP), recording the mass of each sample to the nearest milligram. Add 100 mL of deionized water to each flask, and swirl to dissolve the KHP. Add 5 drops of 0.1% thymol blue indicator to the dissolved KHP.

Fill a 25- or 50-mL buret (see the Introduction at the front of the lab manual) with approximately 0.10 M NaOH, taking care to expel all bubbles from the buret tip before beginning the titration. Record the initial volume reading of the buret, and titrate the first solution to the appearance of the first permanent blue color. Then record the final volume reading of the buret. Repeat the titration with the remaining two samples of KHP.

Calculations: From the mass and formula weight of the KHP, calculate the number of millimoles of KHP in the sample. This will be equal to the number of millimoles of NaOH titrant. Divide the number of millimoles of NaOH by the volume of titrant added, in milliliters, to get the molar concentration of the NaOH titrant in units of moles/Liter. Average the three values, which should be in good agreement (relative standard deviation < 5%).

2. TITRATION OF VINEGAR.

Record in your notebook the brand name and percentage of acetic acid in the vinegar that is specified on the manufacturer's label.

Microscale: Obtain 5 mL of standardized 1.00 M NaOH and 5 mL of vinegar in two clean, dry, stoppered 13 × 100 mm test tubes. Obtain three clean 25-mL Erlenmeyer flasks to contain three vinegar samples. Using a 1.00 mL serological pipet and pipet-filling device, transfer exactly 1.00 mL of vinegar to each of the three Erlenmeyer flasks.

Alternatively, put the empty flask on the pan of the balance, and push the tare bar so that the balance reads zero (or record the mass of the empty flask). Using a polyethylene pipet filled with vinegar, drop about 1.00 g of vinegar into the Erlenmeyer flask. Record the exact mass of the sample to the nearest 0.001 g. Repeat until you have three weighed samples of vinegar.

Add 10 mL of deionized water and 3 drops of 0.1% thymol blue indicator to each of the vinegar samples. Fill the 1-mL microburet with 1.00 M NaOH, taking care to remove any bubbles in the buret. (**Caution:** If you are using a 1-mL syringe as a microburet, take special care when filling the syringe or expelling bubbles not to get titrant on your skin or squirt yourself or your neighbor with the NaOH in the syringe.)

Record the initial volume reading of the buret. Titrate the sample of vinegar, swirling the flask to stir the sample while adding NaOH drop by drop from the microburet. When the solution begins to turn blue where the NaOH titrant hits the solution, you are nearing the end point. Titrate to the first appearance of a permanent blue color; then read and record the final volume of NaOH in the microburet. Discard the blue solution, and rinse the flask with deionized water. Repeat the titration for the remaining two samples of vinegar.

Macroscale: Obtain in two clean, dry flasks 100 mL of standardized 0.10 M NaOH and about 15 mL of vinegar. Next obtain three clean 250-mL Erlenmeyer flasks to contain three vinegar samples. Using a 5-mL serological pipet and pipet-filling device, transfer exactly 2.50 mL of vinegar to each of the three flasks. (Alternatively, you may put the flask on an electronic balance, tare to zero, and transfer 2.50 g of vinegar to the flask, using a polyethylene transfer pipet. Record to the nearest milligram the exact mass of vinegar weighed into each flask.)

Add 100 mL deionized water and 5 drops of 0.1% thymol blue indicator to each sample of vinegar. Rinse a 50-mL buret with standardized 0.010 M NaOH, and fill the buret, taking care to expel all bubbles from the tip of the buret. Record the initial volume of the buret to the nearest 0.02 mL. Titrate the first sample of vinegar to the appearance of a permanent blue color. Record the volume reading of the buret. Repeat the titration for the remaining two samples of vinegar.

(Keep your buret, but rinse and refill it with 0.30 M NaOH if you are going to do Part 3, the titration of antacid tablets.)

Calculations: From the concentration and added volume of the NaOH titrant, calculate for each titration the millimoles of NaOH added. This will equal the millimoles of acetic acid contained in the vinegar sample. Then calculate the mass of acetic acid contained in the sample; take as a given that the molar mass of acetic acid to be 60.0 g/mol. If you measured the vinegar by volume to obtain the mass of vinegar, assume that the density of vinegar is about 1.005 g/mL (at 20 °C). Finally, calculate the mass percent of acetic acid in the sample as (g acetic acid/g of vinegar) × 100.

3. TITRATION OF ANTACID TABLETS.

Various brands of antacid tablets will be available, such as Tums Antacid Chewable Tablets, Regular ($CaCO_3$); Phillips Milk of Magnesia tablets ($Mg[OH]_2$), etc. The identity of the active ingredient and the mass of active ingredient per tablet will be found on the manufacturer's label. Record this information in your report.

Analyze one or more brands of tablets as your instructor directs, using the following procedure. Put an antacid tablet in a small beaker, and weigh the beaker and tablet on the analytical balance to the nearest milligram. Record the mass. Then transfer the tablet to a clean 250-mL Erlenmeyer flask. Reweigh the empty beaker, and record the mass so that you can obtain the mass of the tablet.

Put 5.00 mL of 3.00 M HCl in the flask containing the tablet, using a pipet or buret to accurately measure the volume of HCl solution.[1] Add approximately 50 mL deionized water. Heat the flask to boiling on a hot plate or over a Bunsen burner. (**Caution:** If you use a Bunsen burner, clamp the flask so that it cannot be overturned.) Gently boil the solution for about 5 minutes. This expels any carbon dioxide produced by reaction with HCl. Use a glass stirring rod, if necessary, to help break up and disperse the tablet. (**Caution:** Do not use so much force that you risk breaking a hole in the flask with the glass stirring rod.)

Microscale: While the solution is boiling, rinse and fill a 1-mL microburet with standardized 6.0 M NaOH, and record the initial buret reading.

---

[1]This amount of HCl is calculated to neutralize a tablet that will consume 10 mmols of HCl, leaving about 5 mmols of unreacted HCl. If the tablet will consume more than 10 mmols of HCl, the volume of 3.00 M HCl added should be increased proportionately.

Macroscale: Use the same 50-mL buret you used in Part 2, but rinse and refill it with 0.30 M NaOH. Record the initial volume reading of the buret.

Cool the solution containing the antacid to room temperature by immersing it in a container of tap water. When the solution has cooled, add 10 drops of 0.1% thymol blue indicator. The solution should appear red, indicating that the pH is less than 2. Then titrate the solution with 6.00 M NaOH (*microscale 1 mL buret*) or 0.30 M NaOH (*macroscale buret*) until you get a color change from red to yellow. Record the buret reading.

....

NOTE: If you are titrating antacids that contain calcium carbonate or magnesium hydroxide as the active ingredient, the steep rise in pH at the end point, shown in Figure 27-2, means that the color transition from yellow to blue, which takes place in the pH range 8.0 to 9.6, will quickly follow the transition from red to yellow, which takes place in the pH range 1.2 to 2.8. If you are titrating too rapidly and overrun the end point slightly, you might not even see the red to yellow transition. For these samples, you can without serious error take the first appearance of the blue color as the endpoint of the titration.

Repeat the titration with a second sample or with another brand of antacid, if your instructor so directs.

....

Calculations: Calculate the millimoles of HCl added, the millimoles of NaOH added, the millimoles of HCl that were neutralized by the antacid, and the millimoles and milligrams of the active ingredient in the tablet.

Write a balanced chemical reaction for the reaction of HCl with the active ingredient in the antacid tablet. From your titration data calculate the mass of the active ingredient, and compare this with the value given on the manufacturer's label. Compare the mass of the active ingredient you found with the mass of the tablet. The difference represents the mass of inert binders used in making the tablet. Calculate the milligrams of inert binder per tablet and the mass percentage of inert binder.

Consider This

Using a pH meter, the pH of a cranberry juice cocktail was found to be 2.46, which indicates that the $[H_3O^+] = 3.47 \times 10^{-3}$ M. Titration of a 1.0 mL sample of the juice to approximately pH 7 required 0.94 mL of 0.10 M NaOH, indicating a total acid content of 0.094 M. Give a plausible explanation for why these two numbers don't agree.

Carbonated beverages like Coke, Pepsi, and 7-Up contain acids, usually phosphoric or citric acid. Propose a procedure to remove the dissolved carbon dioxide from these beverages and determine their total acid content.

BIBLIOGRAPHY

Batson, W. B., and Laswick, P. H. "Pepsin and Antacid Therapy: A Dilemma," *J. Chem. Educ.,* **1979,** *56:* 484–486.

Breedlove, C. H., Jr. "Quantitative Analysis of Antacid Tablets," *Chemistry,* **1972,** *45:* 27–28.

Fuchsman, W. H., and Garg, S. "Acid Content of Beverages: A General Chemistry Experiment Involving the Resolution of Apparent Anomalies," *J. Chem. Educ.,* **1990,** *67:* 67–68. Titration of a variety of beverages makes clear the distinction between the $[H_3O^+]$ and the total acid content of a beverage.

Hem, S. L. "Physicochemical Properties of Antacids," *J. Chem. Educ.,* **1975,** *52:* 383–385.

Lieu, V. T., and Kalbus, G. "Potentiometric Titration of Acidic and Basic Compounds In Household Cleaners," *J. Chem. Educ.,* **1988,** *65:* 184–185. (Note that the two titration curves in Figures 1 and 2 should be interchanged.)

Murphy, J. "Determination of Phosphoric Acid in Cola Beverages: A Colorimetric and pH Titration Experiment for General Chemistry," *J. Chem. Educ.,* **1983,** *60:* 420–421.

The pH Scale

Acid-Base Titrations

The Titration of Stomach Antacids

NAME

SECTION   LOCKER

INSTRUCTOR   DATE

DATA AND CALCULATIONS

1. Standardization of NaOH with Potassium Hydrogenphthalate (KHP)

Data

	Trial 1	Trial 2	Trial 3
Mass of empty flask (or tare mass)	g	g	g
Mass of flask + KHP	g	g	g
Net mass of KHP	g	g	g
Volume of NaOH titrant			
Initial buret reading	mL	mL	mL
Final buret reading	mL	mL	mL
Net volume of NaOH	mL	mL	mL

For each titration, calculate the concentration of NaOH in mol/L:

$$\frac{g\ KHP}{mL\ NaOH} \times \frac{1\ mol\ KHP}{204.23\ g\ KHP} \times \frac{1\ mol\ NaOH}{1\ mol\ KHP} \times \frac{1000\ mL}{1\ Liter} = \frac{mol\ NaOH}{Liter}$$

$\frac{mol\ NaOH}{Liter}$			

Average value of the NaOH molarity _____ mol/L

2. Titration of Vinegar

Data

Brand name of vinegar _____

Acid content of vinegar specified by the manufacturer _____ %

Molarity of standard NaOH _____ mol/L

Give the following data for each titration.

	Trial 1	Trial 2	Trial 3
Mass of empty flask (or tare mass)	g	g	g
Mass of flask + vinegar	g	g	g
Net mass (or volume) of vinegar	g or mL	g or mL	g or mL
Volume of NaOH titrant			
Initial reading of buret	mL	mL	mL
Final reading of buret	mL	mL	mL
Net volume of NaOH	mL	mL	mL

Calculations

The mass percentage of acetic acid in vinegar is given by the expression

$$\text{mass \% acetic acid} = \frac{g \text{ acetic acid}}{g \text{ vinegar}} \times 100$$

(a) For each titration calculate, using Equation (16), the moles of acetic acid in the sample. (Show one complete calculation here.)

moles of acetic acid: _____ mole _____ mole _____ mole

(b) From the number of moles of acetic acid, calculate the number of grams of acetic acid for each titration. (Show one calculation here.)

mass of acetic acid: _____ g _____ g _____ g

(c) Calculate the mass percentage of acetic acid for each titration. (Show one calculation here.) (If you measured the vinegar by volume rather than by weighing it, assume that the density of vinegar is 1.005 g/mL.)

calculated mass% acetic acid in vinegar: _____ % _____ % _____ %

(d) Calculate the average value for the three trials: _____ %

If you have discarded any values, explain why.

(e) What is the difference between your value and the manufacturer's stated value? Comment on the significance of this difference to you as a consumer and to the manufacturer as a producer.

(f) Assuming that the density of vinegar is 1.005 g/mL, calculate the molarity of acetic acid in vinegar from your average value for the mass percentage of acetic acid in vinegar.

3. Titration of Antacid Tablets

Data and calculations	Tablet 1	Tablet 2
Brand name of tablet		
Active ingredient		
Milligrams of active ingredient (from the manufacturer's label)	mg	mg
Mass of tablet + beaker	g	g
Mass of empty beaker	g	g
Net mass of tablet	g	g
Molarity of HCl solution	M	M
Volume of HCl added to tablet	mL	mL
Molarity of NaOH titrant	M	M
Volume of NaOH titrant		
Final buret reading	mL	mL
Initial buret reading	mL	mL
Net volume of NaOH	mL	mL
Millimoles of HCl solution added to tablet	mmol	mmol
Millimoles of NaOH solution required to neutralize the excess HCl	mmol	mmol
Millimoles of HCl neutralized by the antacid tablet	mmol	mmol

Write the balanced chemical equation(s) for the reaction of HCl with the active ingredient in the tablet(s). Calculate from your titration data the millimoles and mass (in milligrams) of active ingredient in the tablet(s). Compare your results with the values given on the manufacturer's label.

Tablet 1: Equation _____

Tablet 2: Equation _____

	Tablet 1	Tablet 2
Calculated mass, m_{calc}, of active ingredient	mg	mg
Milligrams of active ingredient, m_{mfg} (from the manufacturer's label)	mg	mg
Percentage relative difference percentage relative difference $= \dfrac{m_{calc} - m_{mfg}}{m_{mfg}} \times 100\%$	%	%
Mass of inert binder in tablet	mg	mg
Mass percent of inert binder	%	%

☞ Consider This

1. *Using a pH meter, the pH of a cranberry juice cocktail was found to be 2.46, which indicates that the $[H_3O^+] = 3.47 \times 10^{-3}$ M. Titration of a 1.0 mL sample of the juice to approximately pH 7 required 0.94 mL of 0.10 M NaOH, indicating a total acid content of 0.094 M. Give a plausible explanation for why these two numbers don't agree.*

2. *Carbonated beverages like Coke, Pepsi, and 7-Up contain acids, usually phosphoric and/or citric acid. Propose a procedure to remove the dissolved carbon dioxide from these beverages and determine their total acid content.*

3. *In an early TV commercial for Rolaids antacid tablets, the manufacturer claimed that the tablet would neutralize 47 times its own weight of stomach acid. The tablet weighed 1.4 g and contained 0.334 g of the active ingredient, $NaAl(OH)_2CO_3$. Assume that the active ingredient reacts with HCl according to the reaction:*

$$NaAl(OH)_2CO_3 + 4HCl \rightarrow NaCl + AlCl_3 + 3H_2O + CO_2$$

How many moles of HCl would one Rolaids tablet neutralize?

_____ mol

Stomach acid is about 0.14 M HCl and has a density about the same as water. Did a Rolaids tablet neutralize 47 times its own weight of HCl or 47 times its own weight of 0.14 M HCl solution (HCl + water)?

28

Determination of the Molar Mass and Ionization Constant of a Weak Acid

PURPOSE

To determine the molar mass and acid dissociation constant of acetic acid (a weak monoprotic acid) by titration and pH measurement of solutions containing a known mass concentration of acetic acid. To determine the molar mass and acid ionization constant of an unknown weak acid. To learn how to use a pH meter to measure the pH of a solution.

PRE-LAB PREPARATION

Acids and bases are important classes of chemical compounds. They control the pH of living systems and of many reactions carried out in the chemical laboratory. If the pH of the blood of most animals shifts by as little as 0.3 units above or below the normal range of 7.3 to 7.5, severe illness results. Therefore, if we understand the principles of how acids and bases function, we will be better informed about the functioning of biological systems, as well as of purely chemical systems.

In this experiment we will study the properties of weak acids that can ionize (dissociate) by reaction with water to yield one hydronium ion (a protonated water molecule, H_3O^+). Such acids are called weak monoprotic acids. Acetic acid, the active ingredient in vinegar, is a common weak acid belonging to this class. We will determine its molar mass by titration and its ionization constant by making measurements of pure acetic acid solutions and a half-neutralized solution of acetic acid. We can also apply exactly the same principles to determine the molar mass and ionization constant of an unknown weak monoprotic acid.

Acetic acid has the molecular formula

The molecular formula for acetic acid is generally shortened to read CH_3COOH or $HC_2H_3O_2$. These chemical formulas are often further abbreviated to HA, where HA represents acetic acid (or any generic weak monoprotic acid). Then the symbol A^- stands for acetate ion, $CH_3CO_2^-$, the anion of acetic acid (or the anion of any weak monoprotic acid, HA).

The proton that dissociates in acetic acid is the one attached to the oxygen atom. The O—H bond in the acetic acid molecule is appreciably weaker than the C—H bonds, and the polarity of the O—H bond makes it easier for the proton to transfer to a water molecule. In contrast to the behavior of hydrogen atoms bonded to oxygen, there is no significant ionization of the hydrogen atoms bonded to carbon in aqueous solution.

Acetic acid ionizes according to the equation

$$HA + H_2O \rightleftharpoons H_3O^+ + A^- \qquad (1)$$

The equilibrium constant expression is given by

$$K_a = \frac{[H_3O^+][A^-]}{[HA]} \qquad (2)$$

The literature value of the equilibrium constant for this reaction is 1.76×10^{-5} at 25 °C.[1] You are to measure the value of this constant in this experiment and compare your value with the literature value.

A straightforward way to obtain the value of K_a is to measure each of the quantities on the right-hand side of Equation (2) and carry out the indicated arithmetic. For

---

[1] This is the value of K_a, obtained by extrapolating to zero ionic strength. In any real solution of acetic acid, the ionic strength is greater than zero, and the dissociation constant is somewhat larger because of electrostatic interactions between the ions in solution.

solutions containing only the weak acid HA, the value of K_a can be determined by examining the ionization reaction shown in Equation (1) and noting that the concentration of A^- will be equal to that of the H_3O^+ (ignoring the insignificant amount of H_3O^+ that comes from the ionization of water). The HA concentration in the solution is equal to the *stoichiometric* concentration of HA in the solution minus the amount of HA lost by ionization.[2] For acetic acid and most other weak acids, the amount of HA lost by ionization is much smaller than the amount put into solution and thus the equilibrium concentration of HA is only slightly smaller than the stoichiometric concentration.

In solutions that contain both acetic acid (HA) and sodium acetate (Na^+A^-), we will show that the equilibrium concentrations of HA and A^- are approximately equal to their stoichiometric concentrations. (Similar conclusions will apply to any solution containing a mixture of a weak monoprotic acid, HA, and its salt, NA^+A^-.) The stoichiometric concentrations can be readily determined by titration or by knowledge of the quantities of HA and Na^+A^- used to prepare the solutions. If we know the stoichiometric concentrations of HA and A^-, and if the equilibrium concentrations are equal to the stoichiometric concentrations, in order to determine K_a, all that remains is to calculate the $[H_3O^+]$ from the measured pH.

$$pH = -Log[H_3O^+]; [H_3O^+] = 10^{-pH} \qquad (3)$$

MEASURING THE TOTAL ACID CONCENTRATION BY TITRATION

To determine the molar mass of acetic acid, one of the things we need to know is the total amount of acid present in the original sample solution. This can be readily determined by titration of the acetic acid with a strong base of known concentration, such as sodium hydroxide. An acid-base indicator or a pH meter can be used to determine the endpoint of the titration. Figure 28-1 helps to explain how the titration works. It shows how the pH of the solution changes as a strong base, NaOH, is added to the solution containing the weak monoprotic acid.

After an initial small rise in pH, the pH does not change very rapidly as NaOH is added. In the middle of the titration, there is an appreciable concentration of both HA and A^- present, so adding a small amount of base (or acid) does not change the pH very much. For this reason, this is called the *buffering* region.

When an amount of NaOH equivalent to the amount of acid has been added, the concentration of H_3O^+ decreases very sharply with a corresponding sharp increase in the pH. This is the endpoint of the titration, where the indicator changes color. The pH interval of the color change of the indicator must be a good match to the pH at the end point as shown in Figure 28-1.

DETERMINING K_a BY THE METHOD OF HALF-NEUTRALIZATION

At exactly the midpoint of the titration when half of the acid has been neutralized, there will be equal amounts of HA and A^- in solution. This relationship is a good approximation, subject to the condition that [HA] and $[A^-]$ are both greater than 100 K_a, as discussed in greater detail later. From Equation (2) we see that when [HA] = $[A^-]$, the two terms will cancel so that $K_a = [H_3O^+]$ and $pK_a = pH$ at the midpoint of the titration. This provides a simple way to measure the pK_a of a weak acid. If we have a half-neutralized solution of a weak acid, the pH of the solution will be equal to the pK_a of the acid. K_a and pK_a are related like $[H_3O^+]$ and pH:

$$pK_a = -Log(K_a); \qquad K_a = 10^{-pK_a} \qquad (4)$$

FIGURE 28-1

Titration of a weak acid with a strong base. The figure shows the titration of 20 mL of 0.04 M acetic acid with 1.0 M NaOH. Thymol blue is a satisfactory acid-base indicator for this titration.

---

[2]The stoichiometric concentrations of HA and A^- are determined from the amounts of each substance that are put into the solution. However, HA and A^- can each react with water: $HA + HOH \rightleftharpoons HA + OH^-$ and $A^- + HOH \rightleftharpoons HA + OH^-$. Therefore, in general the actual concentrations of HA and A^- at equilibrium (which we will call the equilibrium concentrations) may not be exactly equal to the stoichiometric concentrations.

DETERMINING THE MOLAR MASS OF A WEAK MONOPROTIC ACID

If we know the mass concentration of acid in units of grams HA/L, it is a straightforward procedure to determine the molar mass of the acid (grams HA/mol). We know that the molar mass of HA has units of grams/mol. If we know the concentration of HA in grams/L, then clearly we have to divide this quantity by mol HA/L (or mmol HA/mL) to get the molar mass.

If we take a known volume of HA solution, we can determine the total amount of HA in the sample by titration with a strong base of known concentration. Then, knowing the volume of the original sample (L) and the total amount of HA (moles), we have the necessary information to complete the calculation of the molar mass:

$$\text{Molar mass} = \frac{g\,HA}{Liter} \times \frac{Liter\,HA}{mole\,HA} = \frac{g\,HA}{mole} \quad (5)$$

In a solution of HA and Na^+A^-, the equilibrium concentrations of HA and A^- are approximately equal to their stoichiometric concentrations. By adding sodium hydroxide to a larger amount of a weak acid, we can form a mixture of the weak acid, HA, and its salt, Na^+A^-. To determine the K_a of the weak acid, we must be able to calculate the equilibrium concentrations of HA, A^-, and H_3O^+. In this section we will show how to calculate the stoichiometric concentrations of HA and A^-. We will then show that the equilibrium concentrations are usually equal to the stoichiometric concentrations. In the following section we will discuss how to determine the H_3O^+ concentration by using a pH meter.

To see how we can determine the stoichiometric concentrations of the acid HA and the base A^-, let's suppose that we add a known amount of NaOH to a solution containing a larger (and known) amount of HA. The reaction is

$$HA + Na^+ + OH^- \rightarrow Na^+ + A^- + H_2O \quad (6)$$

(Na^+ ion, which appears on both sides of Equation [6], is simply a spectator ion.)

The *stoichiometric* amount of sodium acetate (Na^+A^-) formed is equivalent to the amount of NaOH added.

$$\text{mol } Na^+A^-_{\text{formed}} = \text{mol NaOH}_{\text{added}} \quad (7)$$

The stoichiometric amount of HA left in solution is obtained easily by subtracting the amount of NaOH added from the original amount of HA present in the solution.

$$\text{mol HA}_{\text{final}} = \text{mol HA}_{\text{original}} - \text{mol NaOH}_{\text{added}} \quad (8)$$

The stoichiometric composition of the solution is precisely the same as would be obtained if we dissolved the same number of moles of acetic acid and sodium acetate as calculated from Equations (5) and (6) and in the same total volume. This is why these concentrations are called stoichiometric concentrations.

Now let's see how the equilibrium concentrations of HA and A^- are related to their stoichiometric concentrations. We will represent these stoichiometric concentrations by a subscript zero: $[HA]_0$ and $[A^-]_0$. We will make use of two equations, one based on a *material balance* and the other based on a *charge balance* condition.

First we write the *material balance* equation that expresses the condition that the sum of the stoichiometric concentrations must equal the sum of the equilibrium concentrations:

$$[HA]_0 + [A^-]_0 = [HA] + [A^-] \quad (9)$$

This equation says that if we put known stoichiometric amounts of HA and A^- in solution, the sum of the stoichiometric concentrations will be exactly equal to the sum of the equilibrium concentrations. This is true because any HA that reacts with water produces A^- and vice versa, as shown by the equations in footnote 2.

The *charge balance* equation expresses the condition that the sum of the positive charges must equal the sum of the negative charges. This is true because the solution as a whole is electrically neutral. (The charge balance equation includes all of the ions in solution, including those that come from the ionization of water.)[3] The charge balance equation for a solution containing HA and Na^+A^- dissolved in water is written as

$$[Na^+] + [H_3O^+] = [A^-] + [OH^-] \quad (10)$$

The stoichiometric concentration of A^- is exactly equal to the concentration of Na^+:

$$[A^-]_0 = [Na^+] \quad (11)$$

This is the same as saying that each mole of NaOH added to a solution containing a larger amount of HA produces a stoichiometric amount of sodium acetate given by Equation (7). Substituting $[A^-]_0$ for $[Na^+]$ in Equation (10) gives

$$[A^-]_0 + [H_3O^+] = [A^-] + [OH^-] \quad (12)$$

---

[3]In the charge balance equation, the concentrations of H_3O^+ and OH^- include the contributions from all sources, both from the ionization of water and the reaction of HA and A^- with water. However, we don't need to worry about keeping separate account of ions coming from water and ions coming from the acid and its salt if they are the same kind of ions.) There is only one equilibrium concentration of H_3O^+ and one equilibrium concentration of OH^- in solution, no matter where they came from.

FIGURE 28-2

The pH meter and its combination pH electrode. (A) The complete apparatus. (B) Detail of the combination pH electrode, which combines a hydrogen ion–sensing glass electrode and reference electrode in a single probe, forming an electrochemical cell. The voltage of the cell is proportional to the logarithm of the ratio of hydrogen ion activities inside and outside the glass bulb.

This result shows that the equilibrium concentration of A^- is equal to the stoichiometric concentration of A^-, provided that the concentrations of H_3O^+ and OH^- are negligible compared to the concentration of A^-. From Equation (9) we see that if $[A^-]_0$ and $[A^-]$ are equal, $[HA]_0$ must equal $[HA]$. This illustrates the general rule that for a solution containing a mixture of HA and Na^+A^-, the equilibrium concentrations of A^- and HA will be equal to the stoichiometric amounts of A^- and HA calculated from Equations (7) and (8) (and divided by the total solution volume) *provided that HA is not too strong an acid* or *its concentration too dilute.* As a good rule of thumb, this rule will apply whenever the stoichiometric concentrations of HA and A^- are both greater than $100K_a$.

USING A pH METER TO MEASURE pH

All pH meters operate by measuring the voltage of an electrochemical cell. The cell is comprised of two half-cell electrodes: a glass-membrane electrode sensitive to the hydrogen ion concentration and a reference electrode.[4] (See Figure 28-2.) The reference electrode is often a sil-

ver-silver chloride electrode. Usually the two half-cell electrodes are combined in a single probe unit called a *combination electrode*. The principle of operation, however, is the same for either type. The silver-silver chloride reference electrode supplies a constant potential ($E° = +0.20$ V vs. the standard hydrogen electrode) determined by the half-reaction

$$AgCl(s) + e^- \rightleftharpoons Ag(s) + Cl^-(3.5\ M) \qquad (13)$$

The $[H_3O^+]$ in the solution determines the potential of the glass electrode. The potential (or voltage) developed across the glass membrane is proportional to the logarithm of the H_3O^+ concentration ratio inside and outside the glass electrode. The pH meter measures the total cell potential across the two half-cell electrodes and displays this measurement on the scale calibrated in pH units (a logarithmic scale, as shown by Equation [3]). When properly standardized with buffers of known pH, a pH meter provides an accurate and simple method for determining the pH of a solution. In Figure 28-2, note the construction details of the combination glass electrode. The thin glass bulb of the glass electrode is very fragile and is often protected with a special plastic guard. Be extremely careful not to hit the glass bulb against the side or bottom of a beaker.

---

[4]Some newer types of pH meters employ a solid state hydrogen ion sensor rather than one based on a glass membrane. The solid state electrodes have the advantage of being more rugged.

EXPERIMENTAL PROCEDURE

Special Supplies: *pH meter, pH 4 and 7 buffers for standardizing the pH meter, 30-mL polyethylene beakers for measuring the pH of solutions, polyethylene transfer pipets.*

Microscale: 1-mL microburet (Air-Tite 1 × 0.01 mL all-polypropylene syringe fitted with a fine tip cut from a fine-tipped polyethylene transfer pipet) or other form of microburet (see the Introduction to the lab manual for a description).

Macroscale: 25- or 50-mL buret.

Chemicals: *2.40 g/L (0.04 M) acetic acid; approximately equivalent molar concentrations of unknown weak monoprotic acids, labeled with their mass concentration (g/L); 0.1% thymol blue indicator (or phenolphthalein).*

Microscale: standardized 1.00 M NaOH.

Macroscale: standardized 0.100 M NaOH.

1. MEASUREMENT OF pH AND TITRATION OF ACETIC ACID SOLUTION.

Your instructor will demonstrate the use of the pH meter. Use 30-mL polyethylene beakers, if available, for holding the samples. Use a wash bottle to rinse the electrode(s) between samples. Excess water can be removed by blotting the bulb with a tissue. (*Do not rub the bulb.*)

Microscale procedure. Obtain 50 mL of 2.40 g/L acetic acid, 6–7 mL of standardized 1.00 M NaOH in a 13 × 100 mm test tube with size 00 rubber stopper, and a 1 mL microburet. Fill the microburet with 1.00 M NaOH and expel any air bubbles. Record the initial volume.

Pour 20mL of the 2.4 g/L acetic acid solution into a clean 30-mL polyethylene beaker. Measure and record the pH. Then transfer the solution to a tared (or weighed) 50-mL Erlenmeyer flask. Using a polyethylene transfer pipet, adjust the mass of acetic acid solution in the flask to 20.0±0.1 g. Record the mass of the sample. Add 5 drops of 0.1% thymol blue indicator. Titrate the sample of acid to the first appearance of a permanent blue color. Record the final volume reading of the microburet.

Repeat the pH measurement and titration with a second 20.0 g sample of 2.4 g/L acetic acid. For each sample calculate the mmole of acid/mL sample. (Assume that each 1.00 g of sample = 1.00 mL.) Average the values for the two samples. If they do not agree within 5 percent, titrate a third sample, and average all three samples. Also average the initial pH values measured for each sample before they were titrated.

Macroscale procedure. Obtain 80 mL of 2.4 g/L acetic acid, 75 mL of standardized 0.10 M NaOH, and a 25- or 50-mL buret. Rinse and fill the buret with 0.10 M NaOH, taking care to expel any bubbles from the tip of the buret. Record the initial volume reading.

Pour 20 mL of the 2.4 g/L acetic acid solution into a clean 30-mL polyethylene beaker. Measure and record the pH. Then transfer the solution to a tared (or weighed) 125-ml Erlenmeyer flask. Add more 2.4 g/L acetic acid and, using a polyethylene transfer pipet, adjust the mass of solution in the flask to 30.0±0.1 g. Record the mass of the sample. Add 5 drops of 0.1% thymol blue indicator. Titrate the sample of acid to the first appearance of a permanent blue color. Record the final volume reading of the buret.

Repeat the pH measurement and titration with a second 30.0-g sample of 2.4 g/L acetic acid. For each sample calculate the mmole of acid/mL sample. (Assume that each 1.00 g of sample = 1.00 mL.) Average the values for the two samples. If they do not agree within 5 percent, titrate a third sample, and average all three samples. Also average the pH values measured for each sample before they were titrated.

Calculations. Calculate the molar concentration of the 2.4 g/L acetic acid (mmol acid/mL of acid) from the average volume of NaOH required to titrate the samples of acetic acid. Calculate the molar mass of acetic acid (g/mol) using Equation (5). Compare with the expected molar mass for acetic acid whose molecular formula is CH_3COOH.

Using the average molar concentration and average pH, calculate the ionization constant for acetic acid. Compare this value with the literature value and with the value measured by the half-neutralization method.

2. MEASUREMENT OF THE pK_a OF ACETIC ACID BY THE HALF-NEUTRALIZATION METHOD.

Weigh into a clean tared Erlenmeyer flask the same size sample of 2.4 g/L acetic acid that you used in Part 1 (*20.0 g for microscale; 30.0 g for macroscale*). Add to the flask exactly half of the average volume of NaOH required to titrate the sample, following the procedure you used in Part 1. Mix thoroughly. This will produce a solution of acetic acid that is exactly half-neutralized. Measure and record the pH of this solution. If the pH meter is properly calibrated, the pH will be equal, to a good approximation, to the pK_a of the acid, as was shown in the Pre-Lab Preparation.

3. THE MOLAR MASS AND IONIZATION CONSTANT OF AN UNKNOWN WEAK MONOPROTIC ACID.

You will be provided with aqueous solutions of an unknown acid, identified only by its mass concentration (g/L). Following the same procedure used in parts 1 and 2, determine the molar mass and ionization constant for the weak monoprotic acid. Assuming the acid is a carboxylic acid of the general formula RCOOH, try to identify the acid from the list of acids given in Table 6 of the Appendix.

☞ Consider This

In the method of half-neutralization for the measurement of the pK_a of an acid, we assume that the equilibrium concentrations of A$^-$ and HA are equal to their stoichiometric concentrations. In order for this to be true, $[H_3O^+]$ and $[OH^-]$ must be negligibly small compared to $[A^-]$. Show that this condition is fulfilled in the measurements you made in Part 2.

By taking the logarithm (base 10) of both sides of Equation (2) and multiplying all of the terms by -1 we can obtain the following equation:

$$-\text{Log}(K_a) = -\text{Log}[H_3O^+] - \text{Log}\frac{[A^-]}{[HA]} \quad (14)$$

which is equivalent to

$$pH = pK_a + \text{Log}\frac{[A^-]}{[HA]} \quad (15)$$

Now if we added 5 percent more than the calculated volume of NaOH required to half-neutralize an acid, the ratio of $[A^-]/[HA]$ would be approximately equal to

1.05/0.95. Show from Equation (15) that this would cause the measured pH to be 0.04 units larger than if we had added the exact amount required. This is probably a fairly realistic estimate of the uncertainty in our measured value of pK_a.

In making the comparison of the ionization constants that you measured with literature values, you should be aware that the interionic attraction effect causes the value of K_a for the solutions used in this experiment to increase (and pK_a to decrease) as the concentration of the ions in solution increases. In brief, the electrostatic attraction of oppositely charged ions acts to decrease the effective concentration (called the activity) of the ions in solution. This causes the ionization constants for weak monoprotic acids of the type we are studying in this experiment to increase by about 15 percent at an ionic strength of 0.02 M. Values of ionization constants tabulated in the literature have usually been determined by extrapolating to zero ionic strength, where interionic attraction effects are negligible.

<table>
<tr><td>REPORT
28</td><td>**Determination of the Molar Mass and Ionization Constant of a Weak Monoprotic Acid**</td><td>NAME

SECTION   LOCKER

INSTRUCTOR   DATE</td></tr>
</table>

DATA AND CALCULATIONS

1. Measurement of pH and Titration of Acetic Acid Solution

Concentration of standardized NaOH titrant _____ mol/L

Mass concentration of acetic acid _____ g/L

	Trial 1	Trial 2	Trial 3
Measured pH of the acetic acid solution			
Mass of acetic acid solution taken for titration	g	g	g
Initial buret reading of NaOH titrant	mL	mL	mL
Final buret reading of NaOH titrant	mL	mL	mL
Net volume of NaOH	mL	mL	mL
Millimoles of NaOH to endpoint of titration	mmol	mmol	mmol
Millimoles of acetic acid in sample	mmol	mmol	mmol
Molar concentration of acetic acid solution	mol/L	mol/L	mol/L
Calculated molar mass of acetic acid	g/mol	g/mol	g/mol

Calculation of the ionization constant for acetic acid from the measured pH of the acetic acid samples

(Average the pH and molar concentration values for the samples of acetic acid you titrated; use the average values in the calculation of the ionization constant.)

(a) Average molar concentration of acetic acid for the samples you titrated: _____ mol/L

Average pH _____

Calculate the corresponding $[H_3O^+]$, compute the concentration of A^- and HA, and calculate the dissociation constant, K_a.

_____ M H_3O^+ _____ M A^- _____ M HA

$$K_a = \frac{[H_3O^+][A^-]}{[HA]} = \text{_____} ; \quad pK_a = \text{_____}$$

2. Measurement of the pK_a of Acetic Acid by the Half-Neutralization Method

Concentration of standardized NaOH titrant _____ mole/L

	Trial 1	Trial 2	Trial 3
Mass of acetic acid solution	g	g	g
Volume of NaOH added to half-neutralize the acetic acid	mL	mL	mL
Measured pH of half-neutralized solution			

Average pH value _____

pK_a = _____ ; K_a = _____

Calculate the $[H_3O^+]$ and the concentrations of A^- and HA, in the half-neutralized solution

(Note: In this solution, $[H_3O^+] \neq [A^-]$)

_____ M H_3O^+ _____ M A^- _____ M HA

3. The Molar Mass and Ionization Constant of an Unknown Weak Monoprotic Acid

Concentration of standardized NaOH titrant _____ mol/L

Mass concentration of unknown weak monoprotic acid _____ g/L

	Trial 1	Trial 2
Measured pH of the unknown acid solution		
Mass of unknown acid solution taken for titration	g	g
Initial buret reading of NaOH titrant	mL	mL
Final buret reading of NaOH titrant	mL	mL
Net volume of NaOH	mL	mL
Millimoles of NaOH to endpoint of titration	mmol	mmol
Millimoles of unknown acid in sample	mmol	mmol
Molar concentration of unknown acid solution	mol/L	mol/L
Calculated molar mass of unknown acid	g/mol	g/mol

Calculation of the ionization constant for acetic acid from the measured pH of the unknown acid samples

(Average the pH and molar concentration values for the samples of unknown acid you titrated; use the average values in the calculation of the ionization constant.)

(a) Average molar concentration of unknown for the samples you titrated _____ mole/L

Average pH _____

Calculate the corresponding $[H_3O^+]$, compute the concentration of A^- and HA, and calculate the dissociation constant, K_a

_____ M H_3O^+ _____ M A^- _____ M HA

$$K_a = \frac{[H_3O^+][A^-]}{[HA]} = \text{_____} ;   pK_a = \text{_____}$$

Measurement of the pK_a of an Unknown Acid by the Half-Neutralization Method

Concentration of standardized NaOH titrant _____ mole/L

	Trial 1	Trial 2
Mass of unknown acid solution	g	g
Volume of NaOH added to half-neutralize the unknown acid	mL	mL
Measured pH of half-neutralized solution		

Average pH value _____

pK_a = _____ ; K_a = _____

Calculate the [H_3O^+] and the concentration of A$^-$ and HA, in the half-neutralized solution

(Note: In this solution, [H_3O^+] $\neq$ [A$^-$])

_____ M H_3O^+ _____ M A$^-$ _____ M HA

Comparing both the calculated molar mass and the ionization constant, which acid in Table 6 of the Appendix is the unknown acid most likely to be?

Give the molar mass and pK_a for the unknown acid that you measured:

Give the molar mass and K_a for the acid listed in Table 6 of the Appendix that you think is the same as your unknown acid.

👉 **Consider This**

In the method of half-neutralization for the measurement of the pK_a of an acid, we assume that the equilibrium concentrations of A$^-$ and HA are equal to their stoichiometric concentrations. In order for this to be true, [H_3O^+] and [OH^-] must be negligibly small compared to [A$^-$]. Show that this condition is fulfilled in the measurements you made in Part 2.

By taking the logarithm (base 10) of both sides of Equation (2) and multiplying all of the terms by -1 we can obtain the following equation:

$$-\text{Log}(K_a) = -\text{Log}[H_3O^+] - \text{Log}\frac{[A^-]}{[HA]} \qquad (14)$$

which is equivalent to

$$pH = pK_a + \text{Log}\frac{[A^-]}{[HA]} \qquad (15)$$

Now if we added 5 percent more than the calculated volume of NaOH required to half-neutralize an acid, the ratio of $[A^-]/[HA]$ would be approximately equal to 1.05/0.95. Show from Equation (15) that this would cause the measured pH to be 0.04 units larger than if we had added the exact amount required. This is probably a fairly realistic estimate of the uncertainty in our measured value of pK_a.

The OH^- concentration of 0.050 M ammonia, NH_3, is 9.5×10^{-4} M.

(a) Write the chemical equation for the dissociation of NH_3 in water.

(b) Write the equilibrium constant expression for the reaction of NH_3 with water.

(c) Calculate the value of K_b using the information given.

Calculate the H_3O^+ concentration and pH of a carbonated beverage that is 0.10 M in dissolved CO_2. *(Essentially all of the H_3O^+ comes from the first stage of dissociation, $CO_2(aq) + 2H_2O \rightleftharpoons H_3O^+ + HCO_3^-$, for which $K_1 = 4.4 \times 10^{-7}$.)*

_____ M

_____ pH

The Chemistry of Natural Waters
Water Hardness by EDTA Titration
Water Softening by Ion Exchange

PURPOSE

To measure water hardness caused by dissolved calcium and magnesium ions in a sample of tap water. To show how water can be softened by an ion exchange process that replaces calcium and magnesium ions with sodium ions. To show how the chemical compositions of river water and seawater are largely determined by geochemistry.

PRE-LAB PREPARATION

It would be hard to imagine life without water. It is the medium in which all of the biological processes of living cells take place. Water is so essential that it can justifiably be called the most valuable of all our resources. Because we breathe oxygen, we might think that it is more essential to life, but that is a human-centered view. Many simple organisms can live without oxygen, but no organism can live and reproduce without water.

There is a lot of water on earth (an estimated 1.4×10^{21} L), but it is mostly seawater. An estimated 97.2 percent is in the oceans, and about 2.1 percent is trapped in the polar icecaps and glaciers, leaving about 0.7 percent of the water in the hydrosphere as the fraction on which we must rely. Our supply of fresh water would soon be exhausted except for the continual renewal of this resource by the solar energy–driven cycle of evaporation, condensation, and precipitation as rain and snow.

THE CHEMICAL COMPOSITION OF NATURAL WATERS

In an area such as the Amazon River basin in South America, far removed from volcanic activity and the smokestacks of industry, rainwater can be very pure, containing a few dissolved substances: mainly carbon dioxide, traces of nitrogen oxides produced by lightning, and

traces of NaCl. Near the sea, the rainfall will contain more NaCl. Although NaCl is not volatile, the interaction of wind and waves produces tiny droplets of seawater (called an *aerosol*) that are carried inland by the wind.

In areas near active volcanoes and in the industrialized countries, precipitation also contains appreciable concentrations of dissolved acids. These arise mostly from the combustion of fossil fuels—for heating, cooking, transportation, industrial processing, and the generation of electricity. Nitrogen oxides and sulfur dioxide (mainly from burning coal that contains sulfur) are produced in the primary combustion process. These are further oxidized to nitric and sulfuric acids in a complex series of reactions that take place in the atmosphere.

Table 29-1 gives the average composition of rainfall in the northeastern United States, for 88 monthly weighted-ion concentrations collected at 17 sampling sites. These data clearly show the presence of substantial quantities of sulfuric and nitric acids. In a region where the soils lack alkaline components that can neutralize the acids, this "acid rain" produces destructive ecological effects, damaging vegetation and killing fish and other aquatic life.

The composition of river water is mainly determined by the reactions of rainwater (and water from melting snow) with rocks and soil as the water makes its way downhill to the sea. *Gypsum* ($CaSO_4 \cdot 2H_2O$) and *calcite* ($CaCO_3$) are commonly found in sedimentary deposits at the earth's surface. Another less stable form of calcium carbonate, *aragonite*, is found in sediments that consist mainly of the mineral skeletons of marine organisms. A relatively common mineral in ancient rocks is *dolomite*, $MgCa(CO_3)_2$. Gypsum is slightly soluble in water, and all of the carbonates can be readily dissolved by water containing traces of acid. Unless the rainfall is acidic from industrial emissions, the main acid present in rainwater is dissolved CO_2, which forms a slightly acidic solution.

TABLE 29-1

The Average Chemical Composition of Natural Waters[a]

	Rainfall, northeastern United States (mmol/L)	Rivers of North America (mmol/L)	Seawater (mmol/L)
Cations			
Ca^{2+}	—	0.53	10.6
Mg^{2+}	—	0.21	54.5
Na^+	0.009	0.39	479.
K^+	—	0.036	10.2
H^+	0.072	—	—
NH_4^+	0.016	—	—
Anions			
HCO_3^-	—	1.1	2.3
SO_4^{2-}	0.028	0.21	28.9
Cl^-	0.012	0.23	546.
Br^-	—	—	0.85
F^-	—	0.008	0.07
NO_3^-	0.026	0.017	0.0001
$H_2SiO_4^{2-}$	—	0.15	—

[a]Quoted by J. N. Butler, *Carbon Dioxide Equilibria and Their Applications,* Addison-Wesley, Reading, MA, 1982, Chapter 5.

$$CO_2(g) + 2\,H_2O(l) \rightleftharpoons H_3O^+(aq) + HCO_3^-(aq) \quad (1)$$

In turn, the acid formed reacts with rocks containing calcium (and magnesium) carbonates.

$$H_3O^+(aq) + CaCO_3(s) \rightleftharpoons$$
$$Ca^{2+}(aq) + HCO_3^-(aq) + H_2O(l) \quad (2)$$

The net reaction is the dissolution of calcium carbonate as given by Equation (3), which is the sum of Equations (1) and (2).

$$CO_2(g) + H_2O(l) + CaCO_3(s) \rightleftharpoons$$
$$Ca^{2+}(aq) + 2\,HCO_3^-(aq) \quad (3)$$

These geochemical processes largely account for the fact that calcium and magnesium ions are the most abundant cations and bicarbonate ion is the most abundant anion in river water, as shown in Table 29-1.

By far the most abundant ionic solute in seawater is sodium chloride (see Table 29-1). This should not be too surprising, because sodium chloride and other sodium salts are both abundant and very soluble. Operating over eons of time, the natural hydrologic cycle has by now washed into the sea most of the soluble sodium salts that lie near the surface of the earth.

WATER HARDNESS AND WATER SOFTENING

In most regions of the United States, water for domestic use comes mainly from rivers, lakes, and wells and contains some dissolved salts, mostly calcium bicarbonate resulting from the reaction shown in Equation (3). Water also contains lesser amounts of sodium, chloride, and sulfate ions. Both Ca^{2+} and Mg^{2+} ions react with soaps (the carboxylate anions of long-chain fatty acids) to form an insoluble curdy scum.

$$Ca^{2+} + 2CH_3(CH_2)_{16}\overset{\overset{\displaystyle O}{\|}}{C}O^- \longrightarrow$$

$$[CH_3(CH_2)_{16}\overset{\overset{\displaystyle O}{\|}}{C}O^-]_2Ca^{2+} \quad (4)$$

This precipitate reduces the cleaning efficiency of soap and is responsible for the ring of scum on the tub when you take a bath. Water that contains calcium and magnesium ions is called hard water, and if the water hardness is too great, it's desirable to "soften" the water by decreasing its content of calcium and magnesium.

Water softening can be done in several ways. Before the advent of synthetic detergents (which do not form insoluble calcium and magnesium salts), washing soda ($Na_2CO_3 \cdot 10H_2O$) was often added to laundry water to remove calcium ions by the reaction

$$Ca^{2+}(aq) + CO_3^{2-}(aq) \rightarrow CaCO_3(s) \quad (5)$$

Another method, called *ion exchange,* replaces calcium ion with sodium ion by using a natural or synthetic solid ion exchanger. The most efficient materials for this purpose are synthetic organic resins that have a hydrocarbon framework to which charged functional groups are chemically bound (see Figure 29-1). If the bound groups are negatively charged, the negative charge is counterbalanced by positive ions (remember the electroneutrality principle). Because the positive ions (called *cations*) are not chemically bound to the resin, they can be replaced by cations having an equivalent positive charge. Such a resin is called a *cation-exchange resin.* If the bound functional groups have positive charges, the counterions will have a negative charge, and the resin is called an *anion-exchange resin.* The following equations show the ion-exchange equilibria for cation- and anion-exchange reactions.

Cation exchange (resin originally in the H^+ form):

$$2RZ{-}SO_3^-H^+ + Ca^{2+} \rightleftharpoons (RZ{-}SO_3^-)_2Ca^{2+} + 2H^+ \quad (6)$$

Anion exchange (resin originally in the OH^- form):

$$RZ{-}N(CH_3)_3OH^- + Cl^- \rightleftharpoons$$
$$RZ{-}N(CH_3)_3Cl^- + OH^- \quad (7)$$

(In these equations the symbol RZ represents the organic polymer matrix [or resin] to which the charged groups are bound.)

If equivalent amounts of cation-exchange resin (in H^+ form) and anion-exchange resin (in the OH^- form) are mixed together, the combination will remove all of the dissolved salts, because the H^+ produced by cation exchange exactly neutralizes the OH^- produced by anion exchange. The resulting water that emerges from the column or tank is pure *deionized* water. In many areas, deionization is a simpler and less costly alternative to distillation. Deionization also is widely used for purifying water used in chemical laboratories.

In domestic water softening by ion exchange, total deionization is not necessary. Only the cations responsible for water hardness need be removed. So only a cation-exchange resin is used, usually in the Na^+ form. The ion-exchange reaction is given by

$$2RZ{-}SO_3^-Na^+ + Ca^{2+} \rightleftharpoons$$
$$(RZ{-}SO_3^-)_2Ca^{2+} + 2Na^+ \quad (8)$$

The ion-exchange reactions are reversible. In domestic water softening systems, NaCl is used to regenerate the cation-exchange resins because it is readily available and inexpensive. The resin is flooded with the NaCl brine, which drives the equilibrium shown in Equation (8) to the left. The calcium and magnesium and excess brine are then flushed out with water, leaving the resin in the Na^+ form.

To regenerate a mixed cation-anion–exchange deionization system, it is first necessary to separate the two resins. They are then regenerated separately, the cation-exchange resin being returned to the H^+ form by treatment with a strong acid (such as H_2SO_4), and the anion resin being returned to the OH^- form by treatment with a strong base (such as NaOH).

To be suitable for use in drinking, cooking, bathing, and watering plants, water should not contain more than 500 parts per million (ppm) of dissolved solids (1 ppm = 1 mg/L). The total dissolved solids can be measured by evaporating a known volume of water to dryness and weighing

FIGURE 29-1

Deionization of a salt solution using a mixture of cation- and anion-exchange resins. The cation-exchange resin is originally in the H^+ form, and the anion-exchange resin in the OH^- form. In this example, $CaCl_2$ solution is poured into the mixed resin bed at the top. Pure water flows out at the bottom. Each Ca^{2+} ion displaces two H^+ ions and each Cl^- displaces one OH^- ion. Exactly equal numbers of H^+ and OH^- ions are produced, which combine to form water. The zigzag lines represent the organic polymer matrix of the resin (most often a styrene-divinylbenzene copolymer) to which the sulfonate ($^-SO_3^-$) and quaternary ammonium [$^-N + (CH_3)_3$] groups are chemically bonded.

the residue. (An approximate value can be obtained more quickly by measuring the electrical conductivity of the water, since most of the dissolved solids are ionic.)

DETERMINING WATER HARDNESS BY EDTA TITRATION

The water hardness (sum of calcium and magnesium ions) is most often measured by titration, using the anion of a polyaminocarboxylic acid called ethylenediaminetetraacetic acid. This name is quite a mouthful, so it is usually abbreviated to EDTA. EDTA is a tetraprotic acid that we will represent by the symbol H_4Y. Usually this substance is employed as the soluble disodium salt dihydrate ($Na_2H_2Y \cdot 2H_2O$), which is available in high purity. The anion, H_2Y^{2-}, reacts with Ca^{2+} or Mg^{2+} (or almost any metal ion with charge $+2$ or greater) to form a complex ion.

$$H_2Y^{2-} + Ca^{2+} \rightleftharpoons CaY^{2-} + 2\,H^+ \qquad (9)$$

The structures of the anion H_2Y^{2-} and the complex with a metal ion, M^{2+}, are shown in Figure 29-2. Think of the EDTA anion as surrounding a metal ion like the large claw of a crab grasping its prey. The technical term for such a complex is *chelate*, which comes from the Greek word meaning "a pincerlike claw."

To determine the end point of the titration, an indicator dye is added that also complexes Ca^{2+} and Mg^{2+}, forming a colored indicator–metal ion complex. As the titrant (disodium EDTA) is added, it complexes the metal ions, but so long as the metal ion is in excess, the indicator-metal ion complex is still present. When precisely enough EDTA has been added to react with all of the metal, the EDTA displaces the indicator from the metal-indicator complex. This happens because the equilibrium constant for the formation of the EDTA-metal ion complex is much larger than that for the indicator-metal ion complex. This produces a reasonably sharp color change as the indicator is converted from the color of the Mg^{2+}-indicator complex (red) to that of the free (or unbound) indicator (blue). In pH 10 buffer the net reaction is described by the equation

$$H_2Y^{3-} + MgIn^- \underset{\text{red}}{\overset{\text{pH 10}}{\rightleftharpoons}} MgY^{2-} + HIn^{2-} \qquad (10)$$

where In^{3-} represents the indicator anion.

In this experiment you will determine the hardness of a sample of tap water by titration with EDTA. You will then treat a sample of the water with a cation exchange resin in the sodium ion form and titrate a sample of the soft water to see how much hardness remains. You will also observe the effects of untreated and softened water on the foaming properties of a soap.

FIGURE 29-2

Structures of ethylenediaminetetraacetic acid (EDTA) and metal-EDTA chelates: (A) structure of the dianion of EDTA, H_2Y^{2-}; (B) structure of an M^{2+}—EDTA chelate, MY^{2-}.

EXPERIMENTAL PROCEDURE

Special Supplies: 1-mL microburet, two polyethylene transfer pipets (25 drops/mL); two 10-mL Erlenmeyer flasks, 5-mL serological or volumetric pipet

Chemicals: 0.0100 M $Na_2EDTA \cdot 2H_2O$[1], 0.01 M $Na_2MgEDTA$/0.1 M NH_3 solution, 1.5 M NH_3/0.3 M NH_4Cl buffer, 2.0% ascorbic acid solution (1 g/50 mL, freshly made each week), Calmagite indicator (0.1 g per 100 mL), 20- to 50-mesh strongly acidic (sulfonated polystyrene) cation-exchange resin in the Na^+ form (Dowex 50WX4-50 or Amberlite IR-120), 0.1% soap solution (1 g/L of Ivory flake or bar soap).

1. DETERMINATION OF WATER HARDNESS BY EDTA TITRATION.

(a) Rinse and fill the 1-mL microburet with 0.0100 M EDTA ($Na_2EDTA \cdot 2H_2O$, not the 0.01 M MgEDTA solution). Add *in order* to a 25-mL Erlenmeyer flask: 5.0 ± 0.1 mL of tap water (this is the sample), 1.0 mL (measured with a polyethylene transfer pipet) of 1.5 M NH_3_ 0.3 M NH_4Cl buffer (the reaction

---

[1] $Na_2EDTA \cdot 2H_2O$ is ethylenediaminetetraacetic acid disodium salt dihydrate. It is available in purity greater than 99 percent and can be weighed out to make a solution whose concentration is known within 1 percent. If greater accuracy is desired, it can be standardized against pure $CaCO_3$ by the procedure described in Part 1. EDTA salts, Calmagite indicator, and Dowex ion exchange resin are available from Aldrich Chemical Co. and other suppliers.

produces H^+, so the buffer is necessary to keep the H^+ concentration at the optimum value), about 0.1 mL (2 drops) of 2% ascorbic acid solution (to prevent oxidation of the indicator), 1.0 ml (measured with a transfer pipet) of 0.01 M $Na_2MgEDTA$/0.1 M NH_3 solution,[2] and 3 drops of 0.1% Calmagite indicator.

Record the initial buret reading. Titrate the solution in the flask with 0.0100 M Na_2EDTA until the last traces of red are gone and the solution is a pure blue color. The color change near the endpoint typically requires about 1–2 drops of 0.0100 M Na_2EDTA. Record the final buret reading. Repeat the titration until you have three titrations that agree within a range of 10 percent of the titrant volume.

(b) Determine the indicator blank by repeating the titration procedure, substituting 5.0 mL of deionized water for the tap water. Record the initial buret reading. Titrate the solution with 0.0100 M Na_2EDTA (it should require only a few drops). Record the volume of the blank as zero if the solution is already blue before any 0.01 M EDTA is added. Record the final buret reading. Subtract the volume required for the indicator blank from the volumes required to titrate the tap water samples to determine the net volume of 0.0100 M Na_2EDTA required to titrate the tap water samples.

Calculate the water hardness in units of millimoles per liter. This quantity is the sum of the calcium and magnesium ion concentrations.

Water quality engineers typically express water hardness in units equivalent to parts per million of $CaCO_3$ (equal to milligrams of $CaCO_3$ per liter). Express your results in these units by assuming that the ions titrated by the EDTA were derived from $CaCO_3$.

$$\frac{\text{mg } CaCO_3}{L} = \frac{\text{mmol } Ca^{2+} + Mg^{2+}}{L} \times$$

$$\frac{1 \text{ mol } CaCO_3}{\text{mol } Ca^{2+} + Mg^{2+}} \times \frac{100 \text{ g } CaCO_3}{\text{mol } CaCO_3} \quad (11)$$

In most natural waters some magnesium ion is present, so expressing water hardness in units of milligrams of $CaCO_3$ per liter is an arbitrary convention that has been adopted for the sake of convenience (or tradition).

---

[2]Because the indicator complexes magnesium ion more strongly than it does calcium ion, the color change at the endpoint will be sharper if Mg^{2+} is present. Although most natural waters will contain some magnesium ion, we deliberately add some to ensure that an adequate amount of magnesium ion will be present. It is added in the form of the Mg-EDTA complex so that no extra EDTA will be required to titrate the added magnesium ion.

[3]The exchange capacity is usually found on the label of the bottle of resin, expressed in milliequivalents (meq) per gram of dry resin or in milliequivalents per milliliter of resin bed. A typical value for the resin you will be using is about 1.7 meq/mL, which means that 1 mL of a packed slurry of the resin can exchange about 1.7 mmol of charge (1.7 mmol of Na^+ ion or 0.85 mmol of Ca^{2+} ion).

2. WATER SOFTENING BY ION EXCHANGE.

Ion-exchange processes are usually carried out by flowing a sample down a bed of ion-exchange resin contained in a column. This procedure gives the most complete exchange because as the sample solution moves down the column it is continually contacting fresh resin, which drives the equilibrium shown in Equation (8) completely to the right.

However, we will use a batch procedure because of its simplicity. To obtain near-quantitative ion exchange, we will use a quantity of resin whose exchange capacity is 50 times larger than the quantity of ions to be exchanged.[3]

Another factor that tends to drive the exchange reaction (Equation [8]) toward completion: Ca^{2+} ions have a charge greater than that of Na^+ ions, so Ca^{2+} ions are more strongly attracted to the negatively charged groups bonded to the resin.

To proceed with the ion exchange, put 50 mL of tap water in a clean 125-mL Erlenmeyer flask. Add 2 g of dry cation-exchange resin (or 3 g if the resin is wet). Swirl the mixture in the flask for 8 to 10 min to allow time for the ion exchange to take place. Save the soft water (supernatant solution) for the following experiments.

When you have finished parts 2a and 2b, pour the resin into a bottle labeled "Used Resin." *Do not pour the resin down the drain or back into the original container.* It is expensive and can be reused after regeneration.

(a) *Soap Sudsing in Hard and Soft Water.* Transfer 5 mL of the soft water to a clean 20 × 150 mm test tube. Measure 5 mL of tap water into a second test tube and 5 mL of deionized water into a third tube. Add exactly 5 mL of 0.1% soap solution to each of the three tubes; stopper and shake them vigorously for a few seconds. Compare the solutions for the amount and stability of the foam and for the appearance of any insoluble calcium soap scum. Record your observations.

(b) *Efficiency of Ion Exchange.* Titrate two 5.0-mL samples of the soft water, using the procedure described in Part 1 for the titration of tap water. (Add the 0.0100 M Na_2EDTA solution drop by drop. You should not have to add much, because most or all of the calcium and magnesium ions have been removed.) Calculate the percentage of water hardness removed by ion exchange.

percentage of hardness removed equals

$$\frac{\frac{\text{mmol}}{L}(Ca + Mg)_{tap} - \frac{\text{mmol}}{L}(Ca + Mg)_{soft}}{\frac{\text{mmol}}{L}(Ca + Mg)_{tap}} \times 100\% \quad (12)$$

where the subscripts *tap* and *soft* refer to the concentrations of calcium + magnesium ions (in millimoles per liter) in tap water and soft water, respectively.

Consider This

The standard EDTA solution that you used in this lab exercise is a general-purpose metal-ion titrant. Develop procedures to determine the metal-ion concentration of solutions of interest, such as milk, soft drinks, creek water, or the water entering and exiting a water softener. Do your values match those on the food labels or the water company's annual report?

About once a month, take water samples from a nearby creek or from your home or school tap throughout the year. Analyze these samples for hardness. Do the values fluctuate in a predictable way? Propose a hypothesis for your local behavior.

EDTA can be found as an ingredient in cosmetics, shampoos, and such. Research the reason for this addition and the mechanism that allows EDTA to act as a preservative in these settings.

Use the Merck Index or Web sources to discover how EDTA is used medicinally. Is it also toxic? What is the mechanism of EDTA's medicinal properties? What is the mechanism of its toxicity?

BIBLIOGRAPHY

Belcher, R., Close, R. A., and West, T. S. "The Complexometric Titration of Calcium in the Presence of Magnesium. A Critical Study," *Talanta*, **1958**, *1:* 238–244 .

Lindstrom, F., and Diehl, H. "Indicator for the Titration of Calcium plus Magnesium with (Ethylenedinitrilo) tetraacetate," *Analytical Chemistry*, **1960**, *32:* 1123–1127.

McCormick, P. G. "Titration of Calcium and Magnesium in Milk with EDTA," *J. Chem. Educ.*, **1973**, *50:* 136–137.

Chemistry of Natural Waters

Water Hardness by EDTA Titration

Water Softening by Ion Exchange

NAME	
SECTION	LOCKER
INSTRUCTOR	DATE

DATA AND CALCULATIONS

1. Determination of Water Hardness by EDTA Titration

Data	Trial 1	Trial 2	Trial 3
(a) Volume of tap water sample	mL	mL	mL
Volume of 0.0100 M Na₂EDTA titrant Final buret reading	mL	mL	mL
Initial buret reading	mL	mL	mL
Net volume of EDTA solution	mL	mL	mL
(b) Volume of deionized water (indicator blank)	mL	mL	mL
Volume of 0.0100 M Na₂EDTA titrant Final buret reading	mL	mL	mL
Initial buret reading	mL	mL	mL
Net volume of EDTA solution (Indicator blank)	mL	mL	mL

Calculate the concentration of $Ca^{2+} + Mg^{2+}$ in the tap water sample in millimoles per liter. Be sure to subtract the net volume of the indicator blank measured in Part 1b from the net volume measured in Part 1a. Show a sample calculation in the space below.

_____ mmol/L   _____ mmol/L   _____ mmol/L

Average _____ mmol/L

Using the average value of the $Ca^{2+} + Mg^{2+}$ concentration in millimoles per liter, calculate the water hardness in units of milligrams of $CaCO_3$ per L (parts per million of $CaCO_3$). (See Equation [11].)

Water hardness _____ mg $CaCO_3$/L(ppm)

2. Water Softening by Ion Exchange

(a) *Soap Sudsing in Hard and Soft Water*

Compare the amount and stability of the foam produced by shaking soap solution with each of the following samples.

  Tap water

  Soft water

  Deionized water

(b) *Efficiency of Ion Exchange*

Data	Trial 1	Trial 2	Trial 3
(a) Volume of soft water sample			
Volume of 0.0100 M Na$_2$EDTA titrant Final buret reading			
Initial buret reading			
Net volume of Na$_2$EDTA solution			

Calculate the concentration of Ca^{2+} + Mg^{2+} in the soft water sample in millimoles per liter. Be sure to subtract the net volume of the indicator blank measured in Part 1b from the net volume measured in Part 2b. Show a sample calculation in the space below.

_____ mmol/L   _____ mmol/L   _____ mmol/L

Average _____ mmol/L

Calculate the percentage of water hardness removed. (See Equation [12].)

Percentage of water hardness removed _____ %

EXERCISES

1. One 10-min shower using a restricted-flow shower head (2.5 gal/min) requires 25 gal of water, and it has been estimated that each person uses about 60 gal of water a day for personal use (mostly bathing and washing). Calculate the volume of cation-exchange resin (in liters) that would be required to remove all the water hardness in the water used by one person in 30 days at 60 gal/day. Assume that the exchange capacity of the resin is 1.7 meq/mL of resin and that the water hardness is equivalent to 200 mg $CaCO_3$ per liter. (Remember that 2 meq of exchange capacity are required per millimole of $CaCO_3$.) 1 gal = 3.785 L.

2. Describe how a cation-exchange resin in the H^+ form and a standardized solution of NaOH might be used to determine the cation content of a solution containing a mixture of NaCl and KNO_3. (**Hint:** if you pass the NaCl and KNO_3 solutions through a column containing a cation exchange resin in the H^+ form, all of the cations will be exchanged for H^+. What will be the composition of the resulting solution?)

☞ **Consider This**

The standard EDTA solution that you used in this lab exercise is a general-purpose metal-ion titrant. Develop procedures to determine the metal-ion concentration of solutions of interest, such as milk, soft drinks, creek water, or the water entering and exiting a water softener. Do your values match those on the food labels or the water company's annual report?

About once a month, take water samples from a nearby creek or from your home or school tap throughout the year. Analyze these samples for hardness. Do the values fluctuate in a predictable way? Propose a hypothesis for your local behavior.

EDTA can be found as an ingredient in cosmetics, shampoos, and such. Research the reason for this addition and the mechanism that allows EDTA to act as a preservative in these settings.

Use the Merck Index *or Web sources to discover how EDTA is used medicinally. Is it also toxic? What is the mechanism of EDTA's medicinal properties? What is the mechanism of its toxicity?*

The Solubility Product Constant of Calcium Iodate, $Ca(IO_3)_2$

PURPOSE

To understand the relation between the molar solubility and the solubility product constant of a sparingly soluble salt. To measure the molar solubility of calcium iodate in pure water and to determine the solubility product constant. To investigate the common ion effect by measuring the molar solubility of calcium iodate in a solution containing added potassium iodate.

PRE-LAB PREPARATION

When soluble ionic compounds are dissolved in water, the solution usually contains just the ions that were present in the solid salt. For example, in a saturated solution of KNO_3, the solution contains only K^+ ions and NO_3^- ions, and the dissolved salt is completely dissociated.[1]

$$KNO_3(s) \rightleftharpoons K^+(aq) + NO_3^-(aq) \qquad (1)$$

When a salt is sparingly soluble, it suggests that the attractive forces between the ions in the solid are comparable to the attractive forces between the ions and the solvent (usually water). Many sparingly soluble salts do not completely dissociate into their constituent ions but interact in solution with one another to form soluble aggregates, called *complexes,* or react with water to form new species.

For example, in a saturated solution of calcium sulfate ($CaSO_4$), in addition to $Ca^{2+}(aq)$ and $SO_4^{2-}(aq)$ ions, a significant fraction of the dissolved calcium sulfate exists

as molecular $CaSO_4(aq)$. This soluble form of $CaSO_4$ may be a molecule with polar covalent bonding or a soluble ion pair, $Ca^{2+}SO_4^{2-}(aq)$. Thus, in this system, it takes two equilibria (and two equilibrium constant expressions) to adequately describe the system:

$$CaSO_4(s) \rightleftharpoons Ca^{2+}(aq) + SO_4^{2-}(aq) \qquad (2)$$

$$CaSO_4(s) \rightleftharpoons CaSO_4(aq) \qquad (3)$$

Here's a second example. A saturated solution of $CaCO_3$ contains appreciable concentrations of HCO_3^- and OH^-, as well as Ca^{2+} and CO_3^{2-} ions. In this system, carbonate ion (CO_3^{2-}) reacts with water to produce bicarbonate ion and hydroxide ion. As a result, it takes two equilibria (and two equilibrium constant expressions) to describe the system:

$$CaCO_3(s) \rightleftharpoons Ca^{2+}(aq) + CO_3^{2-}(aq) \qquad (4)$$

$$CO_3^{2-}(aq) + H_2O \rightleftharpoons HCO_3^-(aq) + OH^-(aq) \qquad (5)$$

The behavior shown in Equation (5) is typical of all salts (such as carbonates, sulfides, and phosphates) in which the anion can react with H_2O to form a weak acid that is only partially ionized (or dissociated).

The two examples whose solubility equilibria are described by Equations (2) through (5) show that it is not always safe to assume that only the ions produced by complete dissociation of the salt are present. The solubility behavior of every salt must be determined by an experimental investigation.

To summarize, we emphasize two important points about solubility equilibria: (a) *The identity of all of the solution species in equilibrium with the solid must be determined by experiment.* It is not safe to assume that all salts

---

[1] In solid KNO_3 each ion is surrounded by ions of the opposite charge. This is till largely true when the KNO_3 is dissolved in water, but the ions are farther apart, and each ion is surrounded by a cluster of water molecules that are bound to the ion. The ions in solution are therefore said to be "hydrated" or "aquated." (The generic term is "solvated.")

dissociate completely into their constituent ions; (b) *The solubility product constant can be calculated directly from the molar solubility* only *if the dissolved salt completely dissociates into its constituent ions and if these ions neither react with one another to form complexes nor react with water to form weak acids or bases.*

For the two example salts we described ($CaSO_4$, $CaCO_3$), it requires two equilibria to describe each system. A measurement of just the molar solubility does not provide enough information to calculate two equilibrium constants. Fuller discussions of solubility equilibria can be found in the references listed in the bibliography.

THE RELATION BETWEEN MOLAR SOLUBILITY AND THE SOLUBILITY PRODUCT CONSTANT

Calcium iodate, a salt whose solubility you will study in this experiment, has been shown to completely dissociate in water into calcium ions, Ca^{2+}, and iodate ions, IO_3^-.[2]

$$Ca(IO_3)_2(s) \rightleftharpoons Ca^{2+}(aq) + 2IO_3^-(aq) \qquad (6)$$

So, a single equilibrium describes the system, and the value of the equilibrium constant can be calculated from the molar solubility.

Equation (6) shows the reaction for the dissociation of solid $Ca(IO_3)_2$ in a saturated aqueous solution. The equilibrium constant expression for this reaction (often called the *solubility product* expression) is written as shown in Equation (7), observing the convention that the activity of the pure solid is taken to be equal to 1.00 so that it does not appear in the equilibrium constant expression.[3]

$$K_{sp} = [Ca^{2+}][IO_3^-]^2 \qquad (7)$$

For any salt that dissolves to give just the ions originally present in the salt, there is a simple relation between the molar solubility of the salt and the solubility product constant of the salt. The concentration of each ion will be equal to the *molar solubility* or to some multiple of it.

What do we mean by the molar solubility of the salt? This quantity is just the concentration of the dissolved salt, expressed in units of moles per liter. In the calcium iodate example, the calcium ion concentration is equal to the molar solubility, and the iodate ion concentration is equal to *two* times the molar solubility. This is true because each mole of calcium iodate that dissolves gives one mole of calcium ions and two moles of iodate ions.

If we let the symbol s represent the molar solubility of calcium iodate (in units of moles per liter), the concentrations of Ca^{2+} and IO_3^- ions will be related to the molar solubility, s, by the following equations:

$$[Ca^{2+}] = s \qquad (8)$$

$$[IO_3^-] = 2s \qquad (9)$$

If we substitute for the concentration of each ion its equivalent in units of molar solubility, s, by substituting Equations (8) and (9) into Equation (7), we obtain an equation that shows that there is a simple relation between the K_{sp} and the molar solubility of the salt.

$$K_{sp} = [Ca^{2+}][IO_3^-]^2 = s(2s)^2 = 4s^3 \qquad (10)$$

A frequently asked question about this expression is, "Why are you doubling the concentration of IO_3^- and then squaring it?" The short answer is, "We aren't doubling the concentration of IO_3^-. Its concentration is equal to 2s."

THE SOLUBILITY OF $Ca(IO_3)_2$ IN KIO_3 SOLUTION: THE COMMON ION EFFECT

From Le Châtelier's principle we would predict that the molar solubility of calcium iodate would be smaller in a solution of potassium iodate, KIO_3, which is a strong electrolyte that completely dissociates in water. The hypothesis is that the addition of KIO_3 would shift the equilibrium shown in Equation (6) toward the left, decreasing the amount of calcium iodate that dissolves. Such a decrease in solubility that occurs when a salt is dissolving in a solution that already contains one of the salt's ions is called the *common ion effect.*

To test this hypothesis you will measure the solubility of calcium iodate in 0.01 M KIO_3. Under these conditions the concentrations of the ions will be related to the molar solubility of $Ca(IO_3)_2$ in the following way (compare Equations [11] and [12] with Equations [8] and [9]):

$$[Ca^{2+}] = s \qquad (11)$$

$$[IO_3^-] = 0.01 + 2s \qquad (12)$$

Note that all of the calcium ion must come from dissolved calcium iodate. However, the iodate ion comes from both KIO_3 and dissolved $Ca(IO_3)_2$. As before, we get *two* moles of iodate ions and *one* mole of calcium ions for every mole of calcium iodate that goes into solution.

---

[2] The actual composition of the solid that is in equilibrium with the ions in solution is the solid hexahydrate, $Ca(IO_3)_2 \cdot 6H_2O$.

[3] The effective concentration, or "activity," of a pure solid that participates in a chemical reaction is constant. (Adding more solid salt to the system does not change the concentration of ions in the solid.) By thermodynamic convention the pure solid is chosen as the thermodynamic reference state, which is equivalent to assigning the value 1.00 to the activity of the solid.

The relation between the solubility product constant and the molar solubility will also be changed (compare Equation [13] with Equation [10]).

$$K_{sp} = [Ca^{2+}][IO_3^-]^2 = s(0.01 + 2s)^2 \quad (13)$$

If we know the concentration of potassium iodate, a single measurement of the total iodate concentration allows us to calculate the molar solubility of calcium iodate and the solubility product constant.

DETERMINATION OF THE MOLAR SOLUBILITY OF CALCIUM IODATE

When calcium iodate dissolves in solution, both the solid and the solution remain electrically neutral.[4] This is called the *electroneutrality* principle and means that in the case of calcium iodate, two IO_3^- ions will go into solution for each Ca^{2+} ion that goes into solution. So molar solubility can be determined by measuring either the concentration of calcium ion or the concentration of iodate ion. We will use a procedure for measuring the IO_3^- concentration that makes use of the fact that IO_3^- oxidizes iodide ion.

$$IO_3^- + 5I^- + 6H^+ \rightarrow 3I_2 + 3H_2O \quad (14)$$

The I_2 produced is in turn titrated with sodium thiosulfate.

$$I_2 + 2S_2O_3^{2-} \rightarrow 2I^- + S_4O_6^{2-} \quad (15)$$

Note from the overall stoichiometry that each mole of IO_3^- will produce enough I_2 to consume 6 moles of $S_2O_3^{2-}$. Therefore the concentration of IO_3^- ion will be given by

$$[IO_3^-], \frac{mol}{L} = \frac{mL\ S_2O_3^{2-}}{mL\ IO_3^-} \times$$

$$\frac{mol\ S_2O_3^{2-}}{L} \times \frac{1\ mol\ IO_3^-}{6\ mol\ S_2O_3^{2-}} \quad (16)$$

EXPERIMENTAL PROCEDURE

Special Supplies: *Microscale:* 1-mL microburet (see Introduction at the front of the lab manual), 1-mL serological pipet (also a 2-ml pipet, if students are to standardize the sodium thiosulfate titrant), pipet-filling devices, fine-tipped (50 drops/mL) polyethylene transfer pipets; polyethylene transfer pipets with 1-mL calibration marks. *Macroscale:* 25- or 50-mL buret, two 10-mL pipets or graduated cylinders.

Chemicals: *Microscale:* standardized 0.05 M sodium thiosulfate (Na$_2$S$_2$O$_3$) solution. *Macroscale:* standardized 0.15 M sodium thiosulfate (Na$_2$S$_2$O$_3$) solution.

KI(s), 1 M HCl, 0.1% starch indicator solution, saturated solution of Ca(IO$_3$)$_2$ · 6H$_2$O in pure water, saturated solution of Ca(IO$_3$)$_2$ · 6H$_2$O in 0.0100 M KIO$_3$. The saturated solutions should be prepared a week in advance; see Notes to Instructor.

NOTES TO INSTRUCTOR: Instructions are given for both *microscale* titrations, using a 1-mL microburet, and *macroscale* titrations, using a 25- or 50-mL buret. (Direct students to the Introduction in the front of the lab manual for instructions in the use of burets and pipets.)

Microscale: The iodate-containing sample volumes may be measured using 1- and 2-ml serological pipets or by weighing the samples on a balance with milligram accuracy (1.00 g = 1.00 mL). Use 0.15 M sodium thiosulfate (Na$_2$S$_2$O$_3$) as the titrant.

Macroscale: If you prefer to use a 25- or 50-mL buret, increase the volume of iodate-containing samples to 10 mL in Parts 1, 2, and 3, and use 0.05 M sodium thiosulfate (Na$_2$S$_2$O$_3$) as the titrant.

Sodium thiosulfate pentahydrate (Na$_2$S$_2$O$_3$ · 5H$_2$O) that has been kept in a well-sealed bottle is normally pure enough so that a solution made up by accurately weighing the salt will be within 1 percent of the target concentration, so standardization of the solution is not absolutely necessary. As an option, students may standardize the nominal 0.15 M (or 0.05 M for macroscale) Na$_2$S$_2$O$_3$ if they are supplied with a solution of 0.0100 M KIO$_3$. The titration is performed as described to determine the molar solubility of calcium iodate.

The solid in equilibrium with the saturated solution of calcium iodate is the hexahydrate. We have found, however, that using anhydrous calcium iodate to prepare the saturated solutions gives the same results. Although anhydrous calcium iodate is available from a commercial supplier of laboratory chemicals (G. Frederick Smith Chemical Co.), a quantity of the hexahydrate sufficient for 100 students is easily prepared. Dissolve 0.5 mol (112 g) of KIO$_3$ in 600 mL of hot water in a liter beaker. To the hot solution, slowly add with stirring 0.25 mol of Ca(NO$_3$)$_2$ (or CaCl$_2$) dissolved in 200 mL water. Allow to cool. Filter off the solid on a large Büchner funnel, and wash the solid with three 50-mL portions of cold water. Allow the solid to air-dry, or use the moist cake to prepare the saturated solutions.

The saturated solution in pure water is prepared by adding 8 g Ca(IO$_3$)$_2$ · 6H$_2$O per liter of water. The saturated solution in 0.0100 M KIO$_3$ is prepared by adding 2.14 g KIO$_3$ + 4 g Ca(IO$_3$) · 6H$_2$O per liter of water. Stir the solutions for at least 24 h; allow to stand for 48 h before use. Ideally, the solutions should be thermostatted at 25 °C, but if this is not possible, a temperature correction can be made to estimate the molar solubility at 25 °C, using data at a different temperature (see Question 3 in the Consider This section of the

---

[4] Separation of positive and negative ions on a large scale would produce such a huge force of attraction between the positive and negative charges that any separation is prevented, except on a microscopic scale, where the distance of separation is on the order of atomic dimensions.

Report form). Use a siphon, or centrifuge individual 5-mL samples so that students receive samples free of particles of solid calcium iodate.

1. STANDARDIZATION OF THE SODIUM THIO-SULFATE SOLUTION (OPTIONAL).

Microscale: Follow the microscale procedure described in Part 2, substituting 2.0-mL samples of 0.0100 M KIO$_3$ for the 1.0-mL samples of saturated Ca(IO$_3$)$_2$ solution. (The volume of 0.15 M sodium thiosulfate required is between 0.7 and 0.9 mL.)

Macroscale: Follow the macroscale procedure described in Part 2, using 10-mL samples of 0.0100 M KIO$_3$. (The volume of 0.05 M sodium thiosulfate required is between 10 and 14 mL.)

Titrate at least two samples. Calculate the exact molar concentration of sodium thiosulfate, keeping in mind that one mole of KIO$_3$ produces enough I$_2$ to consume six moles of sodium thiosulfate. The calculated molarities for the two samples should agree to within 5 to 10 percent.

2. THE MOLAR SOLUBILITY OF Ca(IO$_3$)$_2$ IN PURE WATER.

If you did not perform the standardization procedure in Part 1, record the exact concentration of the sodium thiosulfate (Na$_2$S$_2$O$_3$) titrant. Also record the temperature of the saturated calcium iodate solution. Read the pages of the Introduction describing how to use burets and pipets.

Microscale: Fill a 1-mL microburet with the 0.15 M sodium thiosulfate solution, taking care to expel all bubbles from the buret. Record the initial buret reading. Put 5 mL of deionized water in a 10- or 25-mL Erlenmeyer flask. Add 0.2 g KI, and swirl to dissolve.

Obtain a 5-mL sample of saturated calcium iodate solution in a 13 × 100 mm test tube. (If the sample shows any visible solid particles, it should be centrifuged to settle them to the bottom of the tube.) Then, using a pipet-filling device to fill the pipet, transfer 1.00 mL of the saturated solution into the Erlenmeyer flask. (A 1.00-mL serological pipet may be used or, if one is not available, weigh out 1.00 ± 0.02 g [1 g = 1 mL] of sample directly into the Erlenmeyer flask, using a fine-tipped polyethylene pipet to dispense the sample solution.) Then add 1 mL of 1 M HCl, using a calibrated polyethylene transfer pipet. The solution should turn brown. Without delay, titrate the solution with sodium thiosulfate until the solution is yellow. Then add 1 mL of 0.1% starch indicator. The solution should turn blue-black, the color of the starch-iodine complex. Continue the titration until you get a sharp change from blue to a colorless solution. Record the final buret reading.

Repeat the titration procedure with a second sample. The volumes should agree to within 5 to 10 percent. (A possible source of error is the air oxidation of I$^-$ to give I$_2$. If the solution is allowed to stand for too long before it is titrated, the oxidation will produce values that are too high.)

Macroscale: Fill a 25- or 50-mL buret with 0.050 M sodium thiosulfate solution, taking care to rinse the buret and to expel all bubbles from the tip of the buret. Obtain about 30 mL of saturated calcium iodate solution in a clean, dry flask. (The sample must not show any visible solid particles of undissolved calcium iodate.) Put 50 mL deionized water and 2 g KI into a 250-mL Erlenmeyer flask, and swirl to dissolve the KI. Then, using a pipet-filling device to fill the pipet, transfer 10 mL of the saturated calcium iodate solution into the 250-mL flask, and add 10 mL 1.0 M HCl. (10-mL graduated cylinders may be used to measure the saturated calcium iodate and 1.0 M HCl solutions.) When the HCl is added, the solution should turn brown. Without delay, titrate the solution with 0.05 M sodium thiosulfate until the solution is yellow. Then add 5 mL of 0.1% starch indicator. The solution should turn blue-black, the color of the starch-iodine complex. Continue the titration until you get a sharp change from blue to a colorless solution. Record the final buret reading.

Repeat the titration procedure with a second sample. The volumes should agree to within 5 percent. (A possible source of error is the air oxidation of I$^-$ to give I$_2$. If the solution is allowed to stand for too long before it is titrated, the oxidation will produce values that are too high.)

Calculations: Calculate the concentration of iodate ion in the sample of saturated solution (see Equation [16]). The molar solubility of calcium iodate will be equal to one half the iodate concentration (see Equations [8] and [9]). Calculate the solubility product constant for calcium iodate (see Equation [10]).

3. THE MOLAR SOLUBILITY OF CALCIUM IODATE IN 0.0100 M POTASSIUM IODATE.

Record the temperature of the saturated solution. Record the initial and final buret readings for each titration.

Microscale: Using the microscale procedure described in Part 2, titrate two 1.00-mL samples of the saturated solution of Ca(IO$_3$)$_2$ in 0.0100 M KIO$_3$.

Macroscale: Using the macroscale procedure described in Part 2, titrate two 10-mL samples of the saturated solution of Ca(IO$_3$)$_2$ in 0.0100 M KIO$_3$.

Calculations: Calculate the total iodate concentration as in Part 2. Subtract 0.0100 M from the total concentration to obtain the concentration of iodate ion that comes from dissolved calcium iodate. Divide this result by 2 to obtain the molar solubility of calcium iodate. Is it smaller, as we predicted from Le Châtelier's principle? Calculate the solubility product constant for calcium iodate (see Equation [13]). Does the calculated value of K_{sp} agree with the value for K_{sp} calculated in Part 2? Note that a 10 percent error in measuring the molar solubility, s, will lead to a larger error in the calculation of K_{sp}, because K_{sp} depends on s^3, and cubing the number magnifies the error.

Consider This

Using Le Châtelier's principle, we predicted that the solubility of calcium iodate would decrease in a solution containing a common ion. Now let's think of how we might increase the solubility of calcium iodate, producing a solution that contains more total calcium and total iodate dissolved in solution.

Let's think first about the anion, iodate ion. If the anion is the conjugate base of a weak acid, we might try increasing the solubility of calcium iodate by adding a strong acid such as HCl or HNO₃ to protonate the IO_3^- ion. This would only work if an appreciable amount of undissociated HIO₃ is formed. What about HIO₃, called hydrogen iodate or iodic acid? Is it a strong acid (one that completely dissociates) or a weak acid that has a measurable ionization constant?

Now let's think about the cation, calcium ion. Could we find something that forms a soluble complex with calcium ion or that otherwise interacts with it, for example, forming an ion pair with calcium ion? 1,2-Diols, such as ethylene glycol (1,2-ethanediol) have been reported to form soluble complexes with calcium ion. The chelating agent EDTA (see Experiment 29, The Chemistry of Natural Waters) is known to form soluble complexes in basic solution with nearly any metal ion having +2 or +3 charge.

Finally, there is another rather unexpected possibility. The addition of any soluble "inert" salt (one that completely dissociates) to a solution of any slightly soluble salt will increase its solubility. This is an ion-ion interaction effect or interionic attraction effect. In this context the added "inert" salt is one in which there is no specific chemical interaction between the ions of the inert salt and the ions of the slightly soluble salt—that is, no complexation or ion-pair formation, etc.

The calcium iodate system is ideal for study of these ion-ion interactions because calcium ion has a +2 charge, and such interaction effects turn out to depend on the square of the charge. As a result, the interaction effect of a calcium ion is four times that of an ion with +1 charge.

How does this effect operate? Adding "inert" ions to the solution increases the total concentration of ions, therefore increasing the ionic strength of the solution. Because ions are charged they all interact through the attractive and repulsive forces caused by their charges, causing each ion to become surrounded by a cluster of ions of the opposite charge. Whenever an ion moves, it must drag along its baggage of ions of opposite charge, restricting its ability to act as an independent particle. This in effect decreases the activity, or effective concentration, of the ion. Taking account of these effects quantitatively is usually done by introducing an activity coefficient, γ, which is a measure of the departure from ideal behavior:

$$\text{Activity (effective concentration) of } M^+ = a_{M+} = \gamma\,[M^+]$$

Interionic attraction effects and activity coefficients are not usually discussed in general chemistry textbooks. How significant is the effect of ionic strength on solubility? Ramette (see the Bibliography) cites measurements that show that the molar solubility of calcium iodate at 25 °C increases by a factor of nearly two in 0.5 M KCl compared to its solubility in pure water, certainly a significant effect.

BIBLIOGRAPHY

Butler, J. N., and Cogley, D. R. *Ionic Equilibrium: Solubility and pH Calculations*, John Wiley & Sons, Inc., New York, 1998. Everything you need or want to know about ionic equilibria.

Cavaleiro, A. M. V. S. V. "The Teaching of Precipitation Equilibrium: A New Approach," *J. Chem. Educ.*, **1996**, 73: 423–425. Case studies of systems where the anion of the salt reacts with water to form a weak acid and hydroxide ion.

Clark, R. W., and Bonicamp, J. M. "The K_{sp}-Solubility Conundrum," *J. Chem. Educ.*, **1998**, 75: 1182–1185. A critique of solubility and K_{sp} data presented in general chemistry textbooks.

Hawkes, S. J. "Salts Are Mostly *NOT* Ionized," *J. Chem. Educ.*, **1996**, 73: 421–423.

Hawkes, S. J. "What Should We Teach Beginners about Solubility and Solubility Products?," *J. Chem. Educ.*, **1998**, 75: 1179–1181.

Meites, L.; Pode, J. S. F.; Thomas, H. C. "Are Solubilities and Solubility Products Related?," *J. Chem. Educ.*, **1996**, 43: 667–672. A classic study of the solubility of calcium sulfate including the effects of ion pairing and interionic attraction (ionic activity effects).

Ramette, R. W. *Chemical Equilibrium and Analysis*, Addison-Wesley Publishing Co., Inc., Reading, MA, 1981, pp. 109–112. A case study of calcium iodate solubility and ionic strength (interionic attraction) effects.

The Solubility Product Constant of Calcium Iodate, Ca(IO$_3$)$_2$

NAME

SECTION   LOCKER

INSTRUCTOR   DATE

DATA AND CALCULATIONS

1. Standardization of Sodium Thiosulfate (Optional)

Volume of 0.0100 M KIO$_3$ samples _____ mL

Data	Trial 1	Trial 2	Trial 3
Volume of Na$_2$S$_2$O$_3$ titrant			
Final buret reading	mL	mL	mL
Initial buret reading	mL	mL	mL
Net volume of Na$_2$S$_2$O$_3$	mL	mL	mL
Calculated concentration of Na$_2$S$_2$O$_3$	M	M	M

For each trial, calculate the concentration of Na$_2$S$_2$O$_3$; show a sample calculation in the space below.

Average Na$_2$S$_2$O$_3$ concentration _____ M

(If you did not perform the standardization procedure in Part 1, record here the concentration of the sodium thiosulfate titrant, Na$_2$S$_2$O$_3$.)

2. The Molar Solubility of Ca(IO$_3$)$_2$ in Pure Water

Temperature of the saturated solution of calcium iodate: _____ °C

Volume (or mass) of saturated calcium iodate solution titrated: _____ mL (or g)

Data	Trial 1	Trial 2	Trial 3
Volume of Na$_2$S$_2$O$_3$ titrant			
Final buret reading	mL	mL	mL
Initial buret reading	mL	mL	mL
Net volume of Na$_2$S$_2$O$_3$	mL	mL	mL
Calculated concentration of IO$_3^-$	M	M	M

For each trial, calculate the concentration of IO$_3^-$; show a sample calculation in the space below.

Average IO$_3^-$ concentration _____ M

Calculate the molar solubility of $Ca(IO_3)_2$ in pure water from the average value of the IO_3^- concentration.

Molar solubility: _____ mol/L

Calculate the solubility product constant, K_{sp} for a saturated solution of $Ca(IO_3)_2$ in water. (See Equation [10].)

$$K_{sp} = \text{_____}$$

3. The Molar Solubility of Calcium Iodate in 0.0100 M Potassium Iodate

Temperature of the saturated solution of calcium iodate: _____ °C

Volume (or mass) of saturated calcium iodate solution titrated: _____ mL (or g)

Data	Trial 1	Trial 2	Trial 3
Volume of $Na_2S_2O_3$ titrant Final buret reading	mL	mL	mL
Initial buret reading	mL	mL	mL
Net volume of $Na_2S_2O_3$	mL	mL	mL
Calculated concentration of IO_3^-	M	M	M

For each trial, calculate the concentration of IO_3^-; show a sample calculation in the space below.

Average IO_3^- concentration _____ M

Subtract the concentration of IO_3^- ion that came from the KIO_3 from the average value of the total IO_3^- concentration to get the iodate ion concentration that came from dissolved $Ca(IO_3)_2$.

Total IO_3^- concentration _____ M

IO_3^- concentration from KIO_3 _____ M

IO_3^- concentration from dissolved $Ca(IO_3)_2$ _____ M

Calculate the molar solubility of $Ca(IO_3)_2$ in 0.0100 M KIO_3 solution.

Molar solubility _____ mol/L

Compare the molar solubility of $Ca(IO_3)_2$ in 0.0100 M KIO_3 solution with the molar solubility in pure water, determined in Part 2. Is the change consistent with Le Châtelier's principle?

Calculate the solubility product constant, K_{sp} for a saturated solution of $Ca(IO_3)_2$ in 0.0100 M KIO_3. (See Equation [18].)

Considering the magnitude of possible errors in the determination of molar solubility (of the order of 10 percent maximum error in careful work), is there reasonable agreement between the K_{sp}'s calculated in parts 1 and 2?

☞ **Consider This**

Propose the composition of a solution in which you would expect to find that the molar solubility of $Ca(IO_3)_2$ is greater than in pure water. Briefly and succinctly explain why you would expect to find a greater solubility in the solution you propose.

Calculate the molar solubility, s, of calcium iodate in 0.020 M $Ca(NO_3)_2$, a completely dissociated strong electrolyte. (NO_3^- ion does not chemically interact with either Ca^{2+} or IO_3^-.) Assume that K_{sp} for $Ca(IO_3)_2 = 2.0 \times 10^{-6}$.

To set up the problem, we can write the following equations:

$$\text{Material balance for calcium: } [Ca^{2+}] = s + 0.020$$

$$\text{Material balance for iodate: } [IO_3^-] = 2s$$

$$K_{sp} = 2.0 \times 10^{-6} = [Ca^{2+}][IO_3^-]^2 = (s + 0.020)(2s)^2$$

The last equation contains only one unknown, the value of s, which we would like to calculate; but s is not negligible compared to 0.020, which leaves us with a nasty equation that is cubic in s. An approach that often works in a situation like this is to rearrange the equation into a more useful form and obtain s by iteration. Thus, we can write

$$4s^2 = \frac{K_{sp}}{s + 0.020} \qquad or \qquad 8s = \frac{1}{2}\left(\frac{K_{sp}}{s + 0.020}\right)^{1/2}$$

We must find the value of s that makes both sides of the equation equal. Start first by inserting a trial value of s (say, 0.006 M) in the right-hand side of the latter equation and calculating a value of s. You will get s 5 0.00438. For your next trial value, take something about halfway between 0.00438 and 0.006. Continue the iteration process until you calculate a value of s that is nearly the same (within 5 percent as the value you inserted on the right-hand side of the equation).

If your data for solubility were obtained at a temperature different from 25 °C, estimate the solubility at 25 °C from your data. Around 25 °C, the molar solubility of calcium iodate in water increases with temperature about 4.5 percent per degree. Calculate the molar solubility at 25 °C, s_{25}, by correcting your value determined at t °C, which we will call s_t, using the equation

$$s_{25} = s_t \left[\frac{2 - 0.045(t - 25)}{2 + 0.045(t - 25)} \right]$$

where t is the temperature (°C) of the saturated solution that you analyzed. Ramette (see the Bibliography) quotes the molar solubility of $Ca(IO_3)_2$ in pure water at 25 °C as 0.00798 mol/L. Compare your result with the literature value by calculating the percentage relative difference:

$$\text{percentage relative difference} = \frac{s_{25} - 0.00798}{0.00798} \times 100\%$$

387

31

Identification of Silver, Lead, and Mercury(I) Ions

PURPOSE

To provide an overview of a general scheme for separating and identifying ten cations. To introduce the laboratory techniques used in qualitative analysis. To begin the study of the first group of three cations, Ag^+, Pb^{2+}, and Hg_2^{2+} ions, which are separated out as insoluble chloride salts.

PRE-LAB PREPARATION

Qualitative analysis is a branch of analytical chemistry that identifies particular substances in a given sample of material. In the analysis of inorganic substances, this branch involves the analysis of both metallic constituents as cations and nonmetallic constituents as anions. This and the next five experiments constitute a series in qualitative analysis. Experiments 31–33 deal with the identification of ten common cations, including Ag^+, Pb^{2+}, Hg_2^{2+}, Fe^{3+}, Al^{3+}, Zn^{2+}, Ca^{2+}, Ba^{2+}, K^+, and NH_4^+. Experiment 34 explores the identification of nine anions, NO_3^-, Cl^-, Br^-, I^-, SO_4^{2-}, SO_3^{2-}, S^{2-}, PO_4^{3-}, and CO_3^{2-}. Experiment 35 presents an interesting challenge: the identification of solutions contained in unlabeled bottles. The final experiment in this series, Experiment 36, illustrates a more modern separation technique, paper chromatography, by studying the separation of Fe^{3+}, Co^{2+}, Ni^{2+}, and Cu^{2+}.

Qualitative analysis has remained an important part of the laboratory experience in general chemistry for a number of years, even though the analytical methods have been replaced by sophisticated instrumental methods in practical analysis. We believe that qualitative analysis has two useful purposes: (1) It provides an ideal context for the illustration and application of the principles of ionic equilibria, such as acid-base equilibria and solubility and complex ion equilibria, and (2) it provides a logical and systematic framework for the discussion of the descriptive chemistry of the elements. Its utility aside, qualitative analysis is fun.

ORGANIZATION OF THE SCHEME FOR QUALITATIVE ANALYSIS

Of the approximately 80 metallic elements, 24 of the more common cations are usually included in the scheme of analysis. In the abbreviated scheme presented in this and the next two experiments, we will study the separation and identification of only ten cations. This permits an extensive introduction to the more commonly employed separative techniques in only three 3-hour laboratory periods and does not require handling the dangerous and offensive hydrogen sulfide gas, H_2S.

All qualitative schemes begin with the separation of Ag^+, Hg_2^{2+}, and Pb^{2+} ions as the chlorides. This is made possible by their low solubility and small solubility product constants:

Metal Chloride	K_{SP}
AgCl	1.8×10^{-10}
Hg_2Cl_2	1.1×10^{-18}
$PbCl_2$	1.6×10^{-5}

These ions constitute Group 1 in most schemes of qualitative analysis.

We will not analyze for any of the eight elements generally included in Group 2. These divalent or trivalent elements are all located near the amphoteric or semi-metal element dividing line of the periodic table. They are isolated as sulfides from a 0.3 M H_3O^+ solution that has been saturated with H_2S. The solubility product constants of the sulfides range from 10^{-26} to 10^{-93}.

Group 3 is generally separated as a mixture of hydroxides and sulfides by precipitation from a solution containing NH_3 or both NH_3 and NH_4HS. In the abbreviated scheme employed here, Fe^{3+} and Al^{3+} are first precipitated as hydroxides, and Zn^{2+} is subsequently separated as the sulfide by addition of NH_4HS solution.

Finally, groups 4 and 5 are separated by precipitation of the alkaline earth elements, Ba^{2+} and Ca^{2+}, as the carbonates leaving the alkali metal K^+ and the complex ion that behaves like an alkali metal, NH_4^+, in solution for individual detection.

Table 31-1 presents an overall scheme to illustrate this separation into groups. It is important for you to study the flowchart in this table so that you understand the basis for separating the cations into five groups. This will be especially true if you proceed through the analysis scheme to Experiment 35, where you will analyze a general unknown containing elements from all groups.

It happens that just five steps are required for the analyses in each of the three cation experiments (Experiments 31, 32, and 33). These are sequentially numbered to assist you in constructing an overall flowchart and in analyzing a general unknown. In addition, flowcharts are provided for each experiment. The one in this experiment has been completed; those in the next two are left for your study; you will complete them as a part of your laboratory report.

LABORATORY TECHNIQUES FOR QUALITATIVE ANALYSIS

The oft-repeated admonition *keep your lab bench neat and orderly* will pay big dividends in time saved and in fewer mistakes. Study Figure 31-1. Keep the bench top and the area in front of the sink clear of unnecessary equipment. Arrange conveniently the most frequently used items, such as clean test tubes, your wash bottle, and stirring rods, polyethylene transfer pipets, and indicator test paper, laid out on a clean towel. Keep dirty test tubes and other articles in one place. Clean and rinse these at the first opportunity so that a stock of clean equipment is always ready. Label any solutions that are to be kept for more than a minute or two. *Keep your laboratory records up-to-date, and do your thinking as you work.* Rushing through experimental work without thinking about it, with the idea of understanding it later, is an ill-advised attempt at economy of time.

ESTIMATE THE VOLUME OF SOLUTIONS. Use only the small volumes specified. You can *estimate* most volumes with sufficient precision by counting drops (20 to 25 drops per milliliter) or by estimating the height of the solution in the 100-mm test tube (capacity 8 mL). Use the 10-mL graduated cylinder only when more accurately known volumes are required.

THE HANDLING OF SOLUTIONS. Learn to be clean and efficient. Review the sections on basic laboratory equipment and procedures in the Introduction. When mixing solutions, use a stirring rod (Figure 31-2); do not invert the test tube with your unprotected, and possibly contaminated, thumb as the stopper. When heating solutions, avoid loss by bumping (Figure 31-3). Do not try to boil more than 2 mL of solution in a 10-cm test tube; transfer larger quantities to a 15-cm test tube, a small beaker, or a casserole. The safest way to heat a solution in a test tube is to immerse the tube in a beaker of boiling water.

WASHING PRECIPITATES. When it is desirable to isolate a precipitate and free it from interfering ions in the supernatant in order to carry out further tests on it, you must first wash the precipitate. This is done by centrifuging it to the bottom of a test tube, removing the supernatant solution by carefully pouring it off or by drawing it off with a polyethylene transfer pipet, adding a few milliliters of pure water, mixing as shown in Figure 31-2, and recentrifuging the precipitate. The washing can be repeated, if necessary, to remove all interfering ions.

Ag^+, Pb^{2+}, AND Hg_2^{2+} IONS

The precipitating reagent for this group is chloride ion in an acid solution. Only three of all the common metal ions form a precipitate on the addition of this reagent: Ag^+, Pb^{2+}, and Hg_2^{2+}. Notice that the mercury(I) ion, sometimes called mercurous ion, is not the simple ion Hg^+,

TABLE 31-1

The Abbreviated Scheme for the Separation of Ten Metal Ions

Wash bottle

Soiled apparatus
to be washed

Clean work space

Clean test tubes inverted
on filter paper

Indicator papers

Clean towel

Polyethylene transfer pipets and stirring rods

FIGURE 31-1
Keep your laboratory worktable neat and in order, with the "tools of your trade" conveniently arranged.

This

or this

Don't poke out
the bottom.

Agitate sideways.

NEVER this

Heat the upper part of the liquid
in the test tube and agitate
sideways to prevent
the liquid from bumping
violently.

FIGURE 31-2
Learn to use good technique in mixing solutions.

FIGURE 31-3
Use this technique when you heat a solution in a test tube.

but consists of two mercury atoms held together by a covalent bond, with a double positive charge.

Ammonia solution gives different results with each of the three metal chlorides, as shown by the preliminary experiments that follow. The behavior of mercury salts is unusual and needs some explanation. Mercury exhibits three oxidation states: 0 in free mercury, Hg^0; +1 in mercury(I), or mercur*ous* ion, Hg_2^{2+}, and +2 in mercury(II), or mercur*ic* ion, Hg^{2+}. In some situations, the intermediate mercury(I) ion is unstable, because part of the mercury has been reduced to the metal and part has been oxidized to Hg^{2+}, thus:

$$Hg_2^{2+} \rightleftharpoons Hg^0(l) + Hg^{2+}$$

When ammonia is added to mercury(II) chloride solution, a white "ammonolysis" product, $HgNH_2Cl$, is formed. This is analogous to the partial hydrolysis of $HgCl_2$ to form a basic salt. Compare the two equations:[1]

$$HgCl_2(aq) + HOH \rightleftharpoons Hg(OH)Cl(s) + H^+ + Cl^-$$

$$HgCl_2(aq) + HNH_2(aq) \rightleftharpoons Hg(NH_2)Cl(s) + H^+ + Cl^-$$

If mercury(I) chloride, Hg_2Cl_2, is treated with ammonia, part of it is oxidized to the white Hg(II) amidochloride, above, and part is reduced to black, finely divided mercury metal, Hg(0):

$$Hg_2Cl_2(s) + 2NH_3(aq) \rightleftharpoons$$

$$Hg(NH_2)Cl(s) + Hg(l) + NH_4^+ + Cl^-$$

The mixed precipitates appear black or dark gray.

EXPERIMENTAL PROCEDURE

Special Supplies: Centrifuges

Chemicals: 0.1 M sodium chloride, NaCl; 0.1 M lead(II) nitrate, $Pb(NO_3)_2$; 0.1 M mercury(II) chloride, $HgCl_2$; 0.05 M mercury(I) nitrate, $Hg_2(NO_3)_2$; 0.1 M potassium bromide, KBr; 1 M potassium chromate, K_2CrO_4; 0.1 M potassium iodide, KI; 0.1 M silver nitrate, $AgNO_3$; 1 M sodium thiosulfate, $Na_2S_2O_3$; 6 M hydrochloric acid, HCl; 6 M ammonia, NH_3; 6 M sodium hydroxide, NaOH; 6 M nitric acid, HNO_3.

NOTE TO INSTRUCTOR: If your laboratory is not equipped with centrifuges, you may prefer to have your students filter rather than centrifuge the precipitates. In that case, explain that students may substitute the verb *filter* for the verb *centrifuge*.

---

[1]In this and subsequent experiments, we will use H^+ at times to represent the species H_3O^+, present in an acidic, aqueous solution. We do this for the sake of simplicity. In this context, the chemical meaning of the two symbols is the same.

A. TYPICAL REACTIONS OF THE IONS

1. SOLUBILITY OF THE CHLORIDES. Prepare a sample of each of the three chlorides by adding 2 mL of 0.1 M NaCl to 2 mL each of 0.1 M $AgNO_3$, 0.05 M $Hg_2(NO_3)_2$, and 0.1 M $Pb(NO_3)_2$. If a precipitate fails to form in any of these (why?), add 1 mL of 6 M HCl. Let the precipitates settle, then decant (separate the solution from the solid; see Figure 31-4), and discard the supernatant liquid. Add to each about 2 mL of distilled water. Heat each nearly to boiling and shake. Record your results. Cool under the cold water tap. Describe your results. Save for the addition of ammonia.

Centrifuged sample

Remove liquid by decanting or by using a transfer pipet.

To wash, add H_2O, mix, centrifuge again, and discard washings.

FIGURE 31-4

The procedure for washing a precipitate that has been centrifuged.

2. BEHAVIOR WITH AMMONIA. To each of the three precipitates add 1 mL of 6 M NH_3. Shake these solutions to see if the precipitates will dissolve.

Compare the behavior of the Hg_2Cl_2 precipitate with ammonia with that of $HgCl_2$. To do this, add 1 mL of 6 M NH_3 to 1 mL of 0.1 M $HgCl_2$ solution.

Put 1 mL of the ammoniacal solution of AgCl into a clean test tube, reserving the larger (2-mL) portion for the tests of solubility of silver salts and stability of complex ions. To the smaller (1-mL) portion add a slight excess of 6 M HNO_3. What is the result?

3. RELATIVE SOLUBILITY OF SILVER SALTS AND STABILITY OF COMPLEX IONS. To the larger portion of the ammoniacal AgCl solution, add 1 mL of 0.1 M KBr. Describe what happens. Add 1 mL of 1 M $Na_2S_2O_3$ (sodium thiosulfate), and shake. The complex ion formed is $Ag(S_2O_3)_2^{3-}$. (This is the reaction of "hypo" in fixing the developed film in photography.) To this solution add a little 0.1 M KI. From the results you can evaluate the relative Ag^+ concentrations in the following equilibrium situations: 0.1 M $AgNO_3$, saturated AgCl, saturated AgBr, saturated AgI, $Ag(NH_3)_2^+$, $Ag(S_2O_3)_2^{3-}$. Write net ionic equations for the reactions observed, and list each of the preceding species in order, from the greatest to the least, the amount of $Ag^+(aq)$ that would be present at equilibrium.

B. ANALYSIS OF A KNOWN SOLUTION FOR Ag^+, Pb^{2+}, AND Hg_2^{2+}

First, study the following procedure steps for the analysis and the flowchart shown in Table 31-2, which is completed for this first group of ions. Note the precipitates obtained and the exact ions remaining in the solution. The flowchart will be a useful guide for outlining the essential separation steps and final identification tests for each ion.

Second, write in your experiment report the net ionic equations for the reactions at each successive step of the procedure, as numbered in the flowchart. (Your instructor may require this work before issuing you an unknown.)

Third, prepare a known solution containing Ag^+, Pb^{2+}, and Hg_2^{2+} by placing 1 mL each of 0.1 M $AgNO_3$, 0.1 M $PbNO_3$, and 0.05 M $Hg_2(NO_3)_2$ in a test tube. Analyze this according to the procedure for the analysis steps 1 to 5, following your results on the flowchart. Also study the typical analysis summary (Table 31-3), which is completed for the first group and is similar to the one you will prepare as a record of your actual observations in the analysis of your unknown solutions in Part C. (You will conduct steps 6 to 10 in Experiment 32 and steps 11 to 15 in Experiment 33).

TABLE 31-2
Flowchart: Ag^+, Pb^{2+}, Hg_2^{2+}

Key: *Step numbers* in bold numerals; *reagents* on broken line arrows; *precipitates* or *residues* in solid-line boxes; *solutions* of ions in broken-line boxes.

1. PRECIPITATION OF THE GROUP. To 3 mL of the solution to be tested, add 6 M HCl drop by drop, and mix until any precipitation that occurs is complete. Avoid a large excess. (No precipitation indicates absence of Ag^+ and Hg_2^{2+} and of much Pb^{2+}.) Centrifuge (see Figure 31-5), and test the solution for complete precipitation with a drop of HCl. If the test is positive, add more HCl (one drop at a time) to the solution, and watch for the formation of *additional* precipitate. If more solid forms, recentrifuge; then decant the solution (see Figure 31-4) into another test tube. This solution contains the cations whose chlorides are soluble. These ions will be determined in experiments 32 and 33. Wash the precipitate by adding 3 mL of water; mix, recentrifuge, and pour off the solution.

2. SEPARATION OF THE LEAD ION. Add 3 mL of water to the test tube containing the precipitate from Step 1. Place this test tube in a beaker of boiling water. (Use your test tube holder.) After the tube has heated up, stir the contents with a stirring rod, let the precipitate resettle, and pour off the supernatant solution containing Pb^{2+} ions (if any). Wash the remaining solid residue, which contains AgCl and Hg_2Cl_2, by adding 3 mL of water to the test tube.

Heat the tube in the water bath, stir the contents, let the precipitate settle, then pour off and discard the solution.

TABLE 31-3

Analysis Summary of Each Step, Observation, and Conclusion for a Colorless Solution Known to Contain All Three Ions of the Group

Step	Sample	Reagent	Observations	Conclusions
1	Known sample	Dilute	White precipitate	Ag^+, Pb^{2+}, Hg_2^{2+} (one or more) present
2	Precipitate 1	Hot H_2O	Some precipitate remains	Ag^+ and/or Hg_2^{2+} present
3	Hot H_2O filtrate	K_2CrO_4	Yellow precipitate	Pb^{2+} present, confirmed by dissolving in NaOH
4	Precipitate 2, insoluble in hot H_2O	NH_3	Turns black	Hg_2^{2+} present
5	NH_3 supernatant	HNO_3 (excess)	White precipitate	Ag^+ present

Summary: Ag^+, Pb^{2+}, and Hg_2^{2+} present

3. TEST FOR LEAD ION. Add several drops of 1 M K_2CrO_4 to the supernatant solution from Step 2. A yellow precipitate, soluble when 6 M NaOH is added, proves Pb^{2+}.

4. TEST FOR MERCURY(I) ION. To the solid residue from the hot water treatment (Step 2), add 1 mL of 6 M NH_3, then 2 mL of water. Mix and centrifuge. A black-gray solid residue, a mixture of $HgNH_2Cl(s)$ and finely divided

Hg(l), proves Hg_2^{2+}. Pour off the supernatant solution (containing $Ag[NH_3]_{2+}$) for Step 5.

5. TEST FOR SILVER ION. Acidify the solution from Step 4 with a little 6 M HNO_3, and mix. (If in doubt, test for acidity with litmus.) A white precipitate, AgCl, proves Ag^+.

C. ANALYSIS OF UNKNOWN SOLUTIONS FOR Ag^+, Pb^{2+}, AND Hg_2^{2+}

Obtain one or more unknown solutions from your instructor, and analyze them by the preceding procedure. *At the same time,* record each step, including your actual observations for negative as well as positive tests, by completing an analysis summary (with five headings: Step No., Sample, Reagents, Observations, and Conclusions) like the example for the known solution in Table 31-3. This will constitute your report for each unknown.

Typical centrifuge tube

Aluminum shield — Rubber pad inside

Opposite pairs of tubes should be filled with equal amounts of liquid to prevent excessive vibration.

FIGURE 31-5
The components of the centrifuge.

☞ **Consider This**

After considering the observations below, arrange at the right, in order, from greatest to least, the amount of Cu^{2+}(aq) present at equilibrium: the saturated $CuCO_3$, 1 M $Cu(NH_3)_4^{2+}$, saturated CuS, 1 M $CuCl_2$.

(a) Mixing $CuCl_2$ and Na_2CO_3 solutions will yield a precipitate of $CuCO_3$.

(b) Passing H_2S gas into $Cu(NH_3)_4^{2+}$ solution will precipitate CuS.

(c) $CuCO_3$ solid will dissolve in NH_3 solution to form a deep blue solution.

Explain why you arranged the items as you did.

Identification of Silver, Lead, and Mercury(I) Ions

NAME

SECTION   LOCKER

INSTRUCTOR   DATE

DATA

A. Typical Reactions of the Ions

Solubility of the Chlorides: Their Behavior with Ammonia. In the table below, write the formula(s) of the principal substance(s) present (ions or molecules) when the chlorides are treated as indicated. Also indicate colors.

	AgCl	$PbCl_2{}^a$	Hg_2Cl_2
Hot water			
NH_3(excess)			

^{a}Look up the relative solubilities of $PbCl_2$, and $Pb(OH)_2$ to help you decide whether a reaction occurs with OH^- generated by reaction of NH_3 with H_2O.

Relative Solubility of Silver Salts and Stability of Complex Ions. Complete the net ionic equations for the successive reactions that take place when a $AgNO_3$ solution is treated, in turn, with HCl, excess NH_3, KBr, $Na_2S_2O_3$, and KI.

$Ag^+ + Cl^+ \rightarrow$ _____

$AgCl(s) + 2NH_3 \rightarrow$ _____

$Ag(NH_3)_2{}^+ + Br^- \rightarrow$ _____

$AgBr(s) + 2S_2O_3{}^{2-} \rightarrow$ _____

$Ag(S_2O_3)_2{}^{3-} + I^- \rightarrow$ _____

From the data above, list the following equilibrium solutions in the order of *decreasing free* Ag^+ concentration in the mixture: 0.1 M $AgNO_3$, saturated AgCl, saturated AgBr, saturated AgI, $Ag(NH_3)_2{}^+$, and $Ag(S_2O_3)_2{}^{3-}$.

B. Analysis of a Known Solution for Ag^+, Pb^{2+}, and $Hg_2{}^{2+}$

Balance the following net ionic equations for all of the reactions occuring in the systematic procedure for the analysis of the ions in this group.

1. (a) $Ag^+ + Cl^- \rightarrow AgCl(s)$

   (b) $Pb^{2+} + Cl^- \rightarrow PbCl_2(s)$

   (c) $Hg^{2+} + Cl^- \rightarrow Hg_2Cl_2(s)$

2. $PbCl_2(s) \xrightarrow{\text{hot water}} Pb^{2+} + Cl^-$

3. (a) $Pb^{2+} + CrO_4^{2-} \rightarrow PbCrO_4(s)$

(b) $PbCrO_4(s) + OH^- \rightarrow Pb(OH)_3^- + CrO_4^{2-}$

4. (a) $Hg_2Cl_2(s) + NH_3 \rightarrow Hg(NH_2)Cl(s) + Hg(l) + NH_4^+ + Cl^-$

(b) $AgCl(s) + NH_3 \rightarrow Ag(NH_3)_2^+ + Cl^-$

5. $Ag(NH_3)_2^+ + H^+ + Cl^- \rightarrow AgCl(s) + NH_3$

C. Analysis of Unknown Solutions for Ag^+, Pb^{2+}, and Hg_2^{2+}

Unknown no. _____   Appearance _____   Ions found _____

Unknown no. _____   Appearance _____   Ions found _____

Analysis Summary: Report your actual observations, negative as well as positive, for each step of the analysis of each unknown, and record above the ions found.

Step	Sample	Reagent(s)	Observations	Conclusions

☞ **Consider This**

After considering the observed facts listed below, arrange at the right, in order, from greatest to least, the amount of Cu^{2+}(aq) that would be present at equilibrium: saturated $CuCO_3$, 1 M $Cu(NH_3)_4^{2+}$, saturated CuS, 1 M $CuCl_2$.

(a) Mixing $CuCl_2$ and Na_2CO_3 solutions will yield a precipitate of $CuCO_3$.

(b) Passing H_2S gas into $Cu(NH_3)_4^{2+}$ solution will precipitate CuS.

(c) $CuCO_3$ solid will dissolve in NH_3 solution to form a deep blue solution.

Explain why you arranged the items in the order you did.

_____

_____

_____

_____

Identification of Ferric, Aluminum, and Zinc Ions

PURPOSE

To learn how to separate and identify Fe^{3+}, Al^{3+}, and Zn^{2+} ions as insoluble hydroxides or sulfides. To understand the role of buffers in controlling H_3O^+ and OH^- concentrations. To observe the tendency of Zn^{2+} to form soluble complexes with ammonia. To observe the amphoteric behavior of Al^{3+} as it forms an insoluble hydroxide that dissolves in excess OH^- to form a soluble hydroxy complex.

PRE-LAB PREPARATION

REACTIONS OF Fe^{3+}, Al^{3+}, AND Zn^{2+} IONS

The removal and separation of these three metal ions from our selected group of ten cations (see Table 31-1 of Experiment 31), after removal of the insoluble chlorides, depends on their distinctive behavior with NH_3 and with NaOH. Zinc ion is conveniently removed and identified by its white insoluble sulfide in a basic or slightly acid solution. The control of pH is important in all of these separations.

BUFFER ACTION

Buffers are used to control the relative H_3O^+ and OH^- concentrations within desired limits. Buffers are formed from conjugate acid-base pairs of a weak acid and its salt or a weak base and its salt. Thus, in the Al^{3+} test that follows, a mixture of NH_3 and NH_4Cl is used to produce a basic solution that is much less basic than NH_3 alone. Consider the equilibrium equation and the corresponding equilibrium constant expression:

$$NH_3 + H_2O \rightleftharpoons NH_4^+ + OH^-$$

$$K_b = \frac{[NH_4^+][OH^-]}{[NH_3]} = 1.8 \times 10^{-5}$$

Excess NH_3 in the mixture shifts the equilibrium to the left, resulting in a lower OH^- concentration. While 1 M NH_3 alone contains about 0.004 M OH^-, calculation shows that if NH_4Cl is added to make the NH_4^+ and NH_3 concentrations equal, the resulting OH^- concentration then will be only 1.8×10^{-5} M, a 200-fold reduction.

Likewise, if an acid solution with a smaller concentration of H_3O^+ than in aqueous CH_3COOH is desired, the addition of $Na^+CH_3COO^-$ will substantially decrease the H_3O^+ concentration, as is evident from the equilibrium

$$CH_3COOH(aq) + H_2O \rightleftharpoons H_3O^+ + CH_3COO^-$$

SULFIDE PRECIPITATION

In a solution containing the weak acid H_2S, the resulting S^{2-} concentration is very dependent on the H_3O^+ concentration:

$$H_2S(aq) + H_2O \rightleftharpoons H_3O^+ + HS^-$$

$$HS^- + H_2O \rightleftharpoons H_3O^+ + S^{2-}$$

This fact is utilized in more complete qualitative schemes to separate the metal ions into groups by the degree of insolubility of their sulfides. In our simple scheme, precipitation of Zn^{2+} as $ZnS(s)$ in a basic NH_3 solution is satisfactory, since only alkali and alkaline earth ions remain, the sulfides of which are all readily soluble.

EXPERIMENTAL PROCEDURE

Chemicals: 0.1 M aluminum nitrate, $Al(NO_3)_3$; 0.1 M iron(III) in 1 M nitric acid, $Fe(NO_3)_3/HNO_3$; 0.1 M zinc nitrate, $Zn(NO_3)_2$; "aluminon reagent." (1.0 g/L of the ammonium salt of aurin tricarboxylic acid), 0.1 M ammonium hydrogensulfide, NH$_4$HS or saturated hydrogen sulfide, H$_2$S solution (whichever is used should be freshly prepared); 0.1 M potassium thiocyanate, KSCN; 6 M ammonia, NH$_3$; 6 M sodium hydroxide, NaOH; 6 M hydrochloric acid, HCl.

! **SAFETY PRECAUTIONS:** H$_2$S gas, which has a characteristic odor of rotten eggs, is very toxic. Perform the procedure described in Step 10 (Test for Zn^{2+}) in a fume hood. Boiling to remove H$_2$S also should be performed in a fume hood, or you should use an inverted funnel connected to an aspirator (as shown in Experiment 5, Figure 5-1) to remove H$_2$S.

WASTE COLLECTION: None of the solutions containing the ions in this experiment should be disposed of down the drain. Waste containers should be provided for all waste solutions.

A. TYPICAL REACTIONS OF THE IONS

AMMONIA AND HYDROXIDE ION COMPLEXES.
Place 1-mL samples of 0.1 M Al(NO$_3$)$_3$, 0.1 M Fe(NO$_3$)$_3$, and 0.1 M Zn(NO$_3$)$_2$ into separate test tubes. To each add 1 drop of 6 M NH$_3$, then continue to add more, drop by drop, to determine whether the hydroxide precipitate first formed redissolves with excess NH$_3$. (*Note:* The 0.1 M Fe[NO$_3$]$_3$ contains excess 1 M HNO$_3$, which must first be neutralized before any precipitate appears.) Repeat these tests, using 6 M NaOH instead of 6 M NH$_3$. Summarize all the results in a suitable chart in your report.

RED LAKE FORMATION WITH Al(OH)$_3$.
The light flocculent precipitate of aluminum hydroxide is often difficult to observe. The dye called aluminon (ammonium aurin tricarboxylate) is adsorbed by the precipitate to form a characteristic red lake (red pigment), which makes the identification of the aluminum hydroxide easier.

To 1 mL of water add 2 to 3 drops of 0.1 M Al^{3+} solution. Add 2 drops of aluminon reagent, then 3 to 5 drops of 6 M NH$_3$. Note the appearance of the precipitate. Now make the solution acidic with 0.5 mL 6 M HCl and again just basic with 6 M NH$_3$. Let the mixture stand a moment, and note the characteristic color and flocculation of the precipitate (which now should be red). Also note whether the solution is colored. A good red color is not obtained if the solution is too basic. Explain how the preceding treatment forms a buffer that guarantees a slightly basic solution.

B. ANALYSIS OF A KNOWN SOLUTION FOR Fe^{3+}, Al^{3+}, AND Zn^{2+}
First, study the following procedure steps for analysis until you understand each step and the reason for adding each reagent. Then complete a flowchart for these operations, similar to the one in Table 31-2 of Experiment 31. A skeleton flowchart is provided in Part B of the report form: Simply insert the proper formulas for reagents, precipitates or residues, ions in solution, and so on.

Second, in your experiment report, write the net ionic equations for the reactions occurring at each step of the procedure, as numbered in the flowchart. (Your instructor may require that you do this before you receive an unknown.)

Third, prepare a known solution containing Fe^{3+}, Al^{3+}, and Zn^{2+} by adding 1 mL each of 0.1 M Fe(NO$_3$)$_3$, 0.1 M Al(NO$_3$)$_3$, and 0.1 M Zn(NO$_3$)$_2$ to a test tube. Analyze according to the procedure described in the next paragraph, following your results on the flowchart. *At the same time*, complete an analysis summary similar to Table 31-3 of Experiment 31 (a form for this summary is provided in Part B of the report form).

Directions for steps 1 to 5 are given in Experiment 31, directions for steps 11 to 15 in Experiment 33.

6. PRECIPITATION OF Fe(OH)$_3$ AND Al(OH)$_3$.
To 3 mL of the solution to be tested add an excess (about 1 mL) of 6 M NH$_3$. Mix the solution; then centrifuge it. (The use of a centrifuge, if available, is more convenient for most separations of precipitate and solution. See Figures 31-4 and 31-5. *Obtain instructions from your instructor.* If filter paper is used, modify the directions accordingly.) Wash the precipitated Al(OH)$_3$ and Fe(OH)$_3$ with 1 to 2 mL of water, combining this solution with the supernatant centrifugate, which contains any zinc as Zn(NH$_3$)$_4^{2+}$.

7. SEPARATION OF Fe^{3+} AND Al^{3+}.
Treat the mixed precipitate with about 0.5 mL of 6 M NaOH, followed by 1 mL of water.[1] Centrifuge and wash with 1 mL of water, combining the washings with the filtrate of Al(OH)$_4^-$.

8. TEST FOR Fe^{3+}.
Dissolve the Fe(OH)$_3$ precipitate with 1 to 3 drops of 6 M HCl (if necessary to complete the reaction, warm by placing the tube into a beaker of warm water), and add 2 mL of H$_2$O, then 1 to 3 drops of 0.1 M KSCN. A red color of Fe(SCN)$^{2+}$ proves the presence of Fe^{3+}.

9. TEST FOR Al^{3+}.
To the Al(OH)$_4^-$ filtrate from Step 7 add 6 M HCl until neutral; then add about 0.5 mL excess. Add 1 to 2 drops of aluminon reagent, and make just basic again with 6 M NH$_3$. Mix this and let it stand. A flocculent

---

[1] If filter paper is used, add the NaOH and H$_2$O directly on the filter, collecting the filtrate in a test tube. The basic solution may be passed over the filter a second time for more complete solution of any aluminum as Al(OH)$_4^-$.

precipitate, colored a characteristic red by the dye, proves the presence of Al^{3+}. If color and precipitate are indefinite, you may again make the solution acidic with HCl, then basic with NH_3 to build up the NH_4^+ concentration so the solution becomes less basic.

10. TEST FOR Zn^{2+}. To the filtrate from Step 6, which contains any zinc present as $Zn(NH_3)_4^{2+}$, add 1 to 3 drops of 0.1 M NH_4HS or 2 mL of saturated H_2S water. A white precipitate of ZnS proves the presence of Zn^{2+}. (When a general unknown is being analyzed, centrifuge the ZnS precipitate, and save the solution for Experiment 33. Unless this solution is analyzed at once, it should be acidified with HCl and boiled in the fume hood to remove H_2S. This will avoid oxidation to sulfate ion, which would precipitate Ba^{2+} ion as $BaSO_4$.)

C. ANALYSIS OF UNKNOWN SOLUTIONS FOR Fe^{3+}, Al^{3+}, AND Zn^{2+}

Obtain one or more unknown solutions from your instructor, and analyze them by the preceding steps. *At the same time*, keep a record of each step, including your actual observations for negative as well as for positive tests, by completing an analysis summary in Part C of the report form.

Consider This

A solution is 0.50 M in acetic acid, HA, and 0.25 M in sodium acetate, Na^+A^-. (With the excess of A^-, assume that the equilibrium concentration of HA equals its stoichiometric concentration.) (a) Write the chemical reaction for the dissociation of acetic acid in water. (b) Write the equilibrium constant expression for acetic acid (K_a for HA = 1.8×10^{-5}). (c) Calculate the H_3O^+ concentration in this buffer solution.

A solution is 1.8 M in NH_4Cl, and 0.010 M in NH_3. Will the solution be basic, acidic, *or* neutral? *Show your calculations. (With this excess of NH_4^+, assume that the molarity of NH_3 equals its stoichiometric concentration. [K_b for NH_3 = 1.8×10^{-5}])*

Identification of Ferric, Aluminum, and Zinc Ions

NAME		
SECTION		LOCKER
INSTRUCTOR		DATE

DATA

A. Typical Reactions of the Ions

Ammonia and Hydroxide Ion Complexes. In the spaces provided, write the formulas of the precipitates formed, or new ions formed in solution, if any. Also indicate any characteristic colors and the like that result when each ion is treated with the reagent in the left-hand column.

Reagent	Fe^{3+}	Al^{3+}	Zn^{2+}
6 M NH_3			
6 M NH_3 (excess)			
6 M NaOH			
6 M NaOH (excess)			

Red Lake Formation with $Al(OH)_3$

Equation for the reaction of Al^{3+} with NH_3 _____

Explain how the addition, first, of an excess of HCl, then of a slight excess of NH_3, ensures that the solution will not become too basic to form a satisfactory adsorption compound of the dye with the precipitate.

B. Analysis of a Known Solution for Fe^{3+}, Al^{3+}, and Zn^{2+}

Complete the following flowchart and analysis summary for your known mixture of Fe^{3+}, Al^{3+}, and Zn^{2+}.

Flow Chart: Fe^{3+}, Al^{3+}, Zn^{2+}

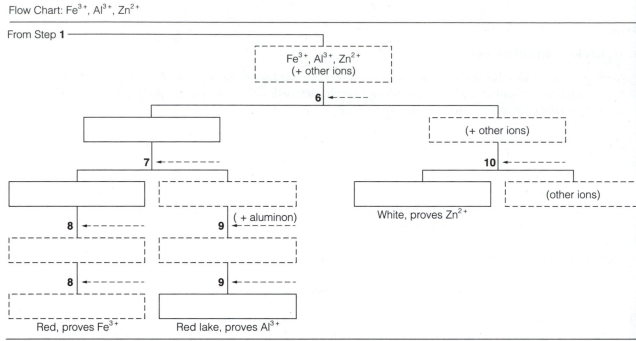

Key: *Step numbers* in bold numerals, *reagents* on broken-line arrows, *precipitates* or *residues* in solid-line boxes, *solutions* in broken-line boxes.

Complete the analysis summary for your known sample, as in Table 31-3.

Step	Sample	Reagent(s)	Observations	Conclusions
6				
7				
8				
9				
10				
Summary conclusions				

Write below, in order, the net ionic equations for all reactions occurring in the systematic analysis of the ions of this group.

C. Analysis of Unknown Solutions for Fe^{3+}, Al^{3+}, and Zn^{2+}

Unknown no. _____   Appearance _____   Ions found _____

Unknown no. _____   Appearance _____   Ions found _____

Analysis Summary: Report your actual observations, negative as well as positive, for each step of the analysis of each unknown, and record above the ions found.

Step	Sample	Reagent(s)	Observations	Conclusions

Consider This

A solution is 0.50 M in acetic acid, HA, and 0.25 M in sodium acetate, Na^+A^-. (With the excess of A^-, assume that the equilibrium concentration of HA equals its stoichiometric concentration.)

(a) Write the chemical reaction for the dissociation of acetic acid in water.

(b) Write the equilibrium constant expression for acetic acid. (K_a for HA $= 1.8 \times 10^{-5}$)

(c) Calculate the H_3O^+ concentration in this buffer solution.

A solution is 1.8 M in NH_4Cl, and 0.010 M in NH_3. Will the solution be basic, acidic, or neutral? *Show your calculations.* (With this excess of NH_4^+, assume that the molarity of NH_3 equals its stoichiometric concentration. [K_b for $NH_3 = 1.8 \times 10^{-5}$])

33

Identification of Alkaline Earth and Alkali Metal Ions

PURPOSE

To study procedures for separating and identifying two alkaline earth cations: Ca^{2+} and Ba^{2+} ions. To learn how to test for the alkali metal cation, K^+, and for ammonium ion, NH_4^+, which is grouped with the alkali metals because its salts have solubilities that resemble the alkali metal salts.

PRE-LAB PREPARATION

The elements making up the alkaline earth and alkali groups are similar and have many properties in common. Their compounds exhibit only one stable oxidation state; they do not form amphoteric hydroxides, being distinctly basic; and they do not readily form complex ions with NH_3 or OH^-. The separation of the alkaline earth elements depends almost entirely on differences in the solubilities of their salts, which show a regular gradation through the periodic table. The salts of the alkali metals are almost all readily soluble. Table 33-1 will be useful in interpreting the analytical procedure.

TABLE 33-1

Solubilities of Alkaline Earth Salts
(g/100 g H_2O, at Room Temperature)

	Mg^{2+}	Ca^{2+}	Sr^{2+}	Ba^{2+}
OH^-	0.001	0.16	1.74	3.89
CO_3^{2-}	0.09	0.0015	0.001	0.0018
SO_4^{2-}	35.5	0.2	0.01	0.00024
CrO_4^{2-}	138.	18.6	0.12	0.00037
$C_2O_4^{2-}$	0.015	0.0007	0.005	0.01

This experiment includes the two alkaline earth ions Ca^{2+} and Ba^{2+}, the alkali ion K^+, and also NH_4^+, whose salts are almost all soluble. We must test for NH_4^+ with a separate portion of the original unknown, since we have used it as a reagent.

PRECIPITATION WITH AMMONIUM CARBONATE

The reagent usually used to precipitate alkaline earth ions is ammonium carbonate $(NH_4)_2CO_3$. Since this is the salt of both a weak base and a weak acid, it reacts in solution to a substantial extent (nearly 80 percent reacted):

$$NH_4^+ + CO_3^{2-} \rightleftharpoons NH_3 + HCO_3^- \ (K = 12)$$

For this reason, we add excess NH_3 to reverse the reaction and increase the concentration of CO_3^{2-} to attain more complete precipitation.

THE SEPARATION OF BARIUM AND CALCIUM IONS

The following analytical procedure takes advantage of the solubility differences listed in Table 33-1. Note that after dissolution of the carbonate precipitate with acetic acid, barium ions can be effectively separated as the chromate, $BaCrO_4$, and then further confirmed by transformation to the still less soluble sulfate, $BaSO_4$. The $BaCrO_4$ dissolves in HCl because the chromate ion can be protonated to form a moderately weak acid, $HCrO_4^-$, which in turn forms dichromate ion, $Cr_2O_7^{2-}$, in acid solution.

$$2CrO_4^{2-} + 2H_3O^+ \rightleftharpoons Cr_2O_7^{2-} + 3H_2O$$

With barium ion removed, calcium ion is easily identified, by precipitation with oxalate ion in NH_3 solution, as CaC_2O_4.

EXPERIMENTAL PROCEDURE

Special Supplies: Nichrome wire, cobalt glass

Chemicals: 6 M acetic acid, CH_3COOH; 3 M ammonium acetate, NH_4CH_3COO; 0.1 M ammonium chloride, NH_4Cl; 3 M ammonium carbonate, $(NH_4)_2CO_3$; 0.1 M barium chloride, $BaCl_2$; 0.1 M calcium chloride, $CaCl_2$; 0.1 M potassium nitrate, KNO_3; 1 M potassium chromate, K_2CrO_4; 1 M potassium oxalate, $K_2C_2O_4$; 0.1 M sodium sulfate, Na_2SO_4; 6 M ammonia, NH_3; 6 M sodium hydroxide, NaOH; 6 M hydrochloric acid, HCl, and 12 M HCl; 1 M sodium acetate, $NaCH_3COO$; 3% sodium tetraphenylborate, $NaB(C_6H_5)_4$.

! SAFETY PRECAUTIONS: Barium is toxic, and chromates are known to be carcinogenic. Try not to spill any Ba^{2+} or CrO_4^{2-} on your skin. If you do, wash it off immediately. Wash your hands thoroughly before you leave the laboratory.

WASTE COLLECTION: None of the solutions or solids containing barium or chromate ions should be disposed of down the drain. Waste containers should be provided for barium and chromate waste solutions. The other solutions can be rinsed down the drain.

A. TYPICAL REACTIONS OF THE IONS

Instead of performing a series of separate, prescribed tests to learn the properties of the ions, proceed at once to the study of the following procedure for the analysis, and at the same time complete the flow chart by filling in the proper formulas for reagents, precipitates or residues, ions in solution, and so on. Note how the procedures for precipitating Ca^{2+} and Ba^{2+} make use of the solubilities of the alkaline earth salts shown in Table 33-1.

On your own initiative, you may wish to try, with the help of your instructor, various test tube experiments, such as the separation of certain ions by a specific reagent, the subsequent precipitation identification tests, and the K^+ flame test in the presence of Na^+ (See Figure 33-1).

B. ANALYSIS OF A KNOWN SOLUTION FOR Ba^{2+}, Ca^{2+}, K^+, AND NH_4^+

Prepare a known solution by adding 1 mL each of 0.1 M $BaCl_2$, 0.1 M $CaCl_2$, 0.1 M KNO_3, and 0.1 M NH_4Cl to a test tube. Analyze this solution according to the following procedure; at the same time, prepare your own flow chart and complete it in accord with your observations. Also complete the analysis summary for this solution (Part B on the report form), outlining the successive steps, reagents, observations, and conclusions.

Directions for steps 1 to 5 are given in Experiment 31, and directions for steps 6 to 10 in Experiment 32.

11. PRECIPITATION OF Ba^{2+} AND Ca^{2+}. Start with a 3-mL sample of the solution to be tested. (If this is a general unknown, evaporate the filtrate from Step 10,

A colored flame indicates a contaminated wire.

A Clean the wire by alternately dipping in 12 M HCl and heating to incandescence.

Observe through cobalt glass.

Look at the flame against a dark background.

Observe the color of the flame when the wire is heated in the edge of the Bunsen flame.

Dip the clean wire into the substance which has been moistened with 12 M HCl.

B

Violet potassium flame (sodium flame obscured)

C

FIGURE 33-1

The flame test for sodium and potassium. (A) shows the procedure for cleaning the wire; (B) and (C) illustrate the steps for observing the flame coloration.

Experiment 32, to a volume of 3 mL.) If your 3-mL sample is not already basic (litmus test), add 6 M NH_3 until it is. Then add, with mixing, 10 drops of 3 M $(NH_4)_2CO_3$. Let the mixture stand 3 to 5 min to complete precipitation; then centrifuge it. Test for completeness of precipitation with further drop-by-drop additions of 3 M $(NH_4)_2CO_3$, until no further precipitation occurs. Centrifuge again, if necessary. Separate the solution, and save it for the K^+ test, Step 14.

12. TEST FOR Ba²⁺.

Wash and centrifuge the preceding precipitate, and discard the wash water. Add 5 drops of 6 M CH_3COOH, warm as needed to dissolve the precipitate, add 1 mL of water, and buffer the solution by adding 5 drops of 3 M NH_4CH_3COO. Add 3 drops of 1 M K_2CrO_4, mix and centrifuge. If the solution is not yellow, add more 1 M K_2CrO_4, drop by drop, but avoid adding over 1 to 3 drops in excess. Wash any precipitate. Yellow $BaCrO_4$ indicates Ba^{2+}. (Save the solution for the Ca^{2+} test, Step 13.) This may be confirmed by adding 2 to 3 drops of 6 M HCl to dissolve the yellow precipitate, then adding 1 mL of 0.1 M Na_2SO_4. Centrifuge, if necessary, to see whether the precipitate is white in the yellow solution. White $BaSO_4$ proves the presence of Ba^{2+}.

13. TEST FOR Ca²⁺.

To the preceding yellow solution from which the $BaCrO_4$ was removed, add 10 drops of 1 M $K_2C_2O_4$; then add 6 M NH_3 to make the solution basic. Let stand 10 min, if a precipitate does not appear before that time. White CaC_2O_4 proves the presence of Ca^{2+}. Further confirmation may be made by decanting the yellow solution, adding to the solid residue a few drops of 6 M HCl and 1 mL of water to dissolve the residue, then adding another drop of 1 M $K_2C_2O_4$, and again making the solution basic with 6 M NH_3. The white precipitate of CaC_2O_4 will reappear.

14. TEST FOR K⁺.

To a portion of the solution from Step 11 (after removal of the Ba^{2+} and Ca^{2+}) add 1 drop of 6 M NaOH to make the solution basic, and gently boil to expel NH_3 (this prevents later interference by NH_4^+ ion).[1] Then acidify with 2 to 3 drops of 6 M HCl, and add 3 mL of 1 M sodium acetate. (The acetate ion reacts with excess H_3O^+ and buffers the solution at about pH 5). Then add 2 to 3 drops of 3% sodium tetraphenylborate reagent, $NaB(C_6H_5)_4$. A white precipitate of potassium tetra-phenylborate, $KB(C_6H_5)_4$, indicates the presence of potassium. Alternatively, a flame test can be used, as outlined in Figure 33-1.

---

[1] Tetraphenylborate ion, $B(C_6H_5)_4^-$, forms a very insoluble precipitate with K^+. However, it also forms precipitates with NH_4^+, Ag^+, Rb^+, Cs^+, and Tl^+, so that these ions must be absent when testing for K^+.

15. TEST FOR NH₄⁺.

This test must always be carried out on the *original solution* or *sample*, since ammonium salts are used as reagents throughout the procedure.

Place 3 mL of the sample in an evaporating dish, and make it basic with 6 M NaOH. Attach a moist strip of red litmus paper to the underside of a watch glass; cover the dish with the watch glass. Warm the solution *very gently* to liberate any ammonia present as the gas. (Avoid boiling, which would contaminate the litmus with spray droplets of the NaOH solution.) An even, unspotted blue color caused by NH_3 gas, not by spattered NaOH, proves ammonium ion.

C. ANALYSIS OF UNKNOWN SOLUTIONS FOR Ba²⁺, Ca²⁺, K⁺, AND NH₄⁺

Obtain one or more unknown solutions from your instructor, and analyze them by the preceding procedure. *At the same time*, keep a record of each step by completing the analysis summary in Part C of the report form, showing your actual observations, negative as well as positive, for each unknown.

☞ Consider This

The variety of colors seen in fireworks is a result of small amounts of metal salts that are added to the explosive mixtures. Make up ~0.1 M solutions of the chloride salts (nitrate salts work less well) of sodium, potassium, barium, copper, calcium, strontium, lithium, rubidium, and cesium. Use the procedure outlined in Figure 33-1 to perform flame tests on these solutions. How many different colors can you make? Can you make new flame colors by combining solutions? For a brightly colored flame like sodium, what are the limits of detection of the salt? (That is, how dilute can you make the solution and still detect sodium?) Make up a series of solutions of a couple of the salts. Can your eyes detect the differences in concentration in a flame test?

Your nose contains a very sensitive ammonia detector. What is the minimum level of NH_3 you can smell in a solution? (See the Introduction, Figure I-1, for the proper technique for a smell test.) Do your classmates have the same limits as you?

From the data contained in Table 33-1, can you propose a statement about the trends in solubilities of the alkaline earth salts? If you check the CRC Handbook for the solubilities of similar beryllium salts, does your statement hold true? Does it work for the nitrate and chloride salts of the alkaline earth elements?

Identification of Alkaline Earth and Alkali Metal Ions

DATA

A. Typical Reactions of the Ions

NOTE: Indicate any experimental tests you have performed personally in order to be certain of the results.

1. Write net ionic equations for the reactions, *if any,* occurring when dilute solutions of the following are mixed. Write "N.R." (for "no reaction") if there is none.

   (a) NH_4Cl, KNO_3, NaOH, heated _____

   (b) $CaCl_2$, $Ba(NO_3)_2$, 1 M NH_3 _____

   (c) $CaCl_2$, $Ba(NO_3)_2$, 1 M NaOH _____

   (d) $Ca(NO_3)_2$, K_2CrO_4 _____

   (e) Mixture d + $BaCl_2$ _____

   (f) Mixture e + HCl _____

   (g) Mixture f + Na_2SO_4 _____

   (h) $BaCl_2$, KNO_3, $ZnCl_2$, H_2S _____

   (i) KCl, $(NH_4)_2CO_3$, $Ba(NO_3)_2$ _____

   (j) $CaCO_3(s)$, NH_4Cl, 6 M $HC_2H_3O_2$ _____

   (k) Mixture j + $K_2C_2O_4$, NH_3 _____

2. On the basis of periodic table relationships, what would you predict about the general solubility (indicate "soluble," "slightly soluble," or "insoluble") of each of the following?

   (a) $Be(OH)_2$ _____   (c) $Ra(OH)_2$ _____

   (b) $BeSO_4$ _____   (d) $RaSO_4$ _____

3. Why does the addition of HCl dissolve such insoluble salts as $BaCrO_4$ and CaC_2O_4?

B. Analysis of a Known Solution for Ba^{2+}, Ca^{2+}, K^+, NH_4^+

1. Complete the following flowchart in accord with your observations.

Flowchart: Ba^{2+}, Ca^{2+}, K^+, NH_4^+

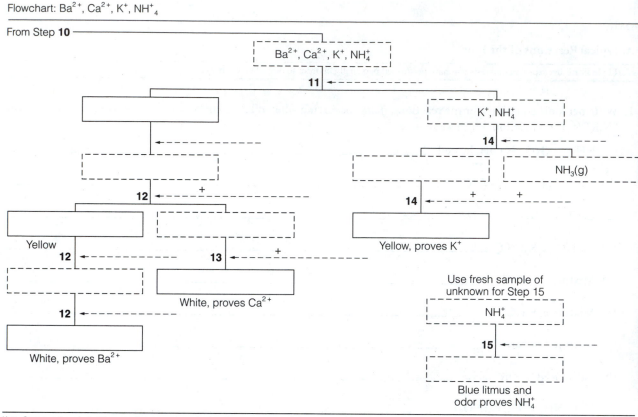

Key: *Step numbers* in bold numerals, *reagents* on broken-line arrows, *precipitates* or *residues* in solid-line boxes, *solutions* in broken-line boxes.

2. Complete the analysis summary for your known sample.

Appearance _____   Ions found _____

Step	Sample	Reagent(s)	Observations	Conclusions
11				
12				
13				
14				
15				
Summary conclusions				

C. Analysis of Unknown Solutions for Ba^{2+}, Ca^{2+}, K^+, and NH_4^+

Unknown no. _____   Appearance _____   Ions found _____

Unknown no. _____   Appearance _____   Ions found _____

Analysis Summary: For your unknown samples, report below your actual observations, negative as well as positive, for each step of each analysis, and record above the ions found.

Step	Sample	Reagent(s)	Observations	Conclusions

☞ **Consider This**

The variety of colors seen in fireworks is a result of small amounts of metal salts that are added to the explosive mixtures. Make up ~0.1 M solutions of the chloride salts (nitrate salts work less well) of sodium, potassium, barium, copper, calcium, strontium, lithium, rubidium, and cesium. Use the procedure outlined in Figure 33-1 to perform flame tests on these solutions. How many different colors can you make? Can you make new flame colors by combining solutions? For a brightly colored flame like sodium, what are the limits of detection of the salt? (That is, how dilute can you make the solution and still detect sodium?) Make up a series of solutions of a couple of the salts. Can your eyes detect the differences in concentration in a flame test?

Your nose contains a very sensitive ammonia detector. What is the minimum level of NH_3 you can smell in a solution? (See the Introduction, Figure I-1, for the proper technique for a small test.) Do your classmates have the same limits as you?

From the data contained in Table 33-1, can you propose a statement about the trends in solubilities of the alkaline earth salts? If you check the CRC Handbook for the solubilities of similar beryllium salts, does your statement hold true? Does it work for the nitrate and chloride salts of the alkaline earth elements?

The Qualitative Analysis of Some Common Anions

PURPOSE

To study specific tests for detecting in an unknown sample the presence of nine common anions: sulfide, sulfate, sulfite, carbonate, chloride, bromide, iodide, phosphate, and nitrate ions.

PRE-LAB PREPARATION

The principles that are employed in the identification of cations can also be applied to the analysis of anions. The qualitative detection of anions in a sample depends on the distinctive solubility properties of particular salts of the ions and specific chemical reactions that are (ideally) unique to a particular ion. In this experiment, we will explore ways to detect the presence of S^{2-}, SO_4^{2-}, SO_3^{2-}, CO_3^{2-}, Cl^-, Br^-, I^-, PO_4^{3-}, and NO_3^-.

Carefully read the following criteria on which the identification of an anion generally depends. (The properties on which these criteria depend are summarized in tables in the Appendix.)

1. Does the anion form a *precipitate* with certain cations, such as Ag^+ or Ba^{2+}, in neutral or in acidic solutions? (See Table 8 in the Appendix for solubility rules.)

2. Is the ion the *anion of a weak or a strong acid*? (See Table 6 in the Appendix for information on the relative strength of acids.)

3. Is the ion the anion of a *volatile acid*?

4. Can the anion act as an *oxidizing agent* or *reducing agent*? What is its relative strength as an oxidizing or reducing agent? (See the Appendix, Table 10, for redox potentials.)

Review the descriptive chemistry of the anions we will study in this experiment as you apply the preceding criteria to the typical reactions and tests for the following anions.

TYPICAL REACTIONS OF COMMON ANIONS

SULFATE ION

Sulfuric acid is a diprotic acid and generally is considered to be a *strong acid;* however, if the concentration of H_3O^+ is 1 M or higher, the sulfate ion is substantially protonated to form HSO_4^-. That is, a 1 M solution of H_2SO_4 has an H_3O^+ concentration of about 1 M, not 2 M as popularly believed. Sulfate ion forms characteristic precipitates with certain metal ions, as with barium ion:

$$Ba^{2+} + SO_4^{2-} \rightleftharpoons BaSO_4(s)$$

Other anions also form insoluble precipitates with barium ion:

$$Ba^{2+} + CO_3^{2-} \rightleftharpoons BaCO_3(s)$$

$$Ba^{2+} + SO_3^{2-} \rightleftharpoons BaSO_3(s)$$

$$3Ba^{2+} + 2PO_4^{3-} \rightleftharpoons Ba_3(PO_4)_2(s)$$

However, these last three anions are salts of *weak acids* and would all dissolve, or fail to precipitate, in an acid solution. In the presence of excess hydronium ion, the concentration of the free anion is reduced to such a low value that the solubility equilibrium is reversed, and the precipitate dissolves or fails to form in the first place. We may therefore use barium ion in an acid solution ($[H_3O^+] \cong 0.1$ M) as a test reagent for the presence of sulfate ion.

SULFITE ION, SULFIDE ION, AND CARBONATE ION

Each of these anions of *weak, volatile acids* forms the free acid when a strong acid is added:

$$SO_3^{2-} + 2H_3O^+ \rightleftharpoons H_2SO_3(aq) + 2H_2O$$

$$\rightleftharpoons SO_2(g) + 3H_2O$$

$$S^{2-} + 2H_3O^+ \rightleftharpoons H_2S(g) + 2H_2O$$

$$CO_3^{2-} + 2H_3O^+ \rightleftharpoons H_2CO_3(aq) + 2H_2O$$

$$\rightleftharpoons CO_2(g) + 3H_2O$$

The sharp odor of sulfur dioxide (SO_2) and the rotten egg smell of hydrogen sulfide (H_2S) are usually sufficient identification. Sulfite ion (SO_3^{2-}) can be oxidized to sulfate ion by bromine water (Br_2) or hydrogen peroxide (H_2O_2), then tested as the sulfate ion is tested. Sulfide ion (S^{2-}) can be further confirmed by allowing the $H_2S(g)$ formed to come into contact with a lead salt solution on filter paper. The result will be the formation of a black deposit of lead sulfide (PbS).

The sulfite ion can be reduced to free sulfur by sulfide ion, which in turn is oxidized also to sulfur, so that a yellow precipitate of sulfur would be the main observable product. The equation is

$$SO_3^{2-} + 2S^{2-} + 6H^+ \rightarrow 3S(s) + 3H_2O$$

These ions, therefore, are not likely to be found together in the same solution, particularly if it is acidic. A solution of sulfite is somewhat unstable because of its ease of oxidation by atmospheric oxygen:

$$2SO_3^{2-} + O_2(g) \rightarrow 2SO_4^{2-}$$

The conventional oxidation number of sulfur in sulfite ion is $+4$, which increases to $+6$ in sulfate ion.

In the test for carbonate, the addition of acid causes the evolution of a colorless gas, having a nose-tickling odor like that of a carbonated beverage. This is an indication, but not proof, that the gas is carbon dioxide. Confirmation that the sample contains carbonate is evidenced by reprecipitation of the carbonate ion as an insoluble carbonate salt, such as calcium carbonate ($CaCO_3$). When the slightly soluble carbon dioxide gas is bubbled into a neutral solution containing calcium ion, the weak acid that is formed, carbonic acid (H_2CO_3),

$$CO_2(g) + H_2O \rightleftharpoons H_2CO_3(aq) \rightleftharpoons 2H^+ + CO_3^{2-}$$

does not furnish a high enough concentration of carbon-

ate ion to precipitate calcium carbonate. But if it is bubbled into an alkaline solution instead of a neutral one, the carbonic acid is deprotonated to form carbonate ion. This is the reason that we use limewater, $Ca(OH)_2$ (although the hydroxide is only slightly soluble), rather than some more soluble calcium salt in the carbonate ion test. The overall net ionic equation is

$$CO_2(g) + Ca^{2+} + 2OH^- \rightleftharpoons CaCO_3(s) + H_2O$$

Sulfite ion, if also present in the unknown, will interfere with the carbonate ion test, since when the sample is acidified, it will liberate sulfur dioxide gas (SO_2) along with the carbon dioxide gas and will form a calcium sulfite ($CaSO_3$) precipitate when the gases are absorbed in the limewater. (Sulfide ion, S^{2-}, would not interfere, since CaS is soluble.) To overcome this interference, it is necessary first to oxidize any sulfite ion to sulfate ion, thus preventing the formation of any sulfur dioxide gas.

CHLORIDE ION, BROMIDE ION, AND IODIDE ION

Like the sulfate ion, chloride, bromide, and iodide ions are the anions of *strong acids*. Their salts with silver ion are insoluble, whereas the silver salts of most other anions, which are also insoluble in neutral solution, will dissolve in acid solution. Silver sulfide (Ag_2S), however, is so very insoluble that the presence of hydrogen ion does not reduce the sulfide ion concentration sufficiently to dissolve it.

We can test for the presence of *chloride* ion in the mixed precipitate of AgCl, AgBr, and AgI. We dissolve the AgCl from the still more insoluble AgBr and AgI by adding ammonia solution, thus forming the complex ion $Ag(NH_3)_2^+$ and Cl^-. When the centrifugate (or filtrate) is again acidified, the AgCl is reprecipitated:

$$AgCl(s) + 2NH_3(aq) \rightleftharpoons Ag(NH_3)_2^+ + Cl^-$$

$$Ag(NH_3)_2^+ + Cl^- + 2H_3O^+ \rightleftharpoons$$
$$AgCl(s) + 2NH_4^+ + 2H_2O$$

It is therefore possible, by carefully controlling the NH_3 and Ag^+ concentrations, to redissolve the AgCl without appreciably redissolving the more insoluble AgBr and AgI. A further distinguishing characteristic is the color of the precipitates: AgCl is white, AgBr is cream-colored, and AgI is light yellow.

The distinctive differences that enable us to separate and test for *bromide ion* and *iodide ion* in the presence of chloride ion depend on differences in their ease of oxidation. Iodide ion is easily oxidized to iodine (I_2) by adding ferric ion (Fe^{3+}), but ferric ion does not affect the others.

After adding hexane (C_6H_{14}) to identify iodine (purple color in hexane) and remove the iodine, we can next oxidize the bromide ion with potassium permanganate or chlorine water solutions, extract the bromine (Br_2) which forms into hexane, and identify it by the brown color produced in the hexane layer.

PHOSPHATE ION

Phosphate ion is the anion of a moderately weak, non-volatile acid—phosphoric acid (H_3PO_4). It is tested for by the typical method of forming a characteristic precipitate. The reagent used is ammonium heptamolybdate solution, to which an excess of ammonium ion is added in order to shift the equilibrium further to the right. An acid solution is necessary. The net ionic reaction is

$$21NH_4^+ + 12\,[Mo_7O_{24}^{6-}] + 7H_3PO_4 + 51H^+ \rightleftharpoons$$
$$7\,(NH_4)_3PO_4 \cdot 12MoO_3(s) + 36H_2O$$

The yellow precipitate is a mixed salt, ammonium phosphomolybdate. Sulfide ion interferes with this test but can be removed first by acidifying the solution with HCl and boiling it.

NITRATE ION

Since all nitrates are soluble, we cannot use a precipitation method to test for nitrate ion. Instead, we use a two-step process. First, nitrate ion is an oxidizing agent that is reduced to nitrite ion in slightly basic $NaHCO_3$ solution by copper-coated zinc metal, a reducing agent:

$$NO_3^- + H_2O + Zn(s) \rightarrow NO_2^- + Zn(OH)_2$$

Second, when acid and excess iodide ion are added, nitrite ion is protonated to form nitrous acid, which oxidizes the iodide ion to form iodine (or triiodide ion):

$$2HNO_2 + 2H_3O^+ + 3I^- \rightarrow 2NO + 4H_2O + I_3^-$$

In the presence of starch indicator, the deep blue color of the starch–triiodide ion complex is formed. Substances such as chlorine, chlorate, sulfite, and sulfide ions interfere, so they must be removed before testing for nitrate.

SUMMARY

Prepare a summary of your results by writing a balanced chemical equation for the reactions that were used to test each anion and the observations that you made to confirm the presence or absence of each anion. Table 34-1 and the Pre-Lab Preparation discussion should serve as valuable resources in identifying the important chemical species. This table should also help you to answer the questions in the report form.

TABLE 34-1
The Behavior of Some Negative Ions with the Metallic Ions Ag^+, Ba^{2+}, Ca^{2+}, and Pb^{2+}.

Negative ions	Metallic ions			
	Ag^+	Ba^{2+}	Ca^{2+}	Pb^{2+}
NO_3^-	Soluble	Soluble	Soluble	Soluble
Cl^-, Br^-, I^-	AgCl, white; AgBr, cream; AgI, yellow. All insoluble in HNO_3. AgCl soluble, AgBr slightly soluble, and AgI insoluble, in NH_3.	Soluble	Soluble	$PbCl_2$ and $PbBr_2$, white, soluble in hot water. PbI_2, yellow, slightly soluble in hot water.
SO_4^{2-}	Moderately soluble	$BaSO_4$, white, insoluble in HNO_3	$CaSO_4$, white, slightly soluble	$PbSO_4$, white, insoluble in HNO_3
SO_3^{2-}	Ag_2SO_3, white, soluble in NH_3 and HNO_3	$BaSO_3$, white, soluble in HNO_3	$CaSO_3$, white, soluble in HNO_3	$PbSO_3$, white, soluble in HNO_3
S^{2-}	Ag_2S, black, soluble in hot, conc. HNO_3	Soluble	Soluble	PbS, black, soluble in HNO_3
PO_4^{3-}	Ag_3PO_4, yellow, soluble in NH_3 and in HNO_3	$Ba_3(PO_4)_2$, white, soluble in HNO_3	$Ca_3(PO_4)_2$, white, soluble in HNO_3	$Pb_3(PO_4)_2$, white, soluble in HNO_3
CO_3^{2-}	Ag_2CO_3, white, soluble in NH_3 and in HNO_3	$BaCO_3$, white, soluble in HNO_3	$CaCO_3$, white, soluble in HNO_3	$PbCO_3$, white, soluble in HNO_3

EXPERIMENTAL PROCEDURE

Special Supplies: One-hole stopper and delivery tube to fit 15-cm test tube as shown in Figure 34-1. Polyethylene transfer pipets with 1-mL calibration marks.

Chemicals: 6 M HCl; 6 M HNO$_3$; 6 M NH$_3$; 0.1 M sodium sulfide, Na$_2$S (freshly made); 0.1 M sodium sulfate, Na$_2$SO$_4$; 0.1 M sodium sulfite, Na$_2$SO$_3$ (freshly made); 0.1 M sodium carbonate, Na$_2$CO$_3$; 0.1 M sodium chloride, NaCl; 0.1 M potassium bromide, KBr; 0.1 M potassium iodide, KI; 1.0 M potassium iodide, KI (used in nitrate ion test); 0.1 M sodium phosphate, Na$_3$PO$_4$; 0.1 M potassium nitrate, KNO$_3$; 0.01 M sodium nitrite, NaNO$_2$; lead acetate paper, Pb(C$_2$H$_3$O$_2$)$_2$; 0.1 M barium chloride, BaCl$_2$; saturated bromine water, Br$_2$(aq) (see Bibliography); 3% hydrogen peroxide, H$_2$O$_2$; 0.02 M calcium hydroxide, Ca(OH)$_2$; mossy zinc metal, Zn(s); 0.1 M silver nitrate, AgNO$_3$; 0.1 M iron(III) chloride, FeCl$_3$; hexane, C$_6$H$_{14}$; saturated chlorine water, Cl$_2$ (see Bibliography); 0.5 M ammonium heptamolybdate (9.8 g/100 mL [NH$_4$]$_6$Mo$_7$O$_{24}$ · 4H$_2$O); 30-mesh zinc metal, Zn(s); 0.03 M CuSO$_4$ in 1 M acetic acid, CH$_3$COOH (0.75 g CuSO$_4$ · 5H$_2$O and 6 g acetic acid in 100 mL of water); 1 M sodium bicarbonate, NaHCO$_3$; 0.5% starch (freshly prepared by boiling for 20 minutes 5 g potato starch / Liter of water); 5 M (55% by weight) tartaric acid (it's OK if the tartaric acid solution is yellow or light-brown).

! **SAFETY PRECAUTIONS:** H$_2$S gas is very toxic and has a characteristic odor of rotten eggs. You should avoid inhaling H$_2$S in Part 1 (Test for Sulfide Ion). The SO$_2$ gas produced in Part 3 (Test for Sulfite Ion) is irritating and toxic and has a strong, suffocating odor. Avoid inhaling SO$_2$ gas. In Part 5 (Test for Chloride Ion), boiling to remove H$_2$S should be done in a fume hood, or you should use an inverted funnel connected to an aspirator (as shown in Figure 5-1) to remove H$_2$S.

Note that 6 M HCl, 6 M HNO$_3$, and 6 M NH$_3$ are corrosive. If they contact your skin, wash them off immediately.

WASTE COLLECTION: While most of the solutions and solids formed in this experiment are nontoxic, some should not get into the environment in the form resulting from this lab. Rather than present an extensive list of hazards, we emphasize that all waste solutions and solids should be collected in an appropriately labeled container. Your instructor may wish to have separate containers for lead waste, silver compounds, and barium products.

First, familiarize yourself with the following test procedures, using 2-mL samples of 0.1 M solutions of each ion to be tested. Any tests that yield unsatisfactory results should be repeated to improve your technique.

Second, answer the review questions and statements in the experiment report before analyzing the unknowns.

Third, obtain one or more unknowns, and perform analyses for each of the ions, using a fresh 2-mL sample for each test.

1. TEST FOR SULFIDE ION.

To about 2 mL of the test solution add a slight excess of 6 M HCl. You may notice an odor of H$_2$S, which is easily detectable in the low parts-per-million range in air. As a second sensitive test, place a piece of moistened lead acetate paper over the mouth of the test tube.[1] Heat the test tube gently in the fume hood. A darkening of the paper indicates S^{2-} in the original solution. (If S^{2-} is found to be present, SO$_3^{2-}$ is not likely to be present—why not?—and the SO$_3^{2-}$ test may then be omitted. Note also the modification of the PO$_4^{3-}$ test when S^{2-} is present.)

2. TEST FOR SULFATE ION.

To 2 mL of the test solution add 6 M HCl drop by drop until the solution is slightly acid. Then add 1 mL of 0.1 M BaCl$_2$ solution, or more as needed to complete the precipitation. A white precipitate of BaSO$_4$ proves the presence of SO$_4^{2-}$. (Save the solution for the sulfite test.)

3. TEST FOR SULFITE ION.

If you noticed a sharp odor of SO$_2$ when the solution from the previous sulfate ion test was made acid with HCl, then SO$_3^{2-}$ is present. If you are in doubt, filter or centrifuge the solution to obtain a clear filtrate, add a drop or more of 0.1 M BaCl$_2$ to be sure all SO$_4^{2-}$ is precipitated, and if necessary add more BaCl$_2$; then refilter or recentrifuge. To the clear solution add 1 to 2 mL of bromine water to oxidize any SO$_3^{2-}$ to SO$_4^{2-}$. Formation of a second white precipitate of BaSO$_4$ now proves the presence of SO$_3^{2-}$.

4. TEST FOR CARBONATE ION.

Fit a 15-cm test tube with a one-hole rubber stopper and bent delivery tube (see Figure 34-1). Place about 3 mL of the test solution in this test tube. If sulfite ion (SO$_3^{2-}$) is present in the unknown, add 1 mL of 3% H$_2$O$_2$ to oxidize it to sulfate ion. Now insert the delivery tube into some clear limewater, 0.02 M Ca(OH)$_2$, in another test tube. When ready, remove the stopper just enough to add a small piece of mossy zinc and 1 mL of 6 M HCl to the test solution. Immediately close the stopper again, and heat the tube gently to boiling to drive any CO$_2$ gas into the limewater. The zinc will react with the HCl to form hydrogen gas, which will help to sweep the CO$_2$(g) into the limewater. Be careful not to let any of the boiling liquid escape through the delivery tube into the limewater. A white precipitate in the limewater indicates CO$_3^{2-}$ or HCO$_3^-$ in the test solution.

5. TEST FOR CHLORIDE ION.

To a 2-mL portion of the test solution, add a few drops of 6 M HNO$_3$, as needed, to make the solution slightly acid. (Test with litmus paper.)

---

[1] If lead acetate paper is not available, a suitable replacement can be made by taking a small strip of filter paper and moistening it with a few drops of 0.1 M lead nitrate solution.

Test solution

Evolved gas

Limewater, Ca(OH)₂

If the evolved gas is CO₂, the limewater will turn milky.

Warm over a low flame.

FIGURE 34-1
The limewater test for carbon dioxide.

Any sulfide ion present can be removed by boiling the solution for a moment in a fume hood. The free sulfur formed in the oxidation of S^{2-} by HNO_3 does not interfere. Add 1 mL of 0.1 M $AgNO_3$. (Here, no precipitate proves the absence of Cl^-, Br^-, or I^-.) Centrifuge the mixture. Test the clear filtrate with 1 drop of 0.1 M $AgNO_3$ for complete precipitation. If necessary, centrifuge again. Discard the filtrate. Wash the precipitate with distilled water to remove excess acid and silver ion. To this precipitate add 3 mL of distilled water, 4 drops of 6 M NH_3, and 12 mL of 0.1 M $AgNO_3$. (The proportions are important, since we want to dissolve only the AgCl from any mixture of AgCl, AgBr, and AgI.) $Ag(NH_3)_2^+$ and Cl^- will form. Shake the mixture well, and centrifuge. Transfer the clear solution to a clean test tube, and acidify with 6 M HNO_3. A white precipitate of AgCl confirms Cl^-.

6. TEST FOR IODIDE ION. To 2 mL of the test solution add 6 M HCl to make the solution acid. If S^{2-} or SO_3^{2-} is present, boil the solution to remove the ion. Add 1 mL of 0.1 M $FeCl_3$ to oxidize any I^- to I_2. (Br^- is not oxidized by Fe^{3+}.) Add 1 mL of hexane (C_6H_{14}), and agitate the mixture. A purple color of I_2 in the hexane layer indicates I^- was present in the original sample. (Save the mixture for the Br^- test.)

7. TEST FOR BROMIDE ION. If no I^- was present in the *preceding* mixture, add 2 mL of chlorine water, and agitate it. A brown color in the hexane layer indicates Br^-. If I^- *was present*, separate, by means of a medicine dropper, as much as possible of the preceding iodide test solu-

tion above the hexane layer that contains the I_2, and place it in a clean test tube. Again extract any remaining I_2 by adding 1 mL of hexane, agitating the mixture, and separating the solution. Repeat this process as many times as necessary to eliminate the violet I_2 color from the hexane layer. The solution may be boiled for a moment to remove any remaining trace of I_2. Then add 2 mL of chlorine water and 1 mL of hexane, and agitate the mixture. A brown color indicates Br^-.

8. TEST FOR PHOSPHATE ION. First mix about 1 mL of 0.5 M $(NH_4)_6Mo_7O_{24}$ reagent with 1 mL of 6 M HNO_3. (If a white precipitate forms, dissolve it by making the solution basic with 6 M NH_3, then reacidify with 6 M HNO_3.) If S^{2-} has been found in the unknown, first make a 2-mL portion of the test sample distinctly acid with 6 M HCl, boil it a moment to remove all H_2S, then add this (or, if no S^{2-} is present, a 2-mL sample of the original unknown) to the clear molybdate solution. If iodide ion, I^-, is found to be present, add 1 mL of 6 M HNO_3 to the test solution, and place the test tube in a boiling water bath for 5 min to remove the iodide. Then add 1 mL of the ammonium molybdate reagent. A yellow precipitate of ammonium phosphomolybdate, $(NH_4)_3PO_4 \cdot 12MoO_3$, appearing at once or after the mixture has been warmed a few minutes to about 40 °C, indicates the presence of PO_4^{3-}.

9. TEST FOR NITRATE ION. Read again the synopsis of the method in the Pre-Lab section. A blank containing unreduced nitrate ion and a control using 1.0×10^{-4} M sodium nitrite solution are run, along with the test solution containing nitrate ion.

Preparing copper-coated zinc metal. Weigh 1.0 g of 30-mesh zinc metal, and place it in a 13 × 100 mm test tube. Add to the zinc metal 2 mL of 0.03 M $CuSO_4$ / 1 M acetic acid solution. Shake the mixture, or use a clean polyethylene transfer pipet to thoroughly mix the solution. (Place the tip of the pipet about one cm above the zinc metal in the bottom of the test tube, draw up the solution into the pipet, and rapidly expel it.) Initially the solution turns cloudy from the evolution of small bubbles of hydrogen gas. After vigorous mixing for about 2 minutes, the solution will become clear, and the blue color of Cu(II) ion will no longer be visible.

Next draw off and discard the solution with a transfer pipet, leaving in the test tube the zinc metal, now coated with copper metal. Add 5 mL of deionized water. Mix thoroughly for a few seconds, then draw off and discard the solution. Repeat the washing procedure two more times with 5 mL portions of deionized water, drawing off and discarding the wash water.

Completing the test. **Note:** The test for nitrate ion is very sensitive. Solutions containing concentrations of nitrate greater than 0.001 M should be diluted with deionized water until the nitrate ion concentration is about 0.001 M. When working with unknown samples containing mixtures of anions, use pH paper or universal indicator to check the pH of a sample of the diluted solution suspected of containing nitrate ion. The pH should be in the range pH 6–8. If not, adjust the pH by adding very small amounts of NaOH or HCl until the solution is approximately neutral.

Prepare 0.001 M KNO_3 solution by diluting 1 mL of 0.1 M KNO_3 with 99 mL of deionized water. Mix this solution thoroughly. Add 2 mL of the 0.001 M KNO_3 solution to the test tube containing the copper-coated zinc metal. Add 2 drops of 1 M $NaHCO_3$ solution to make the solution slightly basic. As before, mix the solution and copper-coated zinc metal vigorously for about 2 minutes. During the mixing some of the nitrate ion is reduced to nitrite ion. After mixing for 2 minutes, draw off 1 mL of the solution and place it into a clean 13 × 100 mm test tube. Add, in the specified order, 2 drops of 1 M KI, 2 drops of 0.5% starch indicator solution, and 5 drops of 5 M tartaric acid solution. Mix the solution thoroughly. In a few seconds after adding the tartaric acid solution, a dark blue color should appear; it will reach its full intensity in 2 to 3 minutes. The formation of the dark blue color confirms the formation of iodine by reaction of iodide with nitrous acid. In the absence of interfering substances, the appearance of the blue color is a positive test for nitrate ion.

Blank and control tests. To run the blank test, place 1 mL of 0.001 M KNO_3 in a clean test tube, add 1 drop of 1 M $NaHCO_3$, 2 drops of 1 M KI, 2 drops of 0.5% starch indicator, and 5 drops of 5 M tartaric acid. Mix thoroughly. No blue color should form, indicating the absence of iodine and nitrite ion. If a faint blue color does form, it may indicate that there has been some air oxidation of the KI solution to form some iodine, or interfering substances previously mentioned may be present (chlorine or the ions chlorate, sulfite, or sulfide).

To run the control test, first prepare a 1 × 10⁻⁴ M $NaNO_2$ solution by diluting 1 mL of 0.01 M $NaNO_2$ solution with 99 mL of deionized water and mixing thoroughly. Place 1 mL of the 1 × 10⁻⁴ M nitrite solution in a clean test tube, then add 1 drop of 1 M $NaHCO_3$, 2 drops of 1 M KI, 2 drops of 0.5% starch indicator, and 5 drops of 5 M tartaric acid, and mix thoroughly. An intense deep blue color should immediately appear. This confirms that nitrite ion is present and that the starch indicator is functioning properly.

Usually the intensity of the blue color in the 1 × 10⁻⁴ M nitrite ion control is greater than that observed with the sample of 0.001 M KNO_3 solution, even though the nitrite control has a tenfold smaller concentration. This suggests that less than 10 percent of the nitrate originally present is reduced to form nitrite ion. Even though nitrate ion is only partially reduced, the test is still very sensitive, easily detecting the presence of 0.001 M nitrate ion.

☞ Consider This

In the test for Cl^- and SO_4^{2-}, explain fully why the addition of acid will dissolve other insoluble silver salts and barium salts, respectively, which would interfere with the tests if the solution were neutral. (See Table 34-1.)

Suppose a white precipitate is obtained on adding $BaCl_2$ reagent to a neutral unknown solution. What might the precipitate be? Write formulas for as many possible substances as you can, considering both positive and negative ions as given in Table 34-1.

A positive test for SO_4^{2-} is to be expected whenever you have SO_3^{2-} in an unknown solution, even when no SO_4^{2-} has been placed in the solution. What could explain this?

Why cannot both S^{2-} and SO_3^{2-} be present in the same solution? Explain, and write the equation.

Both CO_3^{2-} and HCO_3^- respond to the usual carbonate ion test. How could you distinguish between solutions of 1 M $NaHCO_3$ and 1 M Na_2CO_3? (Hint: One way would be to consider the pH of the solution.)

BIBLIOGRAPHY

Baker, Jr., R.C. "Checking Trace Nitrate in Water and Soil Using an Amateur Scientist's Measurement Guide," *J. Chem. Educ.,* **1995,** 72: 57–59.

Feigl, F. *Spot Tests in Inorganic Analysis,* Elsevier Publishing Co., New York, 1958, p. 332.

Hiegel, G. A.; Abdala, M. H.; Burke, S. V.; Beard, D. P. "Methods for Preparing Aqueous Solutions of Chlorine and Bromine for Halogen Displacement Reactions," *J. Chem. Educ.,* **1987,** 64: 156.

Holmes, Jr., L. H. "An Easy Way to Make Chlorine Water," *J. Chem. Educ.,* **1997,** 74: 1326.

The Qualitative Analysis of Some Common Anions

NAME

SECTION   LOCKER

INSTRUCTOR   DATE

SUMMARY OF THE EXPERIMENT

Write balanced chemical equations to describe the reactions which you used to identify the anions in this experiment. For this summary, ignore the complications which may be caused by the presence of other ions. Also describe the observations (i.e., what you actually saw) to confirm the presence or absence of each ion. The entries for phosphate have been filled in as an example (and because it is one of the more complicated reactions).

Species	Equations	Observations
1. Sulfide		
2. Sulfate		
3. Sulfite		
4. Carbonate		
5. Chloride		
6. Iodide		
7. Bromide		
8. Phosphate	$21NH_4^+ + 12\,[Mo_7O_{24}{}^{6-}] + 7H_3PO_4 + 51H+ \rightleftharpoons$ $7\,(NH_4)_3PO_4 \cdot 12MoO_3(s) + 36H_2O$	A yellow precipitate confirms phosphate, PO_4^{3-}
9. Nitrate		

SOME MORE BRAINTEASERS

1. Give the formulas of all new molecules or ions formed when solutions of the following are mixed. Write "N.R." (for "no reaction") if there is none.

(a) Cl^-, Br^-, I^-, Fe^{3+} _____

(b) Fe^{2+}, Br_2 water _____

(c) Cl^-, Br^-, I^-, Cl_2 water _____

(d) $CO_2(g)$, $Ca(OH)_2$ _____

(e) $SO_2(g)$, $CaCl_2$, HCl _____

(f) SO_3^{2-}, H_2O_2 (acid) _____

(g) S^{2-}, SO_3^{2-} (acid) _____

(h) $Pb(C_2H_3O_2)_2$, $H_2S(g)$ _____

(i) Ag^+, Cl^-, NH_3 (excess) _____

(j) Ag^+, I^-, NH_3 (excess) _____

2. An unknown solution (I) is to be tested for the anions, or their derivatives, listed below right. ("Derivatives" are, e.g., PO_4^{3-} or H_3PO_4.) The following tests are performed on *separate* portions of the unknown.

(a) When a test portion is acidified with HCl, and lead acetate is held in the mouth of the test tube, the paper turns black.

(b) A test portion is acidified with HCl, boiled, and cooled; then Cl_2 water and hexane are added. After the mixture is shaken, the hexane layer is a clear, light reddish brown color.

(c) Addition of an excess of HNO_3 produces a colorless gas that forms a white precipitate when bubbled through limewater.

On the basis of these tests considered as a whole, mark, at the right, a plus sign (+) for each ion or its derivative that is definitely present in the original solution, a minus sign (−) if the ion is definitely absent, or a question mark (?) if there is no evidence that the ion is present or absent.

Cl^- _____

Br^- _____

I^- _____

SO_4^{2-} _____

SO_3^{2-} _____

S^{2-} _____

NO_3^- _____

CO_3^{2-} _____

PO_4^{3-} _____

👉 Consider This

In the tests for Cl^- and SO_4^{2-}, explain fully why the addition of acid will dissolve other insoluble silver salts and barium salts, respectively, which would interfere with the tests if the solution were neutral. (See Table 34-1.)

Suppose a white precipitate is obtained on adding $BaCl_2$ reagent to a neutral unknown solution. What might the precipitate be? Write formulas for as many possible substances as you can, considering both positive and negative ions as given in Table 34-1.

A positive test for SO_4^{2-} is to be expected whenever you have SO_3^{2-} in an unknown solution, even when no SO_4^{2-} has been placed in the solution. What could explain this?

Why cannot both S^{2-} and SO_3^{2-} be present in the same solution? Explain, and write the equation.

Both CO_3^{2-} and HCO_3^- respond to the usual carbonate ion test. How could you distinguish between solutions of 1 M $NaHCO_3$ and 1 M Na_2CO_3? (Hint: One way would be to consider the pH of the solution.)

REPORT OF UNKNOWN SOLUTIONS

Write in the correct formula for each ion found, for each unknown. (The ions, if added at all in these unknowns, are present in more than traces.)

Unknown no. _____   Ions found _____

Unknown no. _____   Ions found _____

EXPERIMENT
35

Qualitative Analysis of Unlabeled Solutions:
The N-Solution Problem

PURPOSE

To use resources available to you, including previous experience and literature sources, to identify a set of solutions contained in unlabeled bottles using no other reagents than the solutions themselves.

PRE-LAB PREPARATION

In this experiment we assume that you have been introduced to some of the techniques of qualitative analysis described in Experiments 31 through 34. The separation and identification of ions is generally accomplished by making use of the differences in solubilities of various salts and the tendencies of ions to form complex ions or to be oxidized or reduced. The strategy of developing a qualitative analysis scheme rests first on observing the behavior of cations and anions with various reagents, then attempting to construct a logical scheme of analysis.

This experiment poses a different kind of problem. You will be given N = 7 different unknown solutions, labeled 1, 2, 3, . . . 7. The composition of these solutions, (not in order) is

0.5 M NaOH	0.1 M AgNO₃	0.1 M NaBr
0.1 M KI	0.5 M NH₄Cl	0.5 M Na₂SO₄
0.1 M Fe(NO₃)₃		

Your task will be to identify the contents of each numbered bottle, using no other reagents, indicators, or instruments. One or more unknown solutions must serve as the reagent(s) for identifying another unknown solution. This task may seem impossible, since all the solutions are unknown at the beginning, and it will

indeed be difficult to make all of the identifications correctly if you do not prepare beforehand.

To solve this chemical "crossword" puzzle you will need to make use of your knowledge of the solubilities of salts and the colors of their precipitates, the odors of volatile substances (CO_2, NH_3, H_2S, SO_2, halogens), the possibility of complex ion formation, the interactions of acids and bases, and the possibility of oxidation-reduction reactions.

It is essential that you do some preliminary preparation before coming to the laboratory. Prepare your pre-lab preparation work sheet (Tables 1 and 2 of the report form), showing in tabular form the expected chemical behavior of these solutions when roughly equal volumes are mixed.

For some ions, you can exclude certain types of reactions at the outset. For example, Na^+, K^+, and Ba^{2+} are not likely to be oxidized because they have only one stable oxidation state. Also, they are not likely to be reduced in aqueous solution. In contrast, such ions as I^- or S^{2-} can be oxidized by even moderately strong oxidizing agents such as Fe^{3+}.

In working out the predictions of the chemical reactions that you expect between various ions, you may find it helpful to consult Tables 6 through 10 of the Appendix. (Your instructor may also suggest other reference books on inorganic chemistry or qualitative analysis as sources of information not given in your textbook. For example, you can consult the *CRC Handbook of Chemistry and Physics* to determine the color and solubility of various salts.) If the solubility of a salt is between 0.5 and 10 g per 100 mL, record the figure. If the solubility is less than 0.5 g per 100 mL, you can consider the compound insoluble and assume that it is likely to form a precipitate when its constituent ions are mixed.

427

EXPERIMENTAL PROCEDURE

Special Supplies: Seven 500-mL wash bottles to dispense the seven solutions. Polyethylene transfer pipets (seven per student).

Chemicals: 0.5 M NaOH, 0.1 M AgNO$_3$, 0.1 M NaBr, 0.1 M KI, 0.3 M NH$_4$Cl, 0.5 M Na$_2$SO$_4$, 0.1 M Fe(NO$_3$)$_3$.

•••

NOTE: Prepare beforehand as directed in the Pre-Lab Preparation, and fill in Tables 1 and 2 of the report form. Follow the instructor's directions when you receive your unknown solutions. You might receive them already prepared in test tubes or in dropping bottles, or the seven solutions might be furnished to the entire laboratory section in squeeze bottles from which you take about 5–6 mL of each solution in 13 × 100 mm test tubes (fill them to approximately ⅔ of the test tube capacity). Label each solution with its identification number.

•••

In mixing solutions, place 5–10 drops of a solution in a small test tube, then drop by drop add an equal volume of another unknown solution. Since the order of addition may change the nature of some of the reactions, you might try adding the reagents in reverse order. Record your observations in brief form in Table 3 of the report form, and describe them completely in Table 4.

On the basis of your observations, identify as many of the solutions as you can. With known reagents now at your disposal, devise a logical scheme to identify any remaining unknowns. For example, you may wish to mix three different solutions, or to mix a drop or two of one reagent with a mL of another. If you plan thoughtfully, you may not find it necessary to try all of the possible binary combinations of the unknown solutions.

Your laboratory report should include the results of your pre-lab preparation in Tables 1 and 2, a record of your observations in Tables 3 and 4, and the identification of the unknowns on the answer sheet (Sheet 3) of the report form. In addition to the accuracy of your identifications, part of your grade will be based on the reasoning you adopt in your approach to the problem.

☞ **Consider This**

This experiment may be modified to make a three-lab project in the following way. Divide the lab section into groups of two or three people. Each group is asked to come up with a thought provoking, but not impossible, set of N solutions for an unlabeled bottle experiment. As part of its task, each group is required to

devise and test a strategy that could be used to unambiguously identify the contents of each unlabeled bottle. Two lab periods are allotted for developing and testing the set of N solutions. Then, in the third lab period the groups exchange their set of N unlabeled bottles with another group, and each attempts to solve the other group's set. This project often leads to a considerable amount of lively, even heated discussion.

What should be the value of N? The amount of work required rises rapidly as the value of N increases. If you have N unlabeled solutions, show that the formula for expressing the number of possible combinations, mixing two solutions at a time, is equal to N(N − 1)/2. Here we are distinguishing combinations from permutations. When A and B are mixed together there is only one combination (A + B = B + A), but there are two permutations, which we might use to distinguish the order of addition: A added to B and B added to A (A → B ≠ B → A). It is suggested that a value of N somewhere between six and eight gives a reasonably challenging project, N = 8 requiring about twice as much work as N = 6.

BIBLIOGRAPHY

Finholt, J. E. "The Nine-Bottle Experiment Revisited," *J. Chem. Educ.,* **1984,** *61:* 849. A description of a three-lab project in which groups of students devise N-solution sets and try to solve each other's sets.

MacWood, G. E., Lassettre, E. N., and Breen, G. "A Laboratory Experiment in General Chemistry," *J. Chem. Educ.,* **1940,** *17:* 520. One of the earliest descriptions of the unlabeled bottle problem.

Olander, C. R. "An n-Bottle Lab Exercise with No Hazardous Waste," *J. Chem. Educ.,* **1996,** *73:* 849–850.

Smith, D. G. "The Case of the Unlabeled Bottles," *J. Chem. Educ.,* **1977,** *54:* 701.

Ricketts, J. A. "Laboratory Exercise Emphasizing Deductive Reasoning," *J. Chem. Educ.,* **1960,** *37:* 311. (This article describes a procedure that you might employ for working out a qualitative analysis scheme based on your own observations on the individual ions that might be present in an unknown.)

Stieg, S. "Pattern Recognition in Descriptions of Cation/Anion Reactions for Solving an n-Bottle Puzzle," *J. Chem. Educ.,* **1988,** *65:* 360–361.

Zuehlke, R. W. "The Case of the Unlabeled Bottles," *J. Chem. Educ.,* **1966,** *43:* 601–602. (The author presents a variation of the nine-solution problem.)

---

[1]*CRC Handbook of Chemistry and Physics,* published annually by CRC Press, Inc., Boca Raton, Florida 33431.

Qualitative Analysis of Unlabeled Solutions: The N-Solution Problem

NAME		
SECTION		LOCKER
INSTRUCTOR		DATE

PREPARATION AND OBSERVATIONS

Table 1 (Fill out this table as part of your pre-lab preparation.)

In Table 1 below, write in the net ionic reactions that you would expect for each pair of solutions that are mixed together. Classify each reaction as either a precipitation (P), complex ion formation (CIF), acid-base (neutralization) (AB), or oxidation-reduction (OR) reaction. *For precipitation reactions look up and record the solubility of any insoluble or moderately soluble compounds that might be formed.* Note also any color changes that you expect and indicate whether you anticipate the evolution of a gas with a characteristic odor. If you expect no reaction, write "NR."

Solutions Mixed	Expected Reaction	Classification
$NaOH + AgNO_3$		
$NaOH + NaBr$		
$NaOH + KI$		
$NaOH + NH_4Cl$		
$NaOH + Na_2SO_4$		
$NaOH + Fe(NO_3)_3$		
$AgNO_3 + NaBr$		
$AgNO_3 + KI$		
$AgNO_3 + NH_4Cl$		

Solutions Mixed	Expected Reaction	Classification
$AgNO_3 + Na_2SO_4$	_____	_____
$AgNO_3 + Fe(NO_3)_3$	_____	_____
$NaBr + KI$	_____	_____
$NaBr + NH_4Cl$	_____	_____
$NaBr + Na_2SO_4$	_____	_____
$NaBr + Fe(NO_3)_3$	_____	_____
$KI + NH_4Cl$	_____	_____
$KI + Na_2SO_4$	_____	_____
$KI + Fe(NO_3)_3$	_____	_____
$NH_4Cl + Na_2SO_4$	_____	_____
$NH_4Cl + Fe(NO_3)_3$	_____	_____
$Na_2SO_4 + Fe(NO_3)_3$	_____	_____

TABLES 2 AND 3

Before you go into the laboratory, fill in the squares of Table 2 with a brief description of the results you expect based on your pre-lab preparation summarized in Table 1. For example, if you expect a yellow precipitate you might record "yel. ppt.", or if you expect evolution of NH_3 gas you might write "$NH_3(g)\uparrow$".

As you begin the process of identification in the laboratory, record in the squares of Table 3 a brief description of the results obtained from the mixing of the unknown solutions, penciling in the tentative identification beside the number when you think you have correctly identified an unknown. While you are working and observing, also record in Table 4 a fuller description of your observations. Remember that you may want to mix three or more solutions or to change the order in which you add the solutions before making positive identifications of some unknowns.

Table 2

	AgNO$_3$	NaBr	KI	NH$_4$Cl	Na$_2$SO$_4$	Fe(NO$_3$)$_3$
NaOH →						
AgNO$_3$ →						
NaBr →						
KI →						
NH$_4$Cl →						
Na$_2$SO$_4$ →						

Table 3

1 ↓	2 ↓	3 ↓	4 ↓	5 ↓	6 ↓
2					
3					
4					
5					
6					
7					

Table 4

In the table below, record a complete description of the results obtained from mixing two or more unknown solutions.

Solutions mixed	Observations	Conclusions

ANSWER SHEET

Record below your identification of the contents of each of the seven unlabeled solutions. If you are not confident of an identification, leave it blank.

Solution number   Identity _____

1   _____

2   _____

3   _____

4   _____

5   _____

6   _____

7   _____

In the space below, indicate by writing a few concise sentences, chemical equations, or a flow sheet, the reasoning you have used in making the identifications.

☞ Consider This

This experiment may be modified to make a three-lab project in the following way. Divide the lab section into groups of two or three people. Each group is asked to come up with a thought provoking, but not impossible, set of N solutions for an unlabeled bottle experiment. As part of its task, each group is required to devise and test a strategy that could be used to unambiguously identify the contents of each unlabeled bottle. Two lab periods are allotted for developing and testing the set of N solutions. Then, in the third lab period the groups exchange their set of N unlabeled bottles with another group, and each attempts to solve the other group's set. This project often leads to a considerable amount of lively, even heated discussion.

What should be the value of N? The amount of work required rises rapidly as the value of N increases. If you have N unlabeled solutions, show that the formula for expressing the number of possible combinations, mixing two solutions at a time, is equal to $N(N-1)/2$. Here we are distinguishing combinations from permutations. When A and B are mixed together there is only one combination $(A + B = B + A)$, but there are two permutations, which we might use to distinguish the order of addition: A added to B and B added to A $(A \rightarrow B \neq B \rightarrow A)$. It is suggested that a value of N somewhere between six and eight gives a reasonably challenging project, N = 8 requiring about twice as much work as N = 6.

Paper Chromatography in the Separation of Iron(III), Cobalt(II), Nickel(II), and Copper(II) Ions

PURPOSE

To develop an understanding of how substances can be separated by differential migration, which is the fundamental principle underlying all forms of chromatographic separation. To learn how the chemical properties of transition metal complexes can be adjusted by controlling the acidity and concentration of chloride ion in order to separate the metal complexes on cellulose filter paper. To identify the cations present in an unknown sample containing Fe(III), Co(II), Ni(II), or Cu(II) ions.

PRE-LAB PREPARATION

Experiments 31 through 35 involved the separation and identification of a number of cations and anions by selective precipitation or by selective formation of complex ions with the use of a variety of reagents. It would be ideal if each cation or anion formed a precipitate or colored complex with just one unique reagent. Then no prior separation would be necessary, and adding the proper reagent would detect the sought-for ion. However, this ideal situation has not yet been attained. Consequently, simpler and more efficient methods of separation have been sought to replace the somewhat tedious methods that employ precipitation.

Instruments exist that can identify and quantitatively determine metal ions by their emission spectra, but these instruments are expensive and can be used by only one person at a time. In this experiment we will use the simple yet powerful method of separation called *chromatography*, the history and methods of which were discussed in Experiment 3.

Let's briefly review the important general principle employed in all forms of chromatography: A mixture of solutes in a *moving phase* passes over a selectively adsorbing medium, the *stationary phase*. Separation occurs because the solutes have different affinities for the stationary and moving phases. Solutes that have a greater affinity for the moving phase will spend more time in it and therefore will move along faster and travel farther than solutes that spend more time in the stationary phase. This leads to a *differential rate* of migration of solutes, which means that the solutes move at differing, individual rates and thus become physically separated.

In the moving phase a solute that has no affinity for the stationary phase will move along the medium at the same rate as the molecules of solvent. A solute that has no affinity for the moving phase will not move at all—it will be totally adsorbed by the stationary phase. Ideally we want to adjust the conditions of the separation so that solutes will move along at a rate that is between these two extremes, as shown in Figure 36-1.

There is a variety of ways of achieving a flow around or through a stationary phase (see Experiment 3). Because of its simplicity and low cost, we will use paper chromatography. Paper is mainly cellulose, which contains a large number of hydroxyl ($-OH$) groups; as a result, paper has a great affinity for water (see Figure 36-2).

The solvents employed in paper chromatography must wet the paper so that the mobile phase will move through the paper fibers by capillary attraction. The solute is first applied to the dry paper, not far from one end. Then, during the chromatographic development, the end of the paper nearest the spot is placed in a container of solvent, which begins to flow through the paper. The solutes move along (we hope), and as they do, they continually move back and forth between the moving and stationary phases. The fraction of time they spend in the moving phase determines how fast they move along. (Look again at Figure 36-1.)

FIGURE 36-1

Schematic illustration of separation by differential migration. Parts A through D illustrate the separation of a three-component mixture by differential migration. Note that at the beginning of a separation (A), three components are clustered together at the point of sample application. As the sample migrates across the system (B through D), the three sample components are gradually separated, the fastest-moving component spending the most time in the moving fluid phase. E displays a smoothed plot of the concentration of each substance as a function of the distance from the origin for the distribution shown in D. (After Barry L. Karger et al., *Introduction to Separation Science*, Wiley, New York. Copyright © 1973.)

After the chromatogram has been developed, the positions of all the solutes on the paper are located by spraying the paper with a reagent that reacts with the metal ions to form colored spots. The distance traveled by each solute is measured, as is the distance traveled by the moving solvent front. The ratio of these distances is the retention factor, R_f.

$$R_f = \frac{\text{distance traveled by the solute}}{\text{distance traveled by the solvent front}}$$

Our goal is to arrange conditions so that the solutes will be spread out along the paper, each with a significantly different R_f value.

After the spots are located and their R_f values measured, they are compared with standard known solutes run under the same conditions. If possible, known and unknown substances are run on the same paper. If the known and unknown compounds have identical R_f values in different solvent systems, and if they react in the same way with various chromogenic (color-forming) reagents, it is probable that the compounds are identical.

SEPARATION OF Fe^{3+}, Co^{2+}, Ni^{2+}, AND Cu^{2+}

In separating these transition metal ions, we make use of a mixed solvent system containing water, acetone, and hydrochloric acid. We may envision the stationary phase as a layer of water adsorbed, or hydrogen bonded, to the many hydroxyl groups contained in the cellulose molecules of the paper. The moving phase contains HCl and is rich in the less polar acetone. The metal ions, ranging from water-solvated cations to anionic *chloro*-complexes, will be distributed between the moving and the stationary phases. The metal ions that form weak *chloro*-complexes, or that form no *chloro*-complexes at all, will spend most of their time in the predominantly aqueous stationary phase. The metal ions that form *chloro*-complexes will migrate more rapidly because they will spend most of their time in the acetone-rich phase, either as neutral *chloro*-complexes, MCl_2 or MCl_3 (where M represents a metal ion), or as protonated anionic *chloro*-complexes, such as HMCl or $HMCl_4$ (e.g., $HFeCl_4$).

By adjusting the concentrations of HCl and acetone, we are able to control the composition of the metal ion complexes (and their net charges) and thus control their relative affinities for the aqueous stationary phase and organic-rich moving phase. Such adjustments allow us to optimize the separation of the four metal ions.

FIGURE 36-2

The structure of cellulose.

EXPERIMENTAL PROCEDURE

Special Supplies: Whatman No. 1 paper, cut in 11 × 23.5 cm sheets; thin polyethylene film, such as Handi-Wrap; rubber bands; plastic mm rulers; pencils; glass capillaries of about 0.5-mm internal diameter; ovens or hair dryers are useful.

Chemicals: Solutions of iron(III) chloride, $FeCl_3$; cobalt(II) chloride, $CoCl_2$; nickel(II) chloride, $NiCl_2$; copper(II) chloride, $CuCl_2$ (nitrate salts may also be used); each containing 5 g/L of the ion dissolved in 1 M HCl; you will also be given a solution mixture containing all four metal ions, 5 g/L of each dissolved in 1 M HCl. Finally, you will also be provided with three unknown solutions containing one or more of the metal ions at similar concentrations, 6 M ammonia, NH_3; 0.1 M sodium sulfide, Na_2S; 6 M hydrochloric acid, HCl; acetone, CH_3COCH_3.

! SAFETY PRECAUTIONS: The organic solvent (acetone) used in this experiment is flammable. There must be no open flames in the laboratory.

$NiCl_2$ is recognized as a carcinogen. Avoid skin contact with $NiCl_2$ solution. Wipe up any spills carefully, and wash your hands thoroughly to minimize contact.

Ammonia gas is irritating to the lungs and toxic. Both 6 M NH_3 and 0.1 M Na_2S are corrosive to the skin and to the mucous membranes of the nose and lungs. Spraying these two solutions on the paper chromatograms must be done in a fume hood. Take care not to inhale the mist from the sprayer or to get the spray on your skin. If a hood is not available, place the paper cylinder in a large beaker covered with plastic film. Cut a small 2- to 3-cm hole in the plastic film, insert the nozzle of the sprayer through the hole, and spray the paper cylinder while rotating the beaker.

WASTE COLLECTION: A container should be provided, in the hood if possible, to collect the developing solvent containing HCl and acetone.

Take an 11 × 23.5 cm piece of Whatman No. 1 paper and, use a pencil to draw a line parallel to the long dimension and 2 cm from the edge. Then, pencil in eight small Xs on the line, beginning 3 cm from the edge of the paper and spacing the marks about 2.5 cm apart.

Prepare fresh developing solvent by putting 10.0 mL of 6.0 M HCl and 35 mL of acetone in the bottom of a clean, dry 800-mL beaker; cover the beaker with plastic wrap held in place with a rubber band.

Lay the paper down on a clean paper towel, and carefully spot a tiny drop of each of the four metal ion solutions on four of the Xs, beginning in the lower left corner. Use a glass capillary of about 0.5-mm ID to spot the solutions. Each spot should be no larger than 3 mm. Put a tiny drop of the solution mixture (all four metal ions) on the fifth X and a drop of each of the three unknown solutions on the last three Xs.

You can expedite drying of the spots by gently waving the paper in the air. After they have dried, form the paper into a cylinder and staple the edges together, leaving a gap of 0.5 cm between the edges (see Figure 36-3). Put the paper cylinder in the beaker containing the developing solvent, placing edge nearest the spots at the bottom. Take care that the paper does not touch the wall of the beaker and that the surface of the solvent is below the level of the spots. Immediately seal the beaker with plastic wrap, and allow the solvent to rise to within 1 cm of the top of the paper (about 20 to 30 mins). Observe the development frequently to make sure that the solvent front does not rise all the way to the top of the paper. Also note the colors of the spots as they move along the paper.

When development is finished, remove the paper from the beaker, and *immediately*, before the solvent evaporates, mark the solvent front with a soft pencil. Then dry the paper thoroughly. You can do so in a fume hood by gently waving the paper or more quickly if you put the paper in an oven[1] or use a hair dryer to blow a warm stream of air over the paper. (The laboratory hot-air blower is not recommended, because it will scorch the paper unless it is used very judiciously.)

After the paper has dried, note the colors and locations of any visible spots. Then, in the hood, spray the paper lightly with 6 M NH_3. (Any kind of sprayer providing a fine mist is satisfactory.) If a fume hood is not available, use the procedure described in the safety precautions. The paper should be thoroughly moistened but not dripping wet. You should observe visible changes in the colors

FIGURE 36-3

Preparing the samples for paper chromatography. After the sample solutions have been spotted on the X marks, the paper is rolled into a cylinder and stapled so that it will stand upright. Then it is placed in the beaker containing the developing solvent, and the beaker is sealed with plastic wrap.

---

[1]The temperature should be set at 60 °C. If it is 100 °C, the paper is likely to char and turn dark.

of some of the spots. Note the color and location of any spots that change or become visible. (The presence of a heavy deposit of NH_4Cl salt indicates that the paper was not dried sufficiently to remove most of the HCl before spraying with NH_3.)

Now dry the paper again to remove most of the NH_3, and spray with 0.1 M Na_2S solution. Brown to black spots should immediately form, and each of the four metal ions should produce a visible spot. Locate each of the spots and note its color.

Locate the densest part of the spot for each metal ion, and draw a pencil line through it. Measure the distance from the pencil line at the point of application to the densest part of the spot, and the distance from the point of application to the solvent front. Calculate the R_f values for each spot, and record on your data sheet.

Using the measured R_f values for the known and unknown solutions and for the characteristic colors and intensities of the spots, identify and report the cations that are present in each of the three unknown solutions.

If time permits, run another set of chromatograms to check the reproducibility of the R_f values. Each time you repeat the experiment, make up a fresh developing solvent mixture.

You might also wish to investigate the effect of varying the HCl concentration, keeping the amount of acetone and the total solvent volume constant.

Consider This

Chromatography is an arena in chemistry where both science and art come into play. Use the techniques you developed in the lab to create a chromatographic work of art.

The mechanism of chromatography involves the interplay of the intermolecular forces of the solvent (mobile) phase, the paper (stationary) phase, and the sample. Experiment with the R_f values as functions of the HCl/acetone ratio. Can you correlate your results with the chloro-complex formation constants? The pH?

BIBLIOGRAPHY

Skovlin, D. O. "The Paper Chromatographic Separation of the Ions of Elements 26 through 30," *J. Chem. Educ.*, **1971,** *48:* 274.

Paper Chromatography in the Separation of Iron(III), Cobalt(II), Nickel(II), and Copper(II) Ions

NAME _____

SECTION _____ LOCKER _____

INSTRUCTOR _____ DATE _____

DATA AND RESULTS

1. Record your paper chromatography data in the table below.

   Distance of solvent front from point of application for chromatogram no. 1 _____ mm

   chromatogram no. 2 _____ mm

Sample	Color with solvent	Color with NH_3	Color with Na_2S	Zone distance from point of application (mm)
Fe^{3+}				
Co^{2+}				
Ni^{2+}				
Cu^{2+}				
Mixture containing Fe^{3+} Co^{2+} Ni^{2+} Cu^{2+}				
Unknown no. _____				
Unknown no. _____				
Unknown no. _____				

2. Give your paper chromatography results below. Attach your chromatogram to the report if requested by your instructor.

	Fe^{3+}	Co^{2+}	Ni^{2+}	Cu^{2+}	Mixture	Unknown No. _____	Unknown No. _____	Unknown No. _____
Average R_f values of zones					Fe^{3+} Co^{2+} Ni^{2+} Cu^{2+}			

Unknown no. _____

Unknown no. _____

Unknown no. _____

Cations present: _____

Cations present: _____

Cations present: _____

EXERCISES

1. Write an equation that represents the formation of the tetrachloroiron(III) complex, $FeCl_4^-$ from the aquo complex, $Fe(H_2O)_6^{3+}$.

2. Write an equation that explains the formation of the blue color when Cu(II) is exposed to the vapors of NH_3. [Hints: NH_3 replaces H_2O in the $Cu(H_2O)_4^{2+}$ aquo complex.]

3. Write equations for the reactions of Fe(III), Co(II), Ni(II), and Cu(II) ions with Na_2S.

4. What would you expect to happen to the R_f values of Fe(III), Co(II), and Cu(II) if you decreased the HCl concentration? Why?

5. Why is it important to have clean hands when handling chromatograms?

6. Why is a pencil preferred rather than a pen for marking the points of application?

☞ Consider This

Chromatography is an arena in chemistry where both science and art come into play. Use the techniques you developed in the lab to create a chromatographic artwork.

The mechanism of chromatography involves the interplay of the intermolecular forces of the solvent (mobile) phase, the paper (stationary) phase, and the sample. Experiment with the R_f values as functions of the HCl/acetone ratio. Can you correlate your results with the chloro-complex formation constants? The pH?

Oxidation-Reduction Electron Transfer Reactions

PURPOSE

To study a number of chemical reactions that involve the transfer of electrons from a reducing agent to an oxidizing agent. To establish a *qualitative chemical redox couple table* for seven redox couples by mixing the oxidized form of one redox couple with the reduced form of another couple and observing whether a reaction takes place. The redox couples are Cu(II)/Cu; Zn(II)/Zn; Fe(II)/Fe; H^+/H_2; Fe(III)/Fe(II); Br_2/Br^-; and I_2/I^-.

PRE-LAB PREPARATION

A DEFINITION OF OXIDATION AND REDUCTION

There is a problem in the traditional definition of an oxidation-reduction, or redox, reaction as "a chemical reaction that involves the transfer of electrons from the substance oxidized to the substance reduced" because, strictly speaking, every chemical reaction involves changes in the electron density on atoms and therefore involves charge transfer from one atom to another. As a result, the traditional definition could encompass every chemical reaction.

For this reason, some have suggested (see the Bibliography) that there is no consistent basis for distinguishing so-called oxidation-reduction reactions from other chemical reactions other than by a change in oxidation numbers; and as we have pointed out, oxidation numbers are assigned using arbitrary rules, so these numbers do not accurately reflect the actual distribution of charge in a molecule or polyatomic ion.

This somewhat unsatisfactory state of affairs forces us to conclude that the classification of a reaction as an oxidation reaction is somewhat artificial and arbitrary, based on the purely arbitrary notion of oxidation num-

ber. To be logically consistent, *we must define oxidation as an increase in oxidation number and reduction as a decrease in oxidation number.*

Only for simple reactions involving elements and their ions will there be a direct connection between the changes in oxidation number of the elements and the transfer of electrons. For example, when you put a piece of zinc metal in an aqueous solution of copper(II) sulfate, the zinc reacts with Cu^{2+} to form Zn^{2+} and Cu metal. Thus each Cu^{2+} ion has gained two electrons, and each zinc atom has lost two electrons. In other words, there has been a net transfer of two electrons from a Zn atom to a Cu^{2+} ion.

Likewise, zinc goes from oxidation state 0 to +2, and each Cu(II) ion is reduced from oxidation state +2 to 0. The oxidation number of Zn has increased, so we say it has been oxidized. The oxidation number of Cu has decreased, so we say it has been reduced.

In the reaction of zinc with Cu^{2+}, represented schematically in Figure 37-1, zinc acts as a *reducing agent,* or *reductant,* and Cu^{2+} acts as an *oxidizing agent,* or *oxidant.*

Writing the net stoichiometric reaction as the sum of two half-reactions shows explicitly that the transfer of electrons is involved:

Oxidation half-reaction

$$Zn(s) \rightarrow Zn^{2+}(aq) + 2e^- \tag{1}$$

Reduction half-reaction

$$+ \quad Cu^{2+}(aq) + 2e^- \rightarrow Cu(s) \tag{2}$$

Net reaction

$$Zn(s) + Cu^{2+}(aq) \rightarrow Zn^{2+}(aq) + Cu(s) \tag{3}$$

Solution containing copper ions

FIGURE 37-1
Direct chemical reaction of zinc with Cu^{2+}. As the zinc dissolves, Cu is plated out on the surface of the zinc. A Cu^{2+} ion is reduced by transfer of two electrons from a Zn atom.

Note that *no electrons appear in the overall reaction.* This is why it may sometimes be difficult to look at a chemical reaction and to decide whether it can be classified as an oxidation-reduction reaction. Keep in mind the fact that this decision is arbitrary. Certain conventional rules have been adopted about the definition of oxidation number. We will classify a reaction as an oxidation-reduction reaction *only* if there is a change in oxidation state (or oxidation number) of atoms involved in the reaction according to our previously defined rules. (These rules are summarized in Study Assignment A, together with some examples of how they are applied.)

THE RELATIVE STRENGTH OF OXIDIZING AND REDUCING AGENTS
Oxidation-reduction processes involve a *relative competition* between substances for electrons. The stronger oxidizing agents are substances with greater affinity for additional electrons; the stronger reducing agents are substances with the least attraction for the electrons that they already possess. Thus, silver(I) ion is a stronger oxidizing agent than is copper(II) ion because the reaction

$$2Ag^+ + Cu(s) \rightarrow Cu^{2+} + 2Ag(s) \qquad (4)$$

takes place in the forward direction but not appreciably in the reverse direction. From this fact we may deduce that Ag^+ is a stronger oxidizing agent than Cu^{2+}, meaning that silver(I) ion has a greater affinity for electrons than copper(II) ion does. Another way of looking at this competition is to say that copper metal is a stronger reducing agent than silver metal, because copper atoms give up their electrons more easily than silver atoms do.

Strictly speaking, we need to look at the tendency to transfer electrons from one species to another as a property of each *redox couple*, because (a) the tendency of an Ag(I) ion to accept an electron and become a silver atom depends on the difference in the Gibbs energy between an Ag(I) ion and a silver atom and because (b) the tendency of the copper atom to give up two electrons and become a Cu(II) ion depends on the difference in the Gibbs energy of the copper atom and the Cu(II) ion. In other words, we are really comparing the tendency of redox couples to exchange electrons in an oxidation-reduction reaction. As a result, it is only possible to arrange the various oxidizing and reducing agents as *redox couples* in a table, listing them according to their relative tendencies to transfer electrons. In this experiment, we shall explore such relative tendencies for a limited number of reactions.

Fluorine is one of the strongest oxidizing agents known, and lithium is one of the strongest reducing agents; therefore, these two redox couples would be at opposite ends of the table, with most other redox couples falling somewhere in between. (See Figure 37-2.)

A QUALITATIVE CHEMICAL REDOX SCALE
In this experiment we establish qualitatively the relative position of a limited number of oxidation-reduction couples in a redox couple table. We will start with only three metals and their ions, then expand our study until we have considered seven couples in all.

EXPERIMENTAL PROCEDURE

....
NOTE: Before beginning this experiment, complete the preliminary exercise in your experiment report form. As directed there, indicate the substances oxidized and reduced in the several equations given and the change in oxidation state, if any.
....

Chemicals: *Small pieces of copper, zinc, and iron (in the form of wire or 5×15 mm strips; the iron may be in the form of small, bright nails or clean steel wool [to minimize rusting, nails can be cleaned in 6 M H_2SO_4, then rinsed and dried in an oven]); copper, Cu (turnings); 6 M hydrochloric acid, HCl; 6 M ammonia, NH_3; 0.1 M copper(II) sulfate, $CuSO_4$; 0.1 M zinc sulfate, $ZnSO_4$; iron(II) sulfate, $FeSO_4 \cdot 7H_2O(s)$; bromine water, Br_2 (saturated solution); 0.05 M iodine, I_2, in methanol; hexane, $C_6H_{14}(l)$; 0.1 M iron(II) chloride, $FeCl_3$; 0.1 M potassium bromide, KBr; 0.1 M potassium iodide, KI; 0.1 M potassium ferricyanide, $K_3Fe(CN)_6$; 0.1 M silver nitrate, $AgNO_3$; 6 M sulfuric acid, H_2SO_4.*

! SAFETY PRECAUTIONS: Hexane is volatile and flammable. Carry out the tests requiring hexane in a fume hood away from all open flames.

FIGURE 37-2

Schematic diagram illustrating the strongest and the weakest oxidizing and reducing agents in the table of standard reduction potentials. The strongest oxidizing agent (in the upper left-hand corner) has the most positive standard reduction potential. The strongest reducing agent (in the lower right-hand corner) has the most negative standard reduction potential.

WASTE COLLECTION: *The metal wire or strips should be collected; they can be reused if they are cleaned off. A waste container should be provided, in the hood if possible, for the hexane- and methanol-containing mixtures. Also collect the metal ion solutions in a beaker, neutralize with $NaHCO_3(s)$, and put the solid residue in a waste container.*

1. A QUALITATIVE REDOX SERIES FOR COPPER, ZINC, IRON, AND THEIR IONS.

Explore the behavior of small pieces of Cu, Zn, and Fe metals (if the iron is rusty, it should first be cleaned in 6 M H_2SO_4, then rinsed in water), putting each metal in 3 mL of 0.1 M solutions of the ions of the other metals—that is, Cu with Zn^{2+} and with Fe^{2+}, Zn with Cu^{2+} and with Fe^{2+}, and Fe with Cu^{2+} and with Zn^{2+}—to determine the order in which the redox couples should be placed in a table like that shown in Figures 37-2 and 37-3. The *metal ion/metal* couple that appears lowest in the table will correspond to the metal that visibly reacts with all other metal ions and is therefore the strongest reducing agent.

Sometimes the evidence for reaction is subtle. Keep in mind that when most metal ions are reduced chemically they do not produce a shiny metal surface. More often the metal deposited is in the form of finely divided crystals that appear black or brown. Write equations for the reactions that occur.

Solutions containing Fe^{2+} should be freshly prepared by dissolving 0.1 g of $FeSO_4 \cdot 7H_2O$ in 3 mL of water.

Note that the oxidized form of one redox couple is always mixed with the reduced form of the other couple, or vice versa. (The oxidized form of the couple is always the one with the higher charge or higher oxidation number.) Under Part 1 of the Experimental Data, prepare a table as directed in the report form.

2. THE HYDROGEN ION/HYDROGEN COUPLE.

Test the ability of each metal to reduce H^+ to $H_2(g)$ by placing a small piece of copper, zinc, and iron metal in three separate test tubes; add enough 6 M HCl to cover the metal. (Nitric acid, HNO_3, cannot be used in place of

HCl in this test because nitrate ion in acid solution is a stronger oxidizing agent than H^+; therefore, the HNO_3 would react first and confuse the results.) If there is no immediate reaction visible, warm the solution gently (be careful not to confuse bubbles of water vapor for hydrogen gas). If H^+ reacts with a particular metal, the H^+/H_2 couple should be placed above the *metal ion/metal* couple, as shown in Figure 37-3. Your table should now contain four redox couples.

3. THE OXIDIZING POWER OF THE HALOGENS.

First, if you are not familiar with the colors of the free halogens in hexane, add a little Br_2 water (water saturated with bromine) to 3 mL of H_2O and 1 mL of hexane. (**Caution:** Hexane is flammable.) Mix the liquids together by shaking, and observe the color of the hexane layer floating on top. Dispose of the hexane-solution mixture in the appropriate waste container. Repeat, using 0.05 M I_2 in methanol instead of Br_2 water. Dispose of these mixtures in the appropriate waste containers.

Now, explore the behavior of 3-mL samples of 0.1 M solutions of KBr and KI (with 1 mL of hexane added to each) when each is treated with a little of the other free halogen—that is, mix KBr with a few drops of I_2 in methanol, and KI with a few drops of Br_2 water—to determine the relative order of the Br_2/Br^- and I_2/I^- redox couples. Dispose of these mixtures in the appropriate containers. Write equations for the reactions that take place. Then arrange the two halogens in a separate redox couple table, as directed in the report form.

4. THE IRON(III) ION/IRON(II) ION COUPLE.

Determine whether the Fe^{3+}/Fe^{2+} couple is a stronger or weaker oxidizing couple than the I_2/I^- or Br_2/Br^- couple by adding 1 mL of 0.1 M $FeCl_3$ to 2 mL each of 0.1 M KBr and 0.1 M KI in separate test tubes. (Note that this mixes the oxidized form of the Fe[III]/F[II] couple with the

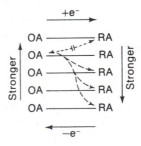

FIGURE 37-3

Tendencies for reaction between oxidizing and reducing agents. The dashed arrows indicate possible redox reactions. An oxidizing agent can react with any reducing agent that lies below it in the table.

reduced forms of the halogens, I^- and Br^-.) Add 1 mL of hexane to each test tube, shake to mix, and from the color of the hexane layer note any evidence for the formation of free halogen. Dispose of these mixtures appropriately. To conduct a separate test for any reduction of Fe^{3+} to Fe^{2+}, prepare two fresh reaction mixtures, omitting the addition of hexane, and add to each a little potassium ferricyanide solution, $K_3Fe(CN)_6$. If Fe^{2+} is present, the deep blue precipitate of $Fe_3(Fe[CN]_6)_2$ will form.

Write equations for any reactions in which iron(III) is reduced. Note that Fe(III) does not go to metallic iron. Place the Fe^{3+}/Fe^{2+} couple in its proper place in your table of redox couples already containing the halogens.

5. THE REACTION OF THE HALOGENS WITH METALS.

Now we want to join the table produced in parts 1 and 2 with the table produced in parts 3 and 4. We will test the ability of the halogens Br_2 and I_2 to oxidize copper metal, as described by the following reactions:

$$Cu(s) + Br_2(aq) \rightarrow Cu^{2+}(aq) + 2Br^-(aq) \qquad (5)$$

and

$$Cu(s) + I_2(methanol) \rightarrow Cu^{2+}(methanol) + \\ 2I^-(methanol) \qquad (6)$$

The tests work like this. Some copper turnings are shaken with Br_2 water. In a separate test tube, copper metal is shaken with I_2 dissolved in methanol. (Methanol is used as the solvent because I_2 is not very soluble in water.) After allowing any precipitate that forms to settle, we will divide the reaction solution in each test tube into two separate portions. One portion of the reaction solution is tested for the presence of Cu^{2+} (formed by oxidation of Cu metal), and the second portion is tested for halide ion (Br^- or I^-), formed by reduction of Br_2 or I_2.

Begin by placing 2 g of copper metal turnings into each of two 15×125 mm test tubes. Add 10 mL of Br_2 water to the first tube and 10 mL of 0.05 M I_2 in methanol to the second tube. Stopper the tubes, and shake each tube vigorously for 30 seconds or until the color of the halogen has faded.

....

NOTE: If a precipitate forms, allow it to settle for a few minutes. The precipitates correspond to formation of Cu(I) bromide (CuBr) or Cu(I) Iodide (CuI) by partial oxidation of Cu to Cu^+, as shown in reactions (7) and (8):

$$2Cu(s) + Br_2 \rightarrow 2CuBr(s) \qquad (7)$$

$$2Cu(s) + I_2 \rightarrow 2CuI(s) \qquad (8)$$

This intermediate ($+1$) oxidation state of copper is stabilized by the formation of the very insoluble salts CuBr and CuI. If Cu^+ is formed as an intermediate oxidation product of Cu metal, it may be immediately precipitated by bromide or by iodide ions if it forms. It is also interesting to note that the existence of this intermediate ($+1$) oxidation state of copper makes it possible for Cu^{2+} and I^- to react with one another to form CuI and I_2, as described by the net reaction

$$Cu^{2+} + 4I^- \rightarrow 2CuI(s) + I_2 \qquad (9)$$

Note that this reaction of Cu^{2+} and I^- is not the reverse of reaction (6), because no copper metal is produced in the reaction. Rather it is a partial reduction of Cu^{2+} that produces a ($+1$) oxidation state of copper, Cu(I); Cu(I) then is stabilized by the formation of the very insoluble CuI(s).
....

Pour off 2–3 mL of the Cu/Br_2 reaction solution into each of two 13×100 mm test tubes. Add to the first tube 5–6 drops of 0.1 M $AgNO_3$. The formation of a precipitate indicates the presence of Br^-, which forms the very insoluble precipitate AgBr(s).

Add to the second portion of the Cu/Br_2 reaction solution 1 mL of 6 M NH_3. Formation of a blue $Cu(NH_3)_4^{2+}$ complex ion indicates the presence of Cu^{2+} ion. If both tests are positive, it is reasonable to conclude that the copper metal reacted with the bromine.

Now divide the Cu/I_2 reaction solution into two portions in 13×100 mm test tubes. Test the solution for the presence of I^- using 0.1 M $AgNO_3$ and for the presence of Cu^{2+} ion with 6 M NH_3 in exactly the same way as you tested the Cu/Br_2 reaction mixture. Were both tests positive? If so, it's reasonable to conclude that the copper metal reacted with the I_2.

Further exploration. Try adding a few drops of 0.1 M Cu^{2+} to 2–3 mL of 0.1 M KI. Do you observe any evidence of reaction like that shown in Reaction (9)?

Considering the results of the tests with copper metal and looking at the table you established for the three *metal ion/metal* redox couples, what would you predict about the ability of other metals, such as Fe and Zn, to be oxidized by bromine or iodine?

6. SUMMARY OF DATA.

Your experimental observations, together with the preceding data, will enable you to combine your two separate redox couple tables into one general redox couple table. The final table will contain all seven redox couples and will show the relative tendencies of the various elements and ions to lose or gain electrons. Note that each redox couple is written so that the change from left to right represents gain of electrons (reduction). In the table in your report form, designate clearly (a) the reducing agents, (b) the oxidizing agents, (c) the end of each column that is the stronger (S), and (d) the end that is the weaker (W).

Such a table can be expanded to include many more oxidation-reduction couples and is useful in predicting the course of many reactions. (Refer to Table 10 in the Appendix.) In reading Table 10, note that *any oxidizing agent (in the left column) has the possibility of reacting with any reducing agent (in the right column) that lies below it in the table* (see Figure 37-3). The table makes no prediction, however, of the *rate* of a given reaction—some reactions are too slow to be practical. Also, you should note that the *concentration* of the ions in a solution has a definite effect on the tendency for reaction, in accordance with the Le Châtelier principle. This factor is considered in Experiments 23 and 39.

Consider This

Before you come to the laboratory to do this experiment, answer the questions posed on the first page of the report form. After you have finished the experiment, answer the questions posed on the last page of the report form.

BIBLIOGRAPHY

Sisler, H. H., and VanderWerf, C. A. "Oxidation-Reduction: An Example of Chemical Sophistry," *J. Chem. Educ.,* **1980,** *57*: 42–44.

Oxidation Reduction
Electron Transfer Reactions

NAME

SECTION   LOCKER

INSTRUCTOR   DATE

☞ Consider This (Before You Do the Experiment)

As a review, in the following equations for familiar reactions, write an "R" over the reducing agent and an "O" over the oxidizing agent. At the right, indicate the change in oxidation number per atom for each element whose oxidation number changes. If there is no change, write in "no change in oxidation number."

Reaction	Element oxidized	Oxidation number increase per atom	Element oxidized	Oxidation number decrease per atom
$2Al(s) + 3Cl_2(g) \rightarrow 2Al^{3+} + 6Cl^-$				
$Ag^+ + Cl^- \rightarrow AgCl(s)$				
$Mg(s) + 2H^+ \rightarrow Mg^{2+} + H_2(g)$				
$Cu^{2+} + H_2S(g) \rightarrow CuS(s) + 2H^+$				
$Ba(s) + 2H_2O \rightarrow Ba^{2+} + 2OH^- + H_2(g)$				
$3Na_2O_2(s) + Cr_2O_3(s) + H_2O \rightarrow 6Na^+ + 2CrO_4^{2-} + 2OH^-$				
$H_2O_2(aq) \rightarrow H_2O + \frac{1}{2} O_2(g)$				
$CO_2(g) + H_2O \rightarrow H_2CO_3(aq)$				
$CO_2(g) + C(s) \rightarrow 2CO(g)$				
$HCl(g) + NH_3(g) \rightarrow NH_4Cl$				
$2ZnS(s) + 3O_2(g) \rightarrow 2ZnO(s) + 2SO_2(g)$				
$4H^+ + 2Cl^- + MnO_2(s) \rightarrow Mn^{2+} + Cl_2(g) + 2H_2O$				
$10H^+ + SO_4^{2-} + 8I^- \rightarrow 4I_2(s) + H_2S(g) + 4H_2O$				

EXPERIMENTAL DATA

1. A Qualitative Redox Table for Copper, Zinc, Iron, and Their Ions

Write net ionic equations for any reactions taking place between the metals and metal ions listed below. If there is no visible evidence of reaction, write "NR" for "no reaction."

Copper and zinc ion: _____

Zinc and copper(II) ion: _____

Iron and copper(II) ion: _____

Iron and zinc ion: _____

Copper and iron(II) ion: _____

Zinc and iron(II) ion: _____

Which metal is the stronger reducing agent, copper or zinc? _____

iron or copper? _____

iron or zinc? _____

In the space at the right, construct a redox couple table for iron, copper, zinc, and their ions. Arrange the three metals in a column at the right-hand side, with the strongest reducing agent at the bottom and the weakest at the top. To the left of each metal symbol make a slash; to the left of the slash write the symbol for the oxidized form of the metal—for example, Cu^{2+}/Cu. You now have a brief "redox couple table."

Which is the strongest oxidizing agent: iron(II) ion, copper(II) ion, or zinc ion? Explain how this table summarizes your observations.

2. The H^+/H_2 Couple

Describe what you observed when you placed iron, copper, and zinc in 6 M HCl. Also, write net ionic equations describing the reactions you observed. Put the H^+/H_2 couple in its proper place in the redox couple table you established for the *metal ion/metal* redox couples in Part 1 by adding it to the redox couples listed in the table.

3. The Oxidizing Power of the Halogens

List your observations, and write plausible net ionic equations for the reactions of the free halogens with the halide ions, in accordance with your experimental data.

Reaction	Observations	Net ionic equation
Bromine and iodide ion		
Iodine and bromide ion		

At the right, construct a redox couple for the halogens and their ions (X_2/X^-), placing the strongest reducing agent at the *bottom* on the *right* of the series, and the strongest oxidizing agent at the *top* and on the *left*.

The strongest oxidizing agent in this series is _____ .

The strongest reducing agent in this series is _____ .

4. The Iron(III) Ion/Iron(II) Ion Couple

Write net ionic equations for the reaction of Fe^{3+} with the halide ions, in accordance with your experimental results. (If there was no evidence of reaction, write "no reaction.")

Iron(III) ion with iodide ion: _____

Iron(III) ion with bromide ion: _____

The Fe^{3+}/Fe^{2+} couple should therefore be placed between the _____ couple

and the _____ couple.

5. The Reaction of the Halogens with Metals

Write net ionic equations describing the results of your experiments on the reaction of Br_2 water and I_2 solution with copper.

What is the experimental evidence for the above reaction products?

6. Summary of Data

In the space at the right, construct a redox couple table for all seven couples studied in this experiment. Write "Oxidizing agents" and "Reducing agents" along the proper sides of the table. Indicate the position of the strongest and weakest oxidizing agents and reducing agents by placing "S" and "W" beside the formulas for these substances.

Consider This (After You Have Finished the Experiment)

NOTE: In answering these questions, refer to the Appendix, Table 10.

1. *In the first parentheses following each formula, write "O" if the substance may be used as an oxidizing agent, then write the formula of the reduced form. (Assume acid solution if the species contains oxygen atoms.) In the second parentheses, write "R" if the substance may be used as a reducing agent, then write the oxidized form. (Note that some substances may be used for either, depending on the substance with which they react.)*

Al ( ) ——————— ( ) ———————   HBr ( ) ——————— ( ) ———————

MnO_4^- ( ) ——————— ( ) ——————— Sn^{2+} ( ) ——————— ( ) ———————

Br_2 ( ) ——————— ( ) ——————— H_2SO_3 ( ) ——————— ( ) ———————

2. *Name a substance for each of the following descriptions:*

   It can oxidize Cd to Cd^{2+} but cannot oxidize Pb to Pb^{2+}. ————————————————

   It can reduce Br_2 to Br^- but cannot reduce I_2 to I^-. ————————————————

   It can oxidize Fe to Fe^{2+} but cannot oxidize Fe^{2+} to Fe^{3+}. ————————————————

   It can reduce Sn^{4+} to Sn^{2+} but cannot reduce Sn^{2+} to Sn. ————————————————

3. *Hypothetical elements A, B, C, and D form the respective divalent ions A^{2+}, B^{2+}, C^{2+}, D^{2+}. The following equations indicate reactions that can, or cannot, occur. Use these data to arrange the metal ion/metal couples into a short redox couple table.*

   $B^{2+} + D \rightarrow D^{2+} + B$

   $B^{2+} + A$ *(will not react)*

   $D^{2+} + C \rightarrow C^{2+} + D$

Redox Titrations

The Oxidizing Capacity of a Household Cleanser or Liquid Bleach

PURPOSE

To show how a redox reaction (the oxidation of iodide ion to iodine) can be used for the quantitative determination of the total amount of oxidizing agent in household cleansers and liquid hypochlorite bleaches.

PRE-LAB PREPARATION

Oxidation-reduction reactions, like acid-base reactions, are widely used as the basis for the analytical determination of substances by titration.[1] In a volumetric titration, a known volume of the *titrant,* usually contained in a buret, is added to the substance being determined, called the *titrand.* The conditions needed for redox titrations are the same as for any titration. The reaction between the titrant and the titrand must be rapid, stoichiometric, and quantitative; that is, the kinetics of the reaction as well as the equilibrium must greatly favor the products. In addition to these basic requirements, the titrant solution must be stable, and there must be some means of determining its concentration accurately. Finally, there must be some means of detecting the end point of the titration reaction.

OXIDIZING AND REDUCING AGENTS

Oxidizing agents available in pure form, such as $K_2Cr_2O_7$, may be weighed out directly to form a titrant solution of known concentration. Potassium permanganate and cerium(IV) salts are often used as oxidizing agents, but they are not pure enough to be weighed out directly, and ordinarily we must titrate them against reducing substances of known high purity (such as As_2O_3) to determine their concentrations accurately. This process is called *standardization.*

The redox potentials of several commonly used oxidizing agents are shown as follows:

$$
\begin{array}{lr}
 & E^0 \; (V) \\
Ce^{4+} + e^- \rightleftharpoons Ce^{3+} & +1.61 \\
MnO_4^- + 8H^+ + 5e^- \rightleftharpoons Mn^{2+} + 4H_2O & +1.51 \\
Cr_2O_7^{2-} + 14H^+ + 6e^- \rightleftharpoons 2Cr^{3+} + 7H_2O & +1.33 \\
I_2(aq) + 2e^- \rightleftharpoons 2I^- & +0.54
\end{array}
$$

Cerium(IV) is one of the strongest oxidizing agents available for use as a titrant, and its solutions are very stable. However, it is more expensive than either potassium permanganate, $KMnO_4$, or potassium dichromate, $K_2Cr_2O_7$. Iodine is a rather weak oxidizing agent and is most often used in indirect procedures in which iodide ion is first oxidized to iodine and then the iodine is titrated with sodium thiosulfate solution.

Most strong reducing agents are easily oxidized by the oxygen in the air, so they are not often employed as titrants. However, they are commonly employed in analysis to reduce a substance to a lower oxidation state before it is titrated with a standard solution of an oxidizing agent. Reducing agents of moderate strength, whose solutions are stable, also exist. One of these is sodium thiosulfate, $Na_2S_2O_3$, which is most often used to titrate I_2.

END-POINT DETECTION IN REDOX TITRATIONS

The most general methods of detecting the end points in redox titrations are like those used in acid-base titrations. When both the oxidized and reduced species are soluble in solution (as MnO_4^-/Mn^{2+}, for example), an inert platinum or gold electrode will respond to the redox

---

[1]Techniques of volumetric analysis are described in the section "Volumetric Measurement of Liquids" in the Introduction.

potential of the system, provided that the electron transfer processes at the electrode surface are reasonably fast so that the equilibrium potential is established. Such an inert electrode may be used in conjunction with a suitable reference electrode in much the same way that the glass electrode is used in acid-base titrations. At the end point of the titration, the redox potential of the system changes abruptly by a few hundred millivolts, and a voltmeter can be used to detect the change. Such a titration is called a *potentiometric redox titration.*

Redox indicators are available that function like acid-base indicators except that they respond to changes in the redox potential of the system rather than to changes in the hydrogen ion concentration. For example, the redox indicator *ferroin* is red in its reduced form, at potentials of less than $+1.12$ V. In titrations in which strong oxidizing agents, such as cerium(IV), are used, the potential jumps abruptly at the end point to potentials of greater than $+1.12$ V, and the indicator changes to its oxidized form, which is blue in color.

In some titrations, a titrant or the substance titrated acts as its own indicator. For example, the permanganate ion, MnO_4^-, is a very deep reddish-purple color, whereas the reduced product of its reaction with reducing agents in acid solution is the colorless manganese(II) ion, Mn^{2+}. Solutions titrated with standard MnO_4^- are usually colorless until the end point, then turn pink when no reducing species are left to react with the added MnO_4^-. Conversely, in the titration of iodine solutions with standard thiosulfate, $S_2O_3^{2-}$, the solutions are brown or yellow until the end point is reached, then become colorless when all of the iodine is titrated. Also, in these titrations, starch is added just before the end point. The starch forms a blue-black colored complex with I_2 and I^- that turns colorless when the I_2 is used up, making the end point sharper and easier to detect.

REDOX TITRATION OF THE OXIDIZING AGENTS IN A HOUSEHOLD CLEANSER OR LIQUID BLEACH

In this experiment, we will make use of two oxidation-reduction reactions to determine the oxidizing capacity of a household cleanser. To a solution containing the sample of cleanser will be added an unmeasured excess of KI in acid solution. Oxidizing agents contained in the cleanser, such as sodium hypochlorite, oxidize the iodide ion to iodine according to the following reaction:

$$
\begin{array}{lr}
 & E^0 \text{ (V)} \\
HClO(aq) + H^+ + 2e^- \rightleftharpoons Cl^- + H_2O & +1.49 \\
2I^- \rightleftharpoons I_2 + 2e^- & -0.54 \\
\hline
HClO(aq) + H^+ + 2I^- \rightleftharpoons I_2 + Cl^- + H_2O & +0.95
\end{array}
$$

The iodine produced in the solution will then be determined by titration with a standardized thiosulfate solution, which reduces iodine stoichiometrically according to the following reaction:

$$
\begin{array}{lr}
 & E^0 \text{ (V)} \\
2S_2O_3^{2-} \rightleftharpoons S_4O_6^{2-} + 2e^- & -0.09 \\
I_2 + 2e^- \rightleftharpoons 2I^- & +0.54 \\
\hline
2S_2O_3^{2-} + I_2 \rightleftharpoons 2I^- + S_4O_6^{2-} & +0.45
\end{array}
$$

Even though the method is indirect, only one standard solution is required.

ESTABLISHING THE END POINT IN THE IODINE-THIOSULFATE TITRATION

Starch (amylose) forms an intensely colored blue complex with traces of iodine (I_2) in the presence of iodide ion. Iodine reacts with I^- to form a complex with the formula I_3^-. This complex has a linear structure ($I—I—I^-$) that fits nicely into the helically coiled starch molecule, as shown in Figure 38-1. The formation of the starch—I_3^- complex is reversible (as is the formation of the deep blue color) and may be used as a sensitive indicator for traces of iodine in solution.

$$\text{starch} + I_2 + I^- \rightleftharpoons \underset{\text{blue}}{(\text{starch}—I_3^-\text{ complex})}$$

Iodine in concentrations as low as 10^{-6} M can be detected, provided that the concentration of iodide ion is 10^{-3} M or greater. Most iodine-containing solutions are titrated with standard thiosulfate to the disappearance of the blue color at the end point. In such titrations, the starch indicator should not be added until just before the end point, when the iodine concentration is low. If it is added too early in the titration, the formation of the blue is not as easily reversible. In practice, therefore, withhold the starch indicator until the last tinge of yellow due to excess iodine has almost disappeared; then add the starch, and quickly complete the titration.

FIGURE 38-1

Starch-triiodide ion complex. The starch (amylose) molecule forms a helix with a cavity large enough to contain the triiodide ions.

Iodide ion is oxidizable by oxygen in air according to the reaction

$$4I^- + O_2 + 4H^+ \rightleftharpoons 2I_2 + 2H_2O$$

Although this reaction is slow in neutral solution, the rate increases with acid concentration. If an iodimetric titration must be carried out in acid solution, exclude the air as much as possible, and perform the titration quickly.

EXPERIMENTAL PROCEDURE

Special Supplies: Several brands of household cleansers (such as Ajax and Comet) or liquid bleach containing approximately 5% sodium hypochlorite, NaOCl; 1-mL microburet (described in the Introduction); 2-mL pipet; two fine-tipped polyethylene transfer pipets.

Chemicals: Potassium iodide, KI(s); 1 M sulfuric acid, H_2SO_4; 0.0100 M potassium iodate, KIO_3; 3% ammonium heptamolybdate, $(NH_4)_6Mo_7O_{24}$ · $4H_2O$, catalyst; 0.15 M sodium thiosulfate, $Na_2S_2O_3$; 0.1% starch indicator freshly made (prepare 1 L by making a paste of 1 g of soluble starch in about 30 mL of water; pour into 1 L of boiling water, and heat about 30 minutes, or until the solution is clear; cool and store in a bottle).

! SAFETY PRECAUTIONS: Liquid bleaches containing 5% sodium hypochlorite are corrosive to the skin and eyes. Your eyes must be protected, as always, by safety goggles. When handling the liquid bleach, take care not to get the solution on your skin. Note that bleach or other hypochlorite-containing substances, such as household cleanser, should never be mixed with ammonia, NH_3; doing so could produce chloramine compounds, such as H_2NCl and $HNCl_2$, which are toxic and volatile.

NOTE TO INSTRUCTOR: The procedures described employ reduced-scale titrations and a 1-mL microburet. If you prefer to use a 50-mL buret instead of a 1-mL microburet, use 0.05 M $Na_2S_2O_3$ as the titrant, and scale up the solution volumes in the standardization procedure of Part 1, using a 25-mL sample of 0.01 M KIO_3 and adding 25 mL of water, 2 g of KI(s), and 10 mL of 1 M H_2SO_4. In Part 2, use a 2-g sample of cleanser in a 250-mL beaker, adding 100 mL of water, 30 mL of 1 M H_2SO_4 added drop by drop, 5 drops of 3% ammonium molybdate catalyst, and 2 g of KI. For liquid bleach, use a 1-g sample in a 250-mL Erlenmeyer flask, adding 50 mL of water, 2 g of KI, 5 drops of 3% ammonium molybdate catalyst, and 10 mL of 1 M H_2SO_4.

1. STANDARDIZATION OF THE SODIUM THIO-SULFATE SOLUTION.
If the approximately 0.15 M $Na_2S_2O_3$ solution has not been standardized, it may be titrated against a solution made up by dissolving pure dry potassium iodate, KIO_3, to a concentration of 0.0100 M.

KIO$_3$ reacts with excess KI in acid solution according to the following reaction:

$$IO_3^- + 5I^- + 6H^+ \rightleftharpoons 3I_2 + 3H_2O$$

Fill a 1-mL microburet with 0.15 M $Na_2S_2O_3$, taking care to eliminate all bubbles. Record the initial volume reading of the buret. To perform the standardization, dispense 2.00 mL of 0.0100 M KIO_3 solution, accurately measured by pipet, into a 25-mL Erlenmeyer flask. Add 3 mL distilled water and 0.2 g of solid KI. Swirl the contents of the flask until the KI dissolves; then add 1 mL of 1 M H_2SO_4, and mix by swirling. A deep brown color should appear, indicating the presence of iodine. Titrate immediately with 0.15 M $Na_2S_2O_3$ solution contained in a 1-mL microburet until the brown fades to a pale yellow. Then add 1 mL of starch indicator solution, and continue the titration until the deep blue color of the starch indicator disappears. The change from blue to a colorless solution is very sharp. Record the volume of sodium thiosulfate solution used in the titration. Repeat the procedure in a duplicate titration. In your report form, calculate the accurate concentration of the $Na_2S_2O_3$ solution.

2. DETERMINATION OF THE OXIDIZING CAPACITY OF A HOUSEHOLD CLEANSER OR LIQUID BLEACH.
Most formulations of household cleanser contain a mild abrasive and cleaning agents such as calcium carbonate and sodium carbonate. They comprise more than 98 percent of the mass of the cleanser. In addition, the cleanser usually contains small amounts of a bleaching (oxidizing agent), anionic detergent, fragrance, and coloring agent. In determining the oxidizing capacity of the cleanser, enough acid must be added to neutralize the carbonates present and provide an acid medium for oxidation of iodide ion by the oxidizing agent. The addition of acid to a carbonate produces a considerable quantity of carbon dioxide gas, causing the solution to foam, so the neutralization step must be carried out slowly, adding the 1 M H_2SO_4 with vigorous stirring to keep the sample from foaming out of its container.

The reduced-scale determination of the oxidizing agent in a powdered household cleanser is carried out by weighing 1.0 g (weighed to nearest milligram) of the cleanser into a 150-mL beaker, adding 50 mL of deionized water, and neutralizing the carbonates in the sample through a drop by drop addition of 15 mL of 1 M H_2SO_4, accompanied by vigorous stirring to disperse the foam. When all of the 1 M H_2SO_4 has been added and the foaming has subsided, add 1 g of KI(s) and 1–2 drops of 3% ammonium heptamolybdate catalyst. Stir to dissolve the KI. The solution should turn brown, indicating the formation of iodine. Immediately titrate the solution with

0.15 M $Na_2S_2O_3$ contained in a 1-mL microburet until the brown color fades to yellow. Then add 2 mL of 0.1% starch indicator, and continue the titration until the purple color of the starch indicator disappears. Record the volume of thiosulfate solution used in the titration. If time permits, make a duplicate titration. Then, if time permits, repeat the procedure, using a second brand of cleanser.

When titrating samples of cleansers that contain powdered scouring agents (and possibly dyes), the end point will be harder to see. It may be helpful to make up a sample of cleanser dispersed in the same volume of water for you to compare with the sample you are titrating, so that the changes can be more easily seen.

A liquid bleach containing about 5% sodium hypochlorite can be titrated by a procedure that closely resembles the standardization titration in Part 1. Using a fine-tipped polyethylene transfer pipet, dispense 0.080–0.090 g of liquid bleach (about 4 drops) directly into a 25-mL Erlenmeyer flask on the pan of a balance. The mass of the sample should be determined to the nearest milligram. (**Caution:** Liquid bleaches are corrosive to the skin and eyes and to balances; wipe up any spills immediately.) Add 5 mL deionized water and 0.2 g KI(s). Swirl to dissolve the KI; then add 1–2 drops of 3% ammonium heptamolybdate catalyst and 1 mL of 1 M H_2SO_4. The solution should turn brown, indicating the formation of iodine. Immediately titrate the solution with 0.15 M $Na_2S_2O_3$ contained in a 1-mL microburet until the brown color fades to yellow. Then add 1 mL of 0.1% starch indicator, and titrate until the blue color of the starch-iodine complex turns colorless. The end point is very sharp. Record the volume of sodium thiosulfate titrant used.

From the concentration and volume of the added sodium thiosulfate solution used to titrate the different brands of cleanser, calculate the weight of oxidizing agent present, assuming it to be sodium hypochlorite, NaOCl. Express the final result as the percentage by weight of sodium hypochlorite: (g NaOCl/g cleanser) × 100%.

If you have the time to compare two cleansers, remember that the effectiveness of a cleanser is influenced by several factors. First, the ability of a cleanser to remove food stains will be related to its oxidizing (bleaching) power. Second, its ability to remove stains from smooth surfaces is also assisted by the abrasive action of its polishing agent, which is usually calcium carbonate. Finally, detergents are also added to cleansers to provide foaming action and to emulsify greases and dirt. In your comparison, you must consider all three of these factors, as well as the cost per unit weight of the cleanser.

3. TITRATION OF THE OXIDIZING CAPACITY OF AN UNKNOWN.

In this optional determination, your instructor will provide you with a sample containing an unknown amount of an oxidizing agent. Using the sample mass specified by the instructor, carry out the titration for the unknown in the same manner as is described in Part 2 for the liquid bleach. Calculate the weight of oxidizing agent present, assuming it to be sodium hypochlorite, NaOCl. Express the final result as the percentage by weight of sodium hypochlorite: (g NaOCl/g sample) × 100%.

☞ Consider This

The oxidizing power of household bleach solutions can be lost when they are exposed to heat, or sunlight, or due to something as simple as leaving the cap off of the bottle. Design experiments to test these storage variables for bleach. Be sure to follow laboratory safety rules as you handle bleach solutions. Test the same variables for cleanser.

The techniques of this experiment should be easily adaptable to measure the concentration of many common oxidizing agents. Develop a method for testing hydrogen peroxide solutions available at the supermarket or drugstore. Devise strategies for testing the stability of these solutions in various storage conditions: temperature, sunlight, etc.

Redox Titrations

The Oxidizing Capacity of a Household Cleanser or Liquid Bleach

NAME

SECTION   LOCKER

INSTRUCTOR   DATE

DATA AND CALCULATIONS

In all of the following calculations, be sure to show the units of each quantity and the proper number of significant figures.

1. Standardization of the Sodium Thiosulfate Solution

	Trial 1	Trial 2	Trial 3
Volume of KIO_3 solution			
Final buret reading (if buret is used)	mL	mL	mL
Initial buret reading	mL	mL	mL
Net volume of KIO_3	mL	mL	mL
Volume of $Na_2S_2O_3$			
Final buret reading (if buret is used)	mL	mL	mL
Initial buret reading	mL	mL	mL
Net volume of $Na_2S_2O_3$	mL	mL	mL
Molarity of the $Na_2S_2O_3$ solution	M	M	M
		Average _____ M	

Show one of your molarity calculations

2. Determination of Oxidizing Capacity of a Household Cleanser or Liquid Bleach

(a) Record the following data for the titration of a household cleanser (or bleach).

Brand name of cleanser or bleach _____

Data	Trial 1	Trial 2	Trial 3
Mass of sample	g	g	g
Volume of $Na_2S_2O_3$ titrant			
Final buret reading	mL	mL	mL
Initial buret reading	mL	mL	mL
Net volume of $Na_2S_2O_3$	mL	mL	mL
Calculation of percentage NaOCl			
Moles of $Na_2S_2O_3$ used	mol	mol	mol
Moles of NaOCl reacted	mol	mol	mol
Mass of NaOCl in sample	g	g	g
Percentage by weight NaOCl (g NaOCl/g sample) × 100%	%	%	%
		Average _____ %	

(b) If you have titrated a sample of another cleanser or bleach, record your data below.

Brand name of cleanser or bleach _____

Data	Trial 1	Trial 2	Trial 3
Mass of sample	g	g	g
Volume of $Na_2S_2O_3$ titrant			
Final buret reading	mL	mL	mL
Initial buret reading	mL	mL	mL
Net volume of $Na_2S_2O_3$	mL	mL	mL
Calculation of percentage NaOCl			
Moles of $Na_2S_2O_3$ used	mol	mol	mol
Moles of NaOCl reacted	mol	mol	mol
Mass of NaOCl in sample	g	g	g
Percentage by weight NaOCl (g NaOCl/g sample) × 100%	%	%	%
		Average _____ %	

3. Titration of the Oxidizing Capacity of an Unknown

Record the following data for the titration of the unknown.

Data for unknown no. _____	Trial 1	Trial 2	Trial 3
Mass of sample	g	g	g
Volume of $Na_2S_2O_3$ titrant			
Final buret reading	mL	mL	mL
Initial buret reading	mL	mL	mL
Net volume of $Na_2S_2O_3$	mL	mL	mL
Calculation of percentage NaOCl			
Moles of $Na_2S_2O_3$ used	mol	mol	mol
Moles of NaOCl reacted	mol	mol	mol
Mass of NaOCl in sample	g	g	g
Percentage by weight NaOCl (g NaOCl/g sample) × 100%	%	%	%
		Average _____	%

EXERCISES

1. Some cleansers may contain bromate salts as oxidizing agents. These salts will react with iodide ion under the conditions we are using according to the following reaction:

$$BrO_3^- + 6H^+ + 6I^- \rightarrow 3I_2 + Br^- + 3H_2O$$

What percentage by weight of $KBrO_3$ would a cleanser have to contain in order to produce an amount of iodine equivalent to that produced by an equal weight of cleanser containing 0.50% NaOCl by weight? (*Hint:* Start with the amount of NaOCl in a 100-g sample of cleanser, and convert it to an equivalent mass of $KBrO_3$.)

2. Many household bleaching solutions are essentially solutions of sodium hypochlorite, NaOCl. Calculate the sample weight of liquid bleach containing 5.0% by weight of NaOCl that would oxidize enough I^- to consume 0.80 mL of 0.15 M sodium thiosulfate.

☞ Consider This

The oxidizing power of household bleach solutions can be lost when they are exposed to heat, or sunlight, or due to something as simple as leaving the cap off of the bottle. Design experiments to test these storage variables for bleach. Be sure to follow laboratory safety rules as you handle bleach solutions. Test the same variables for cleanser.

The techniques of this experiment should be easily adaptable to measure the concentration of many common oxidizing agents. Develop a method for testing hydrogen peroxide solutions available at the supermarket or drugstore. Devise strategies for testing the stability of these solutions in various storage conditions: temperature, sunlight, etc.

EXPERIMENT

39

Electrochemical Cells

PURPOSE

To show how an electrochemical cell can be constructed based on two half-reactions that are separated so that electrons are transferred through an external circuit. To show how cell voltages are related to the concentrations of the components of the cell. To show how measurements of a cell voltage can be used to calculate an equilibrium constant.

PRE-LAB PREPARATION

ELECTROCHEMISTRY AND EVERYDAY LIFE

As you read these words, nerve impulses are traveling along the optic nerves from your eyes to your brain, propagated by electrochemical discharges across the cell walls of the nerves. So important is this electrical activity that one definition of death is the cessation of electrical activity in the brain.

In our everyday lives we use electrochemical cells or their products without thinking about them. When you turn the ignition key of your car, its starter motor is powered by the current from a lead-acid battery. The chromium plate on the car's trim was deposited electrochemically. The aluminum in the engine parts was produced in a cell by electrolysis. Electrochemical reactions are involved in all of these processes, which either use electrical energy to produce chemical substances or vice versa.

THE NATURE OF ELECTROCHEMICAL CELLS

Whenever an oxidation-reduction reaction occurs involving two elements and their ions, electrons are transferred from the substance oxidized to the substance reduced. Thus, when zinc is oxidized by copper(II) ion, the zinc

atom loses two electrons, and the copper(II) ion gains two electrons. We can express this event as two separate half-reactions:

$$Zn(s) \rightarrow Zn^{2+} + 2\,e^- \qquad \text{(oxidation)}$$
$$Cu^{2+} + 2\,e^- \rightarrow Cu(s) \qquad \text{(reduction)}$$

The sum of these two half-reactions gives this net chemical reaction:

$$Zn(s) + Cu^{2+} \rightarrow Zn^{2+} + Cu(s)$$

The net (or total) reaction does not contain any electrons because all of the electrons lost by the zinc are gained by copper(II) ion. An *electrochemical cell* is simply a device used to separate a chemical reaction into two component half-reactions in such a way that the electrons are transferred through an external circuit rather than by direct mixing of the reactants. If the chemical reaction proceeds spontaneously, creating a current flow in the external circuit, we call the cell *self-driven*.

It's interesting that we can construct electrochemical cells based on chemical reactions that we do not ordinarily regard as oxidation-reduction reactions. For example, consider the electrochemical cell based on the following half-reactions:

$$
\begin{aligned}
Ag^+(aq) + e^- &\rightarrow Ag(s) &&\text{(reduction)} \\
Ag(s) + Cl^-(aq) &\rightarrow AgCl(s) + e- &&\text{(oxidation)} \\
\hline
Ag^+(aq) + Cl^- &\rightarrow AgCl(s) &&\text{(net cell reaction)}
\end{aligned}
$$

In this cell, the net reaction is the combination of Ag^+ with Cl^- to form AgCl, which we ordinarily think of as a precipitation reaction, not an oxidation-reduction reaction.

Or consider the cell based on these two half-reactions:

$$Cu^{2+}(1\,M) + 2\,e^- \rightarrow Cu(s) \qquad \text{(reduction)}$$

$$\underline{Cu(s) \qquad\qquad \rightarrow Cu^{2+}\,(0.1\,M) + 2\,e^- \quad \text{(oxidation)}}$$

$$Cu^{2+}\,(0.1\,M) \qquad \rightarrow Cu^{2+}\,(1\,M) \qquad \text{(net cell reaction)}$$

This is called a *concentration cell,* and the net result of its reactions is the transfer of Cu^{2+} from the more concentrated solution to the more dilute solution, a process that we might not even call a chemical reaction.

The message of these two examples is clear: An electrochemical cell can be based on any reaction or process that can be separated into two half-reactions involving the transfer of electrons to or from an external circuit. The net reaction does not have to be a conventionally defined oxidation-reduction reaction. (Keep in mind that the definition of an oxidation-reduction reaction is somewhat arbitrary, as we discussed in Experiment 37, and every chemical reaction involves a redistribution of electronic charge.) In other words, chemical reactions generally involve at least partial *charge transfer* from one atom to another.

An apparatus for carrying out the reaction of zinc with Cu^{2+} in an electrochemical cell is shown in Figure 39-1. In the cell, oxidation takes place at the zinc electrode (the anode),[1] liberating electrons to the external circuit. Reduction takes place at the copper electrode (the cathode),[2] consuming electrons coming from the external circuit. By isolating each half-reaction in its own compartment, we have arranged things so that electron transfer must take place through the external circuit made of metallic wire. It is not possible for the electrons to travel long distances through the solution because they are much too reactive, reacting rapidly with water to reduce it ($e^- + H_2O \rightarrow \frac{1}{2}H_2 + OH^-$).

The current flow in solution consists of positive ions moving in one direction and negative ions moving in the opposite direction. This ionic current flow in solution is a direct result of electron transfer that takes place at the surface of the electrodes. As current flows in the cell, there is a tendency for excess positive charge (in the form of Zn^{2+} ions) to accumulate in solution around the zinc anode as zinc atoms are oxidized. Likewise, excess negative charge (in the form of SO_4^{2-} ions) accumulates around the copper cathode as Cu^{2+} ions are removed from solution by reduction to copper metal. These excess charges create an electric field that causes the ions to *migrate,* positive ions (cations) migrating toward the

cathode and negative ions (anions) migrating toward the anode. This migration of ions between the two compartments of the cell constitutes the cell current in the solution. To summarize, *ions are the charge carriers in solution, and electrons are the charge carriers in the external circuit.*[3]

In Figure 39-1 a porous barrier (e.g., a glass frit) is shown separating the two compartments of the cell. This prevents gross mixing of the solutions in the two compartments by slowing down simple diffusion, but ions can still pass through the porous barrier. If it

FIGURE 39-1

The Daniell cell, a simple electrochemical cell that transforms the energy liberated by a chemical reaction into electrical energy. The electric current in the solution consists of sulfate ions moving toward the left, and of zinc(II) and copper(II) ions moving toward the right.

---

[1]This statement may be regarded as a definition of *anode:* The anode is the electrode at which oxidation takes place.

[2]This statement may be regarded as a definition of *cathode:* The cathode is the electrode at which reduction takes place.

---

[3]In the wires, electrons are the charge carriers, and the flow of current in a wire consists entirely of a flow of negative charge. Since current is conventionally defined as a flow of positive charge, the conventional current flow is opposite to the electron flow. The movement of an electron in one direction in the wire is equivalent to the movement of a hypothetical positive charge in the opposite direction.

became plugged or blocked so that no ions could pass through, current flow in the cell would cease, just as if you had cut the wire in the external circuit. This could happen, for example, if the reaction of ions in the two compartments produced a precipitate, because the precipitate would form right in the porous plug where the two solutions come in contact. Another point to note is that all of the ions in solution migrate under the influence of the electric field, even ions from an added inert electrolyte such as K_2SO_4, which are not involved in the redox reactions. Each ion moves at an average speed determined by the strength of the field and the size of the ion. Big bulky ions move more slowly (and carry a smaller fraction of the current) than small fast-moving ions.

It is important to realize that the net chemical result of the operation of the cell shown in Figure 39-1 is exactly the same as the net stoichiometric reaction: One Cu^{2+} ion is reduced for each zinc atom that is oxidized. But there are some important nonchemical differences between the direct reaction and the reaction carried out in the cell. By using the cell, we are able to convert a fraction of the chemical energy of the reaction directly into electrical energy that can be used to do useful work (like running an electric motor). The direct chemical redox reaction wastes all of the chemical energy as heat—the random thermal motion of the metallic atoms and the ions in solution.

There is another practical difference in the two reactions. In the direct reaction, copper is plated out on the zinc metal as the reaction proceeds, so that after a period of time the zinc atoms get coated with a layer of copper until the reaction slows down and practically stops. In the cell, the reaction can proceed until the cell reaches equilibrium (where the cell voltage is zero). For this reaction, where equilibrium lies far to the right, current would flow either until the Zn anode is practically consumed or until practically all of the Cu^{2+} ions are plated out on the copper cathode.

THE SCHEMATIC REPRESENTATION OF A CELL.

We can always construct, at least in principle, a half-cell that corresponds to a particular redox half-reaction. For example, the redox half-reactions that take place in the Daniell cell are the following:

$$Cu^{2+} + 2\,e^- \rightarrow Cu(s)$$
$$Zn(s) \rightarrow Zn^{2+} + 2\,e^-$$

The sum of these two half-reactions gives the net chemical reaction:

$$Zn(s) + Cu^{2+} \rightarrow Zn^{2+} + Cu(s)$$

We will represent such a cell by the following diagram:

$$Zn|\,ZnSO_4\,\vdots\,CuSO_4|\,Cu$$

in which a vertical bar (|) represents a phase boundary and the dashed vertical bar (⋮) represents the boundary between two miscible ionic solutions (liquid junction). A double dashed vertical bar (⋮⋮) will be used to represent a double liquid junction through an intermediate ionic solution called a *salt* bridge. Salt bridges are used to minimize the liquid junction potential and to prevent mixing of the components of two half-cells.

CELL VOLTAGE.[4] The *volt* is the unit of electrical potential, or driving force. The product of the voltage times the charge of the electron, e, is a measure of the work done when this unit electric charge is transferred from one substance to another. The *voltage of a cell*— sometimes called its *electromotive force* or *potential*—is thus a *quantitative value expressing the tendency of the chemical reaction occurring in the cell to take place.* The magnitude of this voltage depends on the relative strengths of the oxidizing and reducing agents. If the oxidizing agent has an affinity for electrons that is stronger than the tendency of the reducing agent to hold electrons, the electrical potential, or voltage, is correspondingly large.

STANDARD ELECTRODE POTENTIALS. Although we cannot measure a single half-cell potential, we can construct a scale of half-cell potentials by choosing a single *reference* half-cell and measuring the potential of all other half-cells with respect to it. The reference half-cell chosen is based on the half-reaction $2H^+ + 2e^- \leftrightarrow H_2(g)$. This couple is arbitrarily assigned a potential of zero, so that the total cell voltage is ascribed to the other couple. For example, in a cell composed of the Zn^{2+}/Zn half-reaction and the H^+/H_2 half-reaction, where all species are at unit activity,[5] the potential of the cell is found to be $-0.763V$, with the zinc electrode being more negative than the hydrogen electrode. This value, $-0.763V$, is called the *standard electrode potential* for the Zn^{2+}/Zn couple (see Figure 39-2, p. 462). All standard electrode potentials are the values of voltage obtained when all substances in solution are present at unit activity (approximately 1 M), all gases are at unit fugacity

---

[4]The voltmeter used to measure the cell voltage must draw only a small current from the cell, in order not to load the cell and change the concentrations at the electrode surface. The cell voltage under load will be smaller than the open-circuit voltage.

[5]The activity of an ion in solution is usually less than the molar concentration because of ionic interactions. It may be crudely thought of as the "effective concentration."

FIGURE 39-2

On the potential scale, the H^+/H_2 half-reaction is arbitrarily assigned the value zero. The zinc electrode has a voltage of -0.76 V measured against this standard hydrogen electrode (SHE). This value is assigned as the standard electrode potential of the Zn^{2+}/Zn couple. This procedure is analogous to measuring elevation from sea level (rather than from the center of the earth), with sea level being assigned zero in the scale of elevation. The saturated calomel electrode (SCE) is more often used as a practical reference electrode.

FIGURE 39-3

Hydrogen gas, adsorbed on the platinum electrode and in contact with 1 M H^+, forms the reference half-cell. When this half-cell is coupled with a Zn electrode in contact with 1 M Zn^{2+} to form the cell $(Pt)|H_2|H^+ \vdots Zn^{2+}/Zn$, the meter indicates that the zinc electrode is more negative than the hydrogen electrode. The cell voltage will not be the same as the theoretical $E°$ (-0.76 V versus SHE for the Zn^{2+}/Zn half-reaction), because of activity and junction potential effects.

(approximately 1 atm pressure), and the temperature is at a fixed, convenient value, usually 25 °C (see Table 10 of the Appendix).

THE STANDARD HYDROGEN ELECTRODE (SHE) AND PRACTICAL REFERENCE ELECTRODES. A practical hydrogen reference electrode is shown in Figure 39-3, but it is not possible to construct a standard hydrogen electrode (SHE) whose composition corresponds to the arbitrarily chosen reference state.[6] This is because the choice of the standard state for an electrolyte is a solution with a concentration that is 1 molal and in which the ionic activity coefficient is also equal to unity. The standard state thus corresponds to a *hypo-*

thetical solution (one that cannot be made in the laboratory), because the activity coefficients of real solutions are usually less than 1 for 1 molal solutions. In order to obtain a standard electrode potential ($E°$), we extrapolate measured cell voltages to conditions that correspond to the standard reference state. This procedure is too cumbersome for everyday measurements, so in practice we measure cell potentials versus reference half-cells whose potentials have been very accurately determined with respect to the SHE. The saturated calomel electrode (SCE) is a popular reference electrode based on the half-reaction

$$Hg_2Cl_2(s) + 2e^- \rightarrow 2Hg(l) + 2Cl^- \text{ (sat. KCl)}$$

$$Hg|Hg_2Cl_2(s)|\text{sat. KCl} \vdots$$

$$E = 0.244 \text{ V versus SHE at 25 °C}$$

One form of the saturated calomel electrode is shown in Figure 28-1. The saturated KCl electrolyte reduces liquid-junction potentials to a small and reproducible

---

[6]See T. Biegler and R. Woods. "The Standard Hydrogen Electrode—A Misrepresented Concept," *J. Chem. Educ.*, **1973**, *50:* 604; and O. Robbins, Jr. "The Proper Definition of Standard EMF," *J. Chem. Educ.*, **1971**, *48:* 737.

value (a few millivolts or less). Half-cell potentials measured versus the saturated calomel electrode are easily converted to the standard hydrogen electrode scale by adding +0.244 V (see Figure 39-2).

CALCULATING THE STANDARD POTENTIAL OF AN ELECTROCHEMICAL CELL.

When two half-cells are combined to make an electrochemical cell, the standard $E°$ of the cell will equal the difference of the standard half-cell potentials of the two redox couples. For example, the $E°$ for the electrochemical cell formed from the half-cell couples Zn^{2+}/Zn and Cl_2/Cl^- is +2.122 V. Note that this is just the difference between the two standard half-cell potentials on the scale of redox potentials shown in Figure 39-2. The net reaction that occurs when the cell operates spontaneously is

$$Zn(s) + Cl_2(g) \rightarrow Zn^{2+} + 2\,Cl^-$$

ELECTRODE POTENTIALS AND THE PRINCIPLE OF LE CHÂTELIER

THE EFFECT OF CONCENTRATION. We have observed that the tendency for an oxidation-reduction reaction to take place is measured by the voltage created when the reaction takes place in an electrochemical cell. Thus, the electromotive force of 2.122 V created by the cell in the preceding paragraph, for the reaction

$$Zn(s) + Cl_2(g) \rightleftharpoons Zn^{2+} + 2\,Cl^-$$

will be attained when the reactants and products are in their standard states.

According to Le Châtelier's principle, and in accord with observed fact, an increase in the concentration (or pressure) of chlorine gas, $Cl_2(g)$, will increase the voltage of the cell. Conversely, an increase in the concentration of zinc ion, or of chloride ion, will favor the reverse process and therefore decrease the voltage.

THE NERNST EQUATION. In 1889 Walther Nernst (1864–1941) introduced the well-known equation that expresses the quantitative relationship between the voltage of a cell and the concentrations of the reactants and products,

$$E = E° - \frac{2.3\,RT}{nF} \log Q$$

where E is the measured voltage of the cell, $E°$ is the standard potential as calculated from the standard electrode potentials of the half-reactions, R is the gas constant ($8.314\ \mathrm{J \cdot mol^{-1} \cdot K^{-1}}$), T is the temperature (in

kelvins), n is the number of moles of electrons transferred per mole of net chemical reaction, F (called the *Faraday constant*) is the number of coulombs per mole of electrons (96,485), and Q is the product of the activities of the reaction products divided by the product of the activities of the reactants. In determining the value of Q, the activity of each substance is raised to a power equal to its coefficient in the net chemical reaction. (The expression for Q is formulated exactly like the equilibrium constant, K, for the net chemical reaction, but the activities of reactants and products can assume any arbitrary value so that Q is not a constant.) In all of our applications of the Nernst equation, we will make the approximation that the activity coefficients are all equal to 1, replacing *activities* by molar concentrations and the *fugacities* of gases by their *pressures* in atmospheres. It is also convenient to remember that

$$\frac{2.3RT}{F} = 0.059 \text{ at } 25\ °C\ (298\ K)$$

To see the ramifications of the Nernst equation, consider the following. In the preceding reaction, we found that

$$Zn(s) + Cl_2(g,\ 1\ atm) \rightarrow Zn^{2+}(1\ M) + 2\,Cl^-(1\ M)$$

and $E° = 2.122$ V. To calculate the corresponding voltage if the $Cl_2(g)$ were at 4.0 atm, the Zn^{2+} were 0.01 M, and the Cl^- remains at 1 M we would write

$$E = E° - \frac{0.059}{2} \log \frac{(0.010)(1)^2}{(1)(4.0)}$$

$$= 2.122 + 0.077$$

$$= 2.199\ V$$

Both the increase in the $Cl_2(g)$ pressure and the decrease in the Zn^{2+} concentration have a modest effect in increasing the voltage of the cell.

CALCULATING AN EQUILIBRIUM CONSTANT BY MEASURING THE VOLTAGE OF AN ELECTROCHEMICAL CELL

Electrochemical cells can be used to determine the equilibrium constant of chemical reactions. As an example of this, let's consider a cell composed of the following half-reactions:

$$
\begin{aligned}
CuCO_3(s) + 2\,e^- &\rightarrow Cu(s) &&+ CO_3^{2-}\ (1\ M) \\
Cu(s) &\rightarrow Cu^{2+}\ (1\ M) &&+ 2\,e^- \\
\hline
CuCO_3(s) &\rightarrow Cu^{2+}\ (1\ M) &&+ CO_3^{2-}\ (1\ M)
\end{aligned}
$$

$$\text{(net cell reaction)} \qquad (1)$$

We can represent the cell by the following schematic diagram:

$$Cu \mid Cu^{2+} \mathbin{\vdots} CO_3^{2-} \ (1\,M) \mid CuCO_3(s) \mid Cu \qquad (2)$$

The voltage of the cell is given by the Nernst equation:

$$E_{cell} = E° - \frac{2.3RT}{nF} \log Q \qquad (3)$$

Because $Q = 1$ for the cell represented by Equation (1) (neglecting activity coefficients), Equation (3) will reduce to $E_{cell} = E°$:

$$Q = [Cu^{2+}][CO_3^{2-}] = (1)(1) = 1 \qquad (4)$$

If

$$Q = 1 \qquad then \qquad \log Q = 0 \qquad (5)$$

then

$$E_{cell} = E° \qquad (6)$$

If the cell is allowed to operate spontaneously, E_{cell} will gradually decrease until it reaches zero. At this point the chemical reaction has reached equilibrium and $Q = K_{eq}$, the equilibrium constant for the chemical reaction. Therefore, at equilibrium we can rewrite Equation (3) as

$$E_{cell} = 0 = E° - \frac{2.3RT}{nF} \log K_{eq} \qquad (7)$$

which we can rearrange to

$$\log K_{eq} = \frac{nE°}{2.3RT/nF} = \frac{nE°}{0.0059} \qquad (8)$$

where we have introduced the constant $2.3RT/F = 0.059$ at 25 °C. This gives the interesting result that the equilibrium constant for the net chemical reaction of the cell can be calculated if we know $E°$ for the cell.

For the chemical reaction of the cell we have described, the equilibrium constant is the solubility product constant, K_{sp}, for the dissociation of $CuCO_3(s)$ in aqueous solution.

EXPERIMENTAL PROCEDURE

Special Supplies: Voltmeter (see following note); 9-cm lengths of 13-mm-OD glass tubing (7 per student), lightly fire polished on both ends (see Figure 39-4); cellulose dialysis tubing (cut in 3 × 3 cm pieces and stored in water 24 h before use); 12–15-mm lengths of ½ in.-ID by ¹⁄₁₆ in.–wall polyvinyl chloride (Tygon) tubing; wire or 0.5 × 10 cm strips

of Cu, Zn, and Fe metal (or a large iron nail) for use as metal electrodes; 10 cm of 20 to 22 gauge platinum or gold wire (or 0.6 × 10 cm graphite rods) for use as inert electrodes.

Chemicals: 1.0 M copper(II) nitrate, $Cu(NO_3)_2$; 0.10 M copper(II) nitrate, $Cu(NO_3)_2$; 0.10 M zinc nitrate, $Zn(NO_3)_2$; 0.10 M iron(II) sulfate, $FeSO_4$, prepared just at the beginning of the experiment); 0.10 M iron(III) chloride, $FeCl_3$; 1 M potassium nitrate, KNO_3; bromine water, Br_2 (saturated solution); 0.10 M potassium bromide, KBr; 0.05 M iodine, I_2, dissolved in methanol; 0.10 M potassium iodide, KI; 6 M ammonia, NH_3; 1 M sodium sulfide, Na_2S; 1 M sodium carbonate, Na_2CO_3.

NOTES TO INSTRUCTOR: For the voltage measurements in this experiment, a voltmeter with internal resistance of at least 20,000 ohms/V is recommended. The measured voltage will then approach the open-circuit voltage since only a small current (5 µA or less) will be drawn from the cell. A vacuum tube voltmeter (or its solid-state equivalent) is even better. The mV scale of a pH meter also works well for these measurements. The voltmeter should have leads ending in alligator clips convenient for attaching the leads to the electrodes of the cells.

For the construction of the cells in this experiment, the use of glass tubes closed with cellulose dialysis tubing is preferred. These are constructed as shown in Figure 39-4 and as described in Part 1; their use is illustrated in Figure 39-5 (see p. 466). These cells are inexpensive, easily renewed, and require only a small volume of solution.

The simple classic arrangement employing two beakers or test tubes to hold the half-cells and an inverted U-tube salt bridge is also satisfactory. The salt bridge is conveniently filled with 1 M KNO_3 or NH_4NO_3 in a 1% agar gel. The resistance of a 5-mm-ID bridge made in this way is about 1000 ohms, satisfactory for use with a voltmeter with an internal resistance of at least 20,000 ohms/V. If a solid-state voltmeter is used, a 50-fold higher salt-bridge resistance could be tolerated, but it is best to keep cell resistance as low as possible.

An alternate arrangement, recommended in earlier editions of this manual, employs a porous porcelain cup (3 in. tall and 1 in. in diameter), containing one half-cell. This is placed in a 150-mL beaker containing the other half-cell. Although easy to use, these cups are expensive and are difficult to clean when inadvertently clogged with precipitates.

In Part 1 of this experiment you will measure the potentials of several cells composed of redox couples that you studied in Experiment 37. In Part 2 you will relate Le Châtelier's principle to the effect of concentration changes on the voltage of a cell. In Part 3 you will calculate an approximate value for the K_{sp} of $CuCO_3$ from the measured voltage of the cell represented by Equation (2).

Before beginning the experiments, it is necessary to put cellulose on seven half-cell tubes, if this has not already been done. (See Figure 39-4.) Take a 3 × 3 cm

Half-cell tube

FIGURE 39-4
(A) A method of making simple metal/metal ion half-cells. (B) Detail of the construction of the ion-permeable junction.

piece of dialysis tubing that has been stored in water to keep it flexible and center it over the end of the tube. Soften a 12–15 mm length of ½ in.-ID Tygon tubing in hot water, and work it between your fingers until it is pliable. Then force the tubing over the cellulose until the tubing is about two thirds of the way on. Now use a razor blade to trim away excess cellulose around the top of the tubing, and slide the tubing the rest of the way on. At the end of this operation, the cellulose should be stretched tightly and wrinkle free across the end of the tube. Fill the tubes with water to check for leaks (no water should drip through, but the membrane should be moist, since water can diffuse through it). Store the tubes in water until you re ready to use them. When you are through with them t the end of the laboratory period, rinse them out and store them in water.

1. STANDARD CELL POTENTIALS. In your report, rewrite the redox potential series for the couples studied in Experiment 37 (it is Part 6—Summary of the Data—of your report for that experiment). By referring to the Appendix, Table 10, also enter the standard electrode potential ($E°$) of each couple you have listed.

(a) **M^{2+}/M Half-Cells.** For the experimental part of this section, prepare half-cells for Cu, Zn, and Fe by filling three half-cell tubes about three-quarters full of 0.1 M $Cu(NO_3)_2$, 0.1 M $Zn(NO_3)_2$, and 0.1 M $FeSO_4$ (freshly prepared), and placing electrodes of the appropriate metal in each of the three half-cells. (Clean the electrodes with sandpaper or steel wool, then with detergent solution, and rinse with deionized water. A large iron nail, cleaned by immersion in a test tube of 6 M HCl, then rinsed, may be used for the iron electrode.)

To voltmeter

Other redox couple

Tube containing one redox couple

Inert electrolyte salt bridge

Cellulose membranes

FIGURE 39-5

A simple way to make an electrochemical cell. The two redox couples are contained in half-cell tubes that are closed with cellulose membranes permeable to water and ions but that prevent gross mixing of the solutions. Electrical contact in solution is provided by the inert electrolyte salt bridge.

(b) **A Simple Daniell Cell.** To make a simple Daniell cell, $Zn|Zn(NO_3)_2 \vdots Cu(NO_3)_2|Cu$, place the Zn^{2+}/Zn and Cu^{2+}/Cu half-cells you have made in a 150-mL beaker containing 25 ml of 1 M KNO_3, which functions as a salt bridge (see Figure 39-5). Which of these two metals will give up its electrons more readily? To determine the answer, connect the metal electrodes by means of alligator clips and leads to the voltmeter, first with both half-cells in the beaker, then with one half-cell removed from the beaker. Read the meter carefully to the nearest 0.01 V, noting which electrode is the negative terminal (the electron source or anode). (As further proof of this polarity, recall the comparative behavior of Zn metal in Cu^{2+} solution, and of Cu metal in Zn^{2+} solution, from Experiment 37, Part 1 of the Experimental Procedure.) Explain fully all aspects of the operation of the Daniell cell, and complete a diagram of this cell in your report sheet. Be sure you understand the following: (a) What constitutes the electric current in the wire? (b) What constitutes the electric current in the solution? (c) Why must there be actual contact of the two solutions? (d) What are the chemical reactions at each electrode?

In the same manner, put the appropriate pairs of half-cells in the beaker to form the following cells:

$$Fe|FeSO_4 \vdots Cu(NO_3)_2|Cu$$

$$Zn|Zn(NO_3)_2 \vdots FeSO_4|Fe$$

Measure and record the voltage of each cell, noting carefully which electrode is positive. Save the Zn^{2+}/Zn and Cu^{2+}/Cu half-cells for later use.

(c) **Nonmetal Half-Cells.** When the oxidized and reduced form of a redox couple are water soluble, electrical contact is made by placing an inert electrode in the solution. It conducts electrons to and from the external circuit. Most often, a platinum or gold wire or a graphite rod is used. (Graphite is somewhat porous, which presents a problem if the electrode will be transferred from one solution to another; it is hard to rinse off solution that has entered the pores.) Prepare in three small beakers or test tubes the following solutions, which contain equimolar amounts of the oxidized and reduced forms of the redox couple:

$$5 \text{ mL } Br_2 \text{ water } (sat. \; soln.) + 5 \text{ mL } 0.1 \text{ M } KBr$$

$$5 \text{ mL } 0.1 \text{ M } FeCl_3 + 5 \text{ mL } 0.1 \text{ M } FeSO_4$$

$$5 \text{ mL } 0.05 \text{ M } I_2 \text{ in methanol} + 5 \text{ mL } 0.1 \text{ M } KI$$

(Because I_2 reacts almost quantitatively with an excess of I^- to form I_3^-, the last mixture is essentially 0.025 M in I_3^- and 0.025 M in I^-.)

Now pour about 7 mL of each mixture into separate half-cell tubes. A half-cell is completed by inserting an inert electrode, preferably a platinum or gold wire. In the same way as before, a complete cell is formed by placing two half-cells in a 150-mL beaker containing 25 mL of 1 M KNO_3.

Working in this fashion, measure and record the voltages of the following cells, being sure to note which electrode is positive:

$$Zn|Zn(NO_3)_2 \vdots Br_2, KBr|Pt$$

$$Zn|Zn(NO_3)_2 \vdots FeCl_3, FeSO_4|Pt$$

$$Zn|Zn(NO_3)_2 \vdots KI_3, KI|Pt$$

Use the same platinum or gold wire for each half-cell in turn, rinsing it in water between each measurement.

For all six cells for which you made measurements, compare the voltage of each cell (including the sign) with that calculated by combining the $E°$s of the two half-cells that constitute the cell. In each case, the $E°$ of a cell is the algebraic difference of the $E°$ values for the two half-cells. (Why is it necessary to subtract one $E°$ from the other? You will probably notice that the measured cell voltage is not exactly equal to the calculated $E°$ for the cell. The discrepancies may be attributed to differences in the activities of the ions from the hypothetical state of unit activity of the ions that was discussed earlier. However, note that for the seven half-cells studied, the measured values are still sufficiently accurate to establish a redox couple table that is consistent with the order shown in Table 10 of the Appendix.

2. THE EFFECT OF CONCENTRATION.

Place 50 mL of 0.1 M $Cu(NO_3)_2$ (or $CuSO_4$) in a 150-mL beaker, along with a copper electrode. Also place the Zn^{2+}/Zn half-cell you saved from Part 1 into the beaker, and connect the electrodes to a voltmeter. Read and record the voltage; then add, stirring continuously, about 5 mL of 6 M NH_3 to the Cu^{2+} solution, until the deep blue $Cu(NH_3)_4^{2+}$ complex ion is obtained, thus reducing the concentration of $Cu(H_2O)_6^{2+}$. Read and record the voltage. Further reduce the concentration of $Cu(H_2O)_6^{2+}$ by adding, while stirring, an excess (10 mL) of 1 M Na_2S. Again read and record the voltage. Interpret the voltage changes you observe in terms of the Le Châtelier principle.

3. CALCULATING AN EQUILIBRIUM CONSTANT FROM A CELL VOLTAGE MEASUREMENT.

Place 50 mL of 1.0 M Na_2CO_3 and a clean copper strip in a 100-mL beaker. Add 5 drops of 1.0 M $Cu(NO_3)_2$ to form a precipitate of $CuCO_3$, and stir the solution. Fill a half-cell tube (like that shown in Figure 39-4) about two-thirds full of 1.0 M $Cu(NO_3)_2$, and place a clean copper electrode in the tube. Place the tube in the beaker containing the precipitate of $CuCO_3$, connect the voltmeter to the copper strips, and record the voltage. Which electrode is the positive electrode?

Because the concentrations of all the reactants and products in the cell are 1 M, the cell voltage will be equal to $E°_{cell}$, neglecting activity coefficients. (Remember that the activity of a pure solid is defined to be 1.)

From Equation (8) you can calculate K_{sp}, the equilibrium constant for the reaction

$$CuCO_3(s) \rightarrow Cu^{2+}(aq) + CO_3^{2-}(aq)$$

In using Equation (8) you must know the value of n and the magnitude and sign of $E°_{cell}$. What is the value of n that appears in the half-reactions that are summed to give Equation (1)? The cell diagram is shown in Equation (2). By convention, the cell voltage, E_{cell}, is defined as the potential of the right-hand electrode (the Cu strip dipping in 1 M CO_3^{2-}) measured with respect to the left-hand electrode. If the right-hand electrode is the negative electrode, E_{cell} is negative. As defined by this convention, is the $E°_{cell}$ you measured positive or negative?

Consider This

Choose one of the electrochemical cells in this experiment, and study the effect of concentration on the cell voltage. Does it follow the Nernst equation?

How does the cell voltage of these electrochemical cells vary with the temperature? With the proper equipment, study one of them over the range of 5 °C to 50 °C. Show how the slope of the voltage-versus-temperature graph is related to ΔS of the cell reaction.

Prepare a copper or zinc concentration cell, and explore the values of E obtained at a variety of concentrations.

Electrochemical Cells

NAME

SECTION   LOCKER

INSTRUCTOR   DATE

1. Standard Cell Potentials

Complete the diagram at the right by filling in the formulas for the composition of the electrodes and of the ions in the solutions. Place these formulas near the appropriate arrows leading into the solutions. Also indicate at the top which electrode is the anode and which electrode is the cathode.

What constitutes an electric current in a wire?

Indicate the direction of these particles by drawing an arrow over the wire connected to the voltmeter.

What constitutes an electric current in a solution?

Indicate the direction of movement of these particles in the solution by writing their formulas on the proper arrow below the sketch.

e^- flow
$\rightarrow$ / $\leftarrow$?

anode/cathode?

e^-

anode/cathode?

$K^+\ NO_3^-$

Zn

Cu

Why must the two solutions come into actual contact?

Write the half-reaction taking place at the cathode.

Write the half-reaction taking place at the anode.

Write the total cell reaction.

Give the experimental voltage for this cell.

In the right-hand column, enter the standard electrode potential for each half-cell reaction that you studied in Part 1 of the Experimental Procedure, as obtained from Table 10 in the Appendix. In the space below, calculate the $E°$ for each cell whose voltage you measured in Part 1 of the experiment. Write the calculated values in the table below. Also supply the equation for each cell reaction measured in Part 1. Indicate which element is the positive electrode and which the negative, and give the experimental and calculated voltages in each case.

Redox couples	
Couple	$E°$, Volts
Zn^{2+}/Zn	
Fe^{2+}/Fe	
Cu^{2+}/Cu	
I_3^-/I^-	
Fe^{3+}/Fe^{2+}	
Br_2/Br^-	

Cell reaction	Positive electrode	Negative electrode	Experimental voltage	Calculated voltage

2. The Effect of Concentration

Record your observed voltages, and write net ionic equations for the reactions, first of NH_3, and then of Na_2S with Cu^{2+}.

Initial voltage _____ _____

Voltage with
NH_3 added _____ _____

Voltage with
Na_2S added _____ _____

Explain the reasons for the effects on the cell voltage you observed when NH_3 and Na_2S were added to the Cu^{2+}/Cu half-cell.

How would you adjust the concentrations of Cu^{2+} and Zn^{2+} in the cell in order to obtain the maximum voltage possible for this cell?

3. Calculating an Equilibrium Constant from a Cell Voltage Measurement

Give the magnitude of the cell voltage and indicate which electrode is positive for the following cell.

Cell voltage _____ $Cu|\,Cu^{2+}\,|\,CO_3^{2-}\,(1\ M)|\,CuCO_3(s)|\,Cu$

Calculate $\log K_{sp}$ (and K_{sp}) using Equation (8) and your cell voltage data

$$\log K_{sp} =$$

$$K_{sp} =$$

Compare your calculated value with the value for K_{sp} of $CuCO_3$ given in the Appendix, Table 9.

Suggest reasons why the value of K_{sp} that you calculated might be considerably larger than the equilibrium value given in Table 9.

Consider This

Choose one of the electrochemical cells in this experiment, and study the effect of concentration on the cell voltage. Does it follow the Nernst equation?

How does the cell voltage of these electrochemical cells vary with the temperature? With the proper equipment, study one of them over the range of 5 °C to 50 °C. Show how the slope of the voltage-versus-temperature graph is related to ΔS of the cell reaction.

Prepare a copper or zinc concentration cell, and explore the values of E obtained at a variety of concentrations.

Electrochemical Puzzles
Metal Corrosion and Anodic Protection
The Golden Penny Experiment

PURPOSE

To observe the deposition of zinc metal on a penny in contact with metallic zinc and to develop a hypothesis to explain how this happens. To observe how iron may be corroded or protected from corrosion depending on whether it is connected to copper or to zinc.

PRE-LAB PREPARATION

ELECTROCHEMISTRY IS EVERYWHERE

In general, whenever two condensed phases (solid or liquid) are brought into contact, a potential (or voltage) difference develops across the interface. Because the interface region is very thin, even transfer of a small amount of charge across the interface can create a very large electric field. For example, transferring about one picomole (10^{-12} mole) of electron charge per square centimeter of area will typically create a potential difference of approximately 1 volt across an interface layer about one nanometer thick. The electric field in this interface region would be about 10^9 volts/meter. Electric fields this large can cause the transfer of electrons across an interface layer or the transfer of ions between the inside and outside of cells in living organisms. Because contacts between condensed phases are very common in nature, electrochemical phenomena are very common, even though we are often unaware of them. At the cellular level, electrochemical phenomena are crucial to the propagation of nerve impulses, the timing of muscular contractions of the heart, and activity in your brain cells.

Most of the electrochemical technology created by humans involves the simplest kind of chemical change: electron transfer across an interface. Often, the interface is between a good electron conductor, called an *electrode*, and a solution containing molecules or ions. The elec-

trode might be a solid (like platinum or copper metal or graphite), or it could be liquid (like mercury metal). When electrons are transferred from the electrode to a molecule, we say the molecule has been *reduced*. Electron transfer in the opposite sense (from molecule to electrode) is called *oxidation*.

WHAT IS AN ELECTROCHEMICAL CELL?

As we saw earlier, an electrochemical cell is a device used to separate the reactants in a chemical reaction in such a way that the electrons are transferred from one reactant to another through an external circuit rather than by direct mixing of the reactants. We will call the cell a *galvanic*, or *self-driven*, cell if the chemical reaction proceeds spontaneously, creating a current flow in the external circuit. A self-driven electrochemical cell converts the energy of a chemical reaction into electrical energy.

This process can be reversed in an *electrolytic*, or *driven*, electrochemical cell. That is, current can be forced to flow through a cell by using an external power source, such as a battery or direct current (DC) power supply. This process, called *electrolysis*, can create new chemical substances that have greater chemical energy than the reactants, thereby "storing" energy. Upon demand, this chemical energy can be converted back into electrical energy.

The electrolysis of water by means of the cell shown in Figure 40-1 (p. 474) is a familiar example of a driven cell. Less well known is the fact that chemical energy stored in the electrolysis products (hydrogen and oxygen gases) can be directly converted into electrical energy by means of the self-driven fuel cell[1] shown in Figure 40-2 (p. 475).

---

[1] A *fuel cell* is any electrochemical cell in which a net reaction is supported by a supply of reactants external to the cell. Often the oxidant is oxygen from the air.

FIGURE 40-1
Production of hydrogen and oxygen by electrolysis of water in an electrolytic (driven) cell.

Several important industrial chemicals are produced by electrochemical processes. Chlorine is produced by the electrolysis of brine (NaCl) solution. Aluminum is produced by high-temperature electrolysis of alumina (Al_2O_3) dissolved in molten cryolite (Na_3AlF_6). In these examples driven cells are used as "substance-producing" devices. Driven cells are also used in electroplating, in which a thin metal layer is plated out on a conducting substrate, and in the passivation of metals (like aluminum) to produce a thin layer of oxide that provides a protective (and decorative) coating.

FARADAY'S LAWS AND ELECTROLYSIS

Michael Faraday (1791–1867) studied the effects of passing current through cells and published his findings in 1833 and 1834. His experimental conclusions have come to be known as Faraday's laws. Paraphrased in modern language they may be stated as follows: (1) The amount of chemical change produced is proportional to the total amount of charge passed through the cell. (2) The passage of a mole of electric charge produces an equivalent of chemical change. *An equivalent of chemical change* is equal to a mole of chemical change divided by the number of electrons transferred per mole of chemical change.

The amount of electric charge is expressed in coulombs, and can be determined experimentally by measuring the electric current (flow of charge). The current is measured in units of amperes: One ampere is equal to a charge flow of one coulomb per second. If the

FIGURE 40-2

A scheme for obtaining electrical energy from a galvanic (self-driven) hydrogen-oxygen fuel cell. The net chemical reaction in the cell is equivalent to the combustion of hydrogen.

current is constant, the total charge transferred is equal to the current multiplied by the time in seconds:

$$\text{coulombs} = \text{amperes} \left(\text{or} \; \frac{\text{coulombs}}{\text{second}} \right) \times \text{seconds} \quad (1)$$

When a mole of silver (107.87 g) is deposited at a cathode, the half-cell equation

$$Ag^+ + e^- \rightarrow Ag(s) \quad (2)$$

indicates that an Avogadro's number (6.022×10^{23}) of individual silver ions must be reduced by the same number of electrons—"a mole of electrons." The charge on a mole of electrons is

$$6.0221 \times 10^{23} \; \text{electrons} \times 1.6022 \times 10^{-19} \frac{\text{coulombs}}{\text{electron}} =$$

$$96,485 \; \text{coulombs} \quad (3)$$

which is called a *faraday* and given the symbol *F*. When copper is plated out at the cathode, the half-cell equation

$$Cu^{2+} + 2e^- \rightarrow Cu(s) \quad (4)$$

shows that 1 mol of electrons (1 faraday) will produce only ½ mol of copper $\left(\frac{63.54}{2} = 31.77 \; \text{g} \right)$.

To summarize, when 1 faraday (96,485 coulombs of charge) is passed through a cell, chemical change equivalent to transfer of 1 mole of electrons (oxidation at the anode and reduction at the cathode) occurs at each electrode. This quantity of material is called one *electrochemical equivalent.*

Only a few electrochemical reactions are known that give a simple, well-defined reaction conforming exactly to Faraday's laws. More often, an electrochemical reaction is complex, yielding several products. This may occur when more than one electron transfer reaction takes place on the electrode or when the primary products are very reactive in some side reaction.

Electrochemists use the terms *current efficiency* or *yield* and define these terms as the percentage of total charge that is effective in producing the desired substance. It was just these effects that made it difficult to discover these laws, which may seem self-evident to students already familiar with atomic theory. But in 1834, modern theories of atomic structure, chemical bonding, or ionization did not exist, nor was anything known about the existence and properties of electrons. These laws, which eluded many who were studying electricity, should be regarded as a lasting tribute to Faraday's ingenuity and careful experimental work.

ELECTROCHEMISTRY AND CORROSION

When it is not desired, the dissolution of a metal is called *corrosion,* and much money and effort are spent in preventing or combating it. A simple example of corrosion is the dissolution of a metal in acid, where the oxidation-reduction processes are those shown in Figure 40-3.

Corrosion is a common affliction of automobiles in regions with cold climates, where corrosive salts are

FIGURE 40-3

In corrosion, dissolution of the metal liberates electrons. If these electrons can be taken up by some other electrochemical reaction (like H_2 evolution), the metal continues to dissolve.

spread on the roads to melt ice. When a hole or crack in the paint exposes the steel of the auto body, the steel often begins to corrode. This suggests that the corrosion process is primarily electrochemical in nature, two separated half-reactions being mediated by flow of electrons in the steel (see Figure 40-4). Unless the corroded area is cleaned and repainted, corrosion will continue until the metal is rusted through over a large area.

ANODIC PROTECTION.

By coating steel (which is mostly iron metal) with a more active metal like zinc, a process called *galvanizing*, the steel's corrosion can be retarded or entirely prevented. Often, simply making a good electrical connection between a piece of iron and a piece of zinc is sufficient to keep the iron from corroding. We will study the acceleration and the prevention of iron corrosion by connecting it to various metals.

THE GOLDEN PENNY EXPERIMENT

An experiment that simultaneously charms and puzzles is the plating of a penny with zinc metal. First, the penny is immersed in a solution containing 1 M NaOH and granular zinc. Subsequent heating of the penny for a few seconds on a hot plate causes the silver color of the penny to turn a bright golden yellow. Explaining the details of the process presents a challenging chemical puzzle. Some questions that arise may only be answered by doing further experiments, so this experiment can be somewhat open-ended.

EXPERIMENTAL PROCEDURE

Special Supplies: *Part 1:* Nominal 100 × 15 mm disposable polystyrene Petri dishes (three per student group); fine steel wool; approximately one soldering kit for every six students, consisting of a 140-watt soldering iron, rosin-core solder, and one 6 × 6 inch ceramic fiber square (available from Flinn Scientific Inc.); digital multimeters with alligator clip leads and short lengths (3 cm) of Pt or nichrome wire to use as voltage probes.

Part 2: Digital multimeters with alligator clip leads, pennies (preferably clean and bright), 20-gauge copper wire, hot plates, polypropylene or stainless steel forceps; approximately one soldering kit (see the description in Part 1) for every six students.

Chemicals: *Part 1:* Agar (powder), 1% phenolphthalein indicator, 0.1 M potassium ferricyanide [hexacyanoferrate(III)], $K_3Fe(CN)_6$; two zinc metal strips, 6 × 40 mm, cut from 0.01-inch-thick zinc foil; two copper metal strips, 6 × 40 mm, cut from 0.01- (or 0.005-) inch-thick copper foil; ungalvanized finishing nails, 4-penny size, 40 mm long (before use, clean by soaking briefly in 3 M H_2SO_4 acid, rinsing with deionized water, and drying in an oven).

Part 2: 30-mesh zinc metal; zinc metal powder; 6 × 100 mm strips of zinc metal, cut from 0.01-inch-thick zinc foil; 20-gauge copper wire; 1 M sodium hydroxide, NaOH.

NOTES TO INSTRUCTOR: We have profitably spent a whole lab period or two on just Part 2 of this experiment, the golden penny experiment. It presents an interesting challenge to explain the observations and many opportunities for open-ended experimentation. Besides, it's fun. Part 1

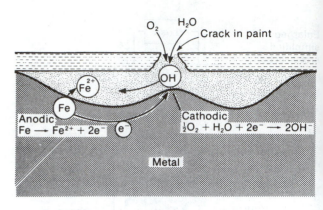

FIGURE 40-4

At a crack in the protective paint, corrosion often develops under the paint. It is the (invisible) part of the steel under the paint that becomes anodic and dissolves. Electron flow in the metal and ionic flow in the corroded area couple the two half-reactions.

---

involves observations over a period of a week or so; as a result, parts 1 and 2, done together, make a good 2-week lab, with time in between for discussion and planning of other experiments to try.

! **SAFETY PRECAUTIONS:** WEAR EYE PROTECTION AT ALL TIMES. Sodium hydroxide is corrosive. You may want to provide latex rubber gloves for handling pennies that have been in contact with 1 M NaOH.

WASTE COLLECTION: Your instructor may direct you to waste containers for NaOH solutions used in this experiment. These substances can be disposed of down the drain only if they are neutralized by sodium bicarbonate.

---

1. METAL CORROSION AND ANODIC PROTECTION.

Obtain two 6 × 40 mm strips of zinc foil, two 6 × 40 mm strips of copper foil, and two 4-penny (40 mm long) ungalvanized iron finishing nails (which should have been previously cleaned by immersion in 3 M H_2SO_4, then rinsed with deionized water, and dried in an oven). Clean the zinc and copper strips with steel wool to produce a clean shiny surface. Use detergent to remove the film of oil on the strips, rinse them with deionized water, and dry them with a tissue.

General Soldering Instructions. Go to a soldering station where you will find a ceramic fiber square, soldering iron, and rosin core solder. When soldering, place the zinc strip on a ceramic fiber square and position the copper strip so the ends of the two strips overlap by about 4–5 mm. Ask a partner to apply pressure to the copper strip, holding it in place while you are soldering the joint. (Your partner may use almost any tool for this—except a bare finger, because the strip will get very hot.) Pull the trigger of the soldering iron to let it heat up. Then tin the tip of the iron with solder, wiping off any excess with a damp sponge. Place the freshly tinned tip on the zinc strip, next to the copper strip, angling the iron to get good thermal contact. Feed the solder into the area between the copper strip and the tip of the soldering iron. When the zinc strip is hot enough, the solder will flow into the joint. Don't dab at the joint with the tip of the iron while soldering. The tip must be kept in continuous good thermal contact with the joint so that the zinc strip heats up. Using steel wool, remove any rosin remaining from the soldering operation; then rinse the soldered joints, and dry the metal strips with a tissue.

Now you should have a Zn/Cu piece, a Cu/Fe piece, and a Zn/Fe piece. (Once these soldered bimetal pieces have been made, they can be cleaned and reused several times, so don't throw them away at the end of the experiment unless your instructor directs you to.)

Add 3 g of agar to 225 mL of boiling deionized water in a 600-mL beaker. At this point it's best to turn down the heat and stir until the agar is dissolved. Be careful not to heat the agar so much that it scorches or boils over. While the agar suspension is hot, add 2 mL of 1% phenolphthalein indicator. Then mix with a stirring rod.

Get three nominal 100×15 mm polystyrene Petri dishes. Put each of the three soldered bimetal pieces in the bottom (taller) half of a Petri dish. Protecting your hands from the beaker containing the hot agar solution, pour the agar over the metal pieces in the Petri dishes. Cover the metal pieces, but do not fill the dish beyond half its depth. There should be no voids or bubbles underneath the metal strips.

After the agar cools and gels, use a fine-tipped polyethylene transfer pipet to place one small (0.02 mL) drop of 0.1 M potassium ferricyanide ($K_3Fe[III][CN]_6$) about 1 cm away from the middle of both iron nails. The ferricyanide salt will diffuse radially outward. If the ferricyanide ions encounter any Fe^{2+} formed by oxidation of Fe, they will react with the Fe(II) ions to form a dark blue compound formulated as $KFe(III)Fe(II)(CN)_6$. (It's reported that you get the same product by mixing Fe^{3+} with $K_4Fe[II][CN]_6$.) Evidently in the final product the iron atoms have exchanged an electron so that the Fe^{2+} ion is oxidized to Fe(III), and the Fe(III) originally in the ferricyanide ion is reduced to Fe(II).

During the course of the afternoon, make periodic observations of the Petri dishes. Look for evidence of formation of hydroxide ion, which will turn the phenolphthalein pink; the formation of insoluble metal ion-hydroxide salts, which will appear as a cloudy band; or the formation of a blue compound in those dishes containing iron nails (with an added drop of 0.1 M potassium ferricyanide, $K_3Fe[CN]_6$), indicating oxidation of Fe to form Fe^{2+}.

Optional. Obtain two short lengths (about 3 cm long) of platinum or nichrome wire and a digital multimeter with leads connected to alligator clips to hold short lengths of wire that will be used as voltage probes. Adjust the voltmeter to its most sensitive voltage range (200 millivolts). First, immerse the probes in the agar about 1 mm from the edge of the two metals soldered together about midway along the length of the metals. (Keep the probes upright, perpendicular to the Petri dish, and make sure the probes do not touch the metal strips.) Note whether there is any voltage difference. Note the polarity. Which metal is nearest the positive (+) end? Any voltage difference indicates an electric field between the two points in the agar, created by the formation of positive and negative ions in the two regions. Considering the polarity of the measured field, what ions do you think might be responsible for the presence of the electric field? Write plausible reactions for the formation of positive ions (metal atom oxidation) and negative ions (reduction of water or oxygen).

Next, touch the probes directly to the two metals at their midpoints and note any voltage reading. (The voltage reading is expected to be zero volts because metals are such good electronic conductors that only a tiny electric field can exist in the two metals together.) The metals soldered together are said to form an *equipotential surface* (a surface where the potential is constant, so that the voltage difference between any two points on the metal is zero.)

Put the top cover on your Petri dish, and tape the top cover in place with two or three short strips of tape. Write your initials or other identifying marks on the tape.

Continue visual observations over a period of several days, looking for evidence of formation of any pink color or any visible precipitates. Make sketches, and write verbal descriptions of the changes you observe.

REACTIONS THAT MIGHT FORM OH^- ION. When a pink color develops around a metal in a gel containing phenolphthalein indicator, it means that the solution next to the metal is basic. In an aqueous gel, the pink color means that some hydroxide ions have been formed.

Although any electrons given up when a reactive metal is oxidized might react at the spot where the oxidation occurs, they can also readily travel to any other spot on the surface of the two joined pieces of metal. That means, it is possible that the point where metal atoms are oxidized could be some distance from the point where hydroxide ions are produced.

Now let's think about what might be most likely to accept these available electrons. Metal atoms typically don't accept electrons to form negatively charged metal ions. Rather, metal ions tend to give up electrons to form positive ions. Things that are easy to reduce have the most positive standard reduction potentials, like the halogens, but we don't have any halogens in our system. The gel surrounding the metal consists mainly of water with about one percent of agar. Although water is not easy to reduce, because water has a negative standard reduction potential in basic solution, this substance can be reduced when the reaction is coupled to the oxidation of Zn metal in basic solution, as shown by the following standard reduction potentials:

$$Zn(OH)_4^{2-} + 2e^- \rightarrow Zn(s) + 4OH^-$$
$$(E° \text{ in } 1 \text{ M OH}^- = -1.28 \text{ volts})$$

$$2H_2O + 2e^- \rightarrow H_2(g) + 2OH^-$$
$$(E° \text{ in } 1 \text{ M OH}^- = -0.80 \text{ volts})$$

Agar is a polysaccharide (like starch), and polysaccharides are not easy to reduce. Finally, we must not forget that the Petri dishes are open to the air, so the agar gel also contains dissolved oxygen, a good acceptor of electrons. At least two reactions involving oxygen deserve serious consideration:

$$O_2(g) + H_2O + 2e^- \rightarrow HO_2^- + OH^-$$
$$(E° \text{ in } 1 \text{ M OH}^- = -0.065 \text{ volts})$$

$$O_2(g) + 2H_2O + 4e^- \rightarrow 4OH^-$$
$$(E° \text{ in } 1 \text{ M OH}^- = +0.40 \text{ volts})$$

The E° for reduction of oxygen in basic solution is considerably more positive than for the reduction of water. So we definitely must consider the possibility that oxygen might be the species that could most easily be reduced, with OH$^-$ (and possibly hydrogen peroxide) being the reduction product.

The reduction of either water or oxygen produces hydroxide ions, but the formation of a pink color with phenolphthalein does not tell us which reaction might be responsible. Thermodynamics (as measured by the standard reduction potentials) favors reduction of oxygen over reduction of water. However, the reduction of oxygen on many metals is known to have a large activation energy, which usually causes the reaction to be slow. Thus, kinetics may favor the reduction of water, particularly because the concentration of water is much greater than the concentration of oxygen in the agar gel.

Can you think of an experiment that might allow you to distinguish if water or oxygen is the major species being reduced?

2. THE GOLDEN PENNY EXPERIMENT. Work in groups of two or three students. Put 8 g of 30-mesh zinc in the bottom of a 400-mL beaker. Tilt and tap the beaker

on the benchtop in order to get the granular zinc to cover about half the bottom of the beaker (one-half of the beaker should be covered, one-half not covered; see Figure 40-5). Carefully pour 200 mL of 1 M NaOH down the side of the beaker, being careful not to disturb the distribution of zinc. Use a stirring rod or spatula to clear any remaining granules so that half of the beaker bottom is completely free of zinc granules. Place the beaker on a hot plate in the fume hood and turn the hot plate to medium heat. The solution should be heated to about 80–90 °C; if it is heated to boiling the distribution of zinc granules will be disturbed.

While waiting for the solution to heat, buff six copper pennies with steel wool until they are shiny. Solder 10-cm lengths of 20-gauge copper wire to two of the pennies, overlapping the wire and penny about 2 to 3 mm from the edge. Solder the free end of one of the copper wires to a 5×100 mm strip of zinc metal, as shown in Figure 40-5. Clean any rosin off the soldered joints with steel wool, and rinse with water.

When soldering, place the penny on a ceramic fiber square, with the end of a copper wire overlapping the penny 2–3 mm. Ask a partner to apply pressure to the wire to hold it in place while you are soldering the joint. (Almost anything will do, but don't use a bare finger

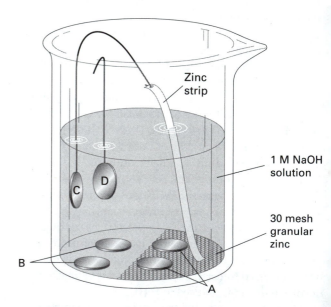

FIGURE 40-5

Four arrangements of copper pennies in the golden penny experiment. The bottom of the beaker is half-covered with 30-mesh zinc metal. (A) Two pennies are lying on top of the 30-mesh zinc. (B) Two pennies are lying on the bottom of the beaker but not in contact with 30-mesh zinc. (C) A penny soldered to copper wire is immersed in solution, the other end of the copper wire being soldered to a strip of zinc metal in contact with 30-mesh zinc on the bottom of the beaker. (D) A penny soldered to copper wire is immersed in solution. The solution in the beaker is 1 M NaOH.

because the wire will get very hot.) Review the General Soldering Instructions in Part 1. Place the freshly tinned tip on the penny, next to the wire, angling the iron to get good thermal contact. Don't dab at the joint with the tip of the iron while soldering.

When the solution has warmed, use forceps to place two copper pennies on top of the granular zinc metal and two pennies in the area that is free of granular zinc (make sure that the pennies on the uncovered side do not contact even one grain of zinc). Bend a small "foot" on the free end of the zinc strip, and place the strip and attached penny in the beaker as shown in Figure 40-5. The "foot" of the zinc strip should rest on the granular zinc so that both are in direct contact. The penny should be completely immersed in solution but should not contact any granular zinc metal on the bottom of the beaker. Finally, hang the last penny (the one with only copper wire soldered to it) over the edge of the beaker so that the penny is completely immersed in solution. See Figure 40-5.

Leave the pennies in the beaker until some of them turn a silvery color. This may take anywhere from 5 to 30 min, depending on the temperature of the solution. (Some of the pennies will never turn silver even after waiting an hour or more.)

Which pennies turn a silvery color? Is it the three pennies that are in contact with the solution and with zinc, either directly or through the copper wire? Or is it the three pennies in contact with the solution but not in direct or indirect contact with zinc metal?

Using a pair of forceps, or wearing latex gloves for protection, remove the pennies that have turned a uniform silvery color, rinse them with water, and put the pennies on a hot plate for a few seconds. Watch what happens to the silver-colored pennies as they heat on the hot plate. Keep the solution warm in the beaker in case you need to repeat some part of the experiment or try some new experiment, as described below.

Further Experiments and Clues to the Puzzle. Solder another length of copper wire to a shiny clean penny. Connect one lead of a voltmeter to a zinc strip and the other lead to the copper wire soldered to the penny. Using the same solution you prepared earlier, immerse the zinc strip and penny in the solution. Is there a voltage difference between the zinc strip and the copper wire soldered to the copper penny? Which metal is the electron source (the negative terminal of this electrochemical cell)?

Do you think a current flows in the copper wire connecting the zinc strip and copper penny when both are immersed in the solution? If so, which direction will electrons flow, and what are the anode and cathode reactions? Put the digital voltmeter into its current measuring mode on its most sensitive (microampere) scale. Then see if any current is flowing when you connect the meter in series between the copper wire soldered to the

penny and the zinc strip—both the penny and the zinc strip should be immersed in the hot 1 M NaOH solution. The series connections should look like this: Penny/ copper wire / (+)ammeter(−) / zinc strip. How large a current flows? Is the current (charge flow) from penny to zinc strip or vice versa?

In wires the charge carriers are electrons. Current (defined as a flow of positive charge) is opposite to the flow of electrons. In which direction are electrons flowing: from penny to zinc strip or vice versa?

Construct a Hypothesis. Describe your observations, and develop a hypothesis to explain what's going on. Why do the copper pennies turn silver? What causes the silver color? Do copper pennies not in contact with zinc turn a silver color? If not, why not? Is this consistent with the observations in Experiment 37, in which you learned that Cu metal does not reduce Zn^{2+} ion? Why do the silver colored pennies turn gold when heated on the hot plate? Is this the realization of the alchemists' dreams of turning base metals into gold? Or is this just a brazen trick?

☞ Consider This

Our goal is to try to use the observations we made to construct a mental image of what the atoms are doing and why they are doing it. First, let's consider the observations and how they provide clues about what's happening at the atomic level.

1. METAL CORROSION AND ANODIC PROTECTION.

(a) Phenolphthalein is an acid-base indicator that turns pink (or red) in basic solution. When it turns pink in solution, that means some hydroxide ions have been formed. The only plausible source of hydroxide ions in the system is the reduction of water and/or oxygen (which could come from the air). Where do you think the electrons required for the reduction came from?

(b) Potassium ferricyanide $[K_3Fe(III)(CN)_6]$ forms a blue color with Fe^{2+}. If a blue color forms at the spot near the iron metal, it's plausible that the iron is being oxidized to form Fe^{2+}.

(c) When freshly made, a dilute $Fe(II)SO_4$ solution is light green, nearly colorless. On standing in contact with air the solution turns yellow, indicating the formation of Fe(III), which must have resulted from oxidation of Fe(II) by oxygen in the air.

(d) If a band of precipitate forms, it seems likely that it would be formed from metal cations reacting with anions (most likely hydroxide ion). The identity of the metals present and the color of the precipitate give us clues about what metal ion might be involved.

(e) Since many metals are easier to oxidize than water, it seems a good bet to assume that if water or oxygen is being reduced to form hydroxide ion, some metal must be getting oxidized. When two different metals are connected together, it's reasonable to assume that the most active metal will get oxidized. The most active metal has the most negative M^{+n}/M standard reduction potential.

(f) One way of looking at the bimetal strips in the Petri dishes is to think of them as an electrochemical cell formed by putting some agar gel that contains phenol-phthalein into a beaker and embedding strips of two different metals in the gel. A voltmeter connected externally to the strips would measure some voltage difference between the two strips. An ammeter connected between the two metal electrodes would measure a small current. (You can easily try this to see if it is true.) Now imagine shorting the strips together in the external circuit by soldering them together. Electrons are able to easily flow from one metal to the other. The electrochemical processes will still go on at the surface of the metal strips, just as when they were connected through the ammeter. Now imagine adding more gel until the electrodes are completely covered. The same electrochemical processes would still go on at the surface of the metals, even though the electrical connection between the metal strips is now "internal" rather than "external." Any electrons that leave the metal would simply cause the reduction of water or oxygen, producing hydroxide ion.

Using this set of hypotheses to assist your thinking, try to construct in your own words an explanation at the atomic level of everything happening in the three Petri dishes.

2. THE GOLDEN PENNY EXPERIMENT.

(a) Assume that the silver coating on the copper penny is zinc metal. From previous experience, we know that zinc metal dissolves in acids. Try putting a few drops of 1 M HCl on a silver-coated penny. What happens?

(b) If the penny is coated with zinc, how is the zinc getting to the penny connected to the copper wire soldered to the zinc strip? Can zinc atoms travel over the wire? Or do they travel through the solution? Do you think they travel from the zinc to the penny as zinc atoms or zinc ions?

(c) If the zinc gets to the copper penny by traveling through solution as zinc ions, where would the zinc ions come from, since the 1 M NaOH solution that we started with did not contain any added zinc salts? Some possible reactions to consider that form zinc ions include:

$$Zn + 2H_2O + 2OH^- \rightarrow H_2(g) + Zn(OH)_4^{2-}$$

$$2Zn + O_2(g) + 2H_2O + 4OH^- \rightarrow 2Zn(OH)_4^{2-}$$

NOTE: Zinc is known to be amphoteric in the sense that $Zn(OH)_2$ is soluble in a solution containing excess hydroxide ion. When a drop of hydroxide ion solution is added to $Zn(NO_3)_2$ solution, a white precipitate of $Zn(OH)_2(s)$ forms. If an excess of hydroxide is added, the precipitate dissolves to give a colorless solution containing the tetrahydroxyzincate complex ion, $Zn(OH)_4^{2-}$.

The first reaction (the reduction of water by zinc in basic solution) would produce hydrogen gas. Did you notice any evidence of evolution of gas that might indicate that hydrogen is being formed? Could the reaction be so slow that the concentration of hydrogen never builds up to the point where bubbles of hydrogen form? To increase the rate of reaction by increasing the surface area of zinc exposed to the 1 M NaOH, try adding powdered zinc to 1 mL of 1 M NaOH in a test tube. Do you see gas evolution?

Is there any way to tell if oxygen is involved in the oxidation of zinc metal? Could both reactions be going on simultaneously?

(d) If the zinc coating on the copper penny is produced by reduction of zinc ion on the surface of the penny, here are two possibilities to consider:

(i) Direct reduction of zinc ion by copper metal:

$$Zn(OH)_4^{2-} + Cu(s) \rightarrow Zn(s) + Cu^{2+}(aq)$$

(ii) Reduction of zinc ion by electrons traveling over the wire. These electrons are produced when zinc atoms are oxidized at the surface of the zinc metal in contact with the solution.

Can copper metal on the surface of the penny directly reduce zinc ions in solution to form zinc atoms on the surface of the copper? If so, would you expect the penny soldered to a copper wire and immersed in solution—but not connected to zinc metal—to get covered with zinc? Did that penny turn a silver color?

A direct test: Immerse a clean shiny strip of copper metal in a solution of 1 M NaOH/0.01 M $Zn(NO_3)_2$ in a test tube. Heat the test tube in a beaker of hot water for 20 minutes. What happens?

(e) What do the measurements with the digital multimeter, operating as a voltmeter and as an ammeter, tell you about the possibility that zinc ion is reduced to zinc metal on the surface of the copper penny by a flow of electrons in the copper wire connecting the zinc metal to the copper penny? If you did the optional tests, did you measure a current? If so, which direction were electrons flowing? From zinc to copper or vice versa?

If this is the process by which zinc ion is reduced on the surface of the copper penny, how long would you expect it to continue? Putting the question another way, when do you think the reaction would stop? How thick would the zinc coating have to get before the surface of the penny began to look (from the viewpoint of the solution) like a piece of zinc metal? Would you expect the process to stop after a single layer (a monolayer) of zinc atoms coated the surface? Or perhaps a thickness of several monolayers is required for the surface of the zinc-coated penny to behave like a zinc metal surface.

(f) *Using Faraday's laws.* Estimate what average constant current would have to flow to produce a monolayer of zinc atoms on the surface of the copper penny in a period of 30 minutes. Make the following assumptions:

Assume that the surface area of a penny is 6.6 cm², and that the diameter of a zinc atom is 2.5×10^{-8} cm. Assuming that a monolayer has a thickness equal to the diameter of a zinc atom, calculate the volume of a monolayer. Calculate the volume of a zinc atom assuming that it is like a little cube of edge length 2.5×10^{-8} cm. (Zinc atoms are presumably spherical, and therefore could be more closely packed as spheres than as cubes, but this is just an approximate, back-of-the-envelope calculation.)

The *ratio: volume of a monolayer/volume of a zinc atom* gives the number of zinc atoms required to form a monolayer, assuming they are packed together like little cubes on the surface of the penny. From the number of atoms, you can calculate the number of moles of zinc. From the moles of zinc, you can calculate the moles of electrons (two moles of electrons must be transferred per mole of Zn^{2+} reduced). From the moles of electrons, you can calculate the total charge transferred in coulombs. The charge in coulombs divided by the total time in seconds gives the average current.

We estimated an average current of about two microamperes, a current that is measurable with most digital multimeters, and the initial current would likely be greater than the average current. If more than a single monolayer were deposited, the average current would be even larger.

(g) What happens when you heat the zinc-coated pennies on the hot plate? Heating could allow the zinc and copper atoms to slip between one another, a process called diffusion. This would form a new phase, a mixture of copper and zinc atoms. Brass is an alloy of copper and zinc, so brass could be formed when the copper and zinc atoms interdiffuse. Would this account for the change in color? What is the color of brass? Unscrupulous alchemists apparently employed this trick in the Middle Ages to fool people into thinking they could make gold from base metals.

BIBLIOGRAPHY

Bonneau, M. C. "Gold Pennies," *J. Chem. Educ.,* **1995,** *72:* 389–390.

Karpenko, V. "Transmutation: The Roots of the Dream," *J. Chem. Educ.,* **1995,** *72:* 383–385.

Shakhashiri, B. Z. "Copper to Silver to Gold," *Chemical Demonstrations,* Vol. 4, The University of Wisconsin Press, 1992, Demonstration 11.33, pages 263–268.

Szczepankiewicz, S. H.; Bieron, J. F.; and Kozik, M. "The Golden Penny Demonstration: An Explanation of the Old Experiment and the Rational Design of the New and Simpler Demonstration," *J. Chem. Educ.,* **1995,** *72:* 386–388. (*Note:* In our hands the new recipe does not work as well as the older one, which we use in the procedure.)

VanderZee, C. and Mosher, M. "Helping Students to Develop an Hypothesis about Electrochemistry," *J. Chem. Educ.,* **1992,** *69:* 924–925. Corrosion and anodic protection.

Electrochemical Puzzles

Metal Corrosion and Anodic Protection

The Golden Penny Experiment

1. Metal Corrosion and Anodic Protection

(a) In the space below make annotated sketches of what you observed in the three Petri dishes containing bimetal strips in agar gel. Also, summarize your observations in the table below.

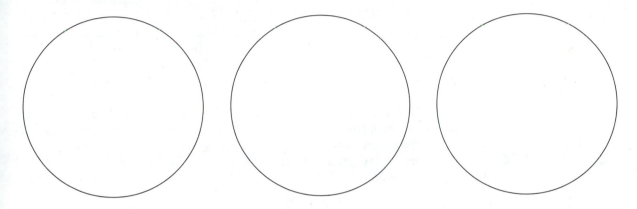

(b) Summarize your observations on the three Petri dishes in the table below:

	Phenolphthalein test: Pink color is an indication of OH⁻	$K_3Fe(CN)_6$ test: Blue color is an indication of Fe^{2+}	Describe the formation of any bands of precipitate	Other observations
Cu/Fe				
Cu/Zn				
Zn/Fe				

(c) If you used a digital voltmeter to probe the voltages in the three Petri dishes, answer the following questions by filling in the table.

When the probes are inserted 1–2 mm from the metal strips, at about the center of the metal strips, do you measure a voltage difference? If so, what is the magnitude of the voltage and next to which metal is there a region of positive charge? What is the magnitude of the voltage when the probes are firmly contacting the metal strips?

	Magnitude of the voltage difference next to the metals, millivolts	The region of positive charge is next to which metal?	Magnitude of the voltage difference when the probes are contacting the metals
Cu/Fe			
Cu/Zn			
Zn/Fe			

(d) Ions migrate in an electric field. So if positive ions are formed at the surface of one metal and negative ions at the surface of the other metal, the ions of opposite charge will migrate toward one another under the influence of the electric field created by the charge separation. How might this explain the formation of bands of precipitate? Describe the color of any bands of precipitate that form. What would be plausible chemical formulas for the precipitates you observed?

(e) Did the iron nail connected to the zinc metal strip corrode? What's your hypothesis (in terms of what the atoms are doing) to explain how the iron nail is protected from corrosion by being connected to zinc metal?

(f) Did the iron nail connected to the copper strip corrode? What's your hypothesis (in terms of what the atoms are doing) to explain this result? Plumbers use a dielectric union when connecting steel water pipes to a copper pipe plumbing system. The union contains an insulating sleeve to prevent a direct electrical contact between the steel and copper pipes. Based on what you observed in the Petri dish, what do you think would happen if you directly connected the steel and copper pipes?

2. The Golden Penny Experiment

General observations

(a) What happens to pennies resting in contact with the 30-mesh zinc on the bottom of the beaker?

(b) What happens to pennies resting on the bottom of the beaker but not in contact with zinc?

(c) What happens to a copper penny soldered to copper wire and immersed in solution when the copper wire is not in contact with zinc metal?

(d) What happens to a penny soldered to copper wire, suspended in solution, when the other end of the copper wire is soldered to a strip of zinc metal which is in turn touching the 30-mesh zinc?

(e) Do you observe the evolution of any gas from the 30-mesh zinc in 1 M NaOH which might come from the following reaction: $Zn + 2H_2O + 2OH^- \rightarrow H_2(g) + Zn(OH)_4^{2-}$?

Does zinc powder react with 1 M NaOH to form a gas?

(f) Is there a voltage difference between a strip of zinc metal and the free end of a copper wire soldered to a penny when both the zinc strip and penny are immersed in solution?

If there is a voltage difference, which is the negative electrode (the electron source)? The copper wire soldered to the penny? Or the strip of zinc metal?

If there is a voltage difference, does it remain constant? Or does it change with time?

Consider This

Constructing a hypothesis to explain the whole process: copper to silver to gold color.

(a) *What do you think makes up the silver-colored coating on the penny? Does it behave in any ways similar to zinc metal?*

(b) *If we assume the silver-colored coating is zinc metal, how do you think the atoms of zinc get from the 30-mesh zinc to the copper penny? By diffusing from the zinc metal to the penny or traveling through the wire connecting the zinc metal to the penny? Or through the solution?*

(c) *If the zinc that coats the copper penny comes from the solution, is the zinc in solution more likely to be in the form of Zn atoms or Zn^{+2} ions? Explain your answer.*

(d) *If you analyzed the solution at the end of the experiment, would you expect to find zinc ions (or $Zn[OH]_4^{2-}$ complex ions) in solution? If so, by what chemical (or electrochemical) reaction could they have been formed?*

(e) *When the silver-plated pennies are heated on the hot plate, they turn a golden color, which is the color of brass, a copper-zinc alloy. Explain how heating could promote the formation of an alloy and cause a color change from silver to gold.*

(f) *Summarize the key observations and your hypothesis for explaining the whole process by which a copper penny acquires a shiny gold color.*

The Chemistry of Some Nonmetals:
Nitrogen, Sulfur, and Chlorine

PURPOSE

To become acquainted with the chemistry of three non-metals (nitrogen, sulfur, and chlorine), the chemistry of which is representative of the elements of Groups 5, 6, and 7. To show that the nonmetals can exist in a variety of formal oxidation states in their compounds with oxygen and hydrogen. To understand how the chemistry of these compounds can be correlated with the electronic structures of the nonmetal elements and their positions in the periodic table.

PRE-LAB PREPARATION

THE OXIDATION STATE OF SOME ELEMENTS AND COMPOUNDS OF THE NONMETALS

The oxidation states of some representative Group 5, 6, and 7 elements are shown in Figure 41-1 (see p. 488). This chart illustrates three important points that we will discuss in the following paragraphs.

1. *Each of the elements in these groups can have many formal oxidation states.* This leads to a rich variety of compounds whose chemistry is complex in comparison with the chemistry of most metals in Groups 1 and 2.

2. *The electronic structures of the elements in the three groups place definite limits on the formal oxidation states accessible to each element.* Nitrogen, whose atoms have five valence electrons, can share these five electrons with electronegative elements like oxygen to produce a formal oxidation state as high as $+5$. It can accept a share of as many as three electrons from less electronegative elements like hydrogen to produce a formal oxidation state as low as -3. In both the $+5$ and -3 oxidation states, the nitrogen atoms have a share of eight valence electrons (like the electron configuration of the inert gases).

Because a neutral sulfur atom has six valence electrons, the formal oxidation state of sulfur ranges from $+6$ in sulfate ion to -2 in sulfide ion. The oxidation states of chlorine range from $+7$ to -1 because a neutral chlorine atom has seven valence electrons.

We must emphasize, however, that the oxidation states are assigned to atoms by formal rules (discussed in Study Assignment A). *The formal oxidation number is not the actual charge on the atom as it exists in the compound.* The actual charge on the atoms of real compounds is usually a fractional charge in the range $+1$ to -1, depending on the electronegativities of the atoms. This is consistent with the electroneutrality principle, which reflects the fact that the distribution of electron density around the nuclei always adjusts itself so that large net charges on adjacent atoms are not formed. As an example, the formal oxidation state we assign to the oxygen atoms in the perchlorate ion (ClO_4^-) is -2. The chlorine atom is therefore assigned oxidation state $+7$. The net charge is given by $4(-2) + 7 = -1$. The actual partial charges have been calculated by Sanderson (see the Bibliography) to be -0.211 on each oxygen atom and -0.155 on the chlorine atom. (Verify that these sum to a net charge of -1 for the perchlorate ion.)

The addition of one electron to a neutral chlorine atom produces stable chloride ion, having an electron configuration like that of argon. (Chloride ion and argon have the same number of electrons; therefore we say they are *isoelectronic.*) In an ion like perchlorate, ClO_4^-, chlorine shares its electrons with the electronegative oxygen atoms, and we can draw Lewis (electron-dot) structures that give each atom a share of eight electrons.

The Lewis structures for some of the compounds mentioned above are shown in Figure 41-2 (see p. 489).

3. *As the formal oxidation state of the nonmetal increases in a compound, the compound will tend to be a stronger oxidizing agent.* Figure 41-1 shows this trend. The

compounds at the top of the table have the largest formal oxidation number assigned to the nonmetal element and are the strongest oxidizing agents. Thus, nitric, sulfuric, and perchloric acids are all strong oxidizing agents, particularly if they are hot and concentrated.

There is also a tendency for the compounds with the most negative oxidation states to be good reducing agents. This is particularly true of NH_3 and hydrazine, N_2H_4, as well as sulfide ion, S^{2-}. However, chloride ion is not a good reducing agent, and in fact all of the chlorine compounds except chloride ion are strong oxidizing agents. This can be understood qualitatively as the effect of the increasing nuclear charge when we go from Group 5 to Group 7 elements. The increased nuclear charge makes Group 7 elements hold on to their electrons more tightly. Therefore Group 7 compounds tend to be stronger oxidizing agents.

THE CHEMISTRY OF REPRESENTATIVE ELEMENTS OF GROUPS 5, 6, AND 7

It would be impossible to explore the chemistry of all the elements in these three groups in the course of one laboratory session; therefore we will choose one member of each group as representative of the chemistry of that particular group. First we will study the chemistry of some of the compounds of nitrogen, a Group 5 element, some compounds of sulfur (Group 6), and of chlorine (Group 7).

Oxygen is so electronegative that it exists primarily in stable compounds in the formal oxidation state -2, with

a few compounds (the peroxides) having oxidation state -1. For these reasons we have chosen sulfur as the representative element of the Group 6 elements (chalcogens).

Similarly, we have chosen chlorine as the representative element of the Group 7 elements (halogens) rather than the first member of the series, fluorine, because fluorine is so electronegative that all stable fluorine compounds contain fluorine with a formal oxidation state of -1.

NITROGEN OXIDES AND SULFUR DIOXIDE IN THE ATMOSPHERE

Nitric oxide is produced in large amounts in steelmaking furnaces and by reaction of nitrogen with oxygen at high temperature in automobile engines. The nitric oxide quickly reacts with oxygen to yield nitrogen dioxide, a brownish gas. Under the action of sunlight, nitrogen dioxide can be photolyzed to give $NO + O$. The free oxygen atom can then react with oxygen to give ozone, O_3. Ozone is a lung irritant and is also very toxic to many species of pine trees. It can also react with unburned hydrocarbons in the atmosphere to yield peroxyacetyl nitrate (PAN), which is very irritating to the eyes and also damages leafy crops such as spinach.

The combustion of coal-containing sulfur produces large quantities of sulfur dioxide by the reaction

$$S + O_2 \rightarrow SO_2$$

Oxidation state	Group 5	Group 6	Group 7
+7			Cl_2O_7, $HClO_4$, ClO_4^-
+6		SO_3, H_2SO_4, HSO_4^-, SO_4^{2-}	Cl_2O_6
+5	$N_2O_5(g)$, HNO_3, NO_3^-		$HClO_3$, ClO_3^-
+4	$NO_2(g)$, $N_2O_4(g)$	SO_2, H_2SO_3, HSO_3^-, SO_3^{2-}	ClO_2
+3	$N_2O_3(g)$, HNO_2, NO_2^-		$HClO_2$, ClO_2^-
+2	$NO(g)$	$S_2O_3^{2-}$	
+1	$N_2O(g)$		Cl_2O, $HClO$, ClO^-
0	$N_2(g)$	S_8	Cl_2
-1	NH_2OH, NH_3OH^+		Cl^-
-2	N_2H_4, $N_2H_5^+$, $N_2H_6^{2+}$	H_2S, HS^-, S^{2-}	
-3	$NH_3(g)$, NH_4^+		

Oxidizing strength →

Reduction / *Oxidation*

FIGURE 41-1
Compounds representing a wide variety of oxidation states of three nonmetals.

In regions in which an appreciable amount of ozone has been formed by a photochemical reaction with nitrogen dioxide, the ozone can further oxidize the SO_2 to sulfur trioxide, SO_3, by the reaction

$$SO_2 + O_3 \rightarrow SO_3 + O_2$$

Sulfur trioxide is the anhydride of sulfuric acid. Thus it can combine with water to produce an aerosol of sulfuric acid mist, or, if ammonia is present (produced in large quantities in agricultural areas by livestock), ammonium sulfate particles can be produced. The haze that is associated with photochemical smog often contains a considerable amount of ammonium sulfate as well as organic matter.

Sulfur dioxide is used in agriculture to kill the wild yeasts that are normally present on grapes used in winemaking. The grapes are then inoculated with a pure strain of yeast that produces a higher quality of wine than the wild strains would. SO_2 is also used in bleaching dried fruit, such as apricots and raisins. One can demonstrate this bleaching action easily by taking a few petals of a red rose, boiling them in 30 mL of ethanol to extract the red pigment, acidifying the extract with a drop of 6 M HCl, and bubbling SO_2 through the solution. The solution will turn pale as the sulfur dioxide reduces the red pigment to a colorless chemical form of the pigment.

SOME CHEMISTRY OF NITROGEN COMPOUNDS

Nitrogen, the first element in Group 5, is an important nonmetal that forms compounds illustrating all the oxidation states from -3 to $+5$. In the zero oxidation state, nitrogen is particularly stable and has the electron-dot structure :N:::N: The commercially important compounds are those in the -3 and the $+5$ oxidation states. Ammonia and its salts are important as soluble fertilizers. Nitric acid is an important oxidizing agent and is used in making explosives.

THE -3 OXIDATION STATE

The nitride ion, N^{3-}, forms with a few Group 1 or 2 metals. Once ignited, magnesium metal burns in air with a brilliant white light (dangerous to look at) forming an ash that is predominantly white MgO. However, the reaction temperature is so high that nitrogen also reacts with hot magnesium, producing a small amount of yellow

FIGURE 41-2
Electron-dot structures of three nonmetallic elements and some of their compounds.

magnesium nitride, Mg_3N_2. Nitrides react with water to form ammonia, NH_3, in which nitrogen has a formal oxidation state of -3 but the bonding is predominantly covalent:

$$Mg_3N_2(s) + 6\,H_2O(l) \rightarrow 6\,NH_3(aq) + 3\,Mg(OH)_2(s)$$

You are probably familiar with ammonia as a household cleaning solution. However, ammonia is actually a gas at room temperature. It is extremely soluble in water, and household ammonia is a solution of about 2 M in NH_3. The volatility of ammonia in this solution makes it easy to detect the characteristic odor of $NH_3(g)$. This odor serves to identify solutions that contain NH_3.

Ammonia is a weak base and forms ammonium ion, its conjugate base, along with hydroxide ion when it is dissolved in water:

$$NH_3(aq) + 6\,H_2O(l) \rightarrow NH_4^+(aq) + OH^-(aq)$$
$$K_b = 1.8 \times 10^{-5}$$

You may still see old texts containing aqueous ammonia labeled ammonium hydroxide (NH_4OH). This formula stems from a notion that ammonia forms a molecule of NH_4OH by combining with a water molecule, and then this weak base, ammonium hydroxide, dissociates slightly to give NH_4^+ and OH^-. There is some controversy over whether the NH_4OH molecule has an independent existence. It is probably more accurate to think of an aqueous ammonia solution as containing mainly NH_3 molecules dissolved in water.

The addition of ammonia to aqueous solutions of certain metal ions causes precipitates to form. These precipitates are usually not nitrogen compounds but hydroxides, formed from the small amounts of hydroxide produced by reaction of ammonia with water. Naturally, these precipitates form only with metal ions that have very insoluble hydroxides ($K_{sp} < 10^{-11}$ for M^{2+} cations). Because of its lone pair of electrons,

$$\ddot{N} \atop {\Large / \; | \; \backslash} \atop H \; H \; H$$

ammonia can act as a Lewis base, forming donor-acceptor complexes with Lewis acids. An example of this behavior is the formation of ammine complexes of the transition metals such as

$$Ag^+(aq) + 2\,NH_3(aq) \rightleftarrows Ag(NH_3)_2^+(aq)$$

and

$$Cu^{2+}(aq) + 4\,NH_3(aq) \rightleftarrows Cu(NH_3)_4^{2+}(aq)$$

This copper complex with its intense violet-blue color is often used in the laboratory as an indicator for the presence of ammonia.

THE -2 OXIDATION STATE

The -2 oxidation state of nitrogen is represented by hydrazine, N_2H_4. This compound can be produced by reacting ammonia with hypochlorite ion in basic solution:

$$2\,NH_3(aq) + ClO^-(aq) \rightarrow N_2H_4(aq) + H_2O(l) + Cl^-(aq)$$

However, hydrazine is very toxic, and this mixture also produces toxic and explosive chloramine compounds, such as $ClNH_2$. This is the reason for the statement on laundry bleach bottles (5% aqueous NaOCl) warning that bleach should not be mixed with ammonia solutions.

THE OXIDES OF NITROGEN

None of the nitrogen oxides form spontaneously at room temperature and pressure. They can form at high temperatures and pressures, with some beneficial and some unpleasant consequences. Nitrogen oxides are formed in lightning discharges, and this serves as a natural form of "fixed" nitrogen fertilizer for plants as it is washed into the soil by rain. Nitrogen(II) oxide, NO (also called nitric oxide), forms at the high temperatures characteristic of a flame or an internal combustion engine. When this $NO(g)$ reacts with atmospheric oxygen, it forms (as we will observe) nitrogen dioxide, NO_2. This reddish-brown gas is an irritating component of the smog found in many large cities.

We also want to examine an important nitrogen acid, nitric acid (HNO_3). Nitric acid is useful because it is a strong acid and because the nitrate ion does not form precipitates with metal ions, since nitrate salts are soluble. In addition, the nitrate ion in acidic solution is an excellent oxidizing agent. Because of this property, nitric acid can dissolve some metals that do not react with other acids such as HCl. Depending on the concentration of the nitric acid and the strength of the reducing agent, nitric acid can be reduced to $NO_2(g)$, $NO(g)$, or all the way to the -3 oxidation state of ammonia.

EXPERIMENTAL PROCEDURE

NITROGEN COMPOUNDS

NOTE TO INSTRUCTOR: The chemicals required for the study of each element (N, S, Cl) are listed separately under the heading for each element.

Special Supplies: pH paper or universal indicator solution or pH meter; No. 4 one-hole and two-hole rubber stoppers, 25 × 250 mm, 15 × 125 mm, 13 × 100 mm test tubes, 6-mm-OD glass tubing bent to the shapes shown in Figures 41-3 and 41-4, ³⁄₁₆-in.-ID rubber tubing, wood splints.

Chemicals: 1 M ammonia, NH_3; 6 M ammonia, NH_3; 1 M ammonium chloride, NH_4Cl; 0.10 M iron(III) nitrate, $Fe(NO_3)_3$; 1 M sodium hydroxide, NaOH; 0.1 M copper(II) sulfate, $CuSO_4$; 0.1 M sodium chloride, NaCl; 0.1 M silver nitrate, $AgNO_3$; 6 M nitric acid, HNO_3; 16 M nitric acid, HNO_3; 3% hydrogen peroxide, H_2O_2; 3 M iron(III) chloride, $FeCl_3$; copper metal, Cu (turnings).

! SAFETY PRECAUTIONS: Nitrogen oxides (NO and NO_2) will irritate eyes and lungs. Work in a fume hood when generating nitrogen(II) oxide, NO. This experiment should not be performed if a fume hood is not available.

Nitric acid is very corrosive and can cause severe burns. Handle it carefully. If spills occur, wash with plenty of water. Wear eye protection at all times in the laboratory.

WASTE COLLECTION: A waste container should be provided for the silver compound formed in Part 1c. Acid and base wastes should be collected separately or rinsed down the drain after neutralization with sodium bicarbonate, $NaHCO_3$.

....

NOTE TO STUDENTS: In your written report of this experiment, include all observations of the properties of the products formed, and write equations for all reactions.

....

FIGURE 41-3

Apparatus for the preparation of nitrogen gas or of the oxides of nitrogen.

1. AMMONIA AND AMMONIUM ION: THE −3 OXIDATION STATE.

(a) **Acid-Base Properties.** Determine the pH of 1.0 M NH_3 and 1.0 M NH_4Cl with pH paper, a few drops of universal indicator solution, or a pH meter.

(b) **Precipitation Reactions.** Place 1 mL of 0.1 M Fe$(NO_3)_3$ in each of three 13 × 100 mm test tubes. To the first tube add 1 mL of 1 M NaOH. Note the precipitate, and describe its characteristics. To the second tube add 1 mL of 6 M NH_3. To the third tube add 1 mL of 1 M NH_4Cl. Note the reaction(s) (if any), and write net ionic equations to describe the reaction(s).

(c) **Ammonia as a Complexing Agent.** Put 1 mL of 0.1 M $CuSO_4$ in a test tube, and add 2 drops of 1 M NH_3. Describe the precipitate. Continue adding 1 M NH_3 with mixing (about 20 drops total). Observe the solution. Is it the same color as the original $CuSO_4$ solution? (See the Pre-Lab Preparation discussion of the −3 oxidation state for the formula of the product.) Put 1 mL of 0.1 M $CuSO_4$ in another test tube, and add 1 mL (20 drops) of 1.0 M NH_4Cl. Do you observe any changes?

Prepare some solid silver chloride by adding 10 drops of 0.10 M NaCl to 10 drops of 0.10 M $AgNO_3$ in a test tube. Then add 3 mL of 1 M NH_3. Mix thoroughly with a glass stirring rod for 30 s. Does the precipitate dissolve? (Again, see the Pre-Lab Preparation for the formula of the product.) Dispose of the product in the silver waste container provided.

2. THE PREPARATION OF NITRIC OXIDE BY REDUCTION OF HNO_3 WITH COPPER: OXIDATION STATES +2, +4, AND +5. Skip this section if a fume hood is not available. If you have access to one, assemble the generator shown in Figure 41-3, which uses a 20 × 200 mm test tube, and place about 3 g of copper turnings in it. Prepare to collect three 15-cm test tubes of nitric oxide by displacement of water. Add 10 mL of 6 M HNO_3 to the generator, replace the delivery tube connection, and warm the test tube gently to initiate the reaction. After the air has been displaced from the apparatus and the gas bubbling through is colorless, collect two full test tubes of the gas and a third test tube about half full. Do not allow the delivery tube to remain under water while the heated test tube cools; if you do, water will be drawn into the test tube. What is the reaction for the reduction of dilute nitric acid by copper?

Test the nitric oxide in one test tube with a glowing splint to see if it supports combustion. Note the colored gas produced when the tube is exposed to the air. Write the equation for the reaction that accounts for this change.

Test the second sample of nitric oxide for solubility in water by swirling the test tube with its mouth under the water to allow contact of fresh water with the gas. Note if the water level in the test tube rises.

Now take the test tube out of the water for a few seconds, and allow the oxygen of the air to react with the gas, as will be evidenced by the formation of a brown gas. Invert the test tube under the water again, and swirl it to note the solubility of the brown gas. Write the equation for the reaction occurring when this gas dissolves in water.

Mark the water level in the third test tube, which is about half full of nitric oxide, with a wax pencil or gummed label. Set up the small oxygen generator shown in Figure 41-4, placing about 10 mL of 3% hydrogen peroxide in the test tube. Draw up a few milliliters of 3 M $FeCl_3$ into the medicine dropper, and replace the two-hole stopper in the mouth of the generator tube. When ready to begin generating oxygen, release a few drops of $FeCl_3$. After a brief period the $FeCl_3$ will catalyze the decomposition of H_2O_2, producing oxygen gas. After the air has been displaced from the generator, place the delivery tube under the marked test tube, and allow 8 to 10 bubbles of oxygen to enter. Note whether the level of the water is lowered by the addition of oxygen. Recall that in the balanced equation for the reaction that is taking place, two volumes of nitric oxide react with one volume of oxygen to produce two volumes of nitrogen dioxide.

FIGURE 41-4
Generator for the preparation of a small amount of oxygen by the reaction of $FeCl_3$ with H_2O_2. This oxygen is then reacted with $NO(g)$.

Now swirl the test tube with its mouth under the water and note what happens to the water level as the NO_2 reacts with the water. Allow more oxygen to bubble into the tube until the gases turn brown. Note the water level, and again allow the gases to dissolve in water. Repeat the process until the water level approaches the top of the tube. Remember that excess oxygen is not soluble in water. What substances are present in the water solution in the test tube? Apply a simple test to verify your answer. What part of one of the commercial processes involved in the production of nitric acid does this experiment illustrate?

SOME CHEMISTRY OF SULFUR COMPOUNDS

In this experiment we study some chemical properties of some common compounds of sulfur in oxidation states -2, $+2$, $+4$, and $+6$. Sulfur is a Group 6 element with the electron configuration $1s^22s^22p^63s^23p^4$. Its electron-dot structure is $\cdot\ddot{S}\cdot$. By sharing two electrons with two hydrogen atoms, it forms the covalent compound H_2S. A saturated aqueous solution of H_2S is weakly acidic and contains small concentrations of H^+ and HS^- ions. In strongly basic solutions the concentrations of HS^- and S^{2-} ions are greater.

By sharing its electrons with more electronegative elements, such as oxygen, sulfur attains positive oxidation states. In sulfur dioxide, four of the sulfur electrons are involved in bonding with oxygen:

$$\ddot{O}:\!\overset{\ddot{}}{\underset{\cdot\cdot}{S}}\!\overset{\times\times}{\underset{}{}}\ + \ H\!:\!\ddot{O}\!:\ \rightleftharpoons\ H\!:\!\ddot{O}\!:\!S\!:\!\ddot{O}\!:\!H$$

SO_2 is the anhydride of sulfurous acid, H_2SO_3, which is a weak acid, and forms some $H^+ + HSO_3^-$ ions. In basic solutions the equilibria

$$H_2SO_3 \rightleftharpoons H^+ + HSO_3^-$$
$$HSO_3^- \rightleftharpoons H^+ + SO_3^{2-}$$

are shifted to the right, to form more bisulfite and sulfite ions and water.

In sulfur trioxide all six of the sulfur electrons are involved in bonding:

$$\ddot{O}\!:\!\ddot{O}\ \overset{+}{\underset{+}{S}}\ + \ H\!:\!\ddot{O}\!:\ \rightleftharpoons\ H\!:\!\ddot{O}\!:\!S\!:\!\ddot{O}$$

Sulfur trioxide is the anhydride of sulfuric acid, H_2SO_4, one of the most important industrial inorganic chemicals. It is a strong acid whose aqueous solutions contain large concentrations of H^+ and HSO_4^- ions. In basic solutions the SO_4^{2-} ion is the predominant species.

The electron-dot structures of sulfite, sulfate, and thiosulfate ions are represented as follows:

sulfite   sulfate   thiosulfate

Note that in the thiosulfate ion, the sulfur atom that replaces the oxygen in the sulfate structure may be assigned a -2 oxidation number and that the central sulfur atom has an oxidation number of $+6$, just as it has in sulfate. The $+2$ oxidation number assigned to sulfur in thiosulfate is obtained by finding the average of $+6$ and -2: $(+6 -2)/2 = +2$.

The experiments you perform will show that H_2SO_4 and SO_4^{2-} are mild oxidizing agents, whereas H_2S, S^{2-}, H_2SO_3, and SO_3^{2-} are reducing agents. In addition, since SO, SO_3^{2-}, and H_2SO_3 represent intermediate oxidation states, they can act as oxidizing agents with a strong reducing agent and as reducing agents with a strong oxidizing agent.

EXPERIMENTAL PROCEDURE

SULFUR COMPOUNDS

Special supplies: Source of H_2S gas.[1]

Chemicals: Crushed roll sulfur, S; iron, Fe (filings); 0.1 M lead(II) nitrate, $Pb(NO_3)_2$; 0.1 M tin(IV) chloride, $SnCl_4$; 0.1 M zinc nitrate, $Zn(NO_3)_2$; 0.05 M iodine, I_2 (0.05 M I_2 in 0.1 M KI); 0.1 M calcium nitrate, $Ca(NO_3)_2$; 0.1 M barium chloride, $BaCl_2$; 0.1 M barium hydroxide, $Ba(OH)_2$; solid sodium sulfite, Na_2SO_3; saturated bromine water, Br_2; sugar (sucrose) crystals; pieces of zinc, Zn, and copper, Cu; sodium chloride, NaCl(s); potassium bromide, KBr(s); lead acetate test paper, $Pb(C_2H_3O_2)_2$; source of hydrogen sulfide, H_2S gas; 6 M hydrochloric acid, HCl; concentrated sulfuric acid, H_2SO_4; 3 M sulfuric acid, H_2SO_4.

! SAFETY PRECAUTIONS: This experiment should not be performed if a fume hood is not available. Hydrogen sulfide has a noxious odor and is very toxic. Part 1 should be carried out in a fume hood.

In Part 3, the addition of concentrated (118 M) sulfuric acid to NaCl, KBr, and KI should be carried out in a fume hood because the products formed (HCl, HBr, and HI) will irritate eyes and lungs.

When you are asked to note the odor of a substance, do so very cautiously. Do not stick your nose into the mouth of a test tube and inhale. Gently fan the vapor toward your nose with your hand.

WASTE COLLECTION: A waste container should be provided for the lead sulfide formed in parts 1b and 3c and a container should be available for the barium compounds in parts 2b and 3c.

NOTE TO STUDENTS: In your written report, include all observations and equations for all reactions.

1. SULFIDES AND HYDROGEN SULFIDE: OXIDATION STATE -2.

(a) **Preparation of a Sulfide.** Mix approximately 3.5 g of iron filings with 2 g of crushed sulfur in a crucible supported on a triangle. Place a lid on the crucible, and heat with a Bunsen burner until the reaction is initiated, removing the burner and lid occasionally to note whether the reaction continues with the evolution of heat. Burn off any excess sulfur, and allow the crucible to cool. Place a small piece of the compound in a small test tube. Add a few milliliters of 6 M HCl, and note cautiously the products of the reaction.

(b) **Hydrogen Sulfide as a Precipitating Agent.** Metallic sulfides, other than those of the alkali and alkaline earth metals, are sparingly soluble in water. In qualitative analysis many metal ions are identified by precipitating them as sulfides. Using the source of H_2S gas available in your laboratory (see footnote 1), saturate 5 mL of each of the following solutions with H_2S gas: 0.1 M $Pb(NO_3)_2$, 0.1 M $Zn(NO_3)_2$, and 0.1 M $SnCl_4$. Record the color and formula of each precipitate. Dispose of the lead sulfide in the container provided.

(c) **Hydrogen Sulfide as a Reducing Agent.** Saturate each of the following solutions with H_2S: (1) 5 mL of warm 3 M HNO_3, (2) 5 mL of 0.05 M I_2, and (3) a freshly prepared solution of H_2SO_3 made by adding a few crystals of Na_2SO_3 and a drop of 6 M H_2SO_4 to 5 mL of water. Test the vapors with moistened lead acetate paper. Note the products, and write balanced equations for each reaction.

2. SULFUR DIOXIDE, SULFUROUS ACID, SULFITE ION: OXIDATION STATE $+4$.

(a) **Preparation of Sulfur Dioxide.** Since sulfur is in an intermediate oxidation state in SO_2, this compound can be prepared by oxidation of sulfides (as is done in metallurgical roasting), by oxidation of elemental sulfur, or by the reduction of hot concentrated H_2SO_4 by copper. It can also be conveniently prepared (without a change in

[1]Consult your instructor about your best source of H_2S gas. A cylinder of the compressed gas (in a fume hood) or a Kipp generator charged with FeS and 5 M HCl (in a fume hood) is often used. A little "Aitch-tu-ess" (a commercial mixture of sulfur, paraffin, and asbestos) heated in a test tube that is fitted with a gas delivery tube is a convenient method.

oxidation state) from metal sulfites by the addition of an acid.

Place about 2 g of Na_2SO_3 in a large test tube. Add 6 M HCl drop by drop until you can smell (**Caution!**) the odor of the gas (SO_2) given off. Now add about 15 mL of water, and stir until all the solid dissolves. Save for Part 2b.

(b) **Chemical Properties of Sulfurous Acid.** Divide the solution equally among three smaller test tubes. To one portion add a few milliliters of 0.1 M $Ba(OH)_2$ until the solution is basic to litmus. What is the precipitate formed? Is it soluble in 6 M HCl added a drop at a time? (Ignore a slight turbidity, which is due to air oxidation of sulfite to sulfate.) Discard the solution in the barium waste container. To another portion add 5 to 6 mL of saturated bromine water, drop by drop. How do you account for the decolorization that takes place? Now add a few milliliters of 0.1 M $Ba(OH)_2$ to this test tube. What is the precipitate formed? Is it soluble in 6 M HCl? Dispose of the solid in the barium waste container.

Add a little of the third portion, a few drops at a time, to a test tube containing 5 mL of 3 M H_2SO_4 and some mossy zinc. Observe the odor of the gas, and note the precipitate formed.

3. SULFURIC ACID, SULFATES, OXIDATION STATE +6.

(a) **Physical and Chemical Properties of Sulfuric Acid.** While stirring, cautiously add a few mL of 18 M H_2SO_4, drop by drop, to 50 mL of tap water in a small beaker. Note the temperature change. To what do you attribute this result?

Place a few drops of 18 M H_2SO_4 on a few crystals of sugar in a small evaporating dish. Repeat the test on a small piece of paper or wood (such as a matchstick). How do you explain the results?

(b) **Effect of Concentrated Sulfuric Acid.** Investigate the oxidizing strength of concentrated (18 M) sulfuric acid by adding in a fume hood 1 mL of concentrated H_2SO_4 to 1 g of each of the following salts in 20 × 150 mm test tubes: NaCl, KBr, and KI. Note the color of any gases evolved. (Place the tube containing KI in a beaker of hot water.) Test for acidity of the gases evolved by holding a piece of moistened blue litmus paper near the mouth of each test tube. Are any halide ions oxidized to the elementary halogens?

(c) **Solubility of Sulfates.** Add 3 mL of 0.1 M $NaSO_4$ to 3 mL each of the following solutions in separate 10-cm test tubes: 0.1 M $Ca(NO_3)_2$, 0.1 M $BaCl_2$, 0.1 M $Pb(NO_3)_2$. Test the solubility of any precipitates in dilute nitric acid by adding 1 mL of 6 M HNO_3 to each precipitate. Dispose of the lead and barium compounds in their respective waste containers.

SOME CHEMISTRY OF CHLORINE COMPOUNDS

Figure 41-5 illustrates the variety of compounds and oxidation states that exist for chlorine, which is a prototype of the halogens.

This chart of the oxidation states of chlorine emphasizes that this element, which in varying degrees is typical of all the Group 7 halogen elements, is *strongly electronegative*. Note that *in all oxidation states except the lowest very stable −1 state, chlorine forms compounds that are strong oxidizing agents.* They have a strong affinity for additional electrons. This tendency is extremely strong for fluorine, which exists only in the oxidation states −1 and 0. The other halogens, in order of their increasing size and atomic mass, are somewhat less electronegative.

All the halogen elements have the same Lewis structures as chlorine. Note the following structures for a chlorine atom, a chlorine molecule, Cl_2, and a chloride ion, Cl^-.

$$:\overset{..}{\underset{..}{Cl}}\cdot \qquad :\overset{..}{\underset{..}{Cl}}:\overset{..}{\underset{..}{Cl}}: \qquad :\overset{..}{\underset{..}{Cl}}:^-$$

| chlorine atom | chlorine molecule | chlorine ion |

In each halogen, the free element can add one more electron to fill its valence orbitals, thus giving the ion the electron structure of the inert noble gas next in order in the periodic table.

Chlorine and its compounds show a marked tendency to undergo self– or auto–oxidation-reduction, in which some molecules or ions of a species are oxidized to a higher state while others are reduced to the stable −1 state. This process is called *disproportionation*.

It is possible to oxidize Cl^- (oxidation state −1) to free chlorine, $Cl_2(g)$ (oxidation state 0), then carry out a series of disproportionation reactions in which the chlorine is successively oxidized to the +1, +5, and finally the +7 oxidation states, as indicated on the flow chart in Figure 41-6. This could be done as follows:

1. $Cl_2(g)$ is passed into a cold basic solution, in which it is auto-oxidized to ClO^-, and auto-reduced to Cl^-. (This is the reaction that occurs in the commercial preparation of 5% NaClO bleaching solution, the bleach sold in grocery markets.)

2. When heated, the ClO^- in basic solution is further oxidized to ClO_3^-, and a portion is reduced back to Cl^-. (Suitable crystallization of the salts from this solution produces the commercially important oxidant $KClO_3$ or the weed killers $NaClO_3$ and $Ca[ClO_3]_2$.)

3. Maintaining $KClO_3$ crystals at a temperature just above their melting point results in further autoxidation of

		+7	(Cl_2O_7), $HClO_4$, ClO_4^-	Cl_2O_7 is unstable; $HClO_4$ is a strong oxidizing agent. Reduced to Cl^-
		+5	$HClO_3$, ClO_3^-	Strong oxidizing agent. Reduced to Cl^-
		+4	ClO_2	Unstable, explosive
O	R	+3	$HClO_2$, ClO_2^-	Good oxidizing agent. Reduced to Cl^-
		+1	Cl_2O, $HClO$, ClO^-	Good oxidizing agent. Reduced to Cl^-
		0	Cl_2	Good oxidizing agent. Reduced to Cl^-
		−1	Cl^-	

FIGURE 41-5

Chlorine compounds.

the ClO_3^- to ClO_4^-, and reduces a portion of it to Cl^-. (Perchlorates are important oxidants in solid rocket fuels.)

THE CHLORIDE IONS AND ITS COMPOUNDS: THE −1 OXIDATION STATE

The −1 oxidation state of the halogens, known collectively as the halides, is by far the most common form of the halogens. We are familiar with chloride ions in the form of salt, both solid NaCl and salt solutions such as seawater and blood. The salt in seawater is predominantly sodium chloride, but bromide and iodide ions are also present in smaller amounts.

Hydrochloric acid, HCl, is one of the common laboratory acids. Hydrochloric acid solutions are formed by dissolving gaseous hydrogen chloride molecules, $HCl(g)$, in water. These acid solutions have powerful, choking odors when they are concentrated due to the volatilization of the gaseous hydrogen halides from these solutions.

Each of the halide ions forms precipitates with Ag^+, Pb^{2+}, and Hg_2^{2+} ions. The solids formed in these reactions are useful in the laboratory for identifying these metal ions in solution or for identifying these halide ions in solution. The silver halide solids serve as the basis for all photographic materials because of their sensitivity to

light. (Light activates the decomposition of these solids into the halogen element and silver metal.)

$$2AgBr(s) \xrightarrow[developer]{light} 2Ag(s) + Br_2(l)$$

The halide ions can also serve as ligands in forming complex ions with metal ions. While the halide ions are not the best ligands, we will observe the formation of some complexes by noting color changes and the dissolution of a solid.

ELEMENTAL CHLORINE: THE 0 OXIDATION STATE

As elements the halogens are identifiable by their color and state: Cl_2 is a pale, greenish-yellow gas at room temperature; Br_2 is a dark, reddish-brown fuming liquid; and iodine is a purple-black solid that gives off a violet vapor. Because these molecules are nonpolar, they are not highly soluble in water, but they are soluble in nonpolar solvents such as hexane, C_6H_{14}. The distinguishing characteristics of the halogens is their oxidizing ability, more commonly recognized as their bleaching action.

The halogens are not such strong oxidizing agents that they can oxidize water to $O_2(g)$, but solutions of

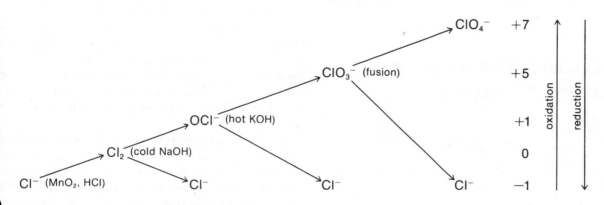

FIGURE 41-6

The interconversions of some ions containing chlorine in several different oxidation states.

the halogens are unstable because of their tendency to disproportionate. For example,

$$Cl_2(aq) + H_2O(l) \rightarrow Cl^-(aq) + HClO(aq) + H^+(aq)$$

Solutions of chlorine water must be freshly made or they will not contain much chlorine in the form Cl_2.

HYPOCHLORITE AND HYPOCHLOROUS ACID: THE +1 OXIDATION STATE

The solutions that we call chlorine bleach do not contain Cl, but are actually solutions of sodium hypochlorite, NaClO, prepared by dissolving chlorine gas into sodium hydroxide solutions:

$$Cl_2(aq) + 2\,NaOH(aq) \rightarrow$$
$$NaClO(aq) + NaCl(aq) + H_2O(l)$$

The hypochlorite ion itself is a good oxidizing agent, so these solutions possess the same desired bleaching action.

Although it is similar to chlorine in this oxidizing power, the hypochlorite ion is very different from the chloride ion. Because hypochlorous acid, HClO, is a weak acid, the hypochlorite ion is a base of moderate strength. This is in contrast to the neutral character of the chloride ion, the conjugate base of the strong acid, HCl.

CHLORATE AND PERCHLORATE: HIGHER OXIDATION STATES

If basic solutions containing ClO^- are heated, the hypochlorite disproportionates to give chlorate and chloride ions.

$$3\,ClO^-(aq) \rightarrow ClO_3^-(aq) + 2\,Cl^-(aq)$$

Chlorate can be further oxidized to the perchlorate ion, ClO_4^-. We will not be exploring the chemistry of these higher oxidation states because of their instability and hazardous nature.

EXPERIMENTAL PROCEDURE

CHLORINE COMPOUNDS

Chemicals: 0.1 M sodium chloride, NaCl; 0.1 M silver nitrate, AgNO₃; 6 M ammonia, NH₃; 6 M nitric acid, HNO₃; 0.1 M copper(II) sulfate, CuSO₄; 12 M hydrochloric acid, HCl; sodium hypochlorite, NaOCl solution (a commercial bleach solution containing 5% NaOCl); red and blue litmus paper; 6 M sodium hydroxide, NaOH; 0.1 potassium iodide, M KI; hexane, C₆H₁₄ (or petroleum ether, b.p. 60–90 °C); 0.1 M potassium bromide, KBr.

! **SAFETY PRECAUTIONS:** Observe the usual precautions, especially eye protection, with the strong acids and bases used in this section as well as the bleach solution. Do not mix NH₃ and NaOCl solutions.

WASTE COLLECTION: A waste container should be provided for the silver compound formed in Part b. A container for the hexane-containing mixtures should be made available, in the hood if possible.

1. THE CHLORIDE ION: OXIDATION STATE −1.

(a) **Solubility and Stability of Chloride Salts.** To 1 mL of 0.1 M NaCl, add 1 mL of 0.1 M AgNO₃. Mix the test tube well. Add an equal volume of 6 M NH₃ (about 3 M in the mixture) to the test tube. Agitate the test tube thoroughly to break up any lumps of precipitate. Finally, add a slight excess of 6 M HNO₃, and mix. Explain the results on your Report Form.

(b) **Transition Metal Complexes with Chloride.** To 2 mL of 0.1 M CuSO₄, add 2 mL of 12 M (concentrated) HCl; dilute this with about 5 mL of water. Write equations, and interpret the color changes you observed. Remember that the copper sulfate solution actually contains $[Cu(H_2O)_6]^{2+}$ ions. Assume that the new complex formed is $CuCl_4^{2-}$.

To 1 mL of 0.1 M AgNO₃, add 3 mL of 12 M (concentrated) HCl; agitate this well for several minutes to redissolve the precipitated AgCl. Now dilute this with about 5 ml of distilled water. Write equations for the reactions, and interpret the changes you observed, assuming that the complex initially formed was $AgCl_2^-$.

2. THE HYPOCHLORITE ION.
Since solid NaOCl cannot be isolated easily without decomposition, we will test portions of the solution of a commercial bleach obtained at the grocery store. It was prepared by passing chlorine into a solution of NaOH.

(a) **Litmus Reaction.** Put several drops of NaOCl solution on red and blue litmus to note its acidity or basicity. Note any bleaching effect.

(b) **Reacting to AgNO₃.** To a 3-mL portion of the NaOCl solution, add 1 mL of 0.1 M AgNO₃. What is the precipitate? (Compare with the behavior of a drop of 6 M NaOH on 0.1 M AgNO₃.) Is it soluble in 6 M HNO₃, and does any other precipitate remain? Explain your observations. Dispose of the silver compounds in the silver waste container.

(c) **Oxidizing Strength.** Place 2 mL of 0.1 M KI and 1 mL of hexane (or petroleum ether) in a test tube. (**Caution:** Work in a fume hood. Hexane is flammable. There must be no open flames nearby.) Add 5% NaOCl solution one drop at a time, shaking the test tube after each drop, and note any color change in the hexane layer. Is there any evidence for formation of I_2? An excess of NaOCl must be avoided because it will remove the color, owing to further oxidation of the initial product to the colorless IO_3^- ion. Dispose of the hexane mixture in the proper waste container.

Repeat this test, using 2 mL of 0.1 M KBr in place of KI. Is there any evidence for the formation of Br_2 detected by a color change in the hexane layer? Where would you place the ClO^- ion with respect to Br_2 and I_2 in a scale of oxidizing strength? Now acidify the test solution with 6 M HCl and shake, noting any color formed in the hexane layer. Does the oxidizing strength of ClO^- change when the solution is acidified? Properly dispose of the hexane mixtures.

Record all of your observations and write equations to explain the reactions.

 Consider This

From thermodynamic tables find the enthalpies of formation of the nitrogen oxides listed in Figure 41-1. Using Le Châtelier's principle, determine whether these oxides are likely to form at high or low temperatures and at high or low pressures. Use this knowledge to discuss the formation of nitrogen oxides in an internal combustion engine and how you would design a catalytic converter to eliminate them from automobile exhaust.

Are the same issues involved in the formation of sulfur oxides if the gasoline (or coal) contains sulfur as it is burned?

BIBLIOGRAPHY

Sanderson, R. T. *Polar Covalence,* Academic Press, New York, 1983, p. 194.

**The Chemistry of Some Nonmetals:
Nitrogen, Sulfur, and Chlorine**

DATA, OBSERVATIONS, AND CONCLUSIONS: NITROGEN COMPOUNDS (GROUP 5)

1. Ammonia and Ammonium Ion: The −3 Oxidation State

(a) What is the pH of 1 M NH_3? _____

What is the pH of 1 M NH_4Cl? _____

Write net ionic equations to describe the reactions that cause these solutions to be acidic or basic.

$NH_3 + H_2O \rightarrow$ _____

$NH_4^+ + H_2O \rightarrow$ _____

(b) *Precipitation reactions*

What is the precipitate formed when $Fe(NO_3)_3$ and NaOH solutions are mixed?

Does the precipitate formed from $Fe(NO_3)_3$ and NH_3 appear to be the same as the previous one?

Write net ionic equations for the reactions of Fe_{3+} with OH^- and with NH_3 in water.

$Fe_{3+} + OH^- \rightarrow$ _____

$Fe_{3+} + NH_3 + H_2O \rightarrow$ _____

(c) *Ammonia as a complexing agent*

What precipitate forms when a small amount of $NH_3(aq)$ is added to $CuSO_4(aq)$?

Net ionic equation _____

What evidence do you have that a reaction has occurred when excess NH_3 is added?

What is the formula of the copper-containing product? (See the Pre-Lab Preparation on the −3 oxidation state of nitrogen.)

Why doesn't a precipitate form when NH_4Cl is added to $Cu^{2+}(aq)$?

Describe what happens when you add an excess of 1 M NH_3 to a precipitate of AgCl.

Net ionic equation _____

2. **The Preparation of Nitric Oxide by Reduction of HNO_3 with Copper: Oxidation States +2, +4, and +5**

   Give the electron-dot structures proposed for NO.

   Why is nitric oxide called an *odd* molecule?

   Write the equation for the reaction of dilute nitric acid with copper.

   _____

   List the properties of NO: color _____; ability to support combustion _____

   What is the colored gas formed when NO is exposed to air? _____

   Write the equation for this reaction.

   _____

   To what extent is NO soluble in water? _____

   Is NO_2 soluble in water? _____

   How do you account for this? _____

   Write the equation for the reaction of NO_2 with H_2O.

   _____

   Write the equation for the production of oxygen from H_2O_2.

   _____

   Describe the changes taking place in the test tube half full of NO gas as oxygen is added.

   color of gases _____

   change in water level _____

   Explain these observations by writing chemical equations for the reactions that are most likely to be taking place.

   _____

   When the test tube is swirled underwater, the water level becomes _____ and the color of the gases becomes _____. Account for these observations.

   What results were obtained when the swirling process was repeated after exposure to the air?

   What substances are present in the aqueous solution in the test tube at the end of the process?

   _____

DATA, OBSERVATIONS, AND CONCLUSIONS: SULFUR COMPOUNDS (GROUP 6)

1. **Sulfides and Hydrogen Sulfide: Oxidation State −2**

   (a) Write the equation for the preparation of a sulfide.

   _____

   Is the reaction endothermic or exothermic? _____

Write the reaction of ferrous sulfide with hydrochloric acid.

_____

(b) *Hydrogen sulfide as a precipitating agent*

Note the formula and color of the insoluble sulfides prepared.

Pb^{2+} _____   _____

Zn^{2+} _____   _____

Sn^{4+} _____   _____

(c) Write the equations for oxidation of H_2S by

3 M HNO_3 _____

0.05 M I_2 _____

H_2SO_3 _____

2. Sulfite Dioxide, Sulfurous Acid, Sulfite Ion: Oxidation State +4

(a) Write the equation for the preparation of SO_2.

_____

(b) Write the equation for reactions of H_2SO_3 with

(i) $Ba(OH)_2$ _____

Solubility of precipitate in dilute HCl

_____

(ii) Saturated bromine water (Br_2)

Reaction of product with $Ba(OH)_2$

_____

Solubility of precipitate in dilute HCl

_____

(iii) Zn and H_2SO_4

_____

3. Sulfuric Acid, Sulfates: Oxidation State +6

(a) *Physical and chemical properties of sulfuric acid*

Temperature effect on dilution _____

Explanation

Effect on sugar _____   on paper _____

Explanation

(b) *Effect of concentrated sulfuric acid*

Note what happens when concentrated sulfuric acid is added to the following.

Substance	Observations	Equations
NaCl		
KBr		
KI		

(c) *Solubility of sulfates*

Formulate a solubility rule for sulfates of the common metals.

DATA, OBSERVATIONS, AND CONCLUSIONS: CHLORINE COMPOUNDS (GROUP 7)

1. The Chloride Ion: Oxidation State −1

(a) *Solubility and Stability of Chloride Salts*

Write net ionic equations to describe the reactions that you observed.

(i) $NaCl(aq)$ was mixed with $AgNO_3(aq)$.

_____

(ii) 6 M NH_3 was added to the precipitate that formed in (i).

_____

(iii) 6 M HNO was added to the solution from (ii).

_____

(b) *Transition Metal Complexes with Chloride*

What evidence do you have that a reaction occurred when 12 M HCl was added to the $CuSO_4$ solution? Did another reaction occur when the mixture was diluted with water?

Write a chemical equation to describe the processes that you observed in Part b.

_____

Write the equation for the reaction that occurred when the concentrated HCl was added to the $AgNO_3$ solution.

_____

Write the equation for the reaction that occurred when distilled water was added to the previous solution.

_____

2. The Hypochlorite Ion

(a) Is the 5% NaOCl solution acidic or basic? _____

Does it have bleaching properties?

(b) Explain and write equations for the reaction with $AgNO_3$.

_____

(c) Write the equation for the reaction of OCl^- in basic solution with I^-.

_____

Write the equation for the reaction of OCl^- in acidic solution with Br^-.

_____

Place the ClO^- (base), ClO^- (acid), I_2, and Br_2 in order of increasing oxidizing strength.

☞ **Consider This**

From thermodynamic tables find the enthalpies of formation of the nitrogen oxides listed in Figure 41-1. Using Le Châtelier's principle, determine whether these oxides are likely to form at high or low temperatures and at high or low pressures. Use this knowledge to discuss the formation of nitrogen oxides in an internal combustion engine and how you would design a catalytic converter to eliminate them from automobile exhaust.

Are the same issues involved in the formation of sulfur oxides if the gasoline (or coal) contains sulfur as it is burned?

Equilibria of Coordination Compounds

PURPOSE

To illustrate the tendency of metal ions to form coordination compounds (metal complexes) with ions and neutral polar molecules that act as electron-pair donors toward metal ions. To study metal ion complexes with ammonia, chloride ion, and hydroxide ion. To determine the equilibrium constant for the dissociation of the diaminesilver(I) complex ion, $Ag(NH_3)_2^+$.

PRE-LAB PREPARATION

THE HYDRATION OF IONS

We have already studied the most common class of coordination compounds. In Experiment 19 we noted that all ions in solution attract polar water molecules to them to form ion-dipole bonds. Thus, the cupric ion, Cu^{2+}, does not exist as a bare ion in water but rather as the hexaaquo ion, $Cu(H_2O)_6^{2+}$. The origin of this attraction lies in the polar nature of the water molecule. Figure 42-1 shows the bent shape of the water molecule, which gives rise to the high dipole moment (a measure of the separation of charge in a molecule) of water.

The bond angle formed by the oxygen atom and the two hydrogen atoms in a water molecule is known to be 105°. This is not far from the 109.5° for ideal tetrahedral coordination, so the oxygen atom in a water molecule can be viewed as having four orbitals arranged in an approximately tetrahedral fashion, two of them forming the two bonds to the hydrogen atoms and the remaining two each containing a lone pair of electrons. It is these lone pairs that are centers of excess electron density. In Figure 42-1, the symbol ⁝ indicates the negative direction of the dipole moment, and the small delta symbols (δ) indicate where the partial charges exist in the water molecule.

The negative side of the water molecule is strongly attracted to cations. In general, the greater the charge on the cation and the larger the cation, the greater the number of coordinated water molecules. Examples are $H(H_2O)^+$, $Be(H_2O)_4^{2+}$, $Cu(H_2O)_6^{2+}$, $Al(H_2O)_6^{2+}$, $Fe(H_2O)_6^{2+}$,

Anions attract the positive side of the water molecule and are also hydrated. However, the attraction is not as large, the hydrates are less stable, and fewer water molecules are coordinated.

COORDINATION COMPOUNDS

These hydrates are one example of the type of substances called coordination compounds. Other neutral, but polar, molecules such as ammonia, NH_3, and also a number of anions, such as OH^-, Cl^-, CN^-, S^{2-}, $S_2O_3^{2-}$, and $C_2O_4^{2-}$, likewise can form similar very stable coordination groupings about a central ion. Such coordination compounds result from the replacement of the water molecule from the hydrated ion by other molecules or ions when they are present in a solution at high concentration, forming a still more stable bond. The resulting coordination compound may be a positively or negatively charged ion (a complex ion), or it may be a neutral molecule, depending on the number and kind of coordinating groups attached to the central ion.

What bonding forces hold the atoms of such a complex coordination compound together? As always, it is the

FIGURE 42-1
Three representations of the water molecule.

mutual attraction of positively charged atom nuclei and negatively charged electrons. Where large differences of electronegativity are present, as in AlF_6^{3-}, the bonding has a large amount of ionic character. In the majority of cases the bonding is mainly covalent, often with partial ionic character. The type of bond orbitals determines the spatial geometry and affects the stability of a given complex. Study the examples in Figure 42-2.

AMMONIA COMPLEX IONS. Some of the important ammonia complexes are the following:

$Ag(NH_3)_2^+$ $Cu(NH_3)_4^{2+}$ $Ni(NH_3)_4^{2+}$

$Au(NH_3)_2^+$ $Cd(NH_3)_4^{2+}$ $Ni(NH_3)_6^{2+}$

$Cu(NH_3)_2^+$ $Zn(NH_3)_4^{2+}$ $Co(NH_3)_6^{2+}$

These complexes are formed by adding ammonia to a solution containing the hydrated cation. Ammonia molecules are bound by the cation one at a time as the concentration of ammonia increases. At low concentrations of ligand, smaller numbers of ammonia molecules may be bound. For instance, the two NH_3 molecules bound to the Ag^+ bind in successive steps. The equilibrium constant is known for each step of the following reaction sequence:

$$Ag^+ + NH_3 \rightleftharpoons Ag(NH_3)^+$$

$$K_1 = \frac{[Ag(NH_3)^+]}{[Ag^+][NH_3]} = 1.6 \times 10^3$$

$$Ag(NH_3)^+ + NH_3 \rightleftharpoons Ag(NH_3)_2^+$$

$$K_2 = \frac{[Ag(NH_3)_2^+]}{[Ag(NH_3)^+][NH_3]} = 6.8 \times 10^3$$

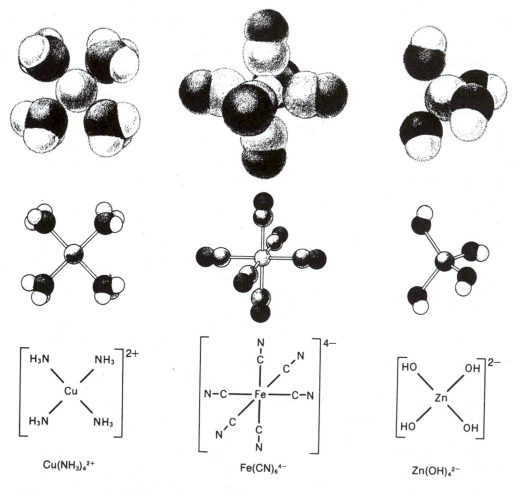

FIGURE 42-2

Diagrams showing the spatial arrangement of the coordinating groups about a central ion in the formation of a complex ion. Above each conventional formula, the "structural formula" represents each covalent electron-pair bond by a single line. The spatial geometry is determined by the type of orbitals represented: dsp^2 (square planar) by $Cu(NH_3)_4^{2+}$, d^2sp^3 (octahedral) by $Fe(CN)_6^{4-}$, and sp^3 (tetrahedral) by $Zn(OH)_4^{2-}$. The ball-and-stick models indicate these geometric patterns more clearly, and the space-filling models in the top row portray the atoms according to their accepted ionic diameters and bond lengths.

The equilibrium constant for the overall reaction can be obtained by the addition of the two reactions and the corresponding multiplication of their equilibrium constants.

$$Ag^+ + 2NH_3 \rightleftharpoons Ag(NH_3)_2^+$$

$$K_{formation} = K_1K_2 = \frac{[Ag(NH_3)^+]}{[Ag^+][NH_3]}$$

$$\times \frac{[Ag(NH_3)_2^+]}{[Ag(NH_3)^+][NH_3]}$$

$$= \frac{[Ag(NH_3)_2^+]}{[Ag^+][NH_3]^2}$$

$$= 1.1 \times 10^7$$

In Part 5 of the Experimental Procedure, you will determine the dissociation constant, K_{diss}, for the $Ag(NH_3)_2^+$ complex ion.[1] Because the $[NH_3]$ is so high under the experimental conditions that are employed, the $Ag(NH_3)_2^+$ species may be neglected.

HYDROXIDE COMPLEX IONS, OR AMPHOTERIC HYDROXIDES.

The hydroxides of most metals are relatively insoluble in water. Thus when sodium hydroxide is added to a metal ion in solution, such as lead ion, a precipitate is formed:

$$Pb(H_2O)_4^{2+} + 2 OH^- \rightleftharpoons Pb(H_2O)_2(OH)_2(s) + 2 H_2O$$

or, using the simple unhydrated metal ion formula,

$$Pb^{2+} + 2 OH^- \rightleftharpoons Pb(OH)_2$$

By Le Châtelier's principle, excess hydroxide ion would give more complete precipitation. Instead, the precipitate dissolves. This is explained by the tendency of lead ion to form a more stable coordination compound[2] with excess hydroxide ion:

$$Pb(H_2O)_2(OH)_2(s) + OH^- \rightleftharpoons Pb(H_2O)(OH)_3^- + 2 H_2O$$

or

$$Pb(OH)_2 + OH^- \rightleftharpoons Pb(OH)_3^-$$

Other ions react similarly. For example, hydrated aluminum ion, $Al(H_2O)_6^{3+}$, reacts to form the hydroxide

---

[1] The overall reaction may be written as a formation or as a dissociation, with $K_{formation} = 1/K_{diss}$.

[2] There is some uncertainty as to whether the lead hydroxide complex (and also the stannous hydroxide complex) will coordinate further to form $Pb(OH)_4^{2-}$ (and $Sn[OH]_4^{2-}$).

precipitate, $Al(H_2O)_3(OH)_3$, or, with excess OH^-, the hydroxide complex ion $Al(H_2O)_2(OH)_4^-$. Traditionally, chemists use the unhydrated formulas in ordinary chemical equations just because they are simpler, except where it is important to emphasize the hydrated structure.

The reactions that form these hydroxide complex ions are entirely reversible. The addition of acid to the above strongly basic $Pb(OH)_3^-$ solution reacts first to re-precipitate the hydroxide,

$$Pb(OH)_3^- + H^+ \rightleftharpoons Pb(OH)_2(s) + H_2O$$

and then, with excess acid,

$$Pb(OH)_2(s) + 2 H^+ \rightleftharpoons Pb^{2+} + 2 H_2O$$

Such metal hydroxides, which may be dissolved by an excess of either a strong acid or a strong base, are called *amphoteric hydroxides*.

The more important metal ions whose hydroxides are amphoteric are given in Table 42-1 (see p. 508).

EXPERIMENTAL PROCEDURE

Chemicals: 1 M ammonium chloride, NH_4Cl; 1 M ammonia, NH_3; 15 M (concentrated) ammonia, NH_3; 6 M ammonia, NH_3; 6 M sodium hydroxide, NaOH; 6 M nitric acid, HNO_3; 12 M hydrochloric acid, HCl; 6 M hydrochloric acid, HCl; copper(II) sulfate pentahydrate, $CuSO_4 \cdot 5H_2O(s)$; 0.1 M copper(II) sulfate, $CuSO_4$; 0.1 M sodium chloride, NaCl; 0.1 M silver nitrate, $AgNO_3$; 0.1 M zinc nitrate, $Zn(NO_3)_2$; 0.1% phenolphthalein, alizarin yellow R, and indigo carmine indicators.

! SAFETY PRECAUTIONS: Concentrated (12 M) HCl and 6 M HCl, 6 M HNO_3, and 6 M NH_3 are lung irritants. These solutions and 6 M NaOH are hazardous to the skin and eyes. If they contact your skin, wash them off immediately. Dispense the solutions in a well-ventilated fume hood. Clean up any spills immediately.

WASTE COLLECTION: Waste containers should be provided for the copper, zinc, and silver compounds formed in this experiment.

1. THE FORMATION OF COMPLEX IONS WITH AMMONIA.

To 3 mL of 0.1 M $CuSO_4$ add a drop of 6 M NH_3. Mix this. (Record your observations and write the equation for the reaction.) Continue to add NH_3 a little at a time, with mixing, until a distinct change occurs. Save this solution. Is this result contrary to the law of Le Châtelier? Obviously the OH^- concentration was increasing while the $Cu(OH)_2$ dissolved. How must the Cu^{2+} concentration have changed? Did it increase or decrease?

To learn which of the substances present in an ammonia solution (NH_4^+, OH^-, NH_3, H_2O) is responsible for

TABLE 42-1

Some Important Amphoteric Hydroxides

Simple ion[a] (acid solution)	Precipitate	Hydroxide complex ion[b] (strongly basic solution)
Pb^{2+}	$Pb(OH)_2$	$Pb(OH)_3^-$, trihydroxyplumbate(II) ion
Zn^{2+}	$Zn(OH)_2$	$Zn(OH)_4^{2-}$, tetrahydroxyzincate(II) ion
Al^{3+}	$Al(OH)_3$	$Al(OH)_4^-$, tetrahydroxyaluminate ion
Cr^{3+}	$Cr(OH)_3$	$Cr(OH)_4^-$, tetrahydroxychromate(III) ion
Sn^{2+}	$Sn(OH)_2$	$Sn(OH)_3^-$, trihydroxystannate(II) ion
Sn^{4+}	$Sn(OH)_4$	$Sn(OH)_6^{2-}$, tetrahydroxystannate(IV) ion

[a]So highly charged an ion as Sn^{4+} probably does not exist as such. In strong HCl solution, tin (IV) salts dissolve as the chloride complex, $SnCl_6^{2-}$.

[b]Formerly these ions were written in the anhydrous form: PbO_2^{2-}, ZnO_2^{2-}, AlO_2^-, CrO_2^-, SnO_2^{2-}, and SnO_3^{2-}. These formulas may be derived from the hydroxide complex ion formulas simply by subtracting the appropriate number of H_2O molecules or H_3O^+ ions.

the change you noted, try the following tests: (a) Add 1 mL of 1 M NH_4Cl to 1 mL of 0.1 M $CuSO_4$. (b) Add 2 drops (an excess) of 6 M $NaOH^3$ to 2 mL of 0.1 M $CuSO_4$. (c) Add ammonia gas by placing several crystals of $CuSO_4 \cdot 5H_2O(s)$ in a small dry beaker. At one side in the beaker place a piece of filter paper moistened with concentrated (15 M) NH_3. Cover with a watch glass, and observe any changes. From this evidence, write an equation to show the formation of this new substance when excess NH_3 is added to Cu^{2+}.

To 1 mL of this cupric ammonia complex ion solution, add 6 M HNO_3 in excess. Explain the result and write the equation for the reaction. Dispose of the solution in the copper waste container.

2. THE FORMATION OF AMPHOTERIC HYDROXIDES.
To 5 mL of 0.1 M $Zn(NO_3)_2$ add 6 M NaOH drop by drop, with mixing, until the precipitate that first forms just redissolves. Avoid undue excess of NaOH. Divide the substance into two portions; test one portion with alizarin yellow R and the other with indigo carmine indicator. Estimate the OH^- concentration (for later comparison in Part 3), using the information on the color changes and pH intervals of the indicators given in Table 5 of the Appendix. Now to one portion add 6 M HCl, drop by drop, until a precipitate forms (what is it?), then redissolves as more HCl is added. Interpret all these changes as related to Le Châtelier's law, and to the relative concentration of the various constituents (the zinc in its various forms, H^+, and OH^-), both in words and in net ionic equations.

3. THE REACTION OF ZINC ION WITH AMMONIA.
When ammonia is added gradually to Zn^{2+}, does the precipitate of zinc hydroxide that first forms redissolve as zincate ion, $Zn(OH)_4^{2-}$, owing to the excess base added, or does it redissolve as $Zn(NH_3)_4^{2+}$, owing to the NH_3 molecules added? To test this point, to 3 mL of 0.1 M $Zn(NO_3)_2$ add 6 M NH_3 drop by drop, with mixing, until the precipitate that first forms just redissolves. Divide this mixture, test one portion with phenolphthalein and the other portion with alizarin yellow R. Estimate the approximate OH^- concentration, and compare this with the corresponding situation in Part 2, where NaOH was used. (See the Appendix, Table 5.) What can you conclude as to the possibility of forming $Zn(OH)_4^{2-}$ by adding NH_3 to a zinc salt solution? Explain. Write the equation for the equilibrium that you have verified.

4. CHLORIDE COMPLEX IONS.
(a) To 2 mL of 0.1 M $CuSO_4$ add 2 mL of 12 M (concentrated) HCl; then dilute the solution with about 5 mL of water. Write equations, and interpret the color changes you observed, assuming that the complex formed is $CuCl_4^{2-}$.

(b) To 1 mL of 0.1 M $AgNO_3$ add 3 mL of 12 M (concentrated) HCl; then agitate the solution well for several minutes to redissolve the precipitated AgCl. Now dilute this mixture with about 5 mL of distilled water. Write equations for the reactions, and interpret the changes you observed, assuming that the complex formed is $AgCl_2^-$. Dispose of the tube contents in the silver waste container.

5. THE EQUILIBRIUM CONSTANT OF AN AMMONIA COMPLEX ION.
The dissociation of silver diamine complex ion is represented by the equilibrium

$$Ag(NH_3)_2^+ \rightleftharpoons Ag^+ + 2\,NH_3 \qquad (1)$$

---

[3]This provides an excess of OH^-, a much stronger base than NH_3. The strong base OH^- shows some amphoteric effect (see Part 2) with copper(II) salts but is far from complete.

and the corresponding equilibrium constant expression

$$\frac{[Ag^+][NH_3]^2}{[Ag(NH_3)_2^+]} = K_{diss} \qquad (2)$$

If you add sufficient Cl^- gradually to an equilibrium mixture of Ag^+ and NH_3, (represented by Equation [1]) so that you can just barely begin precipitation of $AgCl(s)$, a second equilibrium is established simultaneously without appreciably disturbing the first equilibrium. This may be represented by the combined equations

$$Ag(NH_3)_2^+ \rightleftharpoons Ag^+ + 2NH_3$$
$$+$$
$$Cl^-$$
$$\updownarrow$$
$$AgCl(s) \qquad (3)$$

By using a large excess of NH_3, you can shift Equation (1) far to the left, with reasonable assurance that the Ag^+ is converted almost completely to $Ag(NH_3)^{2+}$ rather than to the first step only, $Ag(NH_3)^+$. From the measured volumes of NH_3, Ag^+, and Cl^- solutions used, you can determine the concentrations of the species in Equation (1) and calculate the value of K_{diss}.

To prepare the solution,[4] place 3.0 mL of 0.1 M $AgNO_3$ (measure it accurately in a 10-mL graduate) in a 15 × 150 mm test tube. Add 3.0 mL (also carefully measured) of 1 M NH_3. Now prepare some 0.02 M NaCl by diluting

2.0 mL of 0.1 M NaCl to 10.0 mL in your 10-mL graduate. Mix this thoroughly, and note the exact volume. Then, from a medicine dropper, add it to the mixture of $AgNO_3$ and NH_3, about 1 to 1.5 mL at first, and then drop by drop until a very faint, permanent milky precipitate of AgCl remains. Return any excess NaCl from the medicine dropper to the graduate, and note the exact volume used. From these data, K_{diss} can be calculated. Dispose of the solution in the silver waste container.

👉 Consider This

If a visible spectrophotometer is available (such as a Spectronic 20®), develop a procedure for determining the equilibrium constant of the colored complexes such as $Cu(NH_3)_4^{2+}$.

Cobalt(II) chloride is often used as a humidity meter since its color changes from purple to pink as it forms the hexahydrate form. Estimate the equilibrium constant of this reaction by observing the color of the compound in air at various relative humidities.

---

[4]If desired, some improvement in precision can be obtained by using larger volumes—20.0 mL each of 0.1 M $AgNO_3$ and 1 M NH_3. Then dilute 10.0 mL of 0.1 M NaCl to about 50.0 mL in your 50-mL graduate, mix this well, and note the exact volume. Add first about 15 mL, then very small portions, to the Ag^+—NH_3 mixture, stirring as you do so, until a very faint permanent milky precipitate of AgCl remains. Note the total volume of 0.02 M NaCl used.

REPORT	Equilibria of Coordination Compounds	NAME	
42		SECTION	LOCKER
		INSTRUCTOR	DATE

OBSERVATIONS AND DATA

1. **The Formation of Complex Ions with Ammonia**

    The net ionic equation for the reaction of excess $CuSO_4$ with NH_3, is

    _____

    The predicted effect on the reaction above of adding excess NH_3 (based on Le Châtelier's principle) is

    _____

    Describe the observed results when

    excess NH_3 is added to $CuSO_4$ solution. _____

    NH_4Cl and $CuSO_4$ solutions are mixed. _____

    excess $NaOH$ and $CuSO_4$ solutions are mixed. _____

    $CuSO_4 \cdot 5H_2O(s)$ is exposed to NH_3 gas. _____

    Considering the equilibrium equation $NH_3 + H_2O \rightleftharpoons NH_4^+ + OH^-$, explain which substance (NH_3, NH_4^+, or OH^- causes the deep blue color when ammonia is added to a copper(II) solution, and give the equation for the reaction.

    The observed effect and the net ionic equation for the reaction of HNO_3 on this deep blue solution is

2. **The Formation of Amphoteric Hydroxides**

    The net ionic equation for the reaction of excess $Zn(NO_3)_2$ with $NaOH$ is

    _____

    The predicted effect on the reaction above of adding excess $NaOH$ (based on Le Châtelier's principle) is

    _____

    Explain in your own words why $Zn(OH)_2(s)$ dissolves with excess OH^-, and write the net ionic equation for the reaction.

Color with alizarin	indigo	OH^-
yellow R _____	carmine _____	concentration _____

    Explain the effect of adding a moderate amount of HCl to this strongly basic solution, and give the net ionic equation for the reaction.

What further change occurs when excess HCl is added? (Give the equation and explain.)

3. **The Reaction of Zinc Ion with Ammonia**

Note your observations below on the addition of indicators to the solution formed by adding, one drop at a time, 6 M NH_3 to $Zn(NO_3)_2(aq)$ to first form, then just redissolve the precipitate.

Color with   Alizarin   Estimated OH^-
phenolphthalein _____   yellow R _____   concentration _____

(To estimate the OH^- concentration, use the information on the color changes and pH intervals of the indicators given in Table 5 of the Appendix.)

Which coordination compound, $Zn(OH)_4^{2-}$ or $Zn(NH_3)_4^{2+}$ forms when Zn^{2+} reacts with excess NH_3 solution? Compare with Part 2; explain fully.

Therefore, the equation for the formation of this equilibrium complex ion is as follows:

_____

4. **Chloride Complex Ions**

(a) Explain the successive changes observed when concentrated HCl, then H_2O, is added to a $CuSO_4$ solution; give the equations for the reactions.

(b) Explain the changes observed when concentrated HCl, then H_2O, is added to a $AgNO_3$ solution, and give the equations.

5. **The Equilibrium Constant of an Ammonia Complex Ion**

(Indicate your calculations for each step in the spaces provided.)

Volumes of solutions:

0.1 M	1 M	0.02 M	Total
$AgNO_3$ _____	NH_3 _____	NaCl _____	volume _____

(a) Concentration of $Ag(NH_3)_2^{2+}$

(Assume all the silver to be present as the complex ion, ignoring the trace of free Ag^+ remaining.)

_____ M

(b) Concentration of Cl^-

  (Ignore any trace of Cl^- removed as $AgCl[s]$.)

_____ M

(c) Concentration of Ag^+ (Use the Cl^- concentration above, and the solubility product relationship, $[Ag^+][Cl^-] = 2.8 \times 10^{-10}$.)

_____ M

(d) Concentration of free NH_3

  (First calculate the NH_3 concentration as if none combined with Ag^+, then subtract twice the concentration of $Ag[NH_3]_2^{2+}$ found above.)

_____ M

(e) Use the values found in 5a, 5c, and 5d to calculate the value of the equilibrium constant from Equation (2):

_____

(f) Compare your value for the equilibrium constant with the value given in the Appendix, Table 7.

APPLICATION OF PRINCIPLES

1. Which reagent, $NaOH$ or NH_3, will enable you to precipitate the first-named ion from a solution containing each of the following pairs of ions, leaving the second ion in solution? Give also the formula of the precipitate and the exact formula of the other ion in solution. You will need to refer to Tables 7 and 9 in the Appendix to determine which ions form hydroxide precipitates or hydroxide complexes, and ammonia complexes.

Pair	Reagent	Precipitate	Ion in solution
(a) Al^{3+}, Zn^{2+}			
(b) Cu^{2+}, Pb^{2+}			
(c) Pb^{2+}, Cu^{2+}			
(d) Fe^{3+}, Al^{3+}			
(e) Ni^{2+}, Sn^{2+}			
(f) Sn^{2+}, Ni^{2+}			
(g) Mg^{2+}, Ag^+			

2. When ammonia is added to $Zn(NO_3)_2$ solution, a white precipitate forms, which dissolves on the addition of excess ammonia. But when ammonia is added to a mixture of $Zn(NO_3)_2$ and NH_4NO_3, no precipitate forms at any time. Suggest an explanation for this difference in behavior.

3. Calculate the OH^- concentration in (a) 1 M NH_3, and (b) a solution that is 1 M in NH_3 and also 1 M in NH_4Cl. ($K_b = 1.8 \times 10^{-5}$ for NH_3; see the Appendix, Table 6.)

4. Suppose you are given the following experimentally observed facts regarding the reactions of silver ion.

   (a) Ag^+ reacts with Cl^- to give white $AgCl(s)$.

   (b) Ag^+ reacts with ammonia to give a quite stable complex ion, $Ag(NH_3)_2{}^+$.

   (c) A black suspension of solid silver oxide, $Ag_2O(s)$, shaken with NaCl solution, changes to white $AgCl(s)$.

   (d) $AgCl(s)$ will dissolve when ammonia solution is added, but $AgI(s)$ does not dissolve under these conditions.

   Write equations for any net reactions in the cases above, and then, based on these observations, arrange each of the substances $AgCl$, AgI, Ag_2O, and $Ag(NH_3)_2{}^+$ in such an order that their solutions with water would give a successively decreasing concentration of Ag^+.

   (1) ———————————————————————————————————

   (2) ———————————————————————————————————

   (3) ———————————————————————————————————

   (4) ———————————————————————————————————

Consider This

If a visible spectrophotometer is available (such as a Spectronic 20®), develop a procedure for determining the equilibrium constant of the colored complexes such as $Cu(NH_3)_4{}^{2+}$.

Cobalt(II) chloride is often used as a humidity meter since its color changes from purple to pink as it forms the hexahydrate form. Estimate the equilibrium constant of this reaction by observing the color of the compound in air at various relative humidities.

The Chemistry of Vitamin C

PURPOSE

To show that vitamin C (ascorbic acid) acts as a reducing agent toward iodine and other oxidizing agents. To compare the content of vitamin C in commercial tablets with the content stated on the label by the manufacturer. To determine the vitamin C content of citrus juices or other foods. To illustrate the redox chemistry of the iodate–iodide ion reaction.

PRE-LAB PREPARATION

Until late in the nineteenth century it was common for ships on transoceanic voyages to lose more than half their crews. The deaths were the result of the dread disease scurvy. Of those who survived, many were afflicted with massive hemorrhages, ulcerated gums, diarrhea, and exhaustion. In 1795, the British Navy ordered a daily ration of lime juice for every sailor. (This measure gave rise to the nickname "Limey.") There were no more deaths from scurvy in the British Navy. (The British merchant marine, however, considered the practice quackery, and over the next 70 years hundreds of its seamen died. Finally, the Board of Trade passed a similar regulation, making lime rations mandatory on all vessels.)

The vital antiscurvy factor in lime juice—in fact, in all other citrus fruits, and nearly all fresh vegetables—was not isolated until 1928. The brilliant Hungarian-born scientist Albert Szent-Györgyi made the discovery, and the substance was called *hexuronic acid*. Working about the same time, American biochemist C. Glen King reported the isolation of the antiscurvy factor in crystalline form from lemon juice. In 1932 and 1933 hexuronic acid and the substance isolated by King were found to be identical; it was renamed ascorbic acid and was determined to have the chemical formula $C_6H_8O_6$. Szent-Györgyi received a

Nobel prize in 1937 for his studies of biological oxidations and the discovery of ascorbic acid.

In 1933, the English sugar chemist W. M. Haworth reported the structure of ascorbic acid to be compound I shown in Figure 43-1. (Haworth employed the standard convention for writing the structural formula of sugars with one ring.) A somewhat more realistic representation of the same compound is shown as structure II in Figure 43-1.

For many years public health organizations have recommended a minimum daily intake of ascorbic acid, now also known as *vitamin C*. The present daily dietary allowance recommended by the Food and Nutrition Board of the National Research Council is between 35 and 60 mg/day.

Linus Pauling (1901–1994), winner of the Nobel Prize in Chemistry and the Nobel Peace Prize, compiled much information about vitamin C from published research papers. His small book, *Vitamin C and the Common*

FIGURE 43-1

The structure of ascorbic acid.

Cold,[1] stimulated further work on the subject (and the idea for this experiment). Vitamin C is an essential nutrient because, like certain amino acids, it cannot be synthesized in the body and must be supplied from the food we eat.

It was Pauling's view that an increased intake of vitamin C would not only offer protection against the common cold but would also have a general beneficial effect on the body's natural defense mechanisms against disease. Pauling believed that the ordinary human diet contains a suboptimal amount of vitamin C, which he describes as "a substance that participates in almost all of the chemical reactions that take place in our bodies, and is required for many of them." This view has been disputed by other members of the scientific and medical communities.[2]

It is not our intent to take sides in this dispute; rather, we intend to look at the unique chemical properties of this now famous molecule and determine the amounts of it present in some foodstuffs. Table 43-1 gives you some average values for the vitamin C contents of a variety of foods. This will help you in your selection of foods to study in Part 2 of the Experimental Procedure, and it will also tell you how much you would have to eat of a particular food in order to ingest the amounts recommended by Pauling (at least 2 g/day). He recommends doses of ascorbic acid powder or tablets. It would be difficult to obtain such large amounts of the substance from available foodstuffs—unless you want to consume large quantities of green peppers daily!

The experimental method we will use to determine the amount of vitamin C in a sample exploits one of this vitamin's most distinctive features, its powerful reducing properties. For example, it will reduce Fe(III) to Fe(II), iodine to iodide ion, and the red redox indicator dye 2,6-dichloroindophenol to its colorless form. A method employing the latter dye was developed by C. G. King some years ago but the dye is expensive and somewhat unstable. In this experiment we will employ the reaction with iodine, generated by reaction of iodate ion with iodide ion in acid solution, following a procedure described by R. W. Ramette (see the Bibliography). The procedure works like this: KI, HCl (to supply H^+), and starch solution are added to a sample containing some vitamin C. Potassium iodate, (KIO_3), titrant is added from a buret. As the iodate ion is added, it reacts with I^- and H^+ to form I_2.

$$IO_3^- + 5\,I^- + 6\,H^+ \rightarrow 3\,I_2 + 3\,H_2O \tag{1}$$

Almost as quickly as it is formed, the ascorbic acid reacts with the I_2.

When all the vitamin C has reacted, the concentration of I_2 begins to build up to a level where it will react with I^- to form the linear triiodide ion, I_3^-:

$$I_2 + I^- \rightleftarrows I_3^- \tag{3}$$

Triiodide ion combines with starch (as shown in Figure 38-1) to form the starch-triiodide ion complex, which has a deep blue (almost blue-black) color. So the appearance of the blue color signals the end point of the titration.

TABLE 43-1

Ascorbic Acid Content of Foodstuffs

Ascorbic acid (mg/100 g)	Foodstuffs
100–350	Chili peppers (green and red), sweet peppers (green and red), parsley, turnip greens
25–100	Citrus juices, tomato juice, mustard greens, spinach, brussels sprouts
10–25	Green beans and peas, sweet corn, asparagus, pineapple, cranberries, cucumbers, lettuce
Less than 10	Eggs, milk, carrots, beets, cooked meat

[1]Linus Pauling, *Vitamin C and the Common Cold,* W. H. Freeman and Company, San Francisco, 1970. A revised and somewhat expanded version of the book was published in 1976 under the title *Vitamin C, the Common Cold, and the Flu.*

[2]E. M. N. Hamilton and E. N. Whitney, *Nutrition: Concepts and Controversies,* West Publishing Co., St. Paul, MN, 1979, p. 256.

EXPERIMENTAL PROCEDURE

Special Supplies: 1-mL microburet; 10-mL pipet and pipet bulb, polyethylene transfer pipets; 100-mg vitamin C tablets; glazed weighing paper; mortar and pestle or food blender; cheesecloth; 150 mm–diameter funnel; citrus juices; dried breakfast drink, such as Tang; citrus fruit or other natural foodstuffs containing vitamin C. A pH meter will be required if you are going to measure the pK_a of vitamin C as an optional experiment in Part 3.

Chemicals: *Ascorbic acid tablets, 100 mg; 0.0250 M potassium iodate, KIO₃; potassium iodide, KI(s); 1 M hydrochloric acid, HCl; 0.5% starch indicator solution. If you are going to measure the pKₐ of vitamin C as an optional experiment in Part 3, 1.00 M sodium hydroxide, NaOH, and 1.00 M hydrochloric acid, HCl, will be required (see the article by Hurd in the Bibliography for the procedure).*

NOTE TO INSTRUCTOR: If you prefer to use a 25- or 50-mL buret, dissolve a weighed 100-mg tablet of vitamin C in 50 mL of deionized water, add 1 g of KI(*s*), swirl to dissolve, add 5 mL of 1 M HCl, add 2–3 mL of 0.5% starch solution, and titrate with 0.0100 M KIO₃ contained in a buret.

1. DETERMINATION OF THE VITAMIN C CONTENT OF COMMERCIAL TABLETS.

Rinse and fill a 1-mL buret with 0.0250 M KIO₃ solution, and record the initial buret reading. Place a sheet of glazed weighing paper (or a small beaker) on the pan of an analytical balance; determine and record its mass. Then put a 100-mg vitamin C tablet on the weighing paper and reweigh. Record the mass. Is the difference in masses exactly 100 mg? Think about the answer, and comment on it in your laboratory report. If you don't find anything unusual about your result, think about it again after the next step.

Put 100±1 mL of deionized water in a 250-mL Erlenmeyer flask. Add the weighed tablet to the water, and use a glass stirring rod to crush the tablet and dissolve the vitamin C. Vitamin C is very soluble in water, so if something remains undissolved after 2–3 minutes of stirring, it is probably a binder present in the tablet to make it more coherent. (Does this explain why a 100-mg vitamin C tablet might have a mass greater than 100 mg?) Mix the solution thoroughly before withdrawing a sample in the next step.

Using a 10-mL pipet (fill by using a pipet bulb), transfer 10.0±0.1 mL of the vitamin C solution to a 25-mL Erlenmeyer flask. Add 0.1 g of KI, and swirl to dissolve; then add 1 mL of 1 M HCl, measured with a polyethylene transfer pipet. Add 0.5 ml of 0.5% starch indicator, and titrate with 0.0250 M KIO₃ until the mixture acquires a deep blue color. (As you near the end point, you will get a transient blue color that quickly fades. Keep adding titrant until you get a permanent blue color.) Record the final buret reading.

Repeat the procedure with two more 10-mL samples (or repeat the whole procedure with another vitamin C tablet, as your instructor directs.) For each sample, calculate the mass of ascorbic acid (its molar mass is 176.12 g/mol). (Your instructor may choose to collect a larger set of results from the class in order to carry out a statistical analysis of the data.) Compare the average value for the tablets you analyzed with the content stated on the label by the manufacturer. Keep in mind that you titrated ¹⁄₁₀ of the vitamin C tablet, so you must multiply the calculated vitamin C content of the sample by 10 to obtain the total content of the original tablet.

2. DETERMINATION OF THE VITAMIN C CONTENT OF A FOODSTUFF.

You are to determine the vitamin C content of a foodstuff you have some interest in. (In fact, your instructor may ask you to bring this food to the laboratory for analysis.) Try to obtain a sample of fresh citrus fruit, canned or frozen juices, or a breakfast drink in dry powder form (such as Tang). Keep in mind that if you use a highly colored food such as beets or cranberry juice, the color change at the end point will be modified and may be more difficult to see.

(a) **Juices.** Enough vitamin C for an accurate titration is usually present in 25 mL of a citrus or vegetable juice. Freshly squeezed juices may be strained through cheesecloth to remove the pulp. Obtain a 150 mm–diameter funnel, and support it in an iron ring with the stem of the funnel in the mouth of a 100-mL graduated cylinder. Line the funnel with 2 to 3 layers of cheesecloth, generously cut so that it drapes over the edges of the funnel. Pour the juice and pulp into the funnel. After most of the juice has run through, gather up the edges of the cheesecloth to form a sack, and twist it to squeeze out most of the juice in the pulp.

To determine the vitamin C content of citrus juices, start with a 25-mL sample, adding 0.2 g KI(*s*), 1 mL of 1 M HCl, and 1 mL of 0.5% starch solution. Titrate the sample with 0.025 M KIO₃ to the blue end point. (The sample size should be adjusted to contain 5–10 mg of vitamin C. You may have to do a practice sample if the juice is highly colored.) Record the source and the volume of your sample of juice. Calculate the milligrams of ascorbic acid in the sample, and divide by the sample volume to obtain the ascorbic acid content in milligrams per milliliter of juice.

(b) **Extraction from Solids.** Weigh out a sample of the solid foodstuff sufficient to contain about 10 mg of ascorbic acid. (This would require about 0.8 mL of 0.0250 M KIO₃ titrant.) Record the mass of the sample. Grind the sample in a mortar and pestle or blender, add 25 mL of water. Support a 150 mm–diameter funnel in an iron ring with the stem of the funnel in the mouth of a 100-mL graduated cylinder. Line the funnel with 2 to 3 layers of cheesecloth, generously cut so that it drapes over the edge of the funnel. Add the pulp and juice. Gather up the edges of the cheesecloth to form a sack, and twist it to express the juice from the pulp. If the volume is less than 25 mL, disperse the pulp in a little more water and repeat, to extract as much vitamin C as possible. The final sample should have a volume of about 25 mL.

Record the source and mass of the sample. Titrate the extract of the sample with 0.0250 M KIO₃, using the procedure described in Part 2a for juices. Calculate the milligrams of ascorbic acid in the sample, and divide by the sample mass to obtain the ascorbic acid content in milligrams per gram of sample.

👉 **Consider This**

Compare fresh and reconstituted frozen orange juice for their vitamin C contents.

Compare the vitamin C content of one sample of juice with a second sample of the same juice that has been heated to near boiling for 10 min, then cooled to room temperature.

Compare the vitamin C content of one sample of juice with a second sample of the same juice that has had air (oxygen) bubbled through it for 30 min.

Compare the vitamin C content of tablets that have been "improperly" stored, such as in heated or sunlit areas, to those in a fresh container.

Determine the pK$_a$ of vitamin C. In one interesting article (see the Bibliography), C. D. Hurd discusses possible reasons why ascorbic acid, although lacking a carboxylic acid group (—COOH), is a stronger acid than acetic acid.

BIBLIOGRAPHY

Hurd, C. D. "The Acidities of Ascorbic and Sialic Acids," *J. Chem. Educ.,* **1970,** *47:* 481.

King, C. G. "Chemical Methods for Determination of Vitamin C," *Ind. and Eng. Chem.,* **1941,** *13:* 225.

Ramette, R. W. *Chemical Equilibrium and Analysis,* Addison-Wesley, Reading, MA, 1981, p. 628.

The Chemistry of Vitamin C

NAME	
SECTION	LOCKER
INSTRUCTOR	DATE

DATA AND CALCULATIONS

1. Determination of the Vitamin C Content of Commercial Tablets

Molarity of the potassium iodate (KIO_3) solution _____ M

Data	Trial 1	Trial 2	Trial 3
Mass of the tablet + weighing paper	g	g	g
Mass of the weighing paper	g	g	g
Mass of the tablet	g	g	g
Volume of KIO_3 titrant			
Final buret reading	mL	mL	mL
Initial buret reading	mL	mL	mL
Net volume of KIO_3 solution	mL	mL	mL

Calculate the mass of ascorbic acid (176.12 g/mol) in each tablet, taking account of the reaction stoichiometry shown in Equations (1) and (2). If you titrated a 10-mL aliquot of the tablet dissolved in 100 mL water, be sure to multiply your titration result by 10 to obtain the ascorbic acid content of the original sample. Show a sample calculation below.

Data	Trial 1	Trial 2	Trial 3
Mass of ascorbic acid	mg	mg	mg
		Average _____ mg	

Compare the average content you determined with the content declared on the label by the manufacturer.

Are the tablets pure ascorbic acid, or is there evidence that they contain some inert material? Record your observations made when dissolving the tablets, and compare the mass of a tablet with the mass of ascorbic acid found in the tablet.

2. Determination of the Vitamin C Content of a Foodstuff

Give the name of the foodstuff you analyzed. _____

Data	Trial 1	Trial 2	Trial 3
Mass or volume of foodstuff	g	g	g
Volume of KIO_3 titrant			
Final buret reading	mL	mL	mL
Initial buret reading	mL	mL	mL
Net volume of KIO_3 solution	mL	mL	mL

Calculate the ascorbic acid content of the foodstuff (in milligrams of ascorbic acid per gram or per milliliter of sample). Show sample calculation below.

Milligrams of ascorbic acid per gram (or milliliter) of foodstuff _____

☞ Consider This

Compare fresh and reconstituted frozen orange juice for their vitamin C contents.

Compare the vitamin C content of one sample of juice with a second sample of the same juice that has been heated to near boiling for 10 min, then cooled to room temperature.

Compare the vitamin C content of one sample of juice with a second sample of the same juice that has had air (oxygen) bubbled through it for 30 min.

Compare the vitamin C content of tablets that have been "improperly" stored, such as in heated or sunlit areas, to those in a fresh container.

Determine the pK_a of vitamin C. In one interesting article (see the Bibliography), C. D. Hurd discusses possible reasons why ascorbic acid, although lacking a carboxylic acid group (—COOH), is a stronger acid than acetic acid.

Atmospheric Chemistry of Nitric Oxide Smog and Acid Rain

PURPOSE

To study the reaction between nitric oxide and oxygen in the presence of water. To see how this reaction can lead to the formation of acid rain.

PRE-LAB PREPARATION

The chemistry of nitric oxide is so rich that a study of it could occupy a chemist for a lifetime. It plays a significant role in (1) the formation of photochemical smog and ozone in the troposphere and the destruction of ozone in the stratosphere, (2) the formation of acids in the atmosphere, and (3) the biological chemistry of humans and other animals.

In the troposphere (the layer of the atmosphere closest to the earth's surface), ozone is undesirable because it is a lung irritant, but in the stratosphere ozone is desirable because it shields us from ultraviolet radiation from the sun. In its atmospheric reactions nitric oxide cuts two ways, playing a primary role in the formation of "bad" ozone in the part of the atmosphere where we live and breathe, but catalyzing the decomposition of "good" ozone in the stratosphere, where ozone has a beneficial protective effect.

More recently it has been discovered that nitric oxide plays an important role in biological systems, acting as a neurotransmitter (or messenger molecule) and having profound effects on the vascular system, where it enlarges blood vessels by relaxing the smooth muscles in their walls.

In this experiment we will focus on the role of nitric oxide in the formation of acids in the atmosphere. In rainy weather, these acids can return to earth as acid rain. In dry weather the acids can react with ammonia from dairies and cattle feed lots to form solid particulates such as ammonium nitrate. The tiny solid particles scatter light like the tiny droplets of water in fog, forming a haze that reduces visibility.

THE ROLE OF NITRIC OXIDE IN THE FORMATION OF PHOTOCHEMICAL SMOG AND ACIDS

One of the noxious pollutants that comes out of the tailpipes of autos and trucks is nitric oxide, NO. In fact, any high-temperature combustion that takes place in the presence of air produces some NO, for example, cooking over a gas burner or heating with a gas-fired furnace. The higher the temperature of the combustion, the more NO is produced. It results from the reaction at high temperature of nitrogen and oxygen:

$$N_2(g) + O_2(g) \rightarrow 2NO(g) \tag{1}$$

Although the amount of NO produced is small, in the concentration range of a few parts per million, this small amount of NO has significant environmental effects when it is multiplied in a large urban area by the millions of people driving autos or heating homes and cooking with natural gas. At high concentrations, NO rapidly reacts with oxygen to form nitrogen dioxide (NO_2), a brown noxious gas. The net reaction of NO with oxygen is reported to be

$$2NO(g) + O_2(g) \rightarrow 2\,NO_2(g) \tag{2}$$

Kinetically the reaction is second order in NO, so at the low concentrations found in polluted ambient air the reaction is much slower, taking hours to completely react (Shaaban, 1990).

NO_2 plays a key role in the formation of what is called photochemical (or Los Angeles–type) smog, which contains ozone, a very reactive and irritating gas that is responsible for the painful breathing that you feel if you exercise outdoors during a severe smog alert.

How is the ozone produced? NO_2 has an absorption band centered at 393 nm, near the borderline between the visible and the ultraviolet regions of the spectrum. As the sun shines on a mass of air containing NO_2, some NO_2 is dissociated by absorption of a photon (hv) of light:

$$NO_2(g) + hv \rightarrow NO(g) + O(g) \qquad (3)$$

The oxygen atom produced is very reactive and in the presence of a third molecule, M, which can be a nitrogen or oxygen molecule, it reacts with a dioxygen molecule to form ozone:

$$O(g) + O_2(g) + M(g) \rightarrow O_3(g) + M^*(g) \qquad (4)$$

The molecule M^* (the asterisk means it is in an excited state) carries away excess energy of reaction that would otherwise keep the O atom and the O_2 molecule from forming a stable chemical bond. The NO produced in the photochemical step, Equation (3), can recycle to again form NO_2, so ozone can be formed anytime that sunlight shines on air containing nitrogen oxides (Ainscough and Brodie, 1995).

The brown gas NO_2 is very soluble in water. If you attempt to collect it over water, it is completely absorbed, leaving no measurable volume of gas. Depending on reaction conditions, several different reactions have been reported for the reaction of NO_2 (or NO and O_2) with water. According to some sources, the NO_2 reacts with water to form nitrous and nitric acids by the following net reaction (Bonner and Steadman, 1996; Cotton et al., 1999):

$$2NO_2(g) + H_2O(l) \rightarrow HNO_2(aq) + HNO_3(aq) \qquad (5)$$

It is also reported that, when heated, nitrous acid decomposes to nitric acid and NO (Bonner & Steadman, 1996).

$$3HNO_2 \rightarrow HNO_3 + 2NO + H_2O \qquad (6)$$

The NO formed could then keep recycling and reacting with oxygen, so that when the last bit of nitrous acid, HNO_2, has disappeared, the net overall reaction in the presence of water and excess oxygen can be written as (Hornack, 1990):

$$4NO(g) + 3O_2(g) + 2H_2O \rightarrow 4HNO_3 \qquad (7)$$

Finally, it has been reported that when both NO and O_2 are dissolved in aqueous solution, they react to produce only nitrous acid (or nitrite ion in the presence of

base), corresponding to the following stoichiometry (Bonner and Steadman, 1996; Cotton et al., 1999):

$$4NO(aq) + O_2(aq) + 2H_2O \rightarrow 4HNO_2(aq) \qquad (8)$$

The reactions expressed in Equations (2) + (5), (7) (if HNO_2 readily decomposes), or (8) all ultimately form acids starting with nitric oxide, NO and O_2. If the reactions take place in an area of the atmosphere containing water droplets in a cloud, acid rain would be produced.

SOME POSSIBLE REACTIONS OF NO AND O_2 IN THE PRESENCE OF WATER

Looking at Equations (2) and (5)–(8) we see that there are three distinguishable net chemical reactions that might take place in the presence of water, depending on the reaction conditions:

A: Equation (2) + (5): $2NO(g) + O_2(g) + H_2O \rightarrow$
$$HNO_2(aq) + HNO_3(aq)$$

B: Equation (7): $4NO(g) + 3O_2(g) + 2H_2O \rightarrow 4HNO_3(aq)$

C: Equation (8): $4NO(aq) + O_2(aq) + 2H_2O \rightarrow$
$$4HNO_2(aq)$$

All three reactions predict one mole of total acid produced per mole of NO reacted, so no distinction between the three is possible based solely on the titration of total acid produced. But there are three clearly distinguishable outcomes based on the predicted NO/O_2 reaction stoichiometry and the stoichiometric amounts of HNO_2 and HNO_3 produced, as shown in Table 44-1. In this experiment we will prepare NO and O_2 and make a qualitative and semiquantitative study of their reaction products in the presence of water.

TABLE 44-1

Three Possible Net Reactions of NO and O_2 in the Presence of Water

A: $2NO(g) + O_2(g) + H_2O \rightarrow HNO_2(aq) + HNO_3(aq)$
B: $4NO(g) + 3O_2(g) + 2H_2O \rightarrow 4HNO_3(aq)$
C: $4NO(aq) + O_2(aq) + 2H_2O \rightarrow 4HNO_2(aq)$

Reaction	Mols of O_2 per mol NO	Total mols of acid per mol of NO	Mols of nitrite ion per mol of NO	Mols of nitrate ion per mol of NO
A	0.50	1.0	0.5	0.5
B	0.75	1.0	0.0	1.0
C	0.25	1.0	1.0	0.0

MEASURING THE pH AND TOTAL AMOUNT OF ACID PRODUCED

Nitric acid (HNO_3) and/or nitrous acid (HNO_2) are reported as reaction products when NO_2 reacts with water. Nitric acid is a strong acid; nitrous acid is a weak acid. By measuring both the pH of the reaction products and the total concentration of acid we can, in principle, tell if the solution contains only a strong acid, only a weak acid, or a mixture of weak and strong acids. A properly standardized pH meter is necessary for making the pH measurements.

We can determine the total amount of acid in an aqueous solution by titration with a base. This amounts to adding base to the acid solution until the solution is neutralized. We will add a color indicator that allows us to know when the end point of the titration is reached. The acid/base neutralization is:

$$HNO_2(aq) + OH^-(aq) \rightarrow NO_2^-(aq) + H_2O(l) \quad (9)$$

and/or

$$HNO_3(aq) + OH^-(aq) \rightarrow NO_3^-(aq) + H_2O(l) \quad (10)$$

SEMIQUANTITATIVE TESTING FOR THE PRESENCE OF NITRITE ION

We will use a very sensitive test for nitrous acid, described more fully in the Experimental Procedure. In acid solution, nitrous acid oxidizes iodide ion to iodine, which in the presence of starch indicator forms a deep blue starch-iodine complex. The test easily detects nitrous acid at a concentration of 1×10^{-4} M.

EXPERIMENTAL PROCEDURE

Special Supplies: pH meter; 10-inch length forceps; each student (or student group) will require the following: disposable Luer tip syringes (two 60-mL size and one 10-mL size); two syringe tip caps (available from Flinn Scientific Inc.); two Titeseal polyethylene shell vial caps (Kimble 1-dram shell vials, 15 mm OD × 45 mm length, with Titeseal caps); 3-cm length of 1/8 inch ID rubber tubing; four disposable weighing dishes (1 5/8 inch square); polyethylene transfer pipets with 1-mL calibration marks; 1-mL microburet (Air Tite Normject 1 × 0.01 mL all polypropylene tuberculin syringe).

Chemicals: Active dry baker's yeast (Fleischmann's or other brand, available at supermarkets); sodium nitrite, $NaNO_2(s)$; 1.2 M $FeSO_4$/1.8 M H_2SO_4 (freshly prepared); 0.008 M HNO_3; standardized 0.50 M NaOH; 0.01 M $NaNO_2$; 1 M $NaHCO_3$; 0.5% starch indicator (freshly prepared); 1.0 M KI; ~5 M tartaric acid; 3% hydrogen peroxide; 0.1% thymol blue.

NOTES TO THE INSTRUCTOR: Cylinders of nitric oxide and oxygen can be used to fill the syringes, but NO is available only at relatively low pressure, so a special regulator, compatible with the cylinder fitting, is required. The preparation of NO and O_2 in syringes, as described in the experimental procedure, is convenient and relatively safe,

but the preparation of NO should be done in the fume hood to assure complete safety.

The procedure for detection of nitrous acid, HNO_2, is semiquantitative but can easily be made more quantitative by using a spectrophotometer to measure the absorbance of the sample. This can be done two ways: The first method simply uses a spectrophotometer to measure the absorbance at 592 mm of the deep blue triiodide/starch complex, for both the diluted reaction mixture and a sample of known nitrite ion concentration, as described in Part 4(c) of the experimental procedure. The second method directly measures the absorbance of nitrite ion itself in basic solution at 353 nm. The molar absorptivity of nitrite ion at this wavelength is approximately $23 \text{ L} \cdot \text{mol}^{-1} \cdot \text{cm}^{-1}$, so that a 4.0 millimolar solution of nitrite has an absorbance of about 0.092. (The nitrate ion contributes very little [< 5%] to the total absorbance at this wavelength.) If approximately 3 mL of the aqueous reaction solution is reserved before addition of the thymol blue indicator, the 3 mL of solution can be made basic by the addition of 0.2 mL of 1 M NaOH and the absorbance of the nitrite ion directly measured at 353 nm.

! SAFETY PRECAUTIONS: WEAR EYE PROTECTION AT ALL TIMES. Prepare the NO gas and carry out the reaction of NO with O_2 in a fume hood, if possible.

WASTE COLLECTION: The Fe(II)/Fe(III)/H_2SO_4 reaction products produced in making NO(g) should be neutralized with sodium bicarbonate, $NaHCO_3$, and the solid residue placed in a waste collection bottle. All other solutions may be disposed of in the sink, followed by a water rinse.

GENERAL PROCEDURE FOR GENERATING GASES IN SYRINGES

We will study the reactions of gases prepared in 60-mL disposable Luer-tip syringes, a method that minimizes exposure to noxious and toxic gases. Each gas is generated by mixing two reagents in a syringe. The general arrangement is shown in Figure 44-1. One reagent, generally a solid, is placed in a small vial cap, which is in turn inserted into the syringe. The second reagent is prepared as an aqueous solution and is drawn into the syringe from a small plastic weighing dish. Then the syringe is capped and shaken to mix the reagents and generate the gas. If desired, gases that are not too water soluble can be washed by drawing deionized water into the syringe, capping and shaking the syringe, and expelling the water. Repeating this operation will remove nearly all traces of the original reactants.

Because of the small vial cap present in the syringe, not all of the air originally in the syringe is expelled before drawing in the liquid reagent. As a result, the gases

FIGURE 44-1

A method for generating gases in a syringe. A vial cap containing a solid reagent is placed in the syringe. A solution of a second reagent is drawn into the syringe and the syringe tip is capped. The syringe is then shaken to mix the two reagents and generate the gas.

prepared contain a small amount of air (about 10 percent by volume). For many purposes, this is not a significant drawback.

1. PREPARING O₂ GAS.

Preparing the 60-mL syringe. Remove the plastic cap that sometimes covers the tip of new syringes, and set it aside. Try to move the plunger back and forth. If the plunger movement is stiff and requires a lot of effort, "break in" the syringe by moving the plunger up and down twenty or thirty times. Practice completely removing the plunger, which takes some extra force because of the molded stop designed to prevent the plunger from accidentally coming completely out of the syringe. (Sometimes it also helps to break in a syringe by pulling the plunger completely out and lubricating the plunger tip with a thin film of silicone oil.)

Weighing the solid reagent. Weigh 0.050 g of dry active baker's yeast (Fleischmann's or other brand) on a creased sheet of weighing paper, and transfer the yeast into the cavity of a small polyethylene vial cap (use the cap supplied with a 1-dram Titeseal shell vial). The

amount of yeast is just about enough to cover the bottom of the vial cap when uniformly distributed.

Loading the vial cap into the syringe. Next we want to put the vial cap containing the solid reagent into the syringe. This can be done one of two ways, using forceps or using the flotation technique illustrated in Figure 44-2. With either method, begin by completely removing the plunger from the syringe and setting the plunger aside on a clean towel or sheet of paper.

The vial cap can be loaded using long (10-inch) forceps, grasping the forceps near the top so that you can insert the vial cap all the way to the bottom inside the syringe without your fingers blocking forceps' entry into the syringe. (If you drop the cap, the solid reagent usually spills out, in which case you must start over.) Then, keeping the syringe upright, with the tip pressed against the bench top, insert the plunger, pushing it snugly against the vial cap. Keep the syringe upright (tip down) in a beaker until you are ready to use it.

The flotation method of putting the vial cap in the syringe works well and requires no forceps. Begin by putting your forefinger on the Luer end of the syringe and filling the syringe nearly to the top with deionized water. Then carefully float the vial cap containing the

FIGURE 44-2

Flotation method for loading the vial cap containing the solid reagent. The plunger is removed from the syringe and the syringe is filled with deionized water, sealing the Luer tip with a finger. The vial cap containing the solid reagent is floated on the water, and the water is allowed to slowly drain out.

solid on the water. Controlling the flow with your finger, allow the water to slowly leak out so that the cap rides down to the bottom of the syringe. Perform this operation over the sink, or catch the water in a beaker so you don't spill water all over the bench top. When all of the water has drained out, keep the syringe upright with the tip against the bench top, and insert the plunger, pushing it snugly against the vial cap. Keep the loaded syringe upright in a beaker until you are ready to use it.

Loading the second liquid reagent. Place 6 mL of 3% hydrogen peroxide solution into a small (1 ⅝ inch square) polyethylene weighing dish. (At room temperature each mL of 3% H_2O_2 solution produces about 10–11 mL of oxygen gas.) Draw the hydrogen peroxide solution up into the syringe, taking care not to draw in more than a bubble or two of air as you draw in the last of the solution. Cap the syringe with a syringe tip cap. Shake the syringe to mix the yeast and hydrogen peroxide solution. Oxygen is evolved rapidly by catalytic decomposition of the hydrogen peroxide:

$$H_2O_2(aq) \rightarrow H_2O + \frac{1}{2}O_2(g)$$

As the volume of gas nears the full capacity of the syringe, hold the syringe tip up (directed away from your face), and loosen the cap to relieve the pressure. (If the syringe tip is pointed down when you loosen the cap, the solution will spray out, making a mess.) Tighten the cap, shake, and relieve excess pressure again. Continue until the reaction is complete, and vent any excess pressure one last time. Turn the syringe tip down, and expel the reaction solution into a sink or beaker (it's harmless, being nearly pure water with a small amount of yeast). Keep the syringe capped until you are ready to use the oxygen gas. It isn't necessary to wash the oxygen gas, since the reaction product is essentially pure water. Label the syringe to identify its contents, making sure that the label does not obscure the volume graduations on the syringe.

2. PREPARING NO GAS. For generating $NO(g)$ we will use solid sodium nitrite, $NaNO_2$, which reacts with an excess of Fe(II) solution according to the net ionic reaction:

$$NO_2^- + Fe^{2+}(aq) + 2H^+ \rightarrow$$
$$NO(g) + Fe^{3+}(aq) + H_2O \quad (11)$$

It has been found by trial and error that you can generate 55–60 mL of NO gas by reacting about 0.27 grams of solid $NaNO_2$ with 4 mL of a solution that is 1.2 M in FeSO$_4$ and 1.8 M in H_2SO_4. (This is 50 percent more than the calculated amount of $NaNO_2$, indicating the reaction is only approximately described by Equation [11].)

Weigh 0.27 g of solid $NaNO_2$ on a creased sheet of weighing paper, and load the $NaNO_2$ into the cavity of a small polyethylene vial cap. Using forceps or the flotation method, load the solid in the vial cap into a 60-mL syringe.

If possible, carry out the next steps in a fume hood. Place 4–5 mL of 1.2 M FeSO$_4$/1.8 M H_2SO_4 solution into a small polyethylene weighing dish. Put the tip of the syringe into the solution, and draw up as much of the solution as possible, being careful not to draw in more than a bubble or two of air. Then cap the syringe with a syringe tip cap. In the fume hood, shake the syringe to mix the reagents. The reaction mixture will turn black almost immediately, and NO gas is rapidly evolved. CAUTION!: When the reaction (and the plunger) are nearing the end, turn the syringe upside down, tip end up, before removing the cap to relieve the pressure. (Point the tip away from your face in the hood.) If the tip is pointed down when you remove the cap, the liquid reagent will spray out, making a nasty mess. (If this happens, neutralize the spill with sodium bicarbonate [baking soda], then wipe up the spill.)

Washing the NO gas. After venting excess pressure with the tip up, turn the syringe tip down, remove the tip cap, and expel the black reaction solution into a beaker. (It will be neutralized and collected as waste. Do not pour it down the drain.) If the volume of gas in the syringe is greater than 55 mL, expel some gas (in the fume hood!) until the volume reads 55 mL. In this way you will have some room to draw in wash water. Place some deionized water in a small beaker, and draw up about 5–6 mL of water into the syringe. Immediately cap the syringe, and shake it to wash the gas. Expel the water into the waste collection beaker, draw another portion of water into the syringe, and shake again. Finally, expel the water into the waste collection beaker, and quickly cap the syringe until you are ready to use the NO gas. Put a label on the syringe identifying its contents as NO.

3. REACTING NO WITH O₂. *If possible, carry out all of the operations in a fume hood.* First, gather together on a piece of paper towel or a clean sheet of paper the materials you will need: the two capped 60-mL syringes containing O_2 and NO gas, a clean dry 10-mL disposable syringe, a short (3-cm) length of 1/8-inch ID rubber tubing, and a 100-mL beaker containing 50±1 mL of deionized water. Deionized water is necessary because most tap water contains bicarbonate ion, HCO_3^-, which reacts with acids:

$$HCO_3^- + H^+ \rightarrow H_2O + CO_2(g)$$

Therefore, tap water would spoil the titration by consuming an unknown amount of any acid that we later collect in the beaker.

Move the plunger of the 10-mL syringe back and forth until it moves freely, then push the plunger all the way in until it stops. Now you are going to connect this syringe to the NO supply syringe and transfer a 10-mL sample of NO from the 60-mL syringe to the smaller 10-mL syringe. Uncap the 60-mL NO supply syringe, and push one end of the rubber tubing firmly onto the Luer tip. Flush out the tubing by expelling a small volume (2–3 mL) of NO, stopping at the nearest 5-mL division. Then quickly insert the Luer tip of the empty 10-mL syringe into the free end of the tubing, and slowly pull back the plunger of the 10-mL syringe by exactly 10 mL while at the same time pushing in the plunger of the larger NO syringe by exactly 10 mL. At the end you will have transferred 10 mL of NO, at approximately atmospheric pressure, to the smaller syringe.

Now put the 60-mL O_2 syringe beside the 60-mL NO syringe so that the O_2 syringe can take its place. When ready to make the transfer, remove the 60-mL NO syringe from the tubing and cap the syringe immediately. Uncap the O_2 syringe, and immediately connect it to the free end of the rubber tube (with the 10-mL syringe still on the other end). Immediately push on the plunger of the O_2 syringe until you have transferred exactly 5 mL of oxygen into the smaller 10-mL syringe containing NO. Do *not* move the plunger of the smaller syringe. Keep it at the 10-mL mark. Then quickly disconnect the O_2 syringe from the rubber tubing, and cap the syringe. Use your fingers to pinch off the rubber tubing so that no gas will enter or escape. Do you see any evidence of reaction between NO and O_2?

Put the free end of the rubber tubing into the previously prepared beaker containing 50 mL of deionized water so that the end of the tubing rests on the bottom of the beaker. Draw a little deionized water into the syringe, taking care to keep the free end of the rubber tubing immersed below the surface of the water. Gently shake the gas/water mixture in the syringe. What happens? Continue shaking and drawing water into the syringe until no more gas is absorbed. At this point record the final volume of gas in the syringe.

Now expel the solution containing the reaction products into the beaker. Draw in a few mL of water into the syringe, then expel the water to rinse out the interior of the syringe. Repeat this a couple of times. At the end, the solution in the beaker will contain all of the NO/O_2 reaction products.

Looking at Equation (2) and recalling that gas volumes combine in exactly the ratios of the stoichiometric equation, we see that Equation (2) predicts that 10 mL of NO would combine with 5 mL of O_2 to form 10 mL of NO_2 gas (all at constant atmospheric pressure). That's why we shouldn't have to move the plunger of the 10-mL syringe to accommodate the added volume of O_2 gas.

Record Pressure and Temperature. You will need to measure and record the barometric pressure and the room temperature so that you can calculate how many moles of gaseous reactants were used.

4. ANALYSIS OF THE REACTION SOLUTION

pH Measurements. If you have a pH meter available, measure the pH of the reaction solution that you collected in the beaker. Rinsing the pH electrode with deionized water between each measurement and blotting the electrode with tissue to remove excess water, also measure the pH of 0.008 M HNO_3, then 0.004 M HNO_3 solutions. (The latter solution can be prepared by diluting 0.008 M HNO_3 with an equal volume of water.) *Save the 0.004 M HNO_3 solution to use later in Part 4(c) as a reagent blank in the test for nitrite ion.* The HNO_3 solutions provide reference pH measurements to check the pH calibration of the meter. The pH difference between the two HNO_3 reference solutions should be equal to about Log(2) = 0.30 because HNO_3 is a strong completely ionized acid and the reference solutions differ in concentration by a factor of two.

Titrating the acid. Get a 1-mL microburet (a 1-mL polypropylene syringe equipped with a fine tip cut from a polyetheylene transfer pipet, or whatever form of microburet you usually use). Also get about 6 mL of 0.5 M NaOH in a clean dry 13 × 100 mm test tube equipped with a size 00 rubber stopper. Fill the syringe with 0.5 M NaOH, taking precautions to remove all bubbles from the syringe (and not to squirt yourself or your neighbor with solution from the syringe). Be sure to record the concentration of the NaOH solution and the initial volume reading of the syringe. Add 5 drops of 0.1% thymol blue indicator to the solution in the beaker. Now add the 0.5 M NaOH solution from the microburet, drop by drop, to your acid solution. At the endpoint the solution will turn from yellow to a permanent blue color. (If the endpoint color drifts back toward yellow, add another drop of 0.5 M NaOH; continue until the solution remains blue for at least 30 seconds.) Now read and record the final volume of the added base. *Save the titrated solution for the next test for the presence of nitrite ion.*

Semiquantitative analysis for nitrite ion. In acid solution, nitrous acid oxidizes iodide ion to form iodine (or triiodide ion in the presence of excess iodide ion):

$$2HNO_2 + 2H^+ + 3I^- \rightarrow 2NO + 2H_2O + I_3^-$$

In the presence of starch indicator, a deep blue starch/triiodide ion complex forms. The test easily detects 1 × 10⁻⁷ M nitrous acid, so the reaction solution must be diluted before doing the test.

Dilute the solution that was titrated in Part 4(b) by a factor of 50, using a polyethylene transfer pipet with a 1-mL calibration mark to transfer 1 mL of the titration solution to a clean 125-mL Erlenmeyer flask and adding 49 mL of deionized water. Mix this solution thoroughly.

To perform the tests you will need to get about 2 mL of each of the following reagent solutions: 1 M $NaHCO_3$, 0.5% starch indicator, 1 M KI (freshly prepared), and 5 M tartaric acid (it's OK if the tartaric acid solution is yellow or light brown).

Then, using a clean polyethylene transfer pipet, put 1 mL of the diluted reaction solution into a 13 × 100 mm test tube and add two drops (0.10 mL) 1 M $NaHCO_3$, two drops (0.10 mL) 0.5% starch indicator, two drops (0.10 mL) 1 M KI, and five drops (0.25 mL) 5 M tartaric acid. Shake the test tube to mix the solution thoroughly. Allow the solution to stand for ten minutes for the blue color to develop fully. While waiting, move ahead to do the control and reagent blank tests.

Control and reagent blank tests for nitrite ion. Using the same quantities of the four reagents, perform the same test for nitrite on 1 mL of 1×10^{-4} M sodium nitrite solution (prepared by diluting 1 mL of 0.01 M $NaNO_2$ solution with 99 mL deionized water and thoroughly mixing). We will call this the control solution because it contains a known amount of nitrite ion.

Finally, perform the same test on 1 mL of a solution containing the 0.004 M HNO_3 prepared in Part 4(a). We will call this the *reagent blank test*. The purpose of the blank is to show that a solution containing nitrate ion and the reagents does not give a positive test for nitrite ion. It should be colorless. If it shows a significant blue color, most likely the KI solution contains some iodine produced by air oxidation of the iodide ion or there is nitrite ion contamination from some source.

Estimating the total amount of nitrite produced. After allowing the blue color to fully develop by standing for 10 minutes, compare the intensities of the blue colors of the reaction solution and the control (1×10^{-4} M $NaNO_2$) solution by viewing them against a well-lit white background. (The comparison can be made more quantitative using a spectrophotometer to measure the absorbance at 592 nm.) If the intensities approximately match (or at least appear to be in the same ballpark), you can estimate the total amount of nitrite produced by multiplying by 2500 the amount of nitrite ion in the control. Recall that the total amount of gas (10 mL NO $\cong$ 0.4 mmol NO) was eventually absorbed in 50 mL of deionized water. Then we took 1 mL of this solution and diluted it again to 50 mL, so we took $1/50^{th}$ of the reaction solution and diluted it another factor of 1/50 to give a total dilution of 1/2500.

The amount of nitrite ion in the control solution is equal to 1 mL $\times$ 1×10^{-4} mol/L $= 1 \times 10^{-4}$ mmol. If the blue colors approximately match, this would also be the amount of nitrite ion in the diluted reaction sample. This amount multiplied by 2500 would then give the total amount of nitrite ion produced by reaction of the gas with water (2500 $\times$ 1 $\times$ 10^{-4} mmol $= 0.25$ mmol of nitrite).

You should have enough NO and O_2 left in your syringes to repeat the experiment. If you have time, repeat parts 3 and 4 so that you have two measurements of how much acid and nitrite were formed. Remember, an average of several measurements is always more reliable than a single measurement.

CALCULATIONS AND DATA ANALYSIS

Using the ideal gas law, determine the quantities (in millimoles) of NO and O_2 that were reacted in the syringe. What is the total amount of acid (in units of millimoles) that you would expect to be produced for the given amounts of NO and O_2, considering the three reactions summarized in Table 44-1 (p. 524)?

In your lab notebook, calculate the expected pH for solutions containing 0.008 M HNO_3, 0.004 M HNO_3, and 0.008 M HNO_2 ($K_a = 7.1 \times 10^{-4}$).

From the results of the titration in Part 4(b), calculate the moles of acid formed. Compare this, calculated as a percent yield, with the theoretical number of moles of acid expected. What is the percent yield for the formation of acid in your experiment? Is the result consistent with the stoichiometry shown in Table 44-1 for the three net reactions A, B, and C?

Finally, estimate the amount of nitrite produced (in units of millimoles) as described in Part 4(d).

☞ Consider This

Did you get approximately the expected amount of acid based on the amount of NO gas reacted?

You added 5 mL of O_2 to 10 mL of NO. If these gases reacted as shown in Equation (2) and if the product NO_2 is completely absorbed by water, we would expect that there would be no gas left after shaking the reaction mixture with water. Was there a measurable amount of gas left? If so, how much? What was the color of the gas? If this gas contained NO_2, what color would it be? Explain how this gas might be NO or O_2 or N_2 or a mixture of two gases (N_2 + NO or N_2 + O_2). Where would N_2 have come from? Is it possible from the results of your experiments to say definitely whether the gas left over in the tube is NO, NO_2, O_2, or N_2, or something else, or a mixture of gases? Can you suggest a way to try to determine the composition of the gas left over? What would be your hypothesis about the identity of the gas?

If NO$_2$ simply dissolved in the water without reacting, would you expect the solution to have a brown color? Was the solution brown? What do you think is a good hypothesis to explain the color of the solution?

Most chemists would classify the reactions shown in Equations (1) and (2) as oxidation-reduction reactions. For each reaction, specify the oxidation numbers that you would assign to the nitrogen (N), oxygen (O), and hydrogen (H) atoms on both sides of the equation. Based on this (arbitrary) assignment of oxidation numbers, what was oxidized, and what was reduced in Equations (1) and (2)?

Discussion

In your report, include a discussion of the implications of your experimental results for the broad questions: What human activities lead to the formation of acid rain? Are nitrogen oxides the only source of the acids in acid rain? If not, what are the identities and sources of other important atmospheric pollutants that produce acid rain?

Once formed, why doesn't NO just keep building up to higher and higher concentrations in the troposphere? That is, what processes remove nitrogen oxides from the troposphere?

Is it possible from your pH measurements and the semiquantitative estimate of the amount of nitrite produced to determine if you got exclusively nitric acid, exclusively nitrous acid, or a mixture of nitric and nitrous acids? That is, considering all of the data, which of the three reactions—A, B, or C—is in best agreement with your experimental results?

Note: *The pattern of your pH measurements, when compared with the pattern of calculated pH values for various HNO$_3$ and HNO$_2$ solutions, should give you some important clues. The pattern of measurements is more important than the absolute value of the pHs, since the standardization of the pH meter is often in error by as much as ±0.1 pH unit. For this reason the measured values for 0.008 and 0.004 M HNO$_3$ may give you a more reliable indicator than the calculated pH values.*

BIBLIOGRAPHY

Ainscough, E. W., and Brodie, A. M. "Nitric Oxide—Some Old and New Perspectives," *J. Chem. Educ.,* **1995,** *72:* 686–692.

Alyea, H. N. "Syringe Gas Generators" (Tested Demonstrations), *J. Chem. Educ.,* **1992,** *69:* 65.

Bonner, F. T., and Stedman, G. "The Chemistry of Nitric Oxide and Redox-related Species," in M. Feelisch and J. S. Stamler (eds.), *Methods in Nitric Oxide Research,* John Wiley & Sons, Inc., New York, 1996, Chap. 1.

Cotton, F. A.; Wilkinson, G.; Murillo, C. A.; Bochmann, M. *Advanced Inorganic Chemistry,* 6th ed., John Wiley & Sons, Inc., New York, 1999, pp. 323–333.

Everett, K. G. "The Titration of Air with Nitric Oxide: An Application of Gay-Lussac's Law of Combining Volumes in a General Chemistry Experiment," *J. Chem. Educ.,* **1982,** *59:* 802–803.

Hornack, F. M. "Nitric Oxide Leftovers," *J. Chem. Educ.,* **1990,** *67:* 496.

Mattson, B.; Anderson, M.; Schwennsen, C. *Chemistry of Gases: A Microscale Approach* (available from Flinn Scientific Inc.).

Shaaban, A. F. "The Integrated Rate Equation of the Nitric Oxide-Oxygen Reaction," *J. Chem. Educ.,* **1990,** *67:* 869–871.

Atmospheric Chemistry of Nitric Oxide
Smog and Acid Rain

NAME

SECTION   LOCKER

INSTRUCTOR   DATE

OBSERVATIONS AND DATA

1. Preparing O_2 gas

Describe what you see when yeast cells are mixed with hydrogen peroxide in the syringe. Write a balanced equation for the net chemical reaction.

2. Preparing NO gas

Describe what you see when solid sodium nitrite is mixed with iron(II) sulfate in acid solution. Write a balanced equation for the net chemical reaction.

3. Reacting NO with O_2

Describe what you see when oxygen is mixed with nitric oxide, NO. What is the color of the gas produced? Describe what happens next when deionized water is drawn into the syringe. What is the color of the solution produced?

Barometric Pressure and Room Temperature

Room temperature _____ °C

Barometric pressure _____ torr (mm Hg)

Vapor pressure of water at room temperature _____ torr (mm Hg)

Partial pressure of NO and O_2 gases _____ torr (mm Hg)
(after subtracting the vapor pressure of water at room temperature)

4. Analysis of the Reaction Solution

(a) *pH Measurements*

Calculated pH values assume K_a for $HNO_2 = 7.1 \times 10^{-4}$ and all activity coefficients $= 1.0$

Solution	Measured pH	Calculated pH
NO/O_2 Reaction Solution		XXXX
0.008 M HNO_3		2.10
0.004 M HNO_3 + 0.004 M HNO_2	XXXX	2.34
0.004 M HNO_3		2.40
0.008 M HNO_2	XXXX	2.69

(b) *Titrating the acid produced*

Molarity of NaOH titrant _____ mols/Liter

	Trial 1	Trial 2
Initial volume of NaOH titrant	mL	mL
Final volume of NaOH titrant	mL	mL
Net volume of NaOH titrant	mL	mL
Millimoles of acid titrated	mmol	mmol

(c) *Semiquantitative analysis for nitrite ion*
Describe the nitrite test color comparison of the diluted reaction solution with the control solution
$(1 \times 10^{-4}$ M NaNO$_2)$.

Describe the color of the nitrite test on the blank solution.

(d) Estimate the total amount of nitrite produced in units of millimoles.

CALCULATIONS AND DATA ANALYSIS

Using the ideal gas law and assuming room temperature (in Kelvins) and the partial pressures of NO and O$_2$, calculate the millimoles of NO in 10 mL of NO gas and 5 mL of O$_2$ gas.

Based on the amount (in millimoles) of NO taken, what would you expect to be the total amount (in millimoles) of acid produced?

Calculate the percent yield of acid $= \dfrac{\text{millimoles acid titrated}}{\text{millimoles NO gas reacted}} \times 100$

Consider This

After shaking the NO/O₂ reaction mixture with water, was there some gas left in the 10-mL syringe? How much? What are your thoughts about what might be the identity and source of this gas?

What was the color of the solution resulting from mixing the NO/O₂ gas mixture with water? What is your hypothesis to explain the color of the solution?

Rewrite Equations (1) and (2), assigning oxidation numbers to the nitrogen (N), oxygen (O), and hydrogen (H) atoms on both sides of the equations. What species were oxidized and reduced in Equations (1) and (2)?

Discuss your results, reponding to some of the questions posed in the Discussion paragraph of the Consider This section of the experiment.

Tables of Data

TABLE 1

The International System (SI) of Units and Conversion Factors

Basic SI Units

Physical quantity	Unit	Symbol
Length	meter	m
Mass	kilogram	kg
Time	second	s
Electric current	ampere	A
Temperature	kelvin	K
Amount of substance	mole	mol
Luminous intensity	candela	cd

Common Derived Units

Physical quantity	Unit	Symbol	Definition
Frequency	hertz	Hz	s^{-1}
Energy	joule	J	$kg \cdot m^2 \cdot s^{-2}$
Force	newton	N	$kg \cdot m \cdot s^{-2} = J \cdot m^{-1}$
Pressure	Pascal	Pa	$kg \cdot m^{-1} \cdot s^{-2} = N \cdot m^{-2}$
Power	watt	W	$kg \cdot m^2 \cdot s^{-3} = J \cdot s^{-1}$
Electric charge	coulomb	C	$A \cdot s$
Electric potential difference	volt	V	$kg \cdot m^2 \cdot s^{-3} \cdot A^{-1} = J \cdot A^{-1} \cdot s^{-1}$

Decimal Fractions and Multiples

Factor	Prefix	Symbol	Factor	Prefix	Symbol
10^{-18}	atto	a	10^{-1}	deci	d
10^{-15}	femto	f	10	deca	da
10^{-12}	pico	p	10^2	hecto	h
10^{-9}	nano	n	10^3	kilo	k
10^{-6}	micro	μ	10^6	mega	M
10^{-3}	milli	m	10^9	giga	G
10^{-2}	centi	c	10^{12}	tera	T

Common Conversion Factors

LENGTH

1 angstrom unit (Å) $\equiv 10^{-8}$ cm

2.54 cm $\equiv$ 1 inch (in.)

1 m = 39.37 in.

MASS

453.6 grams (g) = 1 pound (lb)

1 kg = 2.205 lb

28.35 g = 1 ounce (oz, avoirdupois)

VOLUME

1 milliliter (mL) $\equiv$ 1 cubic centimeter (cm^3)

(Note that the mL is now defined precisely equal to 1 cm^3.)

1 liter (L) = 1.057 quarts

28.6 mL = 1 fluid ounce

PRESSURE

1 atmosphere (atm) = 1.013×10^5 Pa (N/m^2)

$= 760$ torr (mm Hg); pressure of a mercury column 760 mm or 29.92 inches high at 0 °C

$= 1.0133$ bar ($dyne/cm^2$)

$= 14.70$ lb/in.^{2}

TEMPERATURE

Absolute zero (K) = -273.15 °C

K = °C = 273.15

Fahrenheit degrees (°F) = 1.8 °C + 32

Celsius degrees (°C) = (°F $-$ 32)/1.8

ENERGY

1 joule $\equiv$ 1 Watt-s $\equiv$ 1 kg $\cdot$ $m^2 \cdot s^{-2}$ = 10^7 erg

1 erg $\equiv$ 1 dyne-cm $\equiv$ 1 g $\cdot$ $cm^2 \cdot s^{-2}$

1 calorie = 4.184 joule (J)

1 electron volt/molecule = 23.06 kcal/mol

TABLE 2

Fundamental Physical and Mathematical Constants
1986 CODATA Recommended values from http://physics.nist.gov

Physical Constants

Symbol	Name	Numerical value
N_A	Avogadro's number	6.0221×10^{23} mol^{-1}
F	Faraday constant	96,485 C per mole of electrons transferred
h	Planck's constant	6.6261×10^{-34} J $\cdot$ s $\cdot$ particle^{-1}
c	Speed of light (*in vacuo*)	2.9979×10^{8} m $\cdot$ s^{-1}
R	The gas constant	0.08206 L $\cdot$ atm $\cdot$ mol^{-1} $\cdot$ K^{-1}
		82.06 mL $\cdot$ atm $\cdot$ mol^{-1} $\cdot$ K^{-1}
		8.3145 J $\cdot$ mol^{-1} $\cdot$ K^{-1}
e	Charge of the electron	1.6022×10^{-19} C
	Volume of 1 mol of an ideal gas	
	at 1 atm 0 °C	22.41 L
	at 1 atm 25 °C	24.46 L

Mathematical Constants

$\pi = 3.14159$

$e = 2.71828$

$Ln(x) = 2.303 \, Log_{10}(x)$

TABLE 3

Vapor Pressure of Water at Different Temperatures

Temperature (°C)	Vapor pressure (torr)
−10 (ice)	1.0
−5 (ice)	3.0
0	4.6
5	6.5
10	9.2
15	12.8
16	13.6
17	14.5
18	15.5
19	16.5
20	17.5
21	18.6
22	19.8
23	21.1
24	22.4
25	23.8
26	25.2
27	26.7
28	28.3
29	30.0
30	31.8
35	42.2
40	55.3
45	71.9
50	92.5
55	118.0
60	149.4
65	187.5
70	233.7
75	289.1
80	355.1
90	525.8
100	760.0
150	3570.5
200	11659.2

TABLE 4

Concentration of Desk Acid and Base Solutions

Reagent	Formula	Molarity	Density	Percent of solute
Acetic acid, glacial	CH_3COOH	17 M	1.05 g/mL	99.5
Acetic acid, dilute		6	1.04	34
Hydrochloric acid, conc.	HCl	12	1.18	36
Hydrochloric acid, dilute		6	1.10	20
Nitric acid, conc.	HNO_3	16	1.42	72
Nitric acid, dilute		6	1.19	32
Sulfuric acid, conc.	H_2SO_4	18	1.84	96
Sulfuric acid, dilute		3	1.18	25
Ammonia solution, conc. (ammonium hydroxide)	NH_3	15	0.90	58
Ammonia solution, dilute (ammonium hydroxide)		6	0.96	23
Sodium hydroxide, dilute	NaOH	6	1.22	20

TABLE 5

Color Changes and pH Intervals of Some Important Indicators

Name of Indicator	pH Interval	Color change	Solvent
Methyl violet	0.2–3.0	Yellow to blue-violet	Water
Thymol blue	1.2–2.8	Red to yellow	Water (+NaOH)
Orange IV (Tropeolin OO)	1.3–3.0	Red to yellow	Water
Benzopurpurin 4B	1.2–4.0	Violet to red	20% alcohol
Methyl orange	3.1–4.4	Red to orange to yellow	Water
Bromophenol blue	3.0–4.6	Yellow to blue-violet	Water (+NaOH)
Congo red	3.0–5.0	Blue to red	70% alcohol
Bromocresol green	3.8–5.4	Yellow to blue	Water (+NaOH)
Methyl red	4.4–6.2	Red to yellow	Water (+NaOH)
Chlorophenol red	4.8–6.8	Yellow to red	Water (+NaOH)
Bromocresol purple	5.2–6.8	Yellow to purple	Water (+NaOH)
Litmus	4.5–8.3	Red to blue	Water
Bromothymol blue	6.0–7.6	Yellow to blue	Water (+NaOH)
Phenol red	6.8–8.2	Yellow to red	Water (+NaOH)
Thymol blue	8.0–9.6	Yellow to blue	Water (+NaOH)
Phenolphthalein	8.3–10.0	Colorless to red	70% alcohol
Thymolphthalein	9.3–10.5	Yellow to blue	70% alcohol
Alizarin yellow R	10.0–12.0	Yellow to red	20% alcohol
Indigo carmine	11.4–13.0	Blue to yellow	50% alcohol
Trinitrobenzene	12.0–14.0	Colorless to orange	70% alcohol

TABLE 6

Equilibrium Constants for the Ionization of Acids and Bases (25 °C)

Compound	Ionization reaction	K_a	pK_a
WATER	$2H_2O \rightleftharpoons H_3O^+ + OH^-$ (25 °C)	1.00×10^{-14}	14.00
	(0 °C)	0.11×10^{-14}	14.94
	(60 °C)	9.61×10^{-14}	13.02
WEAK ACIDS			
Acetic	$CH_3COOH + H_2O \rightleftharpoons H_3O^+ + CH_3COO^-$	1.76×10^{-5}	4.75
Boric	$H_3BO_3 + H_2O \rightleftharpoons H_3O^+ + H_2BO_3^-$	6.0×10^{-10}	9.22
Carbonic (CO_2)	$CO_2 + H_2O \rightleftharpoons H_3O^+ + HCO_3^-$	$K_1 = 4.4 \times 10^{-7}$	6.35
	$HCO_3^- + H_2O \rightleftharpoons H_3O^+ + CO_3^{2-}$	$K_2 = 4.7 \times 10^{-11}$	10.33
Chromic	$H_2CrO_4 + H_2O \rightleftharpoons H_3O^+ + HCrO_4^-$	$K_1 = 2 \times 10^{-1}$	0.7
	$HCrO_4^- + H_2O \rightleftharpoons H_3O^+ + CrO_4^{2-}$	$K_2 = 3.2 \times 10^{-7}$	6.50
Formic	$HCOOH + H_2O \rightleftharpoons H_3O^+ + HCOO^-$	1.8×10^{-4}	3.74
Hydrogen cyanide	$HCN + H_2O \rightleftharpoons H_3O^+ + CN^-$	4×10^{-10}	9.4
Hydrofluoric	$HF + H_2O \rightleftharpoons H_3O^+ + F^-$	6.9×10^{-4}	3.16
Hydrogen peroxide	$H_2O_2 + H_2O \rightleftharpoons H_3O^+ + HO_2^-$	2.4×10^{-12}	11.62
Hydrogen sulfate ion	$HSO_4^- + H_2O \rightleftharpoons H_3O^+ + SO_4^{2-}$	$K_2 = 1.2 \times 10^{-2}$	1.92
Hydrogen sulfide	$H_2S + H_2O \rightleftharpoons H_3O^+ + HS^-$	$K_1 = 1.0 \times 10^{-7}$	7.00
	$HS^- + H_2O \rightleftharpoons H_3O^+ + S^{2-}$	$K_2 = 2 \times 10^{-18}$	17.7
Nitrous	$HNO_2 + H_2O \rightleftharpoons H_3O^+ + NO_2^-$	4.5×10^{-4}	3.50
Oxalic	$H_2C_2O_4 + H_2O \rightleftharpoons H_3O^+ + HC_2O_4^-$	$K_1 = 3.8 \times 10^{-2}$	1.42
	$HC_2O_4^- + H_2O \rightleftharpoons H_3O^+ + C_2O_4^{2-}$	$K_2 = 5.0 \times 10^{-5}$	4.30
Phosphoric	$H_3PO_4 + H_2O \rightleftharpoons H_3O^+ + H_2PO_4^-$	$K_1 = 7.1 \times 10^{-3}$	2.15
	$H_2PO_4^- + H_2O \rightleftharpoons H_3O^+ + HPO_4^{2-}$	$K_2 = 6.3 \times 10^{-8}$	7.20
	$HPO_4^{2-} + H_2O \rightleftharpoons H_3O^+ + PO_4^{3-}$	$K_3 = 4.4 \times 10^{-13}$	12.36
Phosphorous	$H_2HPO_3 + H_2O \rightleftharpoons H_3O^+ + HPO_3^-$	$K_1 = 1.6 \times 10^{-2}$	1.80
Propionic	$CH_3CH_2COOH + H_2O \rightleftharpoons H_3O^+ + CH_3CH_2COO^-$	1.34×10^{-5}	4.87
Sulfurous ($SO_2 + H_2O$)	$H_2SO_3 + H_2O \rightleftharpoons H_3O^+ + HSO_3^-$	$K_1 = 1.2 \times 10^{-2}$	1.92
	$HSO_3^- + H_2O \rightleftharpoons H_3O^+ + SO_3^{2-}$	$K_2 = 5.6 \times 10^{-8}$	7.25
CATION ACIDS—HYDRATED METAL IONS			
Aluminum ion	$Al(H_2O)_6^{3+} + H_2O \rightleftharpoons H_3O^+ + Al(H_2O)_5OH^{2+}$	1.1×10^{-5}	4.9
Chromium(III) ion	$Cr(H_2O)_6^{3+} + H_2O \rightleftharpoons H_3O^+ + Cr(H_2O)_5OH^{2+}$	1.6×10^{-4}	3.80
Iron(III) ion	$Fe(H_2O)_6^{3+} + H_2O \rightleftharpoons H_3O^+ + Fe(H_2O)_5OH^{2+}$	6.7×10^{-3}	2.17
Zinc ion	$Zn(H_2O)_4^{2+} + H_2O \rightleftharpoons H_3O^+ + Zn(H_2O)_3OH^+$	2.5×10^{-10}	9.60

continued on next page

TABLE 6 *continued*

Compound	Ionization reaction	K_a	pK_a
WEAK BASES		K_b	pK_b
Ammonia	$NH_3 + H_2O \rightleftharpoons NH_4^+ + OH^-$	1.8×10^{-5}	4.75
Methylamine	$CH_3NH_2 + H_2O \rightleftharpoons CH_3NH_3^+ + OH^-$	5.0×10^{-4}	3.3
Barium hydroxide	$Ba(OH)_2 \rightleftharpoons BaOH^+ + OH^-$	strong	
	$BaOH^+ \rightleftharpoons Ba^{2+} + OH^-$	$K_2 = 1.4 \times 10^{-1}$	0.85
Calcium hydroxide	$Ca(OH)_2 \rightleftharpoons CaOH^+ + OH^-$	strong	
	$CaOH^+ \rightleftharpoons Ca^{2+} + OH^-$	$K_2 = 3.5 \times 10^{-2}$	1.5

TABLE 7

Equilibrium Constants for the Dissociation of Complex Ions, Amphoteric Hydroxides, and Weak Salts ($\approx 25\ °C$)[a]

Compound	Dissociation reaction	K	pK
AMMINE (AMMONIA) COMPLEX IONS			
Tetraamminecadmium(II)	$Cd(NH_3)_4{}^{2+} \rightleftharpoons Cd^{2+} + 4NH_3$	2×10^{-7}	6.7
Tetraamminecopper(II)	$Cu(NH_3)_4{}^{2+} \rightleftharpoons Cu^{2+} + 4NH_3$	8×10^{-13}	12.1
Diamminesilver(I)	$Ag(NH_3)_2{}^{+} \rightleftharpoons Ag^{+} + 2NH_3$	6×10^{-8}	7.2
Tetraamminezinc(II)	$Zn(NH_3)_4{}^{2+} \rightleftharpoons Zn^{2+} + 4NH_3$	1×10^{-9}	9.0
HYDROXIDE COMPLEX IONS (AMPHOTERIC HYDROXIDES)			
Tetrahydroxoaluminate	$Al(OH)_4{}^{-} \rightleftharpoons Al(OH)_3(s) + OH^{-}$	3×10^{-2}	1.5
Tetrahydroxochromate(III)	$Cr(OH)_4{}^{-} \rightleftharpoons Cr(OH)_3(s) + OH^{-}$	2.5	-0.40
Trihydroxoplumbate(II)	$Pb(OH)_3{}^{-} \rightleftharpoons Pb(OH)_2(s) + OH^{-}$	2×10^{1}	-1.3
Trihydroxostannate(II)	$Sn(OH)_3{}^{-} \rightleftharpoons Sn(OH)_2(s) + OH^{-}$	2.6	-0.41
Tetrahydroxozincate	$Zn(OH)_4{}^{2-} \rightleftharpoons Zn(OH)_2(s) + 2OH^{-}$	4×10^{1}	-1.6
CHLORIDE COMPLEX IONS AND WEAKLY IONIZED SALTS			
Dichlorocadmium	$CdCl_2(aq) \rightleftharpoons Cd^{2+} + 2Cl^{-}$	2.5×10^{-3}	2.60
Tetrachloroaurate(II) ion	$AuCl_4{}^{2-} \rightleftharpoons Au^{3+} + 4Cl^{-}$	5×10^{-22}	21.3
Trichloroiron(III)	$FeCl_3(aq) \rightleftharpoons Fe^{3+} + 3Cl^{-}$	8×10^{-2}	1.9
Dichloroiron(III) ion	$FeCl_2{}^{+}(aq) \rightleftharpoons Fe^{3+} + 2Cl^{-}$	8×10^{-3}	2.9
Chloroiron(III) ion	$FeCl^{2+} \rightleftharpoons Fe^{3+} + Cl^{-}$	3.5×10^{-2}	1.46
Mercury(II) chloride	$HgCl_2(aq) \rightleftharpoons HgCl^{+} + Cl^{-}$	$K_1 = 3.3 \times 10^{-7}$	6.48
Chloromercury(II) ion	$HgCl^{+} \rightleftharpoons Hg^{2+} + Cl^{-}$	$K_2 = 1.8 \times 10^{-7}$	6.74
Tetrachloromercurate(II)	$HgCl_4{}^{2-} \rightleftharpoons Hg^{2+} + 4Cl^{-}$	8.5×10^{-16}	15.07
Tin(II) chloride	$SnCl_2(aq) \rightleftharpoons Sn^{2+} + 2Cl^{-}$	5.7×10^{-3}	2.24
Tetrachlorostannate(II) ion	$SnCl_4{}^{2-} \rightleftharpoons Sn^{2+} + 4Cl^{-}$	3.3×10^{-2}	1.48
Hexachlorostannate(IV) ion	$SnCl_6{}^{2-} \rightleftharpoons Sn^{4+} + 6Cl^{-}$	10^{-4}	4
Dichloroargentate(I) ion	$AgCl_2{}^{-} \rightleftharpoons Ag^{+} + 2Cl^{-}$	5×10^{-6}	5.3
OTHER COMPLEX IONS AND WEAKLY IONIZED SALTS			
Tetracyanocadmate(II) ion	$Cd(CN)_4{}^{2-} \rightleftharpoons Cd^{2+} + 4CN^{-}$	8×10^{-18}	17.1
Thiocyanatoiron(III) ion	$FeSCN^{2+} \rightleftharpoons Fe^{3+} + SCN^{-}$	1×10^{-3}	3.0
Lead(II) acetate	$Pb(CH_3COO)_2(aq) \rightleftharpoons Pb^{2+} + 2CH_3COO^{-}$	1×10^{-4}	4.0
Triacetatoplumbate(II) ion	$Pb(CH_3COO)_3{}^{-} \rightleftharpoons Pb^{2+} + 3CH_3COO^{-}$	2.5×10^{-7}	6.60
Dicyanoargentate(I) ion	$Ag(CN)_2{}^{-} \rightleftharpoons Ag^{+} + 2CN^{-}$	1×10^{-20}	20.0
Dithiosulfatoargentate(I) ion	$Ag(S_2O_3)_2{}^{3-} \rightleftharpoons Ag^{+} + 2S_2O_3{}^{2-}$	4×10^{-14}	13.4

[a]The formation of complex ions undoubtedly occurs by the step-by-step addition of ligands (the coordinated molecules or anions) as the concentration of ligand increases. It is only in the presence of a large excess of the ligand that the overall cumulative ionization constants given here can be used with any measure of quantitative reliability. Furthermore, the total ionic strength of the solution exerts a major influence on the value of the equilibrium constant. For example, as the total ionic strength increases from zero to 0.5 M, the equilibrium constant for dissociation of the $Fe(SCN)^{2+}$ complex increases from 1×10^{-3} to about 7×10^{-3}.

TABLE 8

Solubility of Some Common Salts and Hydroxides at approximately 25 °C
The classification is a summary of observed solubility behavior. The division into three solubility groups is arbitrary.

General rule	Exceptions	
Soluble	Sparingly Soluble	Insoluble
Group 1 (alkali metal salts)		
Nitrates (NO_3^-)		
Fluorides (F^-)		MgF_2, CaF_2, SrF_2, BaF_2, PbF_2
Chlorides, Bromides,	$PbCl_2$, $PbBr_2$	Ag^+, Hg_2^{2+}, PbI_2
Sulfates (SO_4^{2-})[a]	Ca^{2+}, Ag^+, Hg_2^{2+}	Sr^{2+}, Ba^{2+}, Pb^{2+}
Acetates (CH_3COO^-)	Ag^+, Hg_2^{2+}	

General rule	Exceptions	
Insoluble	Sparingly Soluble	Soluble
Carbonates (CO_3^{2-})		Group 1 (alkali metals), NH_4^+
Phosphates (PO_4^{3-})[b]		Group 1 (alkali metals), NH_4^+
Sulfides (S^{2-})[c]		Group 1 (alkali metals), NH_4^+
Oxalates ($C_2O_4^{2-}$)		Group 1 (alkali metals), NH_4^+
Hydroxides (OH^-)	Ca^{2+}, Sr^{2+}	Group 1 (alkali metals), NH_4^+

[a]Many hydrogen sulfates such as $Ca(HSO_4)_2$ and $Ba(HSO_4)_2$ are more soluble.
[b]Many hydrogen phosphates are soluble, such as $Mg(H_2PO_4)_2$ and $Ca(H_2PO_4)_2$.
[c]Sulfides of Al^{3+} and Cr^{3+} hydrolyze and precipitate the corresponding hydroxides.

TABLE 9

Solubility Product Constants (18–25 °C)[a]

For compounds marked with an asterisk(*), the molar solubility calculated from the solubility product constant is in reasonable agreement with the experimentally measured solubility. This suggests that these salts completely dissociate into their constituent ions to produce a solution of low ionic strength and that the ions do not react with water to produce hydrolysis side reactions. (See R. W. Clark and J. M. Bonicamp, "The K_{sp}-Solubility Conundrum," *J. Chem. Educ.,* 1998, *75:* 1182–1185.)

Compound	K_{sp}	Compound	K_{sp}
ACETATES		$Mg(OH)_2$	9×10^{-12}
$Ag(CH_3COO)$	4×10^{-3}	*$Mn(OH)_2$	1.7×10^{-13}
		$Pb(OH)_2$	4×10^{-15}
HALIDES AND CYANIDES		$Sn(OH)_2$	10^{-27}
$AgCN$	10^{-16}	$Zn(OH)_2$	5×10^{-17}
*$AgCl$	1.8×10^{-10}		
*$AgBr$	5×10^{-13}	**OXALATES**	
*AgI	8.5×10^{-17}	*$Ag_2C_2O_4$	1.3×10^{-12}
*CaF_2	4.6×10^{-11}	BaC_2O_4	1.5×10^{-8}
$CuCl$	3.2×10^{-7}	CaC_2O_4	1.3×10^{-9}
Hg_2Cl_2	1.1×10^{-18}	MgC_2O_4	8.6×10^{-5}
*Hg_2Br_2	5.5×10^{-23}	SrC_2O_4	5.6×10^{-8}
*Hg_2I_2	1.1×10^{-28}		
MgF_2	8×10^{-8}	**CARBONATES**	
$PbCl_2$	1.6×10^{-5}	Ag_2CO_3	8×10^{-12}
PbI_2	8.3×10^{-9}	$BaCO_3$	1.6×10^{-9}
		$CaCO_3$	4.8×10^{-9}
SULFATES		$CuCO_3$	2.5×10^{-10}
*Ag_2SO_4	6.2×10^{-5}	$FeCO_3$	2×10^{-11}
*$BaSO_4$	1.2×10^{-10}	$MgCO_3$	4×10^{-5}
$CaSO_4$	2.4×10^{-5}	$MnCO_3$	9×10^{-11}
$PbSO_4$	1.3×10^{-8}	$PbCO_3$	1.5×10^{-13}
$SrSO_4$	7.6×10^{-7}	*$SrCO_3$	1.0×10^{-9}
CHROMATES		**SULFIDES**	
Ag_2CrO_4	2×10^{-12}	Ag_2S	10^{-50}
*$BaCrO_4$	2.3×10^{-10}	CdS	10^{-26}
$PbCrO_4$	2×10^{-16}	CoS	10^{-21}
$SrCrO_4$	3.6×10^{-5}	CuS	10^{-36}
		FeS	10^{-17}
HYDROXIDES		HgS	10^{-50}
$Al(OH)_3$	10^{-33}	MnS	10^{-13}
$Ca(OH)_2$	6.5×10^{-6}	NiS	10^{-22}
$Cr(OH)_3$	10^{-30}	PbS	10^{-26}
$Cu(OH)_2$	2×10^{-19}	SnS	10^{-27}
*$Fe(OH)_2$	8.2×10^{-16}	ZnS	10^{-20}
$Fe(OH)_3$	10^{-37}		

[a]These values are approximate. The solubility is affected by the total ionic strength of the solution, by the temperature, and by the presence of hydronium ion or other ions that may form complex ions or that result in colloidal suspensions. The rate of approach to equilibrium is often slow.

TABLE 10

Standard Reduction Potentials (25 °C)
The reduction potential convention recommended by the IUPAC is used. The half-cell potentials are those that would be obtained in a hypothetical cell with all species at unit activity when measured with respect to the standard hydrogen (H^+/H_2) half-cell. Values in the table are from W. M. Latimer, *Oxidation Potentials, 2nd edition,* Prentice-Hall, Englewood Cliffs, New Jersey, 1952; and A. J. Bard, R. Parsons, and J. Jordan, *Standard Potentials in Aqueous Solution,* Marcel Dekker, Inc., New York, 1985.

Strongest oxidizing agents	Half-reaction	$E°$ (V)	Weakest reducing agents
	$\frac{1}{2} F_2(g) + H^+ + e^- \rightleftharpoons HF(aq)$	+3.06	
	$\frac{1}{2} F_2(g) + e^- \rightleftharpoons F^-$	+2.87	
	$H_2O_2(aq) + 2H^+ + 2e^- \rightleftharpoons 2H_2O$	+1.77	
	$PbO_2(s) + 4H^+ + SO_4^{2-} + 2e^- \rightleftharpoons 2H_2O + PbSO_4(s)$	+1.685	
	$Ce^{4+} + e^- \rightleftharpoons Ce^{3+}$	+1.61	
	$MnO_4^- + 8H^+ + 5e^- \rightleftharpoons 4H_2O + Mn^{2+}$	+1.51	
	$Mn^{3+} + e^- \rightleftharpoons Mn^{2+}$	+1.51	
	$Au^{3+} + 3e^- \rightleftharpoons Au$	+1.50	
	$HClO(aq) + H^+ + 2e^- \rightleftharpoons H_2O + Cl^-$	+1.49	
	$ClO_3^- + 6H^+ + 5e^- \rightleftharpoons 3H_2O + \frac{1}{2} Cl_2(g)$	+1.47	
	$Co^{3+} + e^- \rightleftharpoons Co^{2+}$	+1.45[b]	
	$\frac{1}{2} Cl_2(g) + e^- \rightleftharpoons Cl^-$	+1.358	
	$Cr_2O_7^{2-} + 14H^+ + 6e^- \rightleftharpoons 7H_2O + 2Cr^{3+}$	+1.33	
	$MnO_2(s) + 4H^+ + 2e^- \rightleftharpoons 2H_2O + Mn^{2+}$	+1.23	
	$O_2(g) + 4H^+ + 4e^- \rightleftharpoons 2H_2O$	+1.229	
	$IO_3^- + 6H^+ + 5e^- \rightleftharpoons \frac{1}{2} I_2(aq) + 3H_2O$	+1.195	
	$\frac{1}{2} Br_2(l) + e^- \rightleftharpoons Br^-$	+1.0652	
	$AuCl_4^- + 3e^- \rightleftharpoons 4Cl^- + Au$	+1.00	
	$NO_3^- + 4H^+ + 3e^- \rightleftharpoons 2H_2O + NO(g)$	+0.96	
	$2Hg^{2+} + 2e^- \rightleftharpoons Hg_2^{2+}$	+0.92	
	$ClO^- + H_2O + 2e^- \rightleftharpoons 2OH^- + Cl^-$	+0.89	
	$HO_2^- + H_2O + 2e^- \rightleftharpoons 3OH^-$	+0.88	
	$Cu^{2+} + I^- \rightleftharpoons CuI(s)$	+0.861	
	$Hg^{2+} + 2e^- \rightleftharpoons Hg(l)$	+0.854	
	$O_2(g) + 4H^+(10^{-7} M) + 4e^- \rightleftharpoons 2H_2O$	+0.815	
	$Ag^+ + e^- \rightleftharpoons Ag$	+0.79991	
	$Hg_2^{2+} + 2e^- \rightleftharpoons 2Hg(l)$	+0.789	
	$Fe^{3+} + e^- \rightleftharpoons Fe^{2+}$	+0.771	
	$O_2(g) + 2H^+ + 2e^- \rightleftharpoons H_2O_2(aq)$	+0.682	
	$\frac{1}{2} I_2(aq) + e^- \rightleftharpoons I^-$	+0.621	
	$MnO_4^- + 2H_2O + 3e^- \rightleftharpoons 4OH^- + MnO_2(s)$	+0.60	
	$I_3^- + 2e^- \rightleftharpoons 3I^-$	+0.536	
	$MnO_2(s) + H_2O + NH_4^+ + e^- \rightleftharpoons NH_3(aq) + Mn(OH)_3(s)$	+0.50	

Increasing Oxidizing Strength → / Increasing Reducing Strength →

continued on next page

TABLE 10 *continued*

Half-reaction	$E°$ (V)
$O_2(g) + 2H_2O + 4e^- \rightleftharpoons 4OH^-$	+0.401
$Cu^{2+} + 2e^- \rightleftharpoons Cu$	+0.337
$BiO^+ + 2H^+ + 3e^- \rightleftharpoons H_2O + Bi$	+0.32
$AgCl(s) + e^- \rightleftharpoons Ag + Cl^-$	+0.222
$SbO^+ + 2H^+ + 3e^- \rightleftharpoons H_2O + Sb$	+0.212
$SO_4^{2-} + 4H^+ + 2e^- \rightleftharpoons H_2O + H_2SO_3(aq)$	+0.17
$Sn^{4+} + 2e^- \rightleftharpoons Sn^{2+}$	+0.15
$S + 2H^+ + 2e^- \rightleftharpoons H_2S(g)$	+0.141
$2H^+ + 2e^- \rightleftharpoons H_2(g)$	0.000
$O_2(g) + H_2O + 2e^- \rightleftharpoons OH^- + HO_2^-$	−0.065
$Pb^{2+} + 2e^- \rightleftharpoons Pb$	−0.126
$CrO_4^{2-} + 4H_2O + 3e^- \rightleftharpoons 5OH^- + Cr(OH)_3(s)$	−0.13
$Sn^{2+} + 2e^- \rightleftharpoons Sn$	−0.136
$Ni^{2+} + 2e^- \rightleftharpoons Ni$	−0.250
$Co^{2+} + 2e^- \rightleftharpoons Co$	−0.277
$PbSO_4(s) + 2e^- \rightleftharpoons SO_4^{2-} + Pb$	−0.356
$Cd^{2+} + 2e^- \rightleftharpoons Cd$	−0.403
$Cr^{3+} + e^- \rightleftharpoons Cr^{2+}$	−0.41
$2H^+(10^{-7}\,M) + 2e^- \rightleftharpoons H_2(g)$	−0.414
$Fe^{2+} + 2e^- \rightleftharpoons Fe$	−0.44
$S + 2e^- (1\,M\,OH^-) \rightleftharpoons S^{2-}$	−0.45
$2CO_2(g) + 2H^+ + 2e^- \rightleftharpoons H_2C_2O_4(aq)$	−0.49
$Cr^{3+} + 3e^- \rightleftharpoons Cr$	−0.74
$Zn^{2+} + 2e^- \rightleftharpoons Zn$	−0.763
$2H_2O + 2e^- \rightleftharpoons 2OH^- + H_2(g)$	−0.828
$SO_4^{2-} + H_2O + 2e^- \rightleftharpoons 2OH^- + SO_3^{2-}$	−0.93
$Mn^{2+} + 2e^- \rightleftharpoons Mn$	−1.18
$Al^{3+} + 3e^- \rightleftharpoons Al$	−1.66
$Mg^{2+} + 2e^- \rightleftharpoons Mg$	−2.37
$Na^+ + e^- \rightleftharpoons Na$	−2.714
$Ca^{2+} + 2e^- \rightleftharpoons Ca$	−2.87
$Sr^{2+} + 2e^- \rightleftharpoons Sr$	−2.89
$Ba^{2+} + 2e^- \rightleftharpoons Ba$	−2.90
$Cs^+ + e^- \rightleftharpoons Cs$	−2.92
$K^+ + e^- \rightleftharpoons K$	−2.925
$Li^+ + e^- \rightleftharpoons Li$	−3.045

Increasing Oxidizing Strength

Increasing Reducing Strength

Weakest oxidizing agents	Half-reaction	$E°$ (V)	Strongest reducing agents

TABLE 11

International Atomic Masses (Based on $C^{12} = 12$ Exactly)

Element	Atomic symbol	Atomic number	Mass
Actinium	Ac	89	[227][a]
Aluminum	Al	13	26.9815
Americium	Am	95	[243][a]
Antimony	Sb	51	121.75
Argon	Ar	18	39.948
Arsenic	As	33	74.9216
Astatine	At	85	[210][a]
Barium	Ba	56	137.34
Berkelium	Bk	97	[247][a]
Beryllium	Be	4	9.0122
Bismuth	Bi	83	208.980
Boron	B	5	10.811[b]
Bromine	Br	35	79.909[c]
Cadmium	Cd	48	112.40
Calcium	Ca	20	40.08
Californium	Cf	98	[247][a]
Carbon	C	6	12.01115[b]
Cerium	Ce	58	140.12
Cesium	Cs	55	132.905
Chlorine	Cl	17	35.453[c]
Chromium	Cr	24	51.996[c]
Cobalt	Co	27	58.9332
Copper	Cu	29	63.54
Curium	Cm	96	[247][a]
Dysprosium	Dy	66	162.50
Einsteinium	Es	99	[254][a]
Erbium	Er	68	167.26
Europium	Eu	63	151.96
Fermium	Fm	100	[253][a]
Fluorine	F	9	18.9984
Francium	Fr	87	[223][a]
Gadolinium	Gd	64	157.25
Gallium	Ga	31	69.72
Germanium	Ge	32	72.59
Gold	Au	79	196.967
Hafnium	Hf	72	178.49

continued on next page

TABLE 11 *continued*

Element	Atomic symbol	Atomic number	Mass
Helium	He	2	4.0026
Holmium	Ho	67	164.930
Hydrogen	H	1	1.00797[b]
Indium	In	49	114.82
Iodine	I	53	126.9044
Iridium	Ir	77	192.2
Iron	Fe	26	55.847[c]
Krypton	Kr	36	83.80
Lanthanum	La	57	138.91
Lawrencium	Lw	103	[257][a]
Lead	Pb	82	207.19
Lithium	Li	3	6.939
Lutetium	Lu	71	174.97
Magnesium	Mg	12	24.312
Manganese	Mn	25	54.9380
Mendelevium	Md	101	[256][a]
Mercury	Hg	80	200.59
Molybdenum	Mo	42	95.94
Neodymium	Nd	60	144.24
Neon	Ne	10	20.183
Neptunium	Np	90	[237][a]
Nickel	Ni	28	58.71
Niobium	Nb	41	92.906
Nitrogen	N	7	14.0067
Nobelium	No	102	[254][a]
Osmium	Os	76	190.2
Oxygen	O	8	15.9994[b]
Palladium	Pd	46	106.4
Phosphorus	P	15	30.9738
Platinum	Pt	78	195.09
Plutonium	Pu	94	[242][a]
Polonium	Po	84	[210][a]
Potassium	K	19	39.102
Praseodymium	Pr	59	140.907
Promethium	Pm	61	[147][a]
Protactinium	Pa	91	[231][a]

continued on next page

TABLE 11 *continued*

Element	Atomic symbol	Atomic number	Mass
Radium	Ra	88	[226][a]
Radon	Rn	86	[222][a]
Rhenium	Re	75	186.2
Rhodium	Rh	45	102.905
Rubidium	Rb	37	85.47
Ruthenium	Ru	44	101.07
Samarium	Sm	62	150.35
Scandium	Sc	21	44.956
Selenium	Se	34	78.96
Silicon	Si	14	28.086[b]
Silver	Ag	47	107.870[c]
Sodium	Na	11	22.9898
Strontium	Sr	38	87.62
Sulfur	S	16	32.064[b]
Tantalum	Ta	73	180.948
Technetium	Tc	43	[97][a]
Tellurium	Te	52	127.60
Terbium	Tb	65	158.924
Thallium	Tl	81	204.37
Thorium	Th	90	232.038
Thulium	Tm	69	168.934
Tin	Sn	50	118.69
Titanium	Ti	22	47.90
Tungsten	W	74	183.85
Uranium	U	92	238.03
Vanadium	V	23	50.942
Xenon	Xe	54	131.30
Ytterbium	Yb	70	173.04
Yttrium	Y	39	88.905
Zinc	Zn	30	65.37
Zirconium	Zr	40	91.22

[a]A number in brackets designates the mass number of a selected isotope of the element, usually the one of longest known half-life.

[b]The atomic mass varies because of natural variation in the isotopic composition of the element. The observed ranges are boron, ±0.003; carbon, ±0.00005; hydrogen, ±0.00001; oxygen, ±0.0001; silicon, ±0.001; sulfur, ±0.003.

[c]The atomic mass is believed to have an experimental uncertainty of the following magnitude: bromine ±0.002; chlorine, ±0.001; chromium, ±0.001; iron, ±0.003; silver, ±0.003. For other elements, the last digit given is believed to be reliable to ±0.5.

Atomic masses: Courtesy of the International Union of Pure and Applied Chemistry

Bracketed masses: Courtesy of the National Bureau of Standards (now the National Institute of Science and Technology)

Freeman Laboratory Separates in General Chemistry

Note: When ordering Separates, please use both the separate title and the ISBN. To order please call our Order Department at 1-800-877-5351.

General Reference

0-7167-9409-8	Student Handbook: Introduction and Appendix
0-7167-9410-1	Scientific Measurements
0-7167-9413-6	Observing Chemical Reactions of Household Chemicals. Deductive Chemical Reasoning
0-7167-9457-8	The Language of Chemistry: Chemical Nomenclature. A Study Assignment
0-7167-9452-7	The Chemistry of Some Nonmetals
0-7167-9135-8	The Molar Volume of Oxygen
0-7167-9144-7	The Nomenclature of Inorganic Compounds. A Study Assignment.
0-7167-9145-5	Chemical Stoichiometry. Units of Quantity and Concentration.
0-7167-9148-X	Pressure-Volume-Temperature Relationships in Gases
0-7167-9149-8	Group Relationships in the Periodic Table
0-7167-9150-1	Writing Equations for Ionic Reactions. A Study Assignment
0-7167-9152-8	An Introduction to Chemical Equilibrium. A Study Assignment
0-7167-9154-4	An Introduction to Qualitative Analysis of the Metal Ions. A Study Assignment
0-7167-9161-7	Balancing Oxidation-Reduction Reactions. A Study Assignment
0-7167-9128-5	Recapitulation of Problems on Ionic Equilibria. A Study Assignment
0-7167-9458-6	Writing Lewis Structures. A Study Assignment

The Mole Concept and Stoichiometry

0-7167-9417-9	Determination of a Chemical Formula. The Reaction of Iodine with Zinc
0-7167-9166-8	The Reduction of Copper Oxide to Copper. The Atomic Mass of Copper
0-7167-9165-X	Determination of the Chemical Formula of an Oxide of Bismuth
0-7167-9135-8	The Molar Volume of Oxygen
0-7167-9137-4	Volumetric Analysis. The Titration of Acids and Bases
0-7167-9145-5	Chemical Stoichiometry. Units of Quantity and Concentration

Laboratory Techniques

0-7167-9410-1	Scientific Measurements
0-7167-9411-X	Mass and Volume Relationships
0-7167-9435-7	Determination of an Equilibrium Constant by Spectrophotometry
0-7167-9207-9	An Unknown for Qualitative Analysis
0-7167-9146-3	The Chemical Separation of a Mixture
0-7167-9154-4	An Introduction to the Qualitative Analysis of Metal Ions. A Study Assignment

Fundamental Laws and Principles

0-7167-9414-4	A Cycle of Copper Reactions
0-7167-9416-0	The Chemistry of Oxygen. Basic and Acidic Oxides and the Periodic Table

0-7167-9418-7	Determination of a Chemical Formula by Titration. The Reaction of Calcium with Water
0-7167-9419-5	The Estimation of Avagadro's Number
0-7167-9420-9	The Molar Volume of Dioxygen and Other Gases
0-7167-9421-7	The Molar Mass of a Gas
0-7167-9422-5	The Reactivity of Metals with Hydrochloric Acid
0-7167-9425-X	The Heat Capacity of Metals
0-7167-9426-8	Enthalpy Changes in Chemical Reactions. Hess's Law
0-7167-9428-4	Emission Spectra and the Electronic Structure of Atoms. The Hydrogen Atom Spectrum
0-7167-9229-2	Ionic and Covalent Bonding. Conductivity of Solutions of Ionic and Covalent Compounds
0-7167-9191-9	Writing Lewis Structures. A Study Assignment
0-7167-9430-6	Models of Molecular Shapes. VSEPR Theory and Orbital Hybridization
0-7167-9432-2	Liquids and Solids. The Vapor Pressure and Enthalpy of Vaporization of Water. The Enthalpy of Fusion of Water
0-7167-9433-0	Colligative Properties. The Molar Mass of a Soluble Substance by Freezing Point Depression
0-7167-9131-5	Some Common Physical Properties of Substances
0-7167-9132-3	Some Elementary Chemical Processes of Substances
0-7167-9133-1	The Formula of Compound from Experimental Data
0-7167-9135-8	The Molar Volume of Oxygen
0-7167-9147-1	The Packing of Atoms and Ions in Crystals
0-7167-9148-X	Pressure-Volume-Temperature Relationships in Gases
0-7167-9149-8	Group Relationships in the Periodic Table

Separations and Purifications

0-7167-9412-8	Separations of Food Dyes by Paper Chromatography. Separation of Lycopene and Beta-Carotene by Thin Layer Chromatography
0-7167-9442-X	Identification of Silver, Lead and Mercurous Ions
0-7167-9444-6	Identification of Alkaline Earth and Alkali Metal Ions
0-7167-9445-4	The Qualitative Analysis of Some Common Anions
0-7167-9447-0	Paper Chromatography in the Separation of Iron (III), Cobalt (II), Nickel (II), and Copper (II) Ions
0-7167-9154-4	An Introduction to the Qualitative Analysis of the Metal Ions. A Study Assignment
0-7167-9155-2	The Silver Group